D1053167

SOURCE BOOKS IN THE HISTORY
OF THE SCIENCES
EDWARD H. MADDEN · *General Editor*

A SOURCE BOOK IN PHYSICS

SOURCE BOOKS IN THE HISTORY OF THE SCIENCES

Edward H. Madden · *General Editor*

A SOURCE BOOK IN ASTRONOMY

Harlow Shapley and Helen E. Howarth
Harvard University

SOURCE BOOK IN ASTRONOMY, 1900–1950

Harlow Shapley · *Harvard University*

A SOURCE BOOK IN GEOLOGY

Kirtley F. Mather · *Harvard University*
and Shirley L. Mason

A SOURCE BOOK IN GREEK SCIENCE

Morris R. Cohen · *College of the City of New York
and University of Chicago* and
I. E. Drabkin · *College of the City of New York*

A SOURCE BOOK IN ANIMAL BIOLOGY

Thomas S. Hall · *Washington University*

A SOURCE BOOK IN CHEMISTRY

Henry M. Leicester · *San Francisco College of Medicine
and Surgery* and
Herbert M. Klickstein · *Edgar Fahs Smith Library
in the History of Chemistry, University of Pennsylvania*

A SOURCE BOOK

in

PHYSICS

By

WILLIAM FRANCIS MAGIE

Late Professor of Physics
Princeton University

HARVARD UNIVERSITY PRESS

CAMBRIDGE, MASSACHUSETTS

1965

LIBRARY
APR 20

LIBRARY

APR 29 1966

UNIVERSITY OF THE PACIFIC

149022

Copyright 1935, 1963 by the
President and Fellows of Harvard College

All rights reserved

NINTH PRINTING

Distributed in Great Britain by Oxford University Press, London

Library of Congress Catalog Card Number 63–21307

PRINTED IN THE UNITED STATES OF AMERICA

SOURCE BOOKS IN THE HISTORY
OF THE SCIENCES

General Editor's Preface

This series of Source Books aims to present the most significant passages from the works of the most important contributors to the major sciences during the last three or four centuries. So much material has accumulated that a demand for selected sources has arisen in several fields. Source books in philosophy have been in use for nearly a quarter of a century, and history, economics, ethics, and sociology utilize carefully selected source material. Recently, too, such works have appeared in the fields of psychology and eugenics. It is the purpose of this series, therefore, to deal in a similar way with the leading physical and biological sciences.

The general plan is for each volume to present a treatment of a particular science with as much finality of scholarship as possible from the Renaissance to the end of the nineteenth century. In all, it is expected that the series will consist of eight or ten volumes, which will appear as rapidly as may be consistent with sound scholarship.

In June, 1924, the General Editor began to organize the following Advisory Board:

HAROLD C. BROWN	*Philosophy*	Stanford University
MORRIS R. COHEN	*Philosophy*	College of the City of New York
ARTHUR O. LOVEJOY	*Philosophy*	Johns Hopkins University
GEORGE H. MEAD	*Philosophy*	University of Chicago
WILLIAM P. MONTAGUE	*Philosophy*	Columbia University
WILMON H. SHELDON	*Philosophy*	Yale University
EDWARD G. SPAULDING	*Philosophy*	Princeton University
JOSEPH S. AMES	*Physics*	Johns Hopkins University
FREDERICK BARRY	*Chemistry*	Columbia University
R. T. CHAMBERLIN	*Geology*	University of Chicago
EDWIN G. CONKLIN	*Zoology*	Princeton University
HARLOW SHAPLEY	*Astronomy*	Harvard University
DAVID EUGENE SMITH	*Mathematics*	Columbia University
ALFRED M. TOZZER	*Anthropology*	Harvard University

Each of the scientists on this board, in addition to acting in a general advisory capacity, is chairman of a committee of four or five men, whose business it is to make a survey of their special field and to determine the number of volumes required and the contents of each volume.

In December, 1925, the General Editor presented the project to the Eastern Division of the American Philosophical Association. After some discussion by the Executive Committee, it was approved and the philosophers of the board, with the General Editor as chairman, were appointed a committee to have charge of it. In November, 1927, the Carnegie Corporation of New York granted $10,000 to the American Philosophical Association as a revolving fund to help finance the series. In December, 1927, the American Association for the Advancement of Science approved the project, and appointed the General Editor and Professors Edwin G. Conklin and Harlow Shapley a committee to represent that Association in cooperation with the Advisory Board. In February, 1928, the History of Science Society officially endorsed the enterprise. Endorsements have also been given by the American Anthropological Association, the Mathematical Association of America, the American Mathematical Society, and the American Astronomical Society within their respective fields.

The General Editor wishes to thank the members of the Advisory Board for their assistance in launching this undertaking; Dr. J. McKeen Cattell for helpful advice in the early days of the project and later; Dr. William S. Learned for many valuable suggestions; the several societies and associations that have given their endorsements; and the Carnegie Corporation for the necessary initial financial assistance.

GREGORY D. WALCOTT.

LONG ISLAND UNIVERSITY,
  BROOKLYN, N. Y.
    December, 1928.

A SOURCE BOOK IN PHYSICS

Author's Preface

In accordance with the general plan of the series, this Source Book in Physics contains extracts from important contributions made to that science in the three centuries ending with the year 1900 A.D. The period opens with the introduction of the science of dynamics by Galileo. Throughout the succeeding years the ruling concepts are those of dynamics, developed by Newton, and applied by those who followed him to the explanation by dynamical principles of all the principal physical phenomena. This era, in which the concepts employed were purely dynamical or mechanistic, came to an end precisely in the year 1900, when a new era of development began with the introduction by Planck of the quantum theory of the distribution of energy.

The subject is so vast and ramifies into so many branches, each of which strikes root, like the banyan tree, and has an almost independent growth, that it is impossible to present all the important parts of it in a single volume. Some principles of selection must be adopted, if such can be found, to determine what should be put in and what should be left out. With some misgivings I decided to recognize the distinction between mathematical and experimental physics, to omit the mathematical arguments, and to include only the experimental results and such expositions of the theoretical results as were given in words by their discoverers. The choice of the selections to be made from the vast mass of experiment was determined by considering what would be of interest to a student whose knowledge of physics had been acquired from textbooks. These principles have not been applied with perfect consistency, and I fear that many matters have been omitted which specialists would like to see included. I trust, however, that nothing has been given which is not worthy of a place among the classics of physics.

The prefatory accounts of the lives of the physicists whose works are quoted were taken generally from the histories of physics by Poggendorff, Heller, and Rosenberger, when they were given in these books. The accounts of the men whose lives are not included in them were compiled from various sources. The translations were made by myself except when translations were found in the literature, in which cases these were used and are ascribed to their authors. I thought it best to group the extracts under the general headings into which the science is generally divided in the text-books. Within these groups the chronological order is generally followed, although in some cases cognate matters are grouped together without consideration of their dates. Omissions of parts of the originals are usually indicated by rows of leaders.

Professor Henry Crew, of Northwestern University, and President Joseph S. Ames, of Johns Hopkins University, were kind enough to examine my preliminary list of titles and to aid me by their advice. The limitations of space forced me to omit many subjects which appeared in this first list, and for such omissions I alone am responsible. I often consulted Professor E. P. Adams of Princeton University and received from him many valuable suggestions.

WILLIAM FRANCIS MAGIE.

PRINCETON UNIVERSITY,
  March, 1935.

Contents

xiv   *CONTENTS*

A SOURCE BOOK IN PHYSICS

MECHANICS

GALILEO

Galileo Galilei was born in Pisa on February 15, 1564. His parents were of noble families. His father was a student of music, especially of the mathematical theory. He was without means and felt compelled to train his son for profitable business. The boy's extraordinary talents showed themselves while he was at school and his father finally determined to educate him as a physician. He studied at the University of Pisa, where his interest was excited in the study of mathematics, for which he neglected his medical studies. He left the university without taking a degree. His scientific reputation, however, led to his appointment as a teacher of mathematics in the university in 1589, where he remained three years. It was in this period that he made the fundamental discovery of the laws of falling bodies. These laws he discussed in lectures and collected in a memoir which was not published until two hundred years later. In the later years of his life he wrote the treatise entitled, *Dialoghi della Nuove Scienze*, from which the following extracts are taken. In it he returns to his early studies and presents the connection between the motion of bodies and the forces acting on them in a way which has served as a model for all those who came after him. This treatise may be considered as marking the beginning of the science of dynamics.

In December, 1592, he was appointed as mathematician in the University of Padua for a term of six years. This appointment was renewed and at the end of the second term he was appointed professor for life. He attracted to his lectures students from all over Europe. He invented or improved the telescope and applied it to astronomical observations. His discoveries of Jupiter's moons, of the irregularities of the moon's surface, of sun spots, and of the phases of Mercury and Venus all served to support the Copernican hypothesis and to disprove the Ptolemaic system. On this account a strong feeling against him was raised in the minds of some of the leaders of the church. By his removal in 1610 to Pisa as First Mathematician of the university he became subject to the temporal control of the church. After years of uncertainty he was at length brought before the Inquisition, and was ordered not to publish anything in support of the Copernican system. Nevertheless the relations in which he stood with many of the great dignitaries of the church seemed to him so favorable that he finally ventured to publish a book in his favorite form of a dialogue on the two great systems of the universe. In this book both the Ptolemaic and the Copernican systems were presented, but it was generally admitted that the more convincing argument supported the Copernican system. He was again brought before the Inquisition and

1

finally was so weakened in his old age by fear of what might happen to him that he publicly abjured his belief in the Copernican system. From that time on until his death he was technically a prisoner of the Inquisition, though he was permitted to live in the care of his friends and patrons. It was in this period that he wrote his dialogue on motion. He died on January 8, 1642, the year before that in which Newton was born.

The extracts selected from Galileo's work present his fundamental study of falling bodies and two short passages, one dealing with the pendulum and the other with the composition of motions caused by forces. They are given as they appear in *Two New Sciences*, a translation of Galileo's work published in 1914 by Henry Crew and Alfonso de Salvio.

ACCELERATION AND LAWS OF FALLING BODIES

The following pages contain the essential parts of Galileo's study of accelerated motion and some of the most important propositions deduced from the general laws discussed. The interlocutors are Salviati, who is reading from a manuscript of a certain Academician (Galileo), Sagredo, also a scholar skilled in mechanics, and Simplicio.

SALV. The present does not seem to be the proper time to investigate the cause of the acceleration of natural motion concerning which various opinions have been expressed by various philosophers, some explaining it by attraction to the center, others to repulsion between the very small parts of the body, while still others attribute it to a certain stress in the surrounding medium which closes in behind the falling body and drives it from one of its positions to another. Now, all these fantasies, and others too, ought to be examined; but it is not really worth while. At present it is the purpose of our Author merely to investigate and to demonstrate some of the properties of accelerated motion (whatever the cause of this acceleration may be)—meaning thereby a motion, such that the momentum of its velocity goes on increasing after departure from rest, in simple proportionality to the time, which is the same as saying that in equal time-intervals the body receives equal increments of velocity; and if we find the properties (of accelerated motion) which will be demonstrated later are realized in freely falling and accelerated bodies, we may conclude that the assumed definition includes such a motion of falling bodies and that their speed goes on increasing as the time and the duration of the motion.

SAGR. So far as I see at present, the definition might have been put a little more clearly perhaps without changing the fundamental idea, namely, uniformly accelerated motion is such that its speed increases in proportion to the space traversed; so that, for example, the speed acquired by a body in falling four cubits would be double that acquired in falling two cubits and this latter speed would be double that acquired in the first cubit. Because there is no doubt but that a heavy body falling from the height of six cubits has, and strikes with, a momentum double that it had at the end of three cubits, triple that which it had at the end of one.

SALV. It is very comforting to me to have had such a companion in error; and moreover let me tell you that your proposition seems so highly probable that our Author himself admitted, when I advanced this opinion to him, that he had for some time shared the same fallacy. But what most surprised me was to see two propositions so inherently probable that they commanded the assent of everyone to whom they were presented, proven in a few simple words to be not only false, but impossible.

SIMP. I am one of those who accept the proposition, and believe that a falling body acquires force in its descent, its velocity increasing in proportion to the space, and that the momentum of the falling body is doubled when it falls from a doubled height; these propositions, it appears to me, ought to be conceded without hesitation or controversy.

SALV. And yet they are as false and impossible as that motion should be completed instantaneously; and here is a very clear demonstration of it. If the velocities are in proportion to the spaces traversed, or to be traversed, then these spaces are traversed in equal intervals of time; if, therefore, the velocity with which the falling body traverses a space of eight feet were double that with which it covered the first four feet (just as the one distance is double the other) then the time-intervals required for these passages would be equal. But for one and the same body to fall eight feet and four feet in the same time is possible only in the case of instantaneous (discontinuous) motion; but observation shows us that the motion of a falling body occupies time, and less of it in covering a distance of four feet than of eight feet; therefore it is not true that its velocity increases in proportion to the space.

The falsity of the other proposition may be shown with equal clearness. For if we consider a single striking body the difference of momentum in its blows can depend only upon difference of velocity; for if the striking body falling from a double height were to deliver a blow of double momentum, it would be necessary for this body to strike with a double velocity; but with this doubled speed it would traverse a doubled space in the same time-interval;

observation however shows that the time required for fall from the greater height is longer.

SAGR. You present these recondite matters with too much evidence and ease; this great facility makes them less appreciated than they would be had they been presented in a more abstruse manner. For, in my opinion, people esteem more lightly that knowledge which they acquire with so little labor than that acquired through long and obscure discussion.

SALV. If those who demonstrate with brevity and clearness the fallacy of many popular beliefs were treated with contempt instead of gratitude the injury would be quite bearable; but on the other hand it is very unpleasant and annoying to see men, who claim to be peers of anyone in a certain field of study, take for granted certain conclusions which later are quickly and easily shown by another to be false. I do not describe such a feeling as one of envy, which usually degenerates into hatred and anger against those who discover such fallacies; I would call it a strong desire to maintain old errors, rather than accept newly discovered truths. This desire at times induces them to unite against these truths, although at heart believing in them, merely for the purpose of lowering the esteem in which certain others are held by the unthinking crowd. Indeed, I have heard from our Academician many such fallacies held as true but easily refutable; some of these I have in mind.

SAGR. You must not withhold them from us, but, at the proper time, tell us about them even though an extra session be necessary. But now, continuing the thread of our talk, it would seem that up to the present we have established the definition of uniformly accelerated motion which is expressed as follows:

A motion is said to be equally or uniformly accelerated when, starting from rest, its momentum receives equal increments in equal times.

SALV. This definition established, the Author makes a single assumption, namely,

The speeds acquired by one and the same body moving down planes of different inclinations are equal when the heights of these planes are equal.

By the height of an inclined plane we mean the perpendicular let fall from the upper end of the plane upon the horizontal line drawn through the lower end of the same plane. Thus, to illustrate, let the line *AB* (Fig. 1) be horizontal, and let the planes *CA* and *CD* be inclined to it; then the Author calls the perpendicular *CB* the "height" of the planes *CA* and *CD*; he supposes that the speeds acquired by one and the same body, descending along the planes *CA* and *CD* to the terminal points *A* and *D* are equal since the heights of these planes are the same, *CB*; and also it must be understood that this speed is that which would be acquired by the same body falling from *C* to *B*.

SAGR. Your assumption appears to me so reasonable that it ought to be conceded without question, provided of course there are no chance or outside resistances, and that the planes are hard and smooth, and that the figure of the moving body is perfectly round, so that neither plane nor moving body is rough. All resistance and opposition having been removed, my reason tells me at once that a heavy and perfectly round ball descending along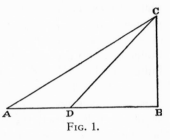

FIG. 1.

the lines *CA*, *CD*, *CB* would reach the terminal points *A*, *D*, *B*, with equal momenta.

SALV. Your words are very plausible; but I hope by experiment to increase the probability to an extent which shall be little short of a rigid demonstration.

Imagine this page to represent a vertical wall, with a nail driven into it; and from the nail let there be suspended a lead bullet of

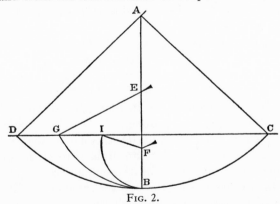

FIG. 2.

one or two ounces by means of a fine vertical thread, *AB*, (Fig. 2) say from four to six feet long, on this wall draw a horizontal line *DC*, at right angles to the vertical thread *AB*, which hangs about two finger-breadths in front of the wall.   Now bring the thread *AB* with the attached ball into the position *AC* and set it free; first it will be observed to descend along the arc *CBD*, to pass the point *B*, and to travel along the arc *BD*, till it almost reaches the horizontal *CD*, a slight shortage being caused by the resistance of the air and the string; from this we may rightly infer that the ball in its descent through the arc *CB* acquired a momentum on reaching *B*, which was just sufficient to carry it through a similar arc *BD* to the same height.   Having repeated this experiment many times, let us now drive a nail into the wall close to the perpendicular *AB*, say at *E* or *F*, so that it projects out some five or six finger-breadths in order that the thread, again carrying the bullet through the arc *CB*, may strike upon the nail *E* when the bullet reaches *B*, and thus compel it to traverse the arc *BG*, described about *E* as center.   From this we can see what can be done by the same momentum which previously starting at the same point *B* carried the same body through the arc *BD* to the horizontal *CD*.   Now, gentlemen, you will observe with pleasure that the ball swings to the point *G* in the horizontal, and you would see the same thing happen if the obstacle were placed at some lower point, say at *F*, about which the ball would describe the arc *BI*, the rise of the ball always terminating exactly on the line *CD*. But when the nail is placed so low that the remainder of the thread below it will not reach to the height *CD* (which would happen if the nail were placed nearer *B* than to the intersection of *AB* with the horizontal *CD*) then the thread leaps over the nail and twists itself about it.

This experiment leaves no room for doubt as to the truth of our supposition; for since the two arcs *CB* and *DB* are equal and similarly placed, the momentum acquired by the fall through the arc *CB* is the same as that gained by fall through the arc *DB*; but the momentum acquired at *B*, owing to fall through *CB*, is able to lift the same body through the arc *BD*; therefore, the momentum acquired in the fall *BD* is equal to that which lifts the same body through the same arc from *B* to *D*; so, in general, every momentum acquired by fall through an arc is equal to that which can lift the same body through the same arc.   But all these momenta which cause a rise through the arcs *BD*, *BG*, and *BI* are equal, since they are produced by the same momentum, gained by fall through *CB*, as experiment shows.   Therefore all the momenta gained by fall through the arcs *DB*, *GB*, *IB* are equal.

Sagr. The argument seems to me so conclusive and the experiment so well adapted to establish the hypothesis that we may, indeed, consider it as demonstrated.

Salv. I do not wish, Sagredo, that we trouble ourselves too much about this matter, since we are going to apply this principle mainly in motions which occur on plane surfaces, and not upon curved, along which acceleration varies in a manner greatly different from that which we have assumed for planes.

So that, although the above experiment shows us that the descent of the moving body through the arc *CB* confers upon it momentum just sufficient to carry it to the same height through any of the arcs *BD, BG, BI*, we are not able, by similar means, to show that the event would be identical in the case of a perfectly round ball descending along planes whose inclinations are respectively the same as the chords of these arcs. It seems likely, on the other hand, that, since these planes form angles at the point *B*, they will present an obstacle to the ball which has descended along the chord *CB*, and starts to rise along the chord *BD, BG, BI*.

In striking these planes some of its momentum will be lost and it will not be able to rise to the height of the line *CD*; but this obstacle, which interferes with the experiment, once removed, it is clear that the momentum (which gains in strength with descent) will be able to carry the body to the same height. Let us then, for the present, take this as a postulate, the absolute truth of which will be established when we find that the inferences from it correspond to and agree perfectly with experiment. The author having assumed this single principle passes next to the propositions which he clearly demonstrates; the first of these is as follows:

Theorem I, *Proposition* I

The time in which any space is traversed by a body starting from rest and uniformly accelerated is equal to the time in which that same space would be traversed by the same body moving at a uniform speed whose value is the mean of the highest speed and the speed just before acceleration began.

Let us represent by the line *AB* (Fig. 3) the time in which the space *CD* is traversed by a body which starts from rest at *C* and is uniformly accelerated; let the final and highest value of the speed gained during the interval *AB* be represented by the line *EB* drawn at right angles to *AB*; draw the line *AE*, then all lines drawn from equidistant points on *AB* and parallel to *BE* will represent the increasing values of the speed, beginning with the instant *A*. Let the point *F* bisect the line *EB*; draw *FG* parallel to *BA*, and *GA* parallel to *FB*, thus forming a parallelogram *AGFB* which will be equal in area to the triangle *AEB*, since the side *GF* bisects the side *AE* at the point *I*; for if the parallel lines in the triangle *AEB* are extended to *GI*, then the sum of all the parallels contained in the quadrilateral is equal to the sum of those con-

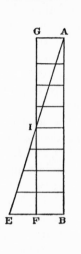

FIG. 3.

tained in the triangle *AEB*; for those in the triangle *IEF* are equal to those contained in the triangle *GIA*, while those included in the trapezium *AIFB* are common. Since each and every instant of time in the time-interval *AB* has its corresponding point on the line *AB*, from which points parallels drawn in and limited by the triangle *AEB* represent the increasing values of the growing velocity, and since parallels contained within the rectangle represent the values of a speed which is not increasing, but constant, it appears, in like manner, that the momenta assumed by the moving body may also be represented, in the case of the accelerated motion, by the increasing parallels of the triangle *AEB*, and, in the case of the uniform motion, by the parallels of the rectangle *GB*. For, what the momenta may lack in the first part of the accelerated motion (the deficiency of the momenta being represented by the parallels of the triangle *AGI*) is made up by the momenta represented by the parallels of the triangle *IEF*.

Hence it is clear that equal spaces will be traversed in equal times by two bodies, one of which, starting from rest, moves with a uniform acceleration, while the momentum of the other, moving with uniform speed, is one-half its maximum momentum under accelerated motion. Q. E. D.

Theorem II, Proposition II

The spaces described by a body falling from rest with a uniformly accelerated motion are to each other as the squares of the time-intervals employed in traversing these distances.

Let the time beginning with any instant *A* be represented by the straight line *AB* (Fig. 4) in which are taken any two time-intervals

AD and *AE*. Let *HI* represent the distance through which the body, starting from rest at *H*, falls with uniform acceleration. If *HL* represents the space traversed during the time-interval *AD*, and *HM* that covered during the interval *AE*, then the space *MH* stands to the space *LH* in a ratio which is the square of the ratio of the time *AE* to the time *AD*; or we may say simply that the distances *HM* and *HL* are related as the squares of *AE* and *AD*.

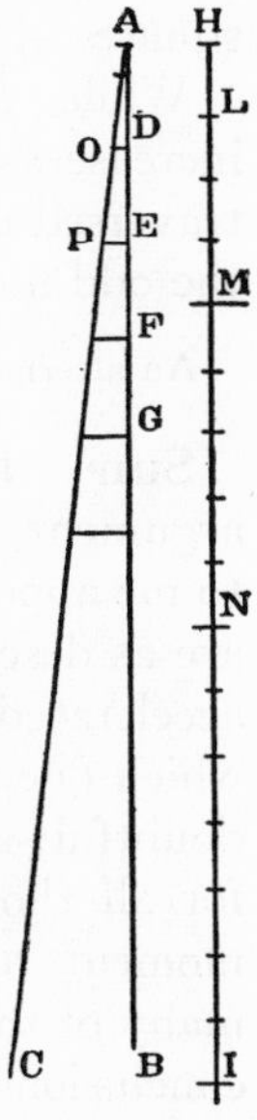

Draw the line *AC* making any angle whatever with the line *AB*; and from the points *D* and *E*, draw the parallel lines *DO* and *EP*; of these two lines, *DO* represents the greatest velocity attained during the interval *AD*, while *EP* represents the maximum velocity acquired during the interval *AE*. But it has just been proved that so far as distances traversed are concerned it is precisely the same whether a body falls from rest with a uniform acceleration or whether it falls during an equal time-interval with a constant speed which is one-half the maximum speed attained during the accelerated motion. It follows therefore that the distances *HM* and *HL* are the same as would be traversed, during the time-intervals *AE* and *AD*, by uniform velocities equal to one-half those represented by *DO* and *EP* respectively. If, therefore, one can show that the distances *HM* and *HL* are in the same ratio as the squares of the time-intervals *AE* and *AD*, our proposition will be proven.

Fig. 4.

But in the fourth proposition of the first book it has been shown that the spaces traversed by two particles in uniform motion bear to one another a ratio which is equal to the product of the ratio of the velocities by the ratio of the times. But in this case the ratio of the velocities is the same as the ratio of the time-intervals (for the ratio of *AE* to *AD* is the same as that of ½*EP* to ½*DO* or of *EP* to *DO*). Hence the ratio of the spaces traversed is the same as the squared ratio of the time-intervals.   Q. E. D.

Evidently then the ratio of the distances is the square of the ratio of the final velocities, that is, of the lines *EP* and *DO*, since these are to each other as *AE* to *AD*.

Corollary I

Hence it is clear that if we take any equal intervals of time whatever, counting from the beginning of the motion, such as *AD*,

DE, EF, FG, in which the spaces *HL, LM, MN, NI* are traversed, these spaces will bear to one another the same ratio as the series of odd numbers, 1, 3, 5, 7; for this is the ratio of the differences of the squares of the lines (which represent time), differences which exceed one another by equal amounts, this excess being equal to the smallest line (viz. the one representing a single time-interval): or we may say (that this is the ratio) of the differences of the squares of the natural numbers beginning with unity.

While, therefore, during equal intervals of time the velocities increase as the natural numbers, the increments in the distances traversed during these equal time-intervals are to one another as the odd numbers beginning with unity.

An alternative demonstration by Sagredo is omitted.

SIMP.   In truth, I find more pleasure in this simple and clear argument of Sagredo than in the Author's demonstration which to me appears rather obscure; so that I am convinced that matters are as described, once having accepted the definition of uniformly accelerated motion.   But as to whether this acceleration is that which one meets in nature in the case of falling bodies, I am still doubtful; and it seems to me, not only for my own sake but also for all those who think as I do, that this would be the proper moment to introduce one of those experiments—and there are many of them, I understand—which illustrate in several ways the conclusions reached.

SALV.   The request which you, as a man of science, make, is a very reasonable one; for this is the custom—and properly so—in those sciences where mathematical demonstrations are applied to natural phenomena, as is seen in the case of perspective, astronomy, mechanics, music, and others where the principles, once established by well-chosen experiments, become the foundations of the entire superstructure.   I hope therefore it will not appear to be a waste of time if we discuss at considerable length this first and most fundamental question upon which hinge numerous consequences of which we have in this book only a small number, placed there by the Author, who has done so much to open a pathway hitherto closed to minds of speculative turn. So far as experiments go they have not been neglected by the Author; and often, in his company, I have attempted in the following manner to assure myself that the acceleration actually experienced by falling bodies is that above described.

A piece of wooden moulding or scantling, about 12 cubits long, half a cubit wide, and three finger-breadths thick, was taken; on its edge was cut a channel a little more than one finger in breadth; having made this groove very straight, smooth, and polished, and having lined it with parchment, also as smooth and polished as possible, we rolled along it a hard, smooth, and very round bronze ball. Having placed this board in a sloping position, by lifting one end some one or two cubits above the other, we rolled the ball, as I was just saying, along the channel, noting, in a manner presently to be described, the time required to make the descent. We repeated this experiment more than once in order to measure the time with an accuracy such that the deviation between two observations never exceeded one-tenth of a pulse-beat. Having performed this operation and having assured ourselves of its reliability, we now rolled the ball only one-quarter the length of the channel; and having measured the time of its descent, we found it precisely one-half of the former. Next we tried other distances, comparing the time for the whole length with that for the half, or with that for two-thirds, or indeed for any fraction; in such experiments, repeated a full hundred times, we always found that the spaces traversed were to each other as the squares of the times, and this was true for all inclinations of the plane, i.e., of the channel, along which we rolled the ball. We also observed that the times of descent, for various inclinations of the plane, bore to one another precisely that ratio which, as we shall see later, the Author had predicted and demonstrated for them.

For the measurement of time, we employed a large vessel of water placed in an elevated position; to the bottom of this vessel was soldered a pipe of small diameter giving a thin jet of water, which we collected in a small glass during the time of each descent, whether for the whole length of the channel or for a part of its length; the water thus collected was weighed, after each descent, on a very accurate balance; the differences and ratios of these weights gave us the differences and ratios of the times, and this with such accuracy that although the operation was repeated many, many times, there was no appreciable discrepancy in the results.

SIMP. I would like to have been present at these experiments; but feeling confidence in the care with which you performed them, and in the fidelity with which you relate them, I am satisfied and accept them as true and valid.

SALV.　Then we can proceed without discussion.

Corollary II

Secondly, it follows that, starting from any initial point, if we take any two distances, traversed in any time-interval whatsoever, these time-intervals bear to one another the same ratio as one of the distances to the mean proportional of the two distances.

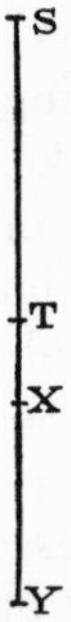

For if we take two distances ST and SY (Fig. 5) measured from the initial point S, the mean proportional of which is SX, the time of fall through ST is to the time of fall through SY as ST is to SX; or one may say the time of fall through SY is to the time of fall through ST as SY is to SX.　Now since it has been shown that the spaces traversed are in the same ratio as the squares of the times; and since, moreover, the ratio of the space SY to the space ST is the square of the ratio SY to SX, it follows that the ratio of the times of fall through SY and ST is the ratio of the respective distances SY and SX.

FIG. 5.

Scholium

The above corollary has been proven for the case of vertical fall; but it holds also for planes inclined at any angle; for it is to be assumed that along these planes the velocity increases in the same ratio, that is, in proportion to the time, or, if you prefer, as the series of natural numbers.

SALV.　Here, Sagredo, I should like, if it be not too tedious to Simplicio, to interrupt for a moment the present discussion in order to make some additions on the basis of what has already been proved and of what mechanical principles we have already learned from our Academician.　This addition I make for the better establishment on logical and experimental grounds, of the principle which we have above considered; and what is more important, for the purpose of deriving it geometrically, after first demonstrating a single lemma which is fundamental in the science of motion.

SAGR.　If the advance which you propose to make is such as will confirm and fully establish these sciences of motion, I will gladly devote to it any length of time.　Indeed, I shall not only be glad to have you proceed, but I beg of you at once to satisfy the curiosity which you have awakened in me concerning your proposition; and I think that Simplicio is of the same mind.

SIMP.   Quite right.

SALV.   Since then I have your permission, let us first of all consider this notable fact, that the momenta or speeds of one and the same moving body vary with the inclination of the plane.

The speed reaches a maximum along a vertical direction, and for other directions diminishes as the plane diverges from the vertical. Therefore the impetus, ability, energy, or, one might say, the momentum of descent of the moving body is diminished by the plane upon which it is supported and along which it rolls.

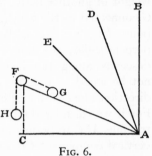

FIG. 6.

For the sake of greater clearness erect the line *AB* (Fig. 6) perpendicular to the horizontal *AC*; next draw *AD, AE, AF*, etc., at different inclinations to the horizontal. Then I say that all the momentum of the falling body is along the vertical and is a maximum when it falls in that direction; the momentum is less along *DA* and still less along *EA*, and even less yet along the more inclined plane *FA*. Finally on the horizontal plane the momentum vanishes altogether; the body finds itself in a condition of indifference as to motion or rest; has no inherent tendency to move in any direction, and offers no resistance to being set in motion. For just as a heavy body or system of bodies cannot of itself move upwards, or recede from the common center toward which all heavy things tend, so it is impossible for any body of its own accord to assume any motion other than one which carries it nearer to the aforesaid common center. Hence, along the horizontal, by which we understand a surface, every point of which is equidistant from this same common center, the body will have no momentum whatever.

This change of momentum being clear, it is here necessary for me to explain something which our Academician wrote when in Padua, embodying it in a treatise on mechanics prepared solely for the use of his students, and proving it at length and conclusively when considering the origin and nature of that marvellous machine, the screw. What he proved is the manner in which the momentum varies with the inclination of the plane, as for instance that of the plane *FA*, one end of which is elevated through a vertical distance *FC*. This direction *FC* is that along which the momentum of a heavy body becomes a maximum; let us discover what ratio this momentum bears to that of the same body moving along the inclined plane *FA*. This ratio, I say, is the inverse of that of the aforesaid lengths. Such is the lemma preceding the theorem which I hope to demonstrate a little later.

It is clear that the impelling force acting on a body in descent is equal to the resistance or least force sufficient to hold it at rest. In order to measure this force and resistance I propose to use the weight of another body. Let us place upon the plane *FA* a body *G* connected to the weight *H* by means of a cord passing over the point *F*; then the body *H* will ascend or descend, along the perpendicular, the same distance which the body *G* ascends or descends along the inclined plane *FA*; but this distance will not be equal to the rise or fall of *G* along the vertical in which direction alone *G*, as other bodies, exerts its force. This is clear. For if we consider the motion of the body *G*, from *A* to *F*, in the triangle *AFC* to be made up of a horizontal component *AC* and a vertical component *CF*, and remember that this body experiences no resistance to motion along the horizontal (because by such a motion the body neither gains nor loses distance from the common center of heavy things) it follows that resistance is met only in consequence of the body rising through the vertical distance *CF*. Since then the body *G* in moving from *A* to *F* offers resistance only in so far as it rises through the vertical distance *CF*, while the other body *H* must fall vertically through the entire distance *FA*, and since this ratio is maintained whether the motion be large or small, the two bodies being inextensibly connected, we are able to assert positively that, in case of equilibrium (bodies at rest) the momenta, the velocities, or their tendency to motion, i.e., the spaces which would be traversed by them in equal times, must be in the inverse ratio to their weights. This is what has been demonstrated in every case of mechanical motion. So that, in order to hold the weight *G* at rest, one must give *H* a weight smaller in the same ratio as the distance *CF* is smaller than *FA*. If we do this, *FA*:*FC* = weight *G*:weight *H*; then equilibrium will occur, that is, the weights *H* and *G* will have the same impelling forces, and the two bodies will come to rest.

And since we are agreed that the impetus, energy, momentum or tendency to motion of a moving body is as great as the force or least resistance sufficient to stop it, and since we have found that the weight *H* is capable of preventing motion in the weight *G*, it follows that the less weight *H* whose entire force is along the perpendicular, *FC*, will be an exact measure of the component of force which the larger weight *G* exerts along the plane *FA*. But the measure of the total force on the body *G* is its own weight, since to prevent its fall it is only necessary to balance it with an equal weight, provided this second weight be free to move vertically; therefore the component of the force of *G* along the inclined plane *FA* will bear to the maximum and total force on this same body *G* along the perpendicular *FC* the same ratio as the weight *H* to the weight *G*. This ratio is, by construction, the same which the height, *FC*, of the inclined plane bears to the length *FA*. We have here the lemma which I proposed to demonstrate and which, as you will see, has been assumed by our Author in the second part of the sixth proposition of the present treatise.

SAGR. From what you have shown thus far, it appears to me that one might infer, arguing *ex aequali con la proportione perturbata*, that the tendencies of one and the same body to move along planes differently inclined, but having the same vertical height, as *FA* and *FI*, are to each other inversely as the lengths of the planes.

SALV. Perfectly right. This point established, I pass to the demonstration of the following theorem:

If a body falls freely along smooth planes inclined at any angle whatsoever, but of the same height, the speeds with which it reaches the bottom are the same.

First we must recall the fact that on a plane of any inclination whatever a body starting from rest gains speed or momentum in direct proportion to the time, in agreement with the definition of naturally accelerated motion given by the Author. Hence, as he has shown in the preceding proposition, the distances traversed are proportional to the squares of the times and therefore to the squares of the speeds. The speed relations are here the same as in the motion first studied [i.e. *vertical motion*], since in each case the gain of speed is proportional to the time.

Let *AB* (Fig. 7) be an inclined plane whose height above the level *BC* is *AC*. As we have seen above, the force impelling a body to fall along the vertical *AC* is to the force which drives the same body along the inclined plane *AB* as *AB* is to *AC*. On the incline *AB*, lay off *AD* a third proportional to *AB* and *AC*; then the force producing motion along AC is to that along AB (i.e., along AD) as the length AC is to the length AD. And therefore the body will traverse the space AD, along the incline AB, in the same time which it would occupy in falling the vertical distance AC, (since the forces are in the same ratio as these distances); also the speed at C is to the speed at D as the distance AC is to the distance AD. But, according to the definition of accelerated motion, the speed at B is to the speed of the same body at D as the time required to traverse AB is to the time required for AD; and, according to the last corollary of the second proposition, the time of passing through the distance AB bears to the time of passing through AD the same ratio as the distance AC (a mean proportional between AB and AD) to AD. Accordingly the two speeds at B and C each bear to the speed at D the same ratio, namely, that of the distances AC and AD; hence they are equal. This is the theorem which I set out to prove.

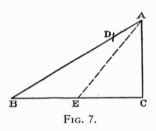

Fig. 7.

From the above we are better able to demonstrate the following third proposition of the Author in which he employs the following principle, namely, the time required to traverse an inclined plane is to that required to fall through the vertical height of the plane in the same ratio as the length of the plane to its height.

For, according to the second corollary of the second proposition, if BA represents the time required to pass over the distance BA, the time required to pass the distance AD will be a mean proportional between these two distances and will be represented by the line AC; but if AC represents the time needed to traverse AD it will also represent the time required to fall through the distance AC, since the distances AC and AD are traversed in equal times; consequently if AB represents the time required for AB then AC will represent the time required for AC. Hence the times required to traverse AB and AC are to each other as the distances AB and AC.

In like manner it can be shown that the time required to fall through AC is to the time required for any other incline AE as the length AC is to the length AE; therefore, ex aequali, the time of fall along the incline AB is to that along AE as the distance AB is to the distance AE, etc.

One might by application of this same theorem, as Sagredo will readily see, immediately demonstrate the sixth proposition of the Author; but let us here end this digression which Sagredo has perhaps found rather tedious, though I consider it quite important for the theory of motion.

SAGR. On the contrary it has given me great satisfaction, and indeed I find it necessary for a complete grasp of this principle.

SALV. I will now resume the reading of the text.

THE PENDULUM

SALV. Let us see whether we cannot derive from the pendulum a satisfactory solution of all these difficulties. And first, as to the question whether one and the same pendulum really performs its vibrations, large, medium, and small, all in exactly the same time, I shall rely upon what I have already heard from our Academician. He has clearly shown that the time of descent is the same along all chords, whatever the arcs which subtend them, as well along an arc of 180° (i.e., the whole diameter) as along one of 100°, 60°, 10°, 2°, ½°, or 4'. It is understood, of course that these arcs all terminate at the lowest point of the circle, where it touches the horizontal plane.

If now we consider descent along arcs instead of their chords then, provided these do not exceed 90°, experiment shows that they are all traversed in equal times; but these times are greater for the chord than for the arc, an effect which is all the more remarkable because at first glance one would think just the opposite to be true. For since the terminal points of the two motions are the same and since the straight line included between these two points is the shortest distance between them, it would seem reasonable that motion along this line should be executed in the shortest time; but this is not the case, for the shortest time— and therefore the most rapid motion—is that employed along the arc of which this straight line is the chord.

As to the times of vibration of bodies suspended by threads of different lengths, they bear to each other the same proportion as the square roots of the lengths of the thread; or one might say the lengths are to each other as the squares of the times; so that if one wishes to make the vibration-time of one pendulum twice that of another, he must make its suspension four times as long. In like manner, if one pendulum has a suspension nine times as long as another, this second pendulum will execute three vibrations during each one of the first; from which it follows that the lengths of the suspending cords bear to each other the (inverse) ratio of the squares of the number of vibrations performed in the same time.

SAGR. Then, if I understand you correctly, I can easily measure the length of a string whose upper end is attached at any height whatever even if this end were invisible and I could see only the lower extremity. For if I attach to the lower end of this string a rather heavy weight and give it a to-and-fro motion, and if I ask a friend to count a number of its vibrations, while I, during the same time-interval, count the number of vibrations of a pendulum which is exactly one cubit in length, then knowing the number of vibrations which each pendulum makes in the given interval of time one can determine the length of the string. Suppose, for example, that my friend counts 20 vibrations of the long cord during the same time in which I count 240 of my string which is one cubit in length; taking the squares of the two numbers, 20 and 240, namely 400 and 57600, then, I say, the long string contains 57600 units of such length that my pendulum will contain 400 of them; and since the length of my string is one cubit, I shall divide 57600 by 400 and thus obtain 144. Accordingly I shall call the length of the string 144 cubits.

SALV. Nor will you miss it by as much as a hand's breadth, especially if you observe a large number of vibrations.

SAGR. You give me frequent occasion to admire the wealth and profusion of nature when, from such common and even trivial phenomena, you derive facts which are not only striking and new but which are often far removed from what we would have imagined. Thousands of times I have observed vibrations especially in churches where lamps, suspended by long cords, had been inadvertently set into motion; but the most which I could infer from those observations was that the view of those who think that such vibrations are maintained by the medium is highly improbable: for, in that case, the air must needs have considerable judgment and little else to do but kill time by pushing to and fro a pendent weight with perfect regularity. But I never dreamed of learning that one and the same body, when suspended from a string a hundred cubits long and pulled aside through an arc of 90° or even 1° or ½°, would employ the same time in passing through the least as through the largest of these arcs; and, indeed, it still strikes me as somewhat unlikely. Now I am waiting to hear how these same simple phenomena can furnish solutions for those acoustical problems—solutions which will be at least partly satisfactory.

SALV. First of all one must observe that each pendulum has its own time of vibration so definite and determinate that it is not possible to make it move with any other period than that which nature has given it. For let any one take in his hand the cord to which the weight is attached and try, as much as he pleases, to increase or diminish the frequency of its vibrations; it will be time wasted. On the other hand, one can confer motion upon even a heavy pendulum which is at rest by simply blowing against it; by repeating these blasts with a frequency which is the same as that of the pendulum one can impart considerable motion. Suppose that by the first puff we have displaced the pendulum from the vertical by, say, half an inch; then if, after the pendulum has returned and is about to begin the second vibration, we add a second puff, we shall impart additional motion; and so on with other blasts provided they are applied at the right instant, and not when the pendulum is coming toward us since in this case the blast would impede rather than aid the motion. Continuing thus with many impulses we impart to the pendulum such momentum that a greater impulse than that of a single blast will be needed to stop it.

SAGR. Even as a boy, I observed that one man alone by giving these impulses at the right instant was able to ring a bell so large that when four, or even six, men seized the rope and tried to stop it they were lifted from the ground, all of them together being unable to counterbalance the momentum which a single man, by properly-timed pulls, had given it.

MOTION OF PROJECTILES

Let us imagine an elevated horizontal line or plane *ab* (Fig. 8) along which a body moves with uniform speed from *a* to *b*. Suppose this plane to end abruptly at *b*; then at this point the body will, on account of its weight, acquire also a natural motion downwards along the perpendicular *bn*. Draw the line *be* along the plane *ba* to represent the flow, or measure, of time; divide this line into a number of segments, *bc, cd, de*, representing equal intervals of time; from the points *b, c, d, e*, let fall lines which are parallel to the perpendicular *bn*. On the first of these lay off any distance *ci*, on the second a distance four times as long, *df*; on the third, one nine times as long, *eh*; and so on, in proportion to the squares of *cb*, *db*, *eb*, or, we may say, in the squared ratio of these same lines. Accordingly we see that while the body moves from *b* to *c* with uniform speed, it also falls perpendicularly through the distance *ci*, and at the end of the time-interval *bc* finds itself at the point *i*. In like manner at the end of the time-interval *bd*, which is the double of *bc*, the vertical fall will be four times the first distance *ci*; for it has been shown in a previous discussion that the distance traversed by a freely falling body varies as the square of the time; in like manner the space *eh* traversed during the time *be* will be nine times *ci*; thus it is evident that the distances

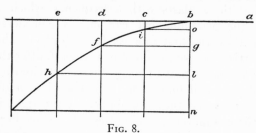

FIG. 8.

eh, *df*, *ci* will be to one another as the squares of the lines *be*, *bd*, *bc*. Now from the points *i*, *f*, *h* draw the straight lines *io*, *fg*, *hl* parallel to *be*; these lines *hl*, *fg*, *io* are equal to *eb*, *db*, and *cb*, respectively; so also are the lines *bo*, *bg*, *bl* respectively equal to *ci*, *df*, and *eh*. The square of *hl* is to that of *fg* as the line *lb* is to *bg*; and the square of *fg* is to that of *io* as *gb* is to *bo*; therefore the points *i*, *f*, *h*, lie on one and the same parabola. In like manner it may be shown that, if we take equal time-intervals of any size whatever, and if we imagine the particle to be carried by a similar compound motion, the positions of this particle, at the ends of these time-intervals, will lie on one and the same parabola.   Q. E. D.

SALV. This conclusion follows from the converse of the first of the two propositions given above. For, having drawn a parabola through the points *b* and *h*, any other two points, *f* and *i*, not falling on the parabola must lie either within or without; consequently the line *fg* is either longer or shorter than the line which terminates on the parabola. Therefore the square of *hl* will not bear to the square of *fg* the same ratio as the line *lb* to *bg*, but a greater or smaller; the fact is, however, that the square of *hl does* bear this same ratio to the square of *fg*. Hence the point *f* does lie on the parabola, and so do all the others.

SAGR. One cannot deny that the argument is new, subtle and conclusive, resting as it does upon this hypothesis, namely, that the horizontal motion remains uniform, that the vertical motion continues to be accelerated downwards in proportion to the square of the time, and that such motions and velocities as these combine without altering, disturbing, or hindering each other, so that as the motion proceeds the path of the projectile does not change into a different curve: but this, in my opinion, is impossible. For the axis of the parabola along which we imagine the natural motion of a falling body to take place stands perpendicular to a horizontal surface and ends at the center of the earth; and since the parabola deviates more and more from its axis no projectile can ever reach the center of the earth or, if it does, as seems necessary, then the path of the projectile must transform itself into some other curve very different from the parabola.

SIMP. To these difficulties, I may add others. One of these is that we suppose the horizontal plane, which slopes neither up nor down, to be represented by a straight line as if each point on this line were equally distant from the center, which is not the case; for as one starts from the middle (of the line) and goes toward either end, he departs farther and farther from the center (of the earth) and is therefore constantly going uphill. Whence it follows that the motion cannot remain uniform through any distance whatever, but must continually diminish. Besides, I do not see how it is possible to avoid the resistance of the medium which must destroy the uniformity of the horizontal motion and change the law of acceleration of falling bodies. These various difficulties render it highly improbable that a result derived from such unreliable hypotheses should hold true in practice.

SALV. All these difficulties and objections which you urge are so well founded that it is impossible to remove them; and, as for me, I am ready to admit them all, which indeed I think our Author would also do. I grant that these conclusions proved in the abstract will be different when applied in the concrete and will be fallacious to this extent, that neither will the horizontal motion be uniform nor the natural acceleration be in the ratio assumed, nor the path of the projectile a parabola, etc. But, on the other hand, I ask you not to begrudge our Author that which other eminent men have assumed even if not strictly true. The authority of Archimedes alone will satisfy everybody. In his Mechanics and in his first quadrature of the parabola he takes for granted that the beam of a balance or steelyard is a straight line, every point of which is equidistant from the common center of all heavy bodies, and that the cords by which heavy bodies are suspended are parallel to each other.

Some consider this assumption permissible because, in practice, our instruments and the distances involved are so small in comparison with the enormous distance from the center of the earth that we may consider a minute of arc on a great circle as a straight line, and may regard the perpendiculars let fall from its two extremities as parallel. For if in actual practice one had to consider such small quantities, it would be necessary first of all to criticise the architects who presume, by use of a plumbline, to erect high towers with parallel sides. I may add that, in all their discussions, Archimedes and the others considered themselves as located at an infinite distance from the center of the earth, in which case their assumptions were not false, and therefore their conclusions were absolutely correct. When we wish to apply our proven conclusions to distances which, though finite, are very large, it is necessary for us to infer, on the basis of demonstrated truth, what correction is to be made for the fact that our distance from the center of the earth is not really infinite, but merely very great in comparison with the small dimensions of our apparatus. The largest of these will be the range of our projectiles—and even here we need consider only the artillery—which, however great, will never exceed four of those miles of which as many thousand separate us from the center of the earth; and since these paths terminate upon the surface of the earth only very slight changes can take place in their parabolic figure which, it is conceded, would be greatly altered if they terminated at the center of the earth.

As to the perturbation arising from the resistance of the medium this is more considerable and does not, on account of its manifold forms, submit to fixed laws and exact description.

STEVINUS

Simon Stevin, whose name is commonly given in the Latinized form, Stevinus, was born in 1548 at Bruges and died at Leyden in 1620. His scientific work covered a wide range and he was distinguished as a civil and military engineer. He was an ardent supporter of William of Orange in the revolt against the Spanish rule.

The extract from Stevinus' work contains an interesting demonstration of the law of equilibrium of a body on an inclined plane and a partial if not a complete statement of the parallelogram of forces. It was first published in Leyden in 1586 and has been translated from the French as it appears in his collected works published at Leyden in 1634.

The Inclined Plane

Up to this point there have been considered the properties of weights acting directly downward; in what follows will be treated the properties and qualities of oblique forces, of which the general foundation is contained in the following theory.

Theorem XI, *Proposition* XIX

If a triangle has its plane perpendicular to the horizon and its base parallel to it; and if on each one of the two other sides there is placed a spherical body of equal weight; then as the right side of the triangle is to the left so is the force of the left hand body to that of the right hand body.

Given. Let *ABC* (Fig. 9) be a triangle with its plane perpendicular to the horizon and its base *AC* parallel to the horizon; and let there be placed on the side *AB* (which is twice the length of *BC*) a globular body *D*, and on *BC* another, *E*, equal in weight and in size.

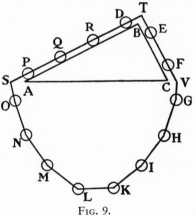

Required. It is to be demonstrated that as the side *AB2* is to the side *BC1* so the force or power of the weight *E* is to that of *D*.

Fig. 9.

Preparation. Let there be arranged about the triangle a set of fourteen spheres, equal in weight and in size, and set at equal distances from each other, such as *D, E, F, G, H, I, K, L, M, N, O, P, Q, R*, strung on a line passing through their centers, so that they can turn about their respective centers, and so placed that there are two spheres on the side *BC* and four on *BA*, then as the one line is to the other line so is the number of the spheres on the one to the number of the spheres on the other; also in *S, T, V*, let there be three rigidly fixed points, over which the line or the thread can slide, so that the two parts of it above the triangle are parallel to the sides *AB*, *BC*; in such a way that the whole string can run freely and without catching over the sides *AB*, *BC*.

Demonstration

If the power of the spheres *D*, *R*, *Q*, *P* were not equal to the power of the two spheres *E*, *F*, one side would be heavier than the other; thus, if it were possible, the four bodies *D*, *R*, *Q*, *P*, will be heavier than the two bodies *E*, *F*; but the four bodies *O*, *M*, *N*, *L*, are equal to the four bodies *G*, *H*, *I*, *K*; wherefore, the set of eight bodies *D*, *R*, *Q*, *P*, *O*, *N*, *M*, *L*, will be heavier, as they are placed, than the six bodies *E*, *F*, *G*, *H*, *I*, *K*, and since the heavier part overpowers the lighter part, the eight spheres will descend and the other six will rise; let this be so, and let *D* come where *O* is at present, and so for the rest; so that *E*, *F*, *G*, *H*, will come where *P*, *Q*, *R*, *D* are at present and *I*, *K*, where *E*, *F* are at present. Nevertheless, the set of spheres will have the same arrangement as before, and by the same reasoning the eight spheres will exceed the others in weight, and in falling will make eight others come into their places, and so this motion will have no end, which is absurd. The demonstration will be the same on the other side; therefore the part of the arrangement *D*, *R*, *Q*, *P*, *O*, *N*, *M*, *L*, will be in equilibrium with the part *E*, *F*, *G*, *H*, *I*, *K*; so that if we take away from the two sides the equal weights which are similarly arranged, as are the four spheres *O*, *N*, *M*, *L*, on the one side and the four *G*, *H*, *I*, *K*, on the other side; the four remaining spheres *D*, *R*, *Q*, *P*, will be and will remain in equilibrium with the two *E*, *F*; wherefore *E* will have a power double of the power of *D*; therefore as the side *BA*2 is to the side *BC*1, so is the power of *E* to the power of *D*.

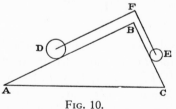

FIG. 10.

Conclusion. Therefore if a triangle has its plane, etc.

Corollary 1.

Let *ABC* (Fig. 10) be a triangle as before, and *AB* twice as great as *BC*, and let *D* be a sphere on *AB* twice as massive as *E*, which is on *BC*; at *F* let there be a rigidly fixed point over which the line *DFE* can slide freely, and so that *DF*, *FE*, are parallel to the sides of the triangle *ABC*, drawn from the centers of the spheres; it is plain that *D*, *E*, will still be in equilibrium, just as before *P, Q, R, D*, were in equilibrium with *E, F*, because as *AB* is to *BC* so is the sphere *D* to the sphere *E*.

Corollary 2.

Further, let one of the sides of the triangle, like *BC* (Fig. 11) (which is half the length of the other side *AB*), be perpendicular to *AC*; the sphere *D* which is twice the mass *E* will still be in equilibrium with *E*, for as the side *AB* is to *BC* so is the sphere *D* to the sphere *E*.

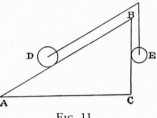

Fig. 11.

In Corollary 3 a pulley is substituted for the fixed point at the top of the plane, and in Corollary 4 the sphere on the plane is replaced by a column or cylinder with its axis perpendicular to the slant face of the plane.

Corollary 5.

Let a perpendicular be drawn from the center of the column *D* (Fig. 12), such as *DK*, cutting the side of it at *L*; then the triangle *LDI* will be similar to the triangle *ABC*, for the angles *ACB*

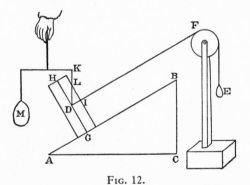

Fig. 12.

and *LID* are right-angles, *LD* is parallel to *BC*, and *DI* to *AB*; wherefore, as *AB* is to *BC* so is *LD* to *DI*; but as *AB* is to *BC* so is the weight of the column to the weight *E*, by the fourth corollary: therefore as *LD* is to *DI* so is the weight of the column to *E*: so that if in *KD* we apply a lifting force *M* in equilibrium with the column, it will be equal in weight to it by the fourteenth proposition: and finally as *LD* is to *DI* so is *M* to *E*.

Corollary 6.

Let us draw the line *BN* (Fig. 13) cutting *AC* produced in *N*, and also *DO*, cutting *LI* produced in *O*, in such a way that the angle

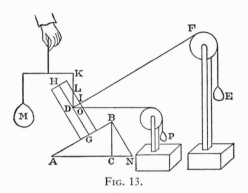

FIG. 13.

IDO is equal to the angle *CBN*, and then let there be applied the direct force *P* along *DO*, holding the column in position (having removed the weights *M*, *E*); then, in as much as *LD* is homologous to *BA* in the triangle *BAC*, and *DI* to *BC*, it follows, since *BA* is

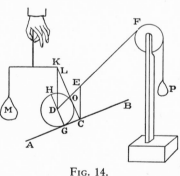

FIG. 14.

to *BC* as the weight on *BA* is to the weight on *BC*, by the second Corollary, that also *DL* is to *DI* as the weight belonging to *DL* is to the weight belonging to *DI*, as *M* to *E*; similarly, the three lines *LD*, *DI*, *DO* being homologous to the three lines *AB*, *BC*, *BN*; then *BA* being to *BN* as the weights pertaining to them so also *LD* will be to *DO* as the weights pertaining to them, that is to say, as *M* to *P*: and the same would be true if *BN* were drawn on the other side of the perpendicular *BC*, that is, between *AB* and *BC*, and similarly *DO* between *DL* and *DI*, and this proportion holds not only when the elevation *DI* is perpendicular to the axis but for every angle.

The above also may be applied to a sphere on the line AB (Fig. 14) as here represented, where we shall say as before: that as LD is to DO so is M to P (provided that CL is at right-angles to AB, that is to say, parallel to the axis HG of the sphere D) and so the weight of the sphere is to P as LD is to DO; but since LD, DO cannot be easily drawn inside the sphere let the perpendicular CE be drawn; then we shall have a triangle CEO outside the solid sphere, similar to LDO: of which the homologous sides are LD, CE, also DO, EO; wherefore as LD is to DO, that is to say as CE is to EO, so the weight of the sphere is to P.

There follows a diagram in which all the lines have been removed except those of the triangle COE, and the result is announced that as CE is to EO so is the weight of the sphere D to the weight P. The author adds here that this relation holds true not only for spheres but for all sorts of solid bodies.

HUYGENS

Christian Huygens was born at the Hague, April 14, 1629. He was the second son of Constantine Huygens, who was a man of wealth and position and also of considerable learning. From him his son received his first instruction in mathematics and mechanics. When sixteen years old Huygens entered the University of Leyden. He studied law there and at Breda. He also pursued studies in mathematics with such success that before his twenty-eighth year he had published several important papers on the subject. He busied himself with the building and improvement of telescopes, and with an instrument of his own construction he discovered a satellite of Saturn and was able to see the rings so clearly that he could describe them correctly. In 1657 he announced his invention of the method of controlling clocks by the pendulum, of which he gave a fuller description in his *Horologium Oscillatorium* in 1673.

In 1665 Huygens was invited on the part of Louis XIV to come to Paris as a member of the Academy which had just been founded. He accepted this invitation and for fifteen years he resided in Paris except for short visits to his native land. He returned to Holland in 1681, in anticipation, as is generally believed, of the revocation of the Edict of Nantes. From then on until his death, which occurred at the Hague on June 8, 1695, he devoted himself to scientific study and to the publication of his works. The most notable product of this period was his treatise on light, in which he laid the foundations of the undulatory theory.

The extract which follows is taken from Huygens' *Horologium Oscillatorium*. It presents theorems on centrifugal force, as Huygens called it, his proofs of which were found among his papers after his death and were published in his collected works.

Theorems on Centrifugal Force

I.

If two equal moving bodies pass over unequal circumferences in equal times, the centrifugal force in the greater circumference will be to that in the smaller, as the circumferences are to each other, or as their diameters.

II.

If two equal moving bodies move with equal velocities in unequal circumferences, their centrifugal forces will be in the inverse ratio of the diameters.

III.

If two equal moving bodies move in equal circumferences with unequal velocities, but each of them with uniform motion, which we are to understand is the case in all these propositions, the centrifugal force of the more rapidly moving body will be to that of the more slowly moving body in the squared ratio of the velocities.

IV.

If two equal moving bodies moving in unequal circumferences have equal centrifugal force, the time of describing the greater circumference will be to the time of describing the smaller circumference in the ratio of the square roots of the diameters.

V.

If a moving body travels in the circumference of a circle with a velocity the same as that which it would acquire by falling from a height equal to a quarter of the diameter, it will have a centrifugal force equal to its weight; that is, it will stretch the string which holds it to the center with the same force as it would if suspended from it.

VI.

If a moving body travels in circumferences parallel to the horizon, drawn on the inner surface of a parabolic conoid, the axis of which is perpendicular, all the circumferences, whether they are small or great, will be described in equal times; and each of these times is equal to the time of a double oscillation of a pendulum, the length of which is half that of the latus rectum of the generating parabola.

VII.

If two moving bodies, suspended by threads of unequal length, swing around so that they traverse circumferences parallel to the horizon, while the other end of the thread remains fixed, and so that the altitudes of the cones which the threads describe in this motion are equal, then the times in which their paths are described are equal.

VIII.

If two moving bodies, as before, swing around so as to describe cones, and are suspended by threads either equal or unequal, and if the altitudes of the cones are unequal, the times of revolution will be in the ratio of the square roots of their altitudes.

IX.

If a pendulum which swings around in a cone describes exceedingly small circuits, the times of describing each of them will have the same ratio to the time of perpendicular fall through a distance equal to twice the length of the pendulum as the circumference of a circle has to its diameter; and further, will be equal to the time of two lateral oscillations of the same pendulum, if they are very small.

X.

If a body moves around in a circumference and completes each circuit in the same time as that in which a pendulum, the length of which is equal to the radius of this circumference, will complete a very small circuit when describing a cone, or two very small lateral oscillations, then it will have a centrifugal force equal to its weight.

XI.

The time of revolution of any pendulum which describes a cone will be equal to the time of perpendicular fall through a height equal to the length of the pendulum, if the angle of inclination of the thread to the horizontal plane is approximately 2 degrees 54 minutes. The exact statement is: If the sine of the angle

described is to the radius as the square inscribed in a circle is to
the square on its circumference.

XII.

If two pendulums equal in weight but with threads of unequal
length swing around as conical pendulums, and the altitudes of
the cones are equal, then the forces by which the threads are
stretched are in the same ratio as the lengths of the threads.

XIII.

If a simple pendulum swings with its greatest lateral oscillation,
that is, if it descends through the whole quadrant of a circle, when
it comes to the lowest point of the circumference it stretches the
string with three times as great a force as it would if it were simply
suspended by it.

NEWTON

Isaac Newton was born in Woolsthorpe, Lincolnshire, on December 25,
1642. He was a posthumous child. His father, also named Isaac, was
the owner of the little manor of Woolsthorpe. When Newton reached his
majority, he inherited this estate, and remained in possession of the manorial
rights until his death.

Newton was prepared for the university at a school in the neighboring
town of Grantham, and in 1661 he was admitted to Trinity College, Cambridge. He received his Bachelor of Arts degree in January 1665, was elected
a Major Fellow and received his Master of Arts degree in 1668. His residence
in Cambridge had been interrupted in 1665 and 1666 by the prevalence of the
plague, which forced him to retire to his home at Woolsthorpe. It was in
those years that his genius developed. In them he invented the fluxional
calculus, discovered the composition of white light, conceived the idea of
universal gravitation, and probably discovered its law. Though he made no
public announcement of these discoveries, yet his abilities in mathematics
and physics were recognized and in 1669, when his teacher and friend Barrow
resigned the Lucasian Professorship of Mathematics, he was appointed in his
stead.

In the years which followed, which were passed quietly in Cambridge,
Newton published much of his work on optics, and developed his thoughts on
gravitation, which were in 1687 announced in *Principia*. The political disturbances of the period led to his serving as a representative of the university
in the Convention Parliament of 1689. In 1692 he suffered from a nervous
breakdown, which showed itself in an attitude of suspicion and dislike of some
of his best friends. From this he seems to have entirely recovered in the

course of a year. He never afterward did any extensive piece of scientific work, devoting most of his thought and work to chronology and theology.

Newton apparently became dissatisfied with his secluded life in the university, and made several efforts to obtain public employment. At last, in 1695, through the good offices of his friend Charles Montague, he was appointed Warden of the Mint, and removed to London. In this office he rendered valuable service in organizing and carrying through the reformation of the coinage. After this special task was accomplished, he was made Master of the Mint, and held this position for life. In 1701 he was sent as the representative of Cambridge University to the short-lived Parliament of that year. In 1703 he was elected to the presidency of the Royal Society and continued in that office by annual reelection, until his death.

Newton died in Kensington on March 20, 1727. He was buried in Westminster Abbey.

The *Principia Mathematica Philosophiae Naturalis*, from which the following extracts are taken, is one of the greatest monuments of the human intellect. In it Newton presents a system of mechanics which rests upon foundations broad enough to serve for all subsequent developments and he applies this system to the discussion of the movements of the heavenly bodies under the law of gravitation. The extracts here given contain, first, the definitions upon which his work is based. His discussion of absolute space and time is inserted because of the interest excited in it by the theory of relativity. The second extract contains the laws of motion and particularly Newton's discussion of the third law, which was his own contribution to the subject. The test of the third law by the collision of swinging pendulums is given as an illustration of Newton's experimental ability.

The translation is by Motte in the edition published in 1803.

Definitions

Definition I

The quantity of matter is the measure of the same, arising from its density and bulk conjunctly.

Thus air of a double density, in a double space, is quadruple in quantity; in a triple space, sextuple in quantity. The same thing is to be understood of snow, and fine dust or powders, that are condensed by compression or liquefaction; and of all bodies that are by any cause whatever differently condensed. I have no regard in this place to a medium, if any such there is, that freely pervades the interstices between the parts of bodies. It is this quantity that I mean hereafter everywhere under the name of body or mass. And the same is known by the weight of each body; for it is proportional to the weight, as I have found by experiments on pendulums, very accurately made, which shall be shewn hereafter.

Definition II

The quantity of motion is the measure of the same, arising from the velocity and quantity of matter conjunctly.

The motion of the whole is the sum of the motions of all the parts; and therefore in a body double in quantity, with equal velocity, the motion is double; with twice the velocity, it is quadruple.

Definition III

The *vis insita*, or innate force of matter, is a power of resisting, by which every body, as much as in it lies, endeavours to persevere in its present state, whether it be of rest, or of moving uniformly forward in a right line.

This force is ever proportional to the body whose force it is; and differs nothing from the inactivity of the mass, but in our manner of conceiving it. A body, from the inactivity of matter, is not without difficulty put out of its state of rest or motion. Upon which account, this *vis insita*, may, by a most significant name, be called *vis inertiae*, or force of inactivity. But a body exerts this force only, when another force, impressed upon it, endeavours to change its condition; and the exercise of this force may be considered both as resistance and impulse; it is resistance, in so far as the body, for maintaining its present state, withstands the force impressed; it is impulse, in so far as the body, by not easily giving way to the impressed force of another, endeavours to change the state of that other. Resistance is usually ascribed to bodies at rest, and impulse to those in motion; but motion and rest, as commonly conceived, are only relatively distinguished; nor are those bodies always truly at rest, which commonly are taken to be so.

Definition IV

An impressed force is an action exerted upon a body, in order to change its state, either of rest, or of moving uniformly forward in a right line.

This force consists in the action only; and remains no longer in the body, when the action is over. For a body maintains every new state it acquires, by its *vis inertiae* only. Impressed forces are of different origins; as from percussion, from pressure, from centripetal force.

Definition V

A centripetal force is that by which bodies are drawn or impelled, or any way tend, towards a point as to a centre.

. . . .

Definition VI

The absolute quantity of a centripetal force is the measure of the same, proportional to the efficacy of the cause that propagates it from the centre, through the spaces round about.

. . . .

Definition VII

The accelerative quantity of a centripetal force is the measure of the same, proportional to the velocity which it generates in a given time.

. . . .

Definition VIII

The motive quantity of a centripetal force is the measure of the same, proportional to the motion which it generates in a given time.

Thus the weight is greater in a greater body, less in a less body; and, in the same body, it is greater near to the earth, and less at remoter distances. This sort of quantity is the centripetency, or propension of the whole body towards the centre, or, as I may say, its weight; and it is always known by the quantity of an equal and contrary force just sufficient to hinder the descent of the body.

. . . .

Scholium

Hitherto I have laid down the definitions of such words as are less known, and explained the sense in which I would have them to be understood in the following discourse. I do not define time, space, place and motion, as being well known to all. Only I must observe, that the vulgar conceive those quantities under no other notions but from the relation they bear to sensible objects. And thence arise certain prejudices, for the removing of which, it will be convenient to distinguish them into absolute and relative, true and apparent, mathematical and common.

I. Absolute, true, and mathematical time, of itself, and from its own nature flows equably without regard to anything external, and by another name is called duration: relative, apparent, and common time, is some sensible and external (whether accurate or unequable) measure of duration by the means of motion, which is commonly used instead of true time; such as an hour, a day, a month, a year.

II. Absolute space, in its own nature, without regard to anything external, remains always similar and immovable. Relative space is some movable dimension or measure of the absolute space; which our senses determine by its position to bodies; and which is vulgarly taken for immovable space; such is the dimension of a subterraneous, an aereal, or celestial space, determined by its position in respect of the earth. Absolute and relative space, are the same in figure and magnitude but they do not remain always numerically the same. For if the earth, for instance, moves, a space of our air, which relatively and in respect of the earth remains always the same, will at one time be one part of the absolute space into which the air passes; at another time it will be another part of the same, and so, absolutely understood, it will be perpetually mutable.

III. Place is a part of space which a body takes up, and is according to the space, either absolute or relative. I say, a part of space; not the situation, nor the external surface of the body. For the places of equal solids are always equal; but their superficies, by reason of their dissimilar figures, are often unequal. Positions properly have no quantity, nor are they so much the places themselves, as the properties of places. The motion of the whole is the same thing with the sum of the motions of the parts; that is, the translation of the whole, out of its place, is the same thing with the sum of the translations of the parts out of their places; and therefore the place of the whole is the same thing with the sum of the places of the parts, and for that reason, it is internal, and in the whole body.

IV. Absolute motion is the translation of a body from one absolute place into another; and relative motion, the translation from one relative place into another. Thus in a ship under sail, the relative place of a body is that part of the ship which the body possesses; or that part of its cavity which the body fills, and which therefore moves together with the ship: and relative rest is the continuance of the body in the same part of the ship, or of its cavity. But real, absolute rest, is the continuance of the body in the same part of that immovable space, in which the ship itself, its cavity, and all that it contains, is moved. Wherefore, if the earth is really at rest, the body, which relatively rests in the ship, will really and absolutely move with the same velocity which the ship has on the earth. But if the earth also moves, the true and absolute motion of the body will arise, partly from the true motion of the earth, in immovable space; partly from the relative motion of the ship on the earth; and if the body moves also relatively in the ship; its true motion will arise, partly from the true motion of the earth, in immovable space, and partly from the relative motions as well of the ship on the earth, as of the body in the ship; and from these relative motions will arise the relative motion of the body on the earth. As if that part of the earth, where the ship is, was truly moved toward the east, with a velocity of 10010 parts; while the ship itself, with a fresh gale, and full sails, is carried towards the west, with a velocity expressed by 10 of those parts; but a sailor walks in the ship towards the east, with 1 part of the said velocity; then the sailor will be moved truly in immovable space towards the east, with a velocity of 10001 parts, and relatively on the earth towards the west, with a velocity of 9 of those parts.

. . . .

The effects which distinguish absolute from relative motion are, the forces of receding from the axis of circular motion. For there are no such forces in a circular motion purely relative, but in a true and absolute circular motion, they are greater or less, according to the quantity of the motion. If a vessel, hung by a long cord, is so often turned about that the cord is strongly twisted, then filled with water, and held at rest together with the water; after, by the sudden action of another force, it is whirled about the contrary way, and while the cord is untwisting itself, the vessel continues for some time in this motion; the surface of the water will at first be plain, as before the vessel began to move; but the vessel, by gradually communicating its motion to the water, will make it begin sensibly to revolve, and recede by little and little from the middle, and ascend to the sides of the vessel, forming itself into a concave figure (as I have experienced), and the swifter the motion becomes, the higher will the water rise, till at last, performing its revolutions in the same times with the vessel, it becomes relatively at rest in it. This ascent of the water shows its endeavour to recede from the axis of its motion; and the

36 *A SOURCE BOOK IN PHYSICS*

true and absolute circular motion of the water, which is here directly contrary to the relative, discovers itself, and may be measured by this endeavour. At first, when the relative motion of the water in the vessel was greatest, it produced no endeavour to recede from the axis; the water showed no tendency to the circumference, nor any ascent towards the sides of the vessel, but remained of a plain surface, and therefore its true circular motion had not yet begun. But afterwards, when the relative motion of the water had decreased, the ascent thereof towards the sides of the vessel proved its endeavour to recede from the axis; and this endeavour showed the real circular motion of the water perpetually increasing, till it had acquired its greatest quantity, when the water rested relatively in the vessel. And therefore this endeavour does not depend upon any translation of the water in respect of the ambient bodies, nor can true circular motion be defined by such translation. There is only one real circular motion of any one revolving body, corresponding to only one power of endeavouring to recede from its axis of motion, as its proper and adequate effect; but relative motions, in one and the same body, are innumerable, according to the various relations it bears to external bodies, and like other relations, are altogether destitute of any real effect, any otherwise than they may perhaps partake of that one only true motion. And therefore in their system who suppose that our heavens, revolving below the sphere of the fixed stars, carry the planets along with them; the several parts of those heavens, and the planets, which are indeed relatively at rest in their heavens, do yet really move. For they change their position one to another (which never happens to bodies truly at rest), and being carried together with their heavens, partake of their motions, and as parts of revolving wholes, endeavour to recede from the axis of their motions.

. . . .

Axioms, or Laws of Motion

Law I

Every body perseveres in its state of rest, or of uniform motion in a right line, unless it is compelled to change that state by forces impressed thereon.

Projectiles persevere in their motions, so far as they are not retarded by the resistance of the air, or impelled downwards by the force of gravity. A top, whose parts by their cohesion are perpetually drawn aside from rectilinear motions, does not cease its rotation, otherwise than as it is retarded by the air. The greater bodies of the planets and comets, meeting with less resistance in more free spaces, preserve their motions both progressive and circular for a much longer time.

Law II

The alteration of motion is ever proportional to the motive force impressed; and is made in the direction of the right line in which that force is impressed.

If any force generates a motion, a double force will generate double the motion, a triple force triple the motion, whether that force be impressed altogether and at once, or gradually and successively. And this motion (being always directed the same way with the generating force), if the body moved before, is added to or subducted from the former motion, according as they directly conspire with or are directly contrary to each other; or obliquely joined, when they are oblique, so as to produce a new motion compounded from the determination of both.

Law III

To every action there is always opposed an equal reaction: or the mutual actions of two bodies upon each other are always equal, and directed to contrary parts.

Whatever draws or presses another is as much drawn or pressed by that other. If you press a stone with your finger, the finger is also pressed by the stone. If a horse draws a stone tied to a rope, the horse (if I may so say) will be equally drawn back towards the stone: for the distended rope, by the same endeavour to relax or unbend itself, will draw the horse as much towards the stone, as it does the stone towards the horse, and will obstruct the progress of the one as much as it advances that of the other. If a body impinge upon another, and by its force change the motion of the other, that body also (because of the equality of the mutual pressure) will undergo an equal change, in its own motion, towards the contrary part. The changes made by these actions are equal, not in the velocities but in the motions of bodies; that is to say, if the bodies are not hindered by any other impediments. For, because the motions are equally changed, the changes of the velocities made towards contrary parts are reciprocally proportional to the bodies. This law takes place also in attractions, as will be proved in the next scholium.

Corollary I

A body by two forces conjoined will describe the diagonal of a parallelogram, in the same time that it would describe the sides, by those forces apart.

If a body in a given time, by the force *M* impressed apart in the place *A*, (Fig. 15) should with an uniform motion be carried from *A* to *B*; and by the force *N* impressed apart in the same place,

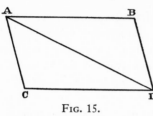

FIG. 15.

should be carried from *A* to *C*; complete the parallelogram *ABCD*, and, by both forces acting together, it will in the same time be carried in the diagonal from *A* to *D*. For since the force *N* acts in the direction of the line *AC*, parallel to *BD*, this force (by the second law) will not at all alter the velocity generated by the other force *M*, by which the body is carried towards the line *BD*. The body therefore will arrive at the line *BD* in the same time, whether the force *N* be impressed or not; and therefore at the end of that time it will be found somewhere in the line *BD*. By the same argument, at the end of the same time it will be found somewhere in the line *CD*. Therefore it will be found in the point *D*, where both lines meet. But it will move in a right line from *A* to *D*, by Law I.

Corollary II

And hence is explained the composition of any one direct force *AD*, out of any two oblique forces *AC* and *CD*; and, on the contrary, the resolution of any one direct force *AD* into two oblique forces *AC* and *CD*: which composition and resolution are abundantly confirmed from mechanics.

. . . .

Corollary III

The quantity of motion, which is collected by taking the sum of the motions directed towards the same parts, and the difference of those that are directed to contrary parts, suffers no change from the action of bodies among themselves.

For action and its opposite re-action are equal, by Law III, and therefore, by Law II, they produce in the motions equal changes towards opposite parts. Therefore if the motions are directed towards the same parts, whatever is added to the motion of the preceding body will be subducted from the motion of that which follows; so that the sum will be the same as before. If the bodies meet, with contrary motions, there will be an equal deduction from the motions of both; and therefore the difference of the motions directed towards opposite parts will remain the same.

· · · · · · · · · · · · · · · · ·

Corollary IV

The common centre of gravity of two or more bodies does not alter its state of motion or rest by the actions of the bodies among themselves; and therefore the common centre of gravity of all bodies acting upon each other (excluding outward actions and impediments) is either at rest, or moves uniformly in a right line.

· · · · · · · · · · · · · · · · ·

Corollary V

The motions of bodies included in a given space are the same among themselves, whether that space is at rest, or moves uniformly forwards in a right line without any circular motion.

· · · · · · · · · · · · · · · · ·

Corollary VI

If bodies, any how moved among themselves, are urged in the direction of parallel lines by equal accelerative forces, they will all continue to move among themselves, after the same manner as if they had been urged by no such forces.

· · · · · · · · · · · · · · · · ·

Scholium

Hitherto I have laid down such principles as have been received by mathematicians, and are confirmed by abundance of experiments. By the first two Laws and the first two Corollaries, Galileo discovered that the descent of bodies observed the duplicate ratio of the time, and that the motion of projectiles was in the curve of a parabola; experience agreeing with both, unless so far as these motions are a little retarded by the resistance of the air. When a body is falling, the uniform force of its gravity acting equally, impresses, in equal particles of time, equal forces upon that body, and therefore generates equal velocities; and in the whole time impresses a whole force, and generates a whole velocity proportional to the time. And the spaces described in proportional times are as the velocities and the times conjunctly; that is, in a duplicate ratio of the times. And when a body is thrown upwards, its uniform gravity impresses forces and takes off velocities proportional to the times; and the times of ascending to the greatest heights are as the velocities to be taken off, and those heights are as the velocities and the times conjunctly, or in the duplicate ratio of the velocities. And if a body be projected in any direction, the motion arising from its projection is compounded with the motion arising from its gravity. As if the body *A* (Fig. 16) by

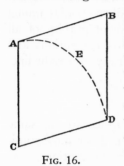

FIG. 16.

its motion of projection alone could describe in a given time the right line *AB*, and with its motion of falling alone would describe in the same time the altitude *AC*; complete the parallelogram *ABCD*, and the body by that compounded motion will at the end of the time be found in the place *D*; and the curve line *AED*, which that body describes, will be a parabola, to which the right line *AB* will be a tangent in *A*; and whose ordinate *BD* will be as the square of the line *AB*. On the same Laws and Corollaries depend those things which have been demonstrated concerning the times of the vibration of pendulums, and are confirmed by the daily experiments of pendulum clocks. By the same, together with the third Law, Sir Christ. Wren, Dr. Wallis, and Mr. Huygens, the greatest geometers of our times, did severally determine the rules of the congress and reflexion of hard bodies, and much about the same time communicated their discoveries to the Royal Society, exactly agreeing among themselves as to those rules. Dr. Wallis, indeed, was something more early in the publication; then followed Sir Christopher Wren, and, lastly, Mr. Huygens. But Sir Christopher Wren confirmed the truth of the thing before the Royal Society by the experiment of pendulums, which Mr. Mariotte soon after thought fit to explain in a treatise entirely upon that subject. But to bring this experiment to an accurate agreement with the theory, we are to have a due regard as well to the resistance of the air as to the elastic force of the concurring

bodies.   Let the spherical bodies *A*, *B* (Fig. 17) be suspended by
the parallel and equal strings *AC*, *BD*, from the centres *C*, *D*.
About these centres, with those intervals, describe the semicircles
EAF, *GBH*, bisected by the radii *CA*, *DB*.   Bring the body *A* to
any point *R* of the arc *EAF*, and (withdrawing the body *B*) let
it go from thence, and after one oscillation suppose it to return to
the point *V*: then *RV* will be the retardation arising from the
resistance of the air.   Of this *RV* let *ST* be a fourth part, situated
in the middle, to wit, so as *RS* and *TV* may be equal, and *RS* may
be to *ST* as 3 to 2: then will *ST* represent very nearly the retardation during the descent from *S* to *A*.   Restore the body *B* to its

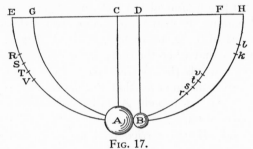

Fig. 17.

place: and, supposing the body *A* to be let fall from the point *S*,
the velocity thereof in the place of reflexion *A*, without sensible
error, will be the same as if it had descended *in vacuo* from the
point *T*.   Upon which account this velocity may be represented
by the chord of the arc *TA*.   For it is a proposition well known
to geometers, that the velocity of a pendulous body in the lowest
point is as the chord of the arc which it has described in its descent.
After reflexion, suppose the body *A* comes to the place *s*, and the
body *B* to the place *k*.   Withdraw the body *B*, and find the place
v, from which if the body *A*, being let go, should after one oscillation return to the place *r*, *st* may be a fourth part of *rv*, so placed
in the middle thereof as to leave *rs* equal to *tv*, and let the chord
of the arc *tA* represent the velocity which the body *A* had in the
place *A* immediately after reflexion.   For *t* will be the true and
correct place to which the body *A* should have ascended, if the
resistance of the air had been taken off.   In the same way we are
to correct the place *k* to which the body *B* ascends, by finding the
place *l* to which it should have ascended *in vacuo*.   And thus
everything may be subjected to experiment, in the same manner
as if we were really placed *in vacuo*.   These things being

done, we are to take the product (if I may so say) of the body A, by the chord of the arc TA (which represents its velocity), that we may have its motion in the place A immediately before reflexion; and then by the chord of the arc tA, that we may have its motion in the place A immediately after reflexion. And so we are to take the product of the body B by the chord of the arc Bl, that we may have the motion of the same immediately after reflexion. And in like manner, when two bodies are let go together from different places, we are to find the motion of each, as well before as after reflexion; and then we may compare the motions between themselves, and collect the effects of the reflexion. Thus trying the thing with pendulums of ten feet, in unequal as well as equal bodies, and making the bodies to concur after a descent through large spaces, as of 8, 12, or 16 feet, I found always, without an error of 3 inches, that when the bodies concurred together directly, equal changes towards the contrary parts were produced in their motions, and, of consequence, that the action and reaction were always equal. As if the body A impinged upon the body B at rest with 9 parts of motion, and losing 7, proceeded after reflexion with 2, the body B was carried backwards with those 7 parts. If the bodies concurred with contrary motions, A with twelve parts of motion, and B with six, then if A receded with 2, B receded with 8; to wit, with a deduction of 14 parts of motion on each side. For from the motion of A subducting twelve parts, nothing will remain; but subducting 2 parts more, a motion will be generated of 2 parts towards the contrary way; and so, from the motion of the body B of 6 parts, subducting 14 parts, a motion is generated of 8 parts towards the contrary way. But if the bodies were made both to move towards the same way, A, the swifter, with 14 parts of motion, B, the slower, with 5, and after reflexion A went on with 5, B likewise went on with 14 parts; 9 parts being transferred from A to B. And so in other cases. By the congress and collision of bodies, the quantity of motion, collected from the sum of the motions directed towards the same way, or from the difference of those that were directed towards contrary ways, was never changed. For the error of an inch or two in measures may be easily ascribed to the difficulty of executing everything with accuracy. It was not easy to let go the two pendulums so exactly together that the bodies should impinge one upon the other in the lowermost place AB; nor to mark the places s, and k, to which the bodies ascended after congress. Nay, and some errors, too, might have happened from the unequal density of the parts of the pendulous bodies themselves, and from the irregularity of the texture proceeding from other causes.

But to prevent an objection that may perhaps be alledged against the rule, for the proof of which this experiment was made, as if this rule did suppose that the bodies were either absolutely hard, or at least perfectly elastic (whereas no such bodies are to be found in nature), I must add, that the experiments we have been describing, by no means depending upon that quality of hardness, do succeed as well in soft as in hard bodies. For if the rule is to be tried in bodies not perfectly hard, we are only to diminish the reflexion in such a certain proportion as the quantity of the elastic force requires. By the theory of Wren and Huygens, bodies absolutely hard return one from another with the same velocity with which they meet. But this may be affirmed with more certainty of bodies perfectly elastic. In bodies imperfectly elastic the velocity of the return is to be diminished together with the elastic force; because that force (except when the parts of bodies are bruised by their congress, or suffer some such extension as happens under the strokes of a hammer) is (as far as I can perceive) certain and determined, and makes the bodies to return one from the other with a relative velocity, which is in a given ratio to that relative velocity with which they met. This I tried in balls of wool, made up tightly, and strongly compressed. For, first, by letting go the pendulous bodies, and measuring their reflexion, I determined the quantity of their elastic force; and then, according to this force, estimated the reflexions that ought to happen in other cases of congress. And with this computation other experiments made afterwards did accordingly agree; the balls always receding one from the other with a relative velocity, which was to the relative velocity with which they met as about 5 to 9. Balls of steel returned with almost the same velocity: those of cork with a velocity something less; but in balls of glass the proportion was as about 15 to 16. And thus the third Law, so far as it regards percussions and reflexions, is proved by a theory exactly agreeing with experience.

In attractions, I briefly demonstrate the thing after this manner. Suppose an obstacle is interposed to hinder the congress of any two bodies A, B, mutually attracting one the other: then if either body, as A, is more attracted towards the other body B, than that other body B is towards the first body A, the obstacle will be more strongly urged by the pressure of the body A than by the pressure of the body B, and therefore will not remain in equilibrio: but the stronger pressure will prevail, and will make the system of the two bodies, together with the obstacle, to move directly towards the parts on which B lies; and in free spaces, to go forward *in infinitum* with a motion perpetually accelerated; which is absurd and contrary to the first Law. For, by the first Law, the system ought to persevere in its state of rest, or of moving uniformly forward in a right line; and therefore the bodies must equally press the obstacle, and be equally attracted one by the other. I made the experiment on the loadstone and iron. If these, placed apart in proper vessels, are made to float by one another in standing water, neither of them will propel the other; but, by being equally attracted, they will sustain each other's pressure, and rest at last in an equilibrium.

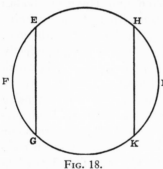

FIG. 18.

So the gravitation betwixt the earth and its parts is mutual. Let the earth FI (Fig. 18) be cut by any plane EG into two parts EGF and EGI, and their weights one towards the other will be mutually equal. For if by another plane HK, parallel to the former EG, the greater part EGI is cut into two parts $EGKH$ and HKI, whereof HKI is equal to the part EFG, first cut off, it is evident that the middle part $EGKH$, will have no propension by its proper weight towards either side, but will hang as it were, and rest in an equilibrium betwixt both. But the one extreme part HKI will with its whole weight bear upon and press the middle part towards the other extreme part EGF; and therefore the force with which EGI, the sum of the parts HKI and $EGKH$, tends towards the third part EGF, is equal to the weight of the part HKI, that is, to the weight of the third part EGF. And therefore the weights of the two parts EGI and EGF, one towards the other, are equal, as I was to prove. And indeed if those weights were not equal, the whole earth floating in the nonresisting aether would give way to the greater weight, and, retiring from it, would be carried off *in infinitum*.

And as those bodies are equipollent in the congress and reflexion, whose velocities are reciprocally as their innate forces, so in the use of mechanic instruments those agents are equipollent, and mutually sustain each the contrary pressure of the other, whose velocities, estimated according to the determination of the forces, are reciprocally as the forces.

So those weights are of equal force to move the arms of a balance; which during the play of the balance are reciprocally as their velocities upwards and downwards; that is, if the ascent or descent is direct, those weights are of equal force, which are reciprocally as the distances of the points at which they are suspended from the axis of the balance; but if they are turned aside by the interposition of oblique planes, or other obstacles, and made to ascend or descend obliquely, those bodies will be equipollent, which are reciprocally as the heights of their ascent and descent taken according to the perpendicular; and that on account of the determination of gravity downwards.

And in like manner in the pully, or in a combination of pullies, the force of a hand drawing the rope directly, which is to the weight, whether ascending directly or obliquely, as the velocity of the perpendicular ascent of the weight to the velocity of the hand that draws the rope, will sustain the weight.

In clocks and such like instruments, made up from a combination of wheels, the contrary forces that promote and impede the motion of the wheels, if they are reciprocally as the velocities of the parts of the wheel on which they are impressed, will mutually sustain the one the other.

The force of tne screw to press a body is to the force of the hand that turns the handles by which it is moved as the circular velocity of the handle in that part where it is impelled by the hand is to the progressive velocity of the screw towards the pressed body.

The forces by which the wedge presses or drives the two parts of the wood it cleaves are to the force of the mallet upon the wedge as the progress of the wedge in the direction of the force impressed upon it by the mallet is to the velocity with which the parts of the wood yield to the wedge, in the direction of lines perpendicular to the sides of the wedge. And the like account is to be given of all machines.

The power and use of machines consist only in this, that by diminishing the velocity we may augment the force, and the contrary: from whence, in all sorts of proper machines, we have the Solution of this problem: *To move a given weight with a given power, or with a given force to overcome any other given resistance.*

For if machines are so contrived that the velocities of the agent
and resistant are reciprocally as their forces, the agent will just
sustain the resistant, but with a greater disparity of velocity will
overcome it.   So that if the disparity of velocities is so great as
to overcome all that resistance which commonly arises either
from the attrition of contiguous bodies as they slide by one another,
or from the cohesion of continuous bodies that are to be separated,
or from the weights of bodies to be raised, the excess of the force
remaining, after all those resistances are overcome, will produce
an acceleration of motion proportional thereto, as well in the
parts of the machine as in the resisting body.   But to treat of
mechanics is not my present business.   I was only willing to show
by those examples the great extent and certainty of the third Law
of motion.   For if we estimate the action of the agent from its
force and velocity conjunctly, and likewise the reaction of the
impediment conjunctly from the velocities of its several parts,
and from the forces of resistance arising from the attrition, cohesion, weight, and acceleration of those parts, the action and
reaction in the use of all sorts of machines will be found always
equal to one another.   And so far as the action is propagated by
the intervening instruments, and at last impressed upon the
resisting body, the ultimate determination of the action will be
always contrary to the determination of the reaction.

VARIGNON

   Pierre Varignon was born in Caen in 1654 and died in 1722 in Paris. He
was professor of mathematics at the Collège Mazarin and later at the Collège
Royal in Paris. He was a member of the French Academy. In 1687 he
laid before the Academy a treatise on the parallelogram of forces. In this
treatise he develops the doctrine of the parallelogram of forces and its connection with the equivalent principle of moments. In a later book, entitled
Nouvelle Mècanique ou Statique, which was published after his death, in 1725, he
applies the parallelogram law to all sorts of statical problems. He also shows
the relation of the parallelogram law to the principle of work, which had been
communicated to him in 1717 by John Bernoulli.
   The extracts are taken from this later work. They cover Varignon's
demonstration of the relation between the parallelogram law and the principle
of moments and his statement of Bernoulli's principle.

VARIGNON'S THEOREM

   Varignon demonstrates the theorem which follows for several cases which
differ in the position of the point about which the moments are taken. It is

not necessary to present all of these.   The first and simplest one will serve as
an example of them all.   Varignon does not state explicitly that the areas
which are connected in his equations represent moments, but the lines of his
parallelogram represent forces and the relation of the areas to the moments
of the forces is sufficiently evident.   The theorem is used in the study of the
lever.

The extract is a part of Lemma XVI, on page 84, Vol. 1 of the *Nouvelle
Mécanique.*

If on the two contiguous sides *AB, AC* (Fig. 19) of any parallelogram *ABDC*, and on its diagonal *AD* which passes through the
angle *BAC* (which I call the *capital* angle) included between these
two sides *AB, AC*, we erect the triangles *ASB, ASC, ASD*, with

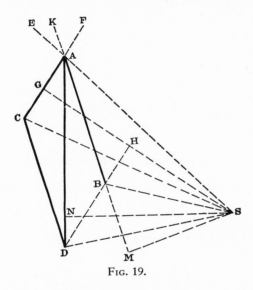

Fig. 19.

a common apex *S* which may be taken anywhere in the plane of the
parallelogram except at the point *A*; then I say,

I. That when the point *S* is in the complement *BAF* or *CAK*
which with the capital angle *BAC* makes two right angles as in the
figure, then the triangle *ASD* constructed on the diagonal *AD* of
the parallelogram *ABDC* will be equal to the sum of the two other
triangles *ASC, ASB*, constructed on the sides *AC, AB*, of the
capital angle *DAC*, that is to say, that in this case we shall always
have *ASD = ASC + ASB*.

Demonstration

Preparation for all the cases.   If from the common apex *S* of the three triangles *ASD, ASB, ASC*, which we are considering, we draw the line *SG* perpendicular at *G* and *H* to the parallel sides *AC*, *BD*, of the parallelogram *ABDC*; we shall have *GS, GH, HS*, for the heights of the triangles *ASC, BAD, BSD*, above their bases *AC, BD* which are by construction perpendicular to these heights. Consequently we shall have their areas $ASC = \frac{1}{2}AC \times GS$, $BAD = \frac{1}{2}BD \times GH = \frac{1}{2}AC \times GH$, $BSD = \frac{1}{2}AC \times HS$; from which we get

I. $BAD + BSD = \frac{1}{2}AC \times GH + \frac{1}{2}AC \times HS = \frac{1}{2}AC \times (GH + HS) = \frac{1}{2}AC \times GS = ASC.$

. . . .

CASE I.   The figure gives $ASD = ASB + BAD + BSD$. . Then in this case, in which the point *S* is in the complement *BAF* of the capital angle *BAC*, as we see it in the figure, we shall always have $ASD = ASB + ASC$, *which was to be proved.*

` . . . .

JEAN BERNOULLI'S PRINCIPLE OF VIRTUAL VELOCITIES

In a letter written from Bâle on January 26, 1717, M. (Jean) Bernoulli, after defining what he means by the word *energy*, in the way that will appear in the following definition, enunciated the principle that in every case of equilibrium of any forces, however they are applied to equilibrate one another, whether immediately or mediately, the sum of the positive energies will be equal to the sum of the negative energies, taken as positive.

This proposition seemed to me so general and so beautiful, that, seeing that I could easily deduce it from the preceding theory, I asked his permission, which he granted, to add it in this place, with the demonstration which my theory furnished of it, and which he did not send me.

. . . .
. . . .

Here is his explanation of what he understands by the word *energy*, in the letter in which he enunciated this beautiful proposition.

Definition XXXII

Conceive (says he) several different forces which are acting along different lines or directions of tendency to maintain in equilibrium a point, a line, a surface, or a body; conceive also that we impress on the whole system of forces a small displacement, either parallel to itself along any direction, or about any fixed point; it is easy to see that by this displacement each of these forces will advance or recede in its direction, unless some one or more of the forces have their directions perpendicular to the direction of the small displacement; in which case the force or the forces will neither advance nor recede: for these advancements or recessions, which I call *virtual velocities*, are nothing other than the amounts by which each line of tendency increases or decreases because of the small displacement; and these increments or decrements are found by drawing a perpendicular from the end of each line of tendency which will cut off from the line of tendency of each force in the neighboring position, to which it has been brought by the small displacement, a small portion which will be the measure of the *virtual velocity* of this force.

FIG. 20.

For example, let P (Fig. 20) be a point in the system of forces which is in equilibrium; let F be one of these forces, which pushes or pulls the point P in the direction FP or PF, let Pp be a small straight line which the point P describes in a small displacement, by which the line of tendency FP takes the position fp, which will either be exactly parallel to FP, if the displacement of the system is so made that all its points move parallel to a given straight line; or, prolonged, will make with FP an infinitely small angle, if the displacement of the system takes place around a fixed point. Now draw PC perpendicular to fp, and you will have Cp representing the *virtual velocity* of the force F, so that $F \times Cp$ is what I call the *energy*. Notice that Cp is either positive or negative with respect to other similar lines: it is positive, if the point P is pushed by the force F and the angle FPp is obtuse; and negative if the angle FPp is acute; but on the contrary, if the point P is pulled by the force, Cp is negative, if the angle FPp is obtuse, and positive, if it is acute. All this being well understood, I lay down (says M. Bernoulli) the following

General Proposition, Theorem XL

In every case of equilibrium of forces, in whatever way they are applied, and in whatever directions they act on one another, either mediately or immediately, the sum of the positive energies will be equal to the sum of the negative energies, taken as positive.

DESCARTES

René Descartes was born in La Haye on March 31, 1596. He was of a family of wealth and position in the country. He received his first education in the school of La Flèche, where he remained eight years. His studies covered a wide range, though he was especially interested in mathematics. After leaving school in 1612 he spent some years in Paris, preparing himself for military life. He there met some of the distinguished mathematicians of his time and was led gradually to withdraw from social life and to devote himself to intensive study. He served for a while in Breda as a volunteer in the army of the Stattholder, Maurice of Orange, and later in the Bavarian army and in other military services. He abandoned the military life and returned home at the age of twenty-five. After various journeys and adventures he went to Holland with the intention of devoting himself to the study of philosophy. The system of philosophy which he developed in the twenty years of his life in Holland had an immense influence upon the thought of the time. In so far as it dealt with the system of the universe it was finally overthrown by the Newtonian system. In 1649 Descartes accepted an invitation from the Queen of Sweden to visit Stockholm and instruct her in his philosophy. The climate was unfavorable for his health and he died in Stockholm on February 11, 1650.

For many years there was a discussion of the question whether the quantity of motion was to be measured by the product of mass and velocity or by the product of mass and the square of velocity. The first view was that of Descartes, the second of Leibnitz. The short extract from Descartes' *Principles of Philosophy*, part II, Section 36, presents Descartes' view that motion in the sense in which he defined it always remains the same or is conserved.

QUANTITY OF MOTION

§ 36. Now that we have examined the nature of motion, we come to consider its cause, and since the question may be taken in two ways, we shall commence by the first and more universal way, which produces generally all the motions which are in the world; we shall consider afterwards the other cause, which makes each portion of matter acquire motion which it did not have before. As for the first cause, it seems to me evident that it is nothing other than God, Who by His Almighty power created matter with motion and rest in its parts, and Who thereafter conserves in the universe by His ordinary operations as much of motion and of rest as He put in it in the first creation. For while it is true that motion is only the behavior of matter which is moved, there is, for all that, a quantity of it which never increases nor diminishes, although there is sometimes more and sometimes less of it in some of its parts; it is for this reason that when a part of matter moves twice as rapidly as another part, and this other part is twice as great as the first part, we have a right to think that there is as much motion in the smaller body as in the larger, and that every time and by as much as the motion of one part diminishes that of some other part increases in proportion. We also know that it is one of God's perfections not only to be immutable in His nature but also to act in a way which never changes: to such a degree that besides the changes that we see in the world, and those that we believe in because God has revealed them, and that we know have happened in nature without any change on the part of the Creator, we ought not to attribute to Him in His works any other changes for fear of attributing to Him inconstancy; from which it follows that, since He set in motion in many different ways the parts of matter when He created them and since He maintained them with the same behavior and with the same laws as He laid upon them in their creation, He conserves continually in this matter an equal quantity of motion.

LEIBNITZ

Gottfried Wilhelm Leibnitz was born on June 21, 1646, in Leipzig. At the age of fifteen he entered the University of Leipzig, where he studied philosophy and law. He took his doctor's degree at the University of Altdorf in 1666. For some years afterward he was largely occupied in philosophical studies and writing. He spent some time in Paris where he became acquainted with many distinguished scientific men. During this period he discovered or invented the differential calculus. By this discovery he was involved in a long and acrimonious contest for priority with Newton and Newton's admirers. After leaving Paris in 1676 he went to Hanover, where he was put in charge of the Ducal Library. He remained in that office for many years and died on November 14, 1716.

The paper presented here is the one in which he introduced the *vis viva* as the measure of quantity of motion. It was published in the *Acta Eruditorum*, 1686.

Quantity of Motion

A short demonstration of a remarkable error made by Descartes and others in that they affirm it to be a law of nature that always the same quantity of motion is conserved by God; which law they make improper use of in applying it to mechanics.

Most mathematicians, when they see, in the cases of the five mechanical powers, that velocity and mass are mutually compensated, generally estimate the motive force by the quantity of motion or by the product of the mass of the body into its velocity. Or to speak more mathematically, the forces of two bodies (of the same sort) which are set in motion, and which act both by reason of their masses and their motions, they say, are in a ratio compounded of the bodies or masses and of the velocities which they possess.   And so it may be agreeable to reason that the same totality of motive power is conserved in nature: and is neither diminished, since we see that no force is lost by a body, but is transferred to some other body; nor increased, because surely perpetual mechanical motion never occurs and no machine or even the world is able to maintain its force without a new external impulse; whence it happens that Descartes, who considered *motive force* and *quantity of motion* as equivalent, affirmed that the same quantity of motion was always conserved by God in the world.

But I, that I may show how much difference there is between these two ideas, assume, *first*, that a body falling from a certain height acquires a force sufficient to raise it to the same height, if it is given the proper direction and no external forces interfere: for example, that a pendulum will return precisely to the height from which it has been released, unless the resistance of the air and other slight obstacles absorb some of its strength, which we need not consider. I assume, *secondly*, that as much force is needed to raise a body *A* weighing one pound to the height *CD* of four ells, as to raise a body *B* weighing four pounds to the height *EF* of one ell. These assumptions are conceded by the Cartesians as well as by other philosophers and mathematicians of our times. Hence it follows that the body *A* (Fig. 21) let fall from the height *CD* acquires exactly as much force as the body *B* let fall from the height *EF*. For the body *A*, after that by its fall from *C* it reaches *D*, there has the force of ascending again to

C, by assumption 1, that is, the force sufficient to raise a body weighing one pound (that is, its own body) to the height of four ells. And similarly the body *B*, after that by its fall *E* it reaches *F*, there has the force of ascending again to *E* by assumption 1, that is, the force sufficient to raise a body weighing four pounds (that is its own body) to the height of one ell. Therefore by assumption 2 the force of the body *A* when it reaches *D* and the force of the body *B* when it reaches *F* are equal.

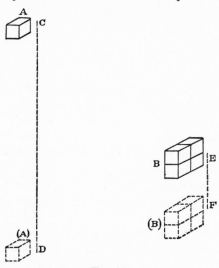

Fig. 21.

Now let us see if the quantity of motion also is the same for both. And here quite unexpectedly a great difference appears. This I show as follows: It has been demonstrated by Galileo that the velocity acquired by the fall *CD* is twice the velocity acquired by the fall *EF*. If therefore we multiply the body *A*, which may be taken as one, by its velocity, which may be taken as two, the product or the quantity of motion will be two. Again if we multiply the body *B*, which may be taken as four, by its velocity, which may be taken as one, the product or the quantity of motion will be four. Therefore the quantity of motion of the body *A* when at the point *D* is half that of the body *B* when at the point *F*, and yet just now the forces of both of these bodies have been found to be equal. And so there is a great difference between motive force and quantity of motion, so that one of these magnitudes cannot be determined from the other; which we have undertaken to show. From this it appears in what way the force should be estimated, from the quantity of the effect which it is able to produce; for example, from the height to which it can lift a heavy body of known magnitude and nature, not from the velocity which it can impress on the body. For there is need not of twice the force but more than that to give twice the velocity to the same body. No one should be surprised that in ordinary machines, such as the lever, the wheel and axle, the pulley, the wedge, the screw, and the like, there is equilibrium when the size of one body is compensated by the velocity of the other, which is brought about by the arrangement of the machine; or when the magnitudes (the same sort of body being assumed) are reciprocally as the velocities, or when the same quantity of motion is produced in any other way, for then it will happen that there will be the same quantity of effect in both the bodies, or the same height of ascent or descent, on whichever side of the equilibrated system you choose to produce the motion. And so by accident it happens in this case that the force can be estimated from the quantity of motion. But other cases occur, such as that which we have previously dealt with, where they are not the same.

While there is nothing more simple than our demonstration, it is strange that it never came into the minds of DesCartes or of his learned followers. But too great confidence in his own ingenuity led him astray, and the others were led astray by their confidence in him. For DesCartes, by an error common to great men, became a little too confident. And I fear that not a few of his followers have been imitating the Peripatetics whom they laugh at, that is, they have been depending upon consulting the books of their master, rather than on right reason and the nature of things.

Therefore we may say that forces are in the compound ratio of the bodies (of the same specific gravity or density) and of the heights productive of velocity, that is, the heights by falling from which they can acquire such velocities; or more generally (since no velocity has really been produced) of the heights that will produce them: but not generally of the velocities themselves, although this seems plausible at first sight, and has so seemed to many; from which many errors have arisen, which are contained in the works on mathematical mechanics of RR.PP. Honoratus Faber and Claudius des Chales, and also of Joh. Alph. Borelli and of other men, otherwise distinguished in such matters. Hence also I think it has happened, that doubts have been thrown by some learned men on the theorem of Huygens about the center of oscillation of a pendulum, which however is certainly valid.

D'ALEMBERT

Jean le Rond d'Alembert was born on November 17, 1717, and died on October 29, 1783. He was an illegitimate child. His mother, Madame de Tencin, exposed him on the steps of the church of St. Jean le Rond. He was placed in charge of a family from which he took his name. At the age of twelve he entered the Collège Mazarin, where he distinguished himself by ability in his studies. In 1741 he was made a member of the French Academy. He collaborated with Diderot in the preparation of the mathematical part of the *Encyclopédie* and was a prominent leader in the stirring intellectual life of the times.

The extract which follows presents d'Alembert's discussion of the measure of the quantity of motion. This question, first proposed by Leibnitz, had since been discussed by many distinguished mathematicians. D'Alembert shows that the question is of no importance for mechanics and is a mere discussion about words. His statements ended the controversy.

The extract is taken from the *Traité de Dynamique*, published in 1743.

QUANTITY OF MOTION

. . . All that we see distinctly in the motion of a body is that the body traverses a certain distance and that it takes a certain time to traverse that distance. It is from this one idea that all the principles of mechanics should be drawn, if we wish to demonstrate them in a clear and accurate way; so no one need be surprised that for this reason I have turned my thought away from causes of motion to consider solely the motions that they produce; and that I have entirely excluded forces inherent in bodies in motion, obscure and metaphysical entities which can only cast shadows on a science that is in itself clear.

It is for this reason that I have thought it best not to take up the consideration of the famous question of *vis viva*. For thirty years mathematicians have been divided in opinion as to whether the force of a body in motion is proportional to the product of the mass by the velocity, or to the product of the mass by the square of the velocity; for example whether a body twice as large as another one, and which has three times as much velocity, has eighteen times as much force or only six times as much. Notwithstanding the disputes to which this question has given rise, its perfect uselessness for mechanics has induced me to make no mention of it in the present work; I believe however that I should not pass over altogether in silence an opinion, which Leibnitz took credit to himself for discovering, which the great Bernoulli has since studied in such a sound and successful way, which MacLaurin has done all he could to overthrow, and about which the writings of a great many illustrious mathematicians have excited public interest. So, without wearying the reader by the detail of all that has been said on this question, it may not be out of place to state briefly the principles by which it can be solved.

When we speak of the force of bodies in motion, either we have no clear idea of what the word means or we can only mean in general the property of moving bodies by which they overcome the obstacles that they encounter, or resist them. It is therefore not by the distance that a body traverses with uniform motion, or by the time that it takes to traverse that distance, or finally by the simple, unique, and abstract consideration of its mass and velocity that we can at once estimate the force; it is solely by the obstacles that a body encounters and by the resistance that these obstacles offer to it. The greater the obstacle that a body can overcome, or that it can resist, the greater may we say is its force, provided that, without meaning to express by this word a hypothetical entity which resides in the body, we use the word only as an abbreviated way of expressing a fact, just as we say that one body has twice as much velocity as another instead of saying that in equal times it traverses twice the distance, without intending to mean by this that the word velocity represents an entity inherent in the body.

This being understood, it is evident that we may oppose to the motion of the body three types of obstacles; impenetrable obstacles, which entirely destroy the motion whatever it may be; or obstacles which have precisely the resistance necessary to destroy the motion of the body and which destroy it instantly, in the case of equilibrium; or finally obstacles which destroy the motion little by little in the case of retarded motion. As the impenetrable obstacles destroy similarly all sorts of motions, they cannot help us to discover the force; it is therefore only in equilibrium or in retarded motion that we should look for the measure of this force. Now everyone agrees that there is equilibrium between two bodies when the products of their masses by their virtual velocities, that is to say, the velocities with which they tend to move, are equal. Therefore in the case of equilibrium the product of the mass by velocity, or what is the same thing, the quantity of motion, may represent the force. Everyone agrees also that in retarded motion the number of obstacles overcome is proportional to the square of the velocity; thus, for example, a body which has compressed a spring with a certain velocity, can with twice that velocity compress either all at once or in succession not two but four springs like the first, nine springs with three times the velocity, etc. From this fact the advocates of *vis viva* conclude that the force of bodies which actually are moving is in general proportional to the product of the mass by the square of the velocity. When we consider the matter, what inconvenience can there be in having the measure of forces different in the two cases of equilibrium and of retarded motion, since, if we think only in terms of clear ideas, the word force should be used to signify only the effect produced by overcoming an obstacle or by resisting it. It must nevertheless be admitted that the opinion of those who consider force as the product of mass by velocity may hold good not only in the case of equilibrium but also in the case of retarded motion, if in the latter case we measure the force, not by the absolute quantity of the obstacles, but by the sum of the resistances of these same obstacles. For there is no doubt that this sum of the resistances is proportional to the quantity of motion, since, as everyone admits, the quantity of motion that a body loses at each instant, is proportional to the product of the resistance by the infinitely short duration of the instant, and that the sum of these products is evidently the total resistance. The whole difficulty is to decide whether we ought to measure force by the absolute quantity of the obstacles or by the sum of their resistances. It seems to be more natural to measure force in this latter way, for an obstacle is only one while it is resisting, and it may properly be said that the sum of the resistances is the obstacle overcome: furthermore, in estimating force in this way we have the advantage of a common measure of force both for equilibrium and for retarded motion; nevertheless as we have no clear and distinct idea of the meaning of the word force, except when we restrict the use of the word to express an effect, I believe that we ought to leave everyone free to make his own choice, and then there will be nothing left in the question except either a futile metaphysical discussion or a dispute about words still more unworthy of the consideration of philosophers.

What we have now said should be enough to make the issue clear to our readers. But another very natural consideration should carry conviction. If a body has a simple tendency to move with a certain velocity, which tendency is arrested by some obstacle; if the body moves really and uniformly with this velocity; or finally if it commences to move with the same velocity, which is gradually consumed and annulled by some cause or other: in these cases the effect produced by the body is different, but the body considered by itself has nothing more in one case than in the other; only the action of the cause which produces the effect is applied differently. In the first case the effect reduces to a simple tendency, which, properly speaking, has no precise measure, since no movement results from it; in the second case, the effect is the distance passed over uniformly in a given time, and this effect is proportional to the velocity; in the third case the effect is the distance passed over before the motion is annulled, and this effect is proportional to the square of the velocity. Now these different effects are evidently produced by one and the same cause; so that those who have said that the force is sometimes proportional to the velocity, and sometimes to its square, can only have been speaking of the effect when they express themselves in that way. These different effects coming all from the same cause may help us, as we may remark in passing, to perceive the lack of precision in the statement so often proposed as an axiom, that causes are proportional to their effects.

Finally even those who are not able to go back to the metaphysical principles of the question of *vis viva* will easily see that it is only a dispute about words if they consider that the two parties to the dispute are in every other way entirely in agreement on the fundamental principles of equilibrium and of motion. If we set the same problem in mechanics before two mathematicians, one of whom is opposed to and the other a partisan of *vis viva*, their solutions of the problem, if they are correct, will always agree; the question of the proper measure of force is therefore entirely useless in mechanics and without any real object. Without doubt it would not have brought to birth so many volumes, if care had been taken to distinguish between the clear and the obscure features of it. If it is taken in that way, no one would have needed more than a few lines to settle the question; but it seems as if most of those who have discussed it have been unwilling to discuss it in a few words.

YOUNG

Thomas Young was born June 13, 1773, at Milverton in Somersetshire, and died in London on May 10, 1829. From his earliest years he showed himself the possessor of unusual talents. While still a schoolboy he acquired not only the learned languages, including Hebrew, but several modern languages as well. He also studied mathematics. At the age of fourteen he served as private tutor to a somewhat younger boy and continued in that function until 1792, when he went to London to pursue the study of medicine. He studied in several universities and finally set up in London as a practising physician. By the death of his uncle in 1797 he became the possessor of considerable means, so that he was able to devote himself largely to the study of physics. In 1801 he was made professor at the Royal Institution. While there he delivered a course of lectures on natural philosophy, from which the following short extract is taken. In it Young gives the name "energy" to the property of matter which Leibnitz called the quantity of motion.

Young's great importance for physics rests upon his discovery of the interference of light, by which he established the undulatory theory. Extracts from his papers on this subject will be given in the proper place.

While he did not abandon his medical work, much of his time and thought was given to other matters. He wrote many articles on scientific subjects for the *Encyclopaedia Britannica* and other publications. By the help of the Rosetta Stone he endeavored to interpret the Egyptian hieroglyphics and, though he was not altogether successful, he was nevertheless a pioneer in this field.

ENERGY

The term energy may be applied, with great propriety, to the product of the mass or weight of a body, into the square of the number expressing its velocity. Thus, if a weight of one ounce moves with a velocity of a foot in a second, we may call its energy 1; if a second body of two ounces have a velocity of three feet in a second, its energy will be twice the square of three, or 18. This product has been denominated the living or ascending force, since the height of the body's vertical ascent is in proportion to it; and some have considered it as the true measure of the quantity of motion; but although this opinion has been very universally rejected, yet the force thus estimated well deserves a distinct denomination. After the considerations and demonstrations which have been premised on the subject of forces, there can be no reasonable doubt with respect to the true measure of motion; nor can there be much hesitation in allowing at once that since the same force, continued for a double time, is known to produce a double velocity, a double force must also produce a double velocity in the same time. Notwithstanding the simplicity of this view of the subject, Leibnitz, Smeaton, and many others, have chosen to estimate the force of a moving body, by the product of its mass into the square of its velocity; and though we cannot admit that this estimation of force is just, yet it may be allowed that many of the sensible effects of motion, and even the advantage of any mechanical power, however it may be employed, are usually proportional to this product, or to the weight of the moving body, multiplied by the height from which it must have fallen, in order to acquire the given velocity. Thus a bullet, moving with a double velocity, will penetrate to a quadruple depth in clay or tallow: a ball of equal size, but of one fourth of the weight, moving with a double velocity, will penetrate to an equal depth: and, with a smaller quantity of motion, will make an equal excavation in a shorter time. This appears at first sight somewhat paradoxical: but, on the other hand, we are to consider the resistance of the clay or tallow as a uniformly retarding force, and it will be obvious, that the motion, which it can destroy in a short time, must be less than that which requires a longer time for its destruction. Thus also when the resistance, opposed by any body to a force tending to break it, is to be overcome, the space through which it may be bent, before it breaks, being given, as well as the force exerted at every point of that space, the power of any body to break it is proportional to the energy of its motion, or to its weight multiplied by the square of its velocity.

In almost all cases of the forces employed in practical mechanics, the labour expended in producing any motion, is proportional, not to the momentum, but to the energy which is obtained; since these forces are seldom to be considered as uniformly accelerating forces, but generally act at some disadvantage, when the velocity is already considerable. For instance, if it be necessary to obtain a certain velocity, by means of the descent of a heavy body from a height, to which we carry it by a flight of steps, we must ascend, if we wish to double the velocity, a quadruple number of steps, and this will cost us nearly four times as much labour. In the same manner, if we press with a given force on the shorter end of a lever, in order to move a weight at a greater distance on the other side of the fulcrum, a certain portion of the force is expended in the pressure which is supported by the fulcrum, and we by no means produce the same momentum, as would have been obtained, by the immediate action of an equal force, on the body to be moved.

LAGRANGE

Joseph-Louis Lagrange was born in Turin on January 25, 1730, and died in Paris on April 10, 1813. He was early interested in the study of mathematics and so rapidly acquired a reputation that when he was scarcely nineteen years old he became professor of mathematics in the Royal Artillery School at Turin. He was soon admitted to the Academy of Turin and contributed largely to its *Proceedings*. In 1766 he was invited to join the Academy at Berlin, to fill the place left vacant by the withdrawal of Euler and the refusal of d'Alembert. He remained in Berlin in this position for twenty years. Dissatisfied by the change of conditions after the death of Frederick the Great he went to Paris, where he was cordially received and given lodgings in the Louvre. While there, he published the *Mécanique Analytique*, his principal contribution to physical science. He passed through the period of the Revolution unhurt. When the École Polytechnique was established he was appointed professor of Mathematics. He was greatly honored by Napoleon.

In the *Mécanique Analytique* Lagrange undertook to lay a foundation for the study of mechanics which would be independent of diagrams and so constructed that any problem could be expressed in equations, the solution of which would offer only mathematical difficulties. His method is in sharp contrast with the synthetic method used by Newton in the *Principia* and by many of Newton's followers. In the extract which follows we have Lagrange's discussion of the principle of virtual velocities, which he makes the foundation of his system. The final outcome of his argument, as embodied in the equations known by his name, would be unintelligible without so long a citation that it cannot be given here.

Principle of Virtual Velocities

The principle of virtual velocities can be expressed generally in the following way:

If any system of as many bodies or points as we please, each acted on by any forces, is in equilibrium, and if this system is given any small displacement, by which each point traverses an infinitely small distance which will be its virtual velocity, the sum of the products obtained by multiplying each of the forces by the distance traversed in the direction of the force by the point at which the force is applied, will be always equal to zero, when we consider positive the distances traversed in the sense of the forces, and negative the distances traversed in the opposite sense.

Jean Bernoulli is the first, so far as I know, who perceived the great generality of this principle of virtual velocities, and its utility for solving the problems of statics. This may be seen in one of his letters to Varignon, dated in 1717, which Varignon has set at the head of the ninth section of his Nouvelle Mécanique, a section devoted entirely to showing by diverse applications the truth and the use of the method of which it treats.

The same principle gave rise to the one which was proposed by Maupertuis in the Mémoires de l'Académie des Sciences de Paris for the year 1740, under the name of the law of repose, which Euler developed still further and rendered more general in the Mémoires de l'Académie de Berlin for the year 1751; also the same principle served as the foundation of that which Courtivon has given in the Mémoires de l'Académie des Sciences de Paris for 1748 and 1749.

In general, I believe that I may assert that all the general principles which may still be discovered in the science of equilibrium, will be only this same principle of virtual velocities, looked at from different points of view, and differing from it only in the form of statement.

But this principle is not only in itself very simple and very general; it has further the great and unique advantage that it may be expressed in a general formula which includes all the problems that can be proposed on the equilibrium of bodies. We shall present this formula in all its range; we shall even attempt to present it in a more general way than has hitherto been done, and to give some novel applications of it.

As to the nature of the principle of virtual velocities, it must be admitted that it is not of itself sufficiently evident to justify erecting it into a primitive principle, but we may regard it as a general expression of the law of equilibrium, deduced from the two principles which we have already presented [principle of moments and the parallelogram of forces]. Thus in the demonstrations that have been given of this principle, it has always been made to depend on them, in a more or less direct way. But there is in statics another general principle, independent of the lever and the composition of forces, although mechanicians commonly refer it to them, which seems to be the natural foundation of the principle of virtual velocities; we may call it the principle of the pulley.

If several pulleys are mounted on the same axle, we call the group a block and the combination of two blocks, one fixed and the other movable, around which is carried a cord, one end of which is fixed to a support while the other is pulled by a force, forms a machine in which the force is to the weight carried by the

movable block as one is to the number of cords that meet this
block, if they are supposed parallel and if we neglect friction and
the stiffness of the cord; for it is evident that, because of the
uniform tension in the whole length of the cord, the weight is
sustained by as many forces equal to that which stretches the
cord, as there are elements of the cord which sustain the movable
block, since these cords are parallel and can even be thought of as
reduced to a single cord, by conceiving, if we please, that the
diameter of the pulleys is made infinitely small.

By increasing the number of the fixed and movable blocks,
and by passing the same cord around them all, as can be done by
the help of extra fixed pulleys, the same force applied to the free
end of the cord can sustain as many weights as there are movable
blocks, each one of which will be to this force as the number of
cords meeting the block will be to one.

For greater simplicity, we will substitute a weight for the force,
after passing over a fixed pulley the free end of the cord which
sustains this weight, which we shall take for a unit weight; and
we will imagine that the different movable blocks, instead of
sustaining weights, are attached to bodies considered as points
and so arranged among themselves as to form any system whatever.
In this way the same weight, by means of the cord which passes
over all the blocks, will set up different forces, which will act on
the different points of the system in the directions of the cords
which end on the blocks attached to these points, and which
will be to the weight as the number of the cords is to one; so that
these forces will be themselves represented by the number of cords
which produce them, because of the tension in the cords.

Now it is evident, in order that the system acted on by these
different forces should remain in equilibrium, that the weight
cannot descend, if there is any infinitely small displacement of
the points of the system; because, since the weight always tends
to descend, if there were any displacement of the system which
would permit it to descend, it would necessarily do so and bring
about this displacement in the system.

Represent by α, β, γ, etc., the infinitely small displacements
which this displacement would bring about in the different points
of the system in the direction of the forces which act on them, and
by P, Q, R, etc., the number of the cords of the blocks applied to
these points to produce these forces; it is evident that the distances
α, β, γ, etc., will also be those by which the movable blocks will

approach the fixed ones which correspond to them, and that these approaches will also diminish the length of the cord which passes around the blocks, by the amounts $P\alpha$, $Q\beta$, $R\gamma$, etc; so that, because the length of the cord is invariable, the weight will descend through the distance $P\alpha + Q\beta + R\gamma + $ etc. Therefore it is necessary for the equilibrium of the forces represented by the numbers P, Q, R, etc., that we have the equation:

$$P\alpha + Q\beta + R\gamma + \text{etc.} = 0$$

which is the analytical expression of the general principle of virtual velocities. It might be thought that, if the quantity $P\alpha + Q\beta + R\gamma + $ etc., instead of being zero, were negative, this condition would be sufficient to bring about equilibrium, because it is impossible that the weight should rise of itself; but it must be noticed that, whatever the connections may be among the points which form the given system, the relations which result among the infinitely small quantities α, β, γ, etc., can be expressed only by differential equations, and consequently by linear equations among these quantities; so that there will be necessarily one or more of them which remain indeterminate and which may be taken either plus or minus; consequently the values of all these quantities will always be such that they may change sign together. Hence it follows that if, in a certain displacement of the system, the value of the quantity $P\alpha + Q\beta + R\gamma + $ etc. is negative, it will become positive by taking the quantities α, β, γ, etc. with opposite signs; and this opposite displacement, being equally possible, will cause the weight to descend and destroy the equilibrium.

Reciprocally, we may prove that if the equation:

$$P\alpha + Q\beta + R\gamma + \text{etc.} = 0$$

holds for all possible infinitely small displacements of the system, the system is necessarily in equilibrium; for since the weight remains motionless in these displacements the forces which act on the system remain in the same state, and there is no reason why they should produce the one rather than the other of the two displacements in which the quantities α, β, γ, etc. have opposite signs. This is the case of the balance, which is in equilibrium because there is no reason why it should incline one way rather than the other.

The principle of virtual velocities, which has thus been demonstrated for commensurable forces, will hold also for incommensur-

able forces; for it is well known that any proposition which has
been demonstrated for commensurable quantities may be demonstrated equally well for incommensurable quantities by a *reductio
ad absurdum*.

POINSOT

Louis Poinsot was born in Paris on January 3, 1777 and died there on
December 15, 1859. He was a pupil of the École Polytechnique and after
his graduation served as engineer for bridges and roads. In 1804 he was made
professor of mathematics at the Lyceum Bonaparte. In 1809 he became
connected with the École Polytechnique as professor of mathematics. He
was chosen member of the Institute in 1813 in place of Lagrange.

His principal contribution to physics is his theory of couples as it was
applied in his *Éléments de Statique*, 1803, and in his *Nouvelle Théorie de la
Rotation des Corps*, 1834. In these works he returned to synthetic methods.
By his methods he greatly clarified the formal presentation of the subject,
though of course he could add nothing to the general treatment of Lagrange.

The extract which follows from the *Éléments de
Statique*, as translated by Thomas Sutton in 1848,
describes some of the properties of couples.

Forces and Couples

68. Let there be any number of forces, P, P',
P'', etc. applied in any manner whatever in
space, to a body or free system.

First consider any one of them, P, for instance
(Fig. 22), which is applied at the point B. Then
at the point A arbitrarily taken in the body, or
without the body (provided we suppose it to be
rigidly connected therewith) apply two opposite

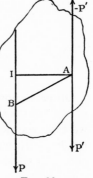

Fig. 22.

forces, P', $-P'$, equal and parallel to the force P. It is clear that
nothing will be altered in the state of the system. But we may
now consider, instead of the force P applied at B, the force P'
applied at A, and the couple $(P, -P')$ acting on the straight line
AB. If, for the sake of greater clearness, we transfer this couple
elsewhere, into any plane whatever, parallel to its own, there will
only remain at the point A the force P', equal and parallel to the
force P, which is nothing more than this force P which we have
transferred parallel to itself from B to A.

If we make the same transformations for all the forces of the
system, with respect to the same point A, it is manifest that all

those forces will be assembled there parallel to themselves, but that there will also be in the system as many couples applied, in consequence of each transformation.   But all the forces applied at A will compound into a single one R, and all the couples into a single couple $(S, -S)$ (Fig. 23) applied to a certain straight line BC.

Hence we learn that any number of forces, applied in any manner to a body, can always be reduced to a single force which passes through any proposed point, and to a single couple, whose plane will be, in general, inclined to the direction of the force.

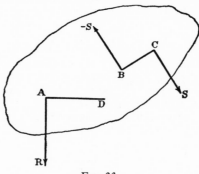

Fig. 23.

Observe, now, that the magnitude, the direction, and the sign of the resultant R will be always the same, in whatever position we take the point A.   By varying the position of this point, the resultant R will only transfer itself parallel to itself to different points in space; but the plane and the magnitude of the resultant couple $(S, -S)$ will necessarily be changed.

But amidst this infinity of reductions, relative to all the points A in space, there is one distinguished from all the rest in that the plane of the resultant couple is perpendicular to the line of action of the resultant.   We may demonstrate this at once in a very ready manner; for all being already reduced to the single force R, and the single couple $(S, -S)$, with respect to some known point A, suppose that we decompose this couple $(S, -S)$ into two others, the one $(T, -T)$, which falls in a plane perpendicular to the line of action of the resultant, and the other $(V, -V)$ in a plane which passes through this line of action AR.   In this plane we find at the same time the couple and the force R; we may therefore transfer this force parallel to itself from A to a point O, so situated, and at such a distance AO as that the couple $(R, -R)$ generated by this translation, may be equal and opposite to the couple $(V, -V)$ and may destroy it: then will there only remain the single force R applied at the new point O, and the couple $(T, -T)$, which is in a plane perpendicular to the direction of this force.   Thus,

Any number of forces are always reducible to a single force, and a single couple whose plane is perpendicular to the direction of the force.   So that there is always in space a certain determinate straight line OR, which may serve at the same time to represent the line of action of the resultant, and the axis of the resultant couple.

This reduction is unique; that is to say, there exists no other position in space at which we could find the resultant couple perpendicular to the resultant force.   For transfer the force R wherever we choose out of the actual position OR, it will produce a couple $(R, -R)$ perpendicular to the couple $(T, -T)$, and these two couples, on being compounded into one, will give the new resultant couple necessarily inclined to the couple $(T, -T)$, and therefore always greater, because the two components are at right angles to one another.   Whence we see not only that the couple $(T, -T)$ is the only one which can be perpendicular to the direction of the resultant, but that it is also the least of all the resultant couples, which can be formed relative to all points in space.   We see, at the same time, that for points taken about OR at equal distances from this straight line, the resultant couples have equal values and are in different planes, but equally inclined to this axis OR, which we may therefore call the central axis of the couples of the system; the further we get from this axis the greater the couples become, and they increase without limit, but they have all this common property, that each of them, resolved upon the plane perpendicular to the constant direction of the force R, gives the same couple $(T, -T)$, whence we see that the value of this minimum couple is always obtained by taking any resultant couple whatever $(S, -S)$ and multiplying it by the cosine of its inclination to the plane of the other.

We only just touch upon this central axis, it gives us so luminous a reduction of all the forces of the system as to throw a light at the same time upon all the other equivalent reductions, and to group them, so to speak, into a single picture, in which we see at once order and mutual dependence.

This theory will be more fully developed in the note upon moments and areas. But we must here confine ourselves to the general corollaries which have more immediate connexion with the elements of statics.

Corollary I

Which contains the laws of the equilibrium of any free system.

69. Since a couple can never be kept in equilibrium by any simple force, directed in any manner in space, it follows, from what we have just said, that there will never be equilibrium in the system, unless both the resultant R of the forces be itself equal to zero, and the resultant couple $(S, -S)$ be at the same time zero.

So that, all the forces applied to the system being transferred parallel to themselves to any point in the system, or in space, must be in equilibrium with one another; and all the couples which they produce by their transference to this point, must be also in equilibrium with one another.

.   .   .   .

PROPERTIES OF MATTER

GALILEO

The following short extract from Galileo's *Two New Sciences* (p. 1) describes an observation from which sprang the invention of the barometer.

RISE OF WATER IN A PUMP

The interlocutors Sagredo and Salviati have been discussing the explanation of tenacity by ascribing it to the existence of a vacuum between the small parts of a body, and the consequent striving of the parts of the body to approach nearer one another, because of the general principle that nature abhors a vacuum. Salviati has said that he accepts this view in part, but that he can prove that other causes are operative, and he describes an apparatus for that purpose.

SAGR. Thanks to this discussion, I have learned the cause of a certain effect which I have long wondered at and despaired of understanding. I once saw a cistern which had been provided with a pump under the mistaken impression that the water might thus be drawn with less effort or in greater quantity than by means of the ordinary bucket. The stock of the pump carried its sucker and valve in the upper part so that the water was lifted by attraction and not by a push as is the case with pumps in which the sucker is placed lower down. This pump worked perfectly so long as the water in the cistern stood above a certain level; but below this level the pump failed to work. When I first noticed this phenomenon I thought the machine was out of order; but the workman whom I called in to repair it told me the defect was not in the pump but in the water which had fallen too low to be raised through such a height; and he added that it was not possible, either by a pump or by any other machine working on the principle of attraction, to lift water a hair's breadth above eighteen cubits; whether the pump be large or small this is the extreme limit of the lift. Up to this time I had been so thoughtless that, although I knew a rope, or rod of wood, or of iron, if sufficiently long, would break by its own weight when held by the upper end, it never occurred to me that the same thing would happen, only much more easily, to a column of water. And really is not that thing which is attracted in the pump a column of water attached at the upper end and stretched more and more until finally a point is reached where it breaks, like a rope, on account of its excessive weight?

SALV. That is precisely the way it works; this fixed elevation of eighteen cubits is true for any quantity of water whatever, be the pump large or small or even as fine as a straw. We may therefore say that, on weighing the water contained in a tube eighteen cubits long, no matter what the diameter, we shall obtain the value of the resistance of the vacuum in a cylinder of any solid material having a bore of this same diameter.

TORRICELLI

Evangelista Torricelli was born of noble parents on October 15, 1608. At the age of twenty he came to Rome, where he studied mathematics under Castelli. He became acquainted with Galileo's work on motion and wrote a memoir on the subject which Castelli submitted to Galileo. This led to Torricelli's coming to Galileo as an assistant. He helped him prepare a work which was published by another of Galileo's followers, Vincenzo Viviani, in 1674. Galileo died three months after Torricelli joined him. The Grand Duke of Tuscany was so struck with the young man's talent that he retained him in Florence as a mathematician and teacher of mathematics. Torricelli died on October 25, 1647, in Florence.

He was a mathematician of considerable importance. As a physicist he is known best for his invention of the barometer and his law of the efflux of liquids from small orifices. The extract which follows contains the announcement of his invention of the barometer and his explanation of its behavior. It is taken from Torricelli's *Collected Works*, published in 1919, Vol. III, p. 186.

THE BAROMETER

To Michelangelo Ricci in Rome.

MOST ILLUSTRIOUS SIR AND   Florence, June 11, 1644
MOST LEARNED PATRON:

Several weeks ago I sent to Sig. Antonio Nardi several of my demonstrations of the areas of cycloids, and asked him that after he had examined them he would send them on at once to yourself or to Sig. Magiotti. I have already called attention to the fact that there are in progress certain philosophical experiments, I do not know just what, relating to vacuum, designed not simply to make a vacuum but to make an instrument which will show the changes in the atmosphere, as it is now heavier and more gross and now lighter and more subtle. Many have said that a vacuum does not exist, others that it does exist in spite of the repugnance of nature and with difficulty; I know of no one who has said that it exists without difficulty and without a resistance from nature.

I argued thus: If there can be found a manifest cause from which the resistance can be derived which is felt if we try to make a vacuum, it seems to me foolish to try to attribute to vacuum those operations which follow evidently from some other cause; and so by making some very easy calculations, I found that the cause assigned by me (that is, the weight of the atmosphere) ought by itself alone to offer a greater resistance than it does when we try to produce a vacuum. I say this because a certain philosopher, seeing that he cannot escape the admission that the weight of the atmosphere causes the resistance which is felt in making a vacuum, does not say that he admits the operation of the heavy air, but persists in asserting that nature also concurs in resisting the vacuum. We live immersed at the bottom of a sea of elemental air, which by experiment undoubtedly has weight, and so much weight that the densest air in the neighborhood of the surface of the earth weighs about one four-hundredth part of the weight of water. Certain authors have observed after twilight that the vaporous and visible air rises above us to a height of fifty or fifty-four miles, but I do not think it is so much, because I can

FIG. 24.

show that the vacuum ought to offer a much greater resistance than it does, unless we use the argument that the weight which Galileo assigned applies to the lowest atmosphere, where men and animals live, but that on the peaks of high mountains the air begins to be more pure and to weigh much less than the four-hundredth part of the weight of water. We have made many vessels of glass like those shown as *A* and *B* (Fig. 24) and with tubes two cubits long. These were filled with quicksilver, the open end was closed with the finger, and they were then inverted in a vessel where there was quicksilver C; then we saw that an empty space was formed and that nothing happened in the vessel where this space was formed; the tube between A and D remained always full to the height of a cubit and a quarter and an inch over. To show that the vessel was entirely empty, we filled the bowl with pure water up to D and then, raising the tube little by little, we saw that, when the opening of the tube reached the water, the quicksilver fell out of the tube and the water rushed with great violence up to the mark E.   It is often said in explanation of the fact that the vessel AE stands empty and the quicksilver, although heavy, is sustained in the tube AC, that, as has been believed hitherto, the force which prevents the quicksilver from falling down, as it would naturally do, is internal to the vessel AE, arising either from the vacuum or from some exceedingly rarefied substance; but I assert that it is external and that the force comes from without.   On the surface of the liquid which is in the bowl there rests the weight of a height of fifty miles of air; then what wonder is it if into the vessel CE, in which the quicksilver has no inclination and no repugnance, not even the slightest, to being there, it should enter and should rise in a column high enough to make equilibrium with the weight of the external air which forces it up?   Water also in a similar tube, though a much longer one, will rise to about 18 cubits, that is, as much more than quicksilver does as quicksilver is heavier than water, so as to be in equilibrium with the same cause which acts on the one and the other.   This argument is strengthened by an experiment made at the same time with the vessel A and with the tube B in which the quicksilver always stood at the same horizontal line AB. This makes it almost certain that the action does not come from within; because the vessel AE, where there was more rarefied substance, should have had a greater force, attracting much more actively because of the greater rarefaction than that of the much smaller space B.   I have endeavored to explain by this principle all sorts of repugnances which are felt in the various effects attributed to vacuum, and I have not yet found any with which I cannot deal successfully.   I know that your highness will perceive many objections, but I hope that if you think them over they will be resolved.   My principal intention I was not able to carry out, that is, to recognize when the atmosphere is grosser and heavier and when it is more subtle and lighter, because the level AB in the instrument EC changes for some other reason (which I would not have believed) especially as it is sensible to cold or heat, exactly as if the vessel *AE* were full of air.

Your devoted and obliged Servant,
V. Torricelli.

PASCAL

Blaise Pascal was born in Clermont on June 19, 1623. His father was a government official and a man of culture. The son showed unusual abilities from his earliest youth and his father devoted himself to his instruction. The boy became interested very early in physical and mathematical studies, although his father attempted for a while to prevent his knowing anything about those subjects for fear that he might neglect his study of languages. It is said that the boy discovered for himself some of the most important propositions of geometry. In his sixteenth year he wrote a paper on conic sections.

His health failed him in his eighteenth year and until his death he was an invalid.

Pascal was one of the founders of the theory of probabilities and contributed to the geometry of cycloids. His principal contribution to physics was in the domain of hydrostatics.

In the later years of his life Pascal became intimately connected with the Jansenists of Port Royal. Under their influence he wrote the *Provincial Letters* and compiled his *Pensées sur la Religion* published after his death. He died on August 19, 1662.

The extracts which follow are, first, a letter from Pascal's brother-in-law, Périer, in which he describes the results of his observations with a barometer at different heights. This experiment had been suggested to him by Pascal and Pascal's comment on the result is given. The second extract is taken from his *Traité de l'Équilibre des Liqueurs*, which was published in 1663 but was probably written ten years earlier.

EXPERIMENTS WITH THE BAROMETER

September 22nd, 1648.

After giving the names of the gentlemen of Clermont who were associated with him, Périer begins the account of the experiments.

We therefore met on that day at eight o'clock in the morning in the garden of the Pères Minimes, which is in almost the lowest part of the town, where the experiment was begun in the following way:

First, I poured into a vessel sixteen pounds of quicksilver, which I had purified during the three preceding days; and taking two tubes of glass of equal size, each about four feet long, hermetically sealed at one end and open at the other, I made with each of them the ordinary experiment of the vacuum in the same vessel, and when I brought the two tubes near each other without lifting them out of the vessel, it was found that the quicksilver which remained in each of them was at the same level, and that it stood in each of them above the quicksilver in the vessel twenty-six inches three lines and a half.   I repeated this experiment twice in the same place, with the same tubes, with the same quicksilver and in the same vessel; and found always that the quicksilver in the tubes was still at the same level and the same height as I found it the first time.

When this had been done, I left one of the two tubes in the vessel, for continual observation: I marked on the glass the height of the quicksilver, and leaving the tube in its place, I begged the Rev. Father Chastin, one of the inmates of the house, a man as pious as he is capable, who thinks very clearly in matters of this sort, to take the trouble to observe it from time to time during the day, so as to see if any change occurred.   And with the other tube and a part of the same quicksilver, I ascended with all these gentlemen to the top of the Puy-de-Dôme, which is higher than the Minimes by about five hundred toises, where, when we made the same experiments in the same way as I had at the Minimes, it was found that there remained in the tube no more than twenty-three inches two lines of quicksilver, whereas at the Minimes there was found in the same tube a height of twenty-six inches, three lines and a half; and so there was between the heights of the quicksilver in these experiments a difference of three inches one line and a half: this result so filled us with admiration and astonishment, and so much surprised us, that for our own satisfaction we wished to repeat it.   I therefore tried the same thing five times more, with great accuracy, at different places on the top of the mountain, once under cover in the little chapel which is there, once exposed, once in a shelter, once in the wind, once in good weather, and once during the rain and the mists which came over us sometimes, having taken care to get rid of the air in the tube every time; and in all these trials there was found the same height of the quicksilver, twenty-three inches two lines, which makes a difference of three inches one line and a half from the twenty-six inches three lines and a half which were found at the Minimes; this result fully satisfied us.

The rest of the letter relates to other experiments made during the descent and reports that the height of the quicksilver in the stationary tube had not changed during the day. It also reports that the height of the quicksilver was found to be slightly lower by ascending to the top of the Cathedral in Clermont.

The following statement is Pascal's comment on this letter:

This account cleared up all my difficulties and I do not conceal the fact that I was greatly delighted with it; and since I noticed that the distance of twenty toises in height made a difference of two lines in the height of the quicksilver, and that six or seven toises made one of about half a line, a fact which it was easy to test in this city, I made the ordinary experiment of the vacuum at the top and at the bottom of the tower of Saint-Jacques de-la-Boucherie, which is from twenty-four to twenty-five toises high: I found a difference of more than two lines in the height of the quicksilver; and then I made the same experiment in a private house, with ninety-six steps in the stairs, where I found very plainly a difference of half a line; which agrees perfectly with the account of Périer.

Fluid Pressure

Chapter I of the *Traité de l'Équilibre des Liqueurs* contains a description of several experiments by which it is proved that the force exerted on a piston or plug of a certain area at the bottom of a column of water is the same, if the vertical height of the column is the same, whether the column itself stands vertically or is inclined, and whether the column has a small or a great cross-section. When Pascal speaks of the weight of the liquid he means the force exerted on the piston or plug of a given area by the weight of the liquid above it.

Chapter II

Why the Weight of Liquids is in Proportion To Their Height.

We see by all these examples that a small filament of water may keep a great weight in equilibrium: it remains to show what is the reason for this increase of force; we proceed to do this by the following experiment (Fig. 25):

If a vessel full of water, otherwise completely closed, has two openings, one of which is one hundred times as large as the other;

by putting in each of these a piston which fits it exactly, a man pushing on the small piston will exert a force equal to that of one hundred men who are pushing on the piston which is one hundred times as large, and will overcome the force of ninety nine men.

Whatever proportion there is between these openings, if the forces which are applied to the pistons are proportional to the openings, they will be in equilibrium. Hence it appears that a vessel full of water is a new principle of mechanics, and a new machine to multiply forces to any degree we please, since a man in this way can lift any load that is given to him.

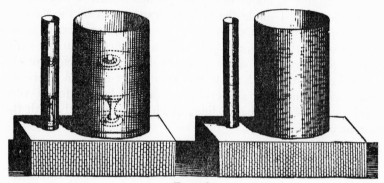

Fig. 25.

And it is truly admirable that there is encountered in this new machine the constant rule which appears in all the older machines, such as the lever, the wheel and axle, the endless screw, etc., which is, that the path is increased in the same proportion as the force; for it is clear that, since one of these openings is one hundred times as large as the other, if the man who pushes on the small piston moves it forward an inch, he will push out the other one only one one-hundredth part of an inch, for, since this movement occurs because of the continuity of the water which acts between the pistons, so that the one of them cannot move without moving the other, it is evident that, when the small piston is moved through an inch of water, which has been pushed forward by it, and which pushes the other piston, since the opening through which it moves is one hundred times as large, it will occupy only one one-hundredth of the height. So that the path is to the path as the force to the force; and this rule we may consider as the true cause of this effect: it is clear that it is the same thing to make one hundred pounds of water move one inch as to make one pound of water move one hundred inches; and thus when a pound of water is so situated in relation to one hundred pounds of water that the one hundred pounds cannot move an inch without making the pound move one hundred inches, they must remain in equilibrium, a pound having as much force to make one hundred pounds move an inch as one hundred pounds have to make a pound move one hundred inches.

It may be added, for greater clearness, that the water is under equal pressure under the two pistons; for if one of them has one hundred times more weight than the other it also touches one hundred times as many parts of the liquid, and so each part is equally pressed; therefore all parts ought to be at rest, because there is no reason why one should yield rather than the other. So that if a vessel has only a single opening, an inch across for example, in which there is a piston loaded with a pound weight, this weight presses against all parts of the vessel generally, because of the continuity and the fluidity of the water: but to determine the pressure of each part we have the following rule: Each part an inch large, like the opening, is subjected to as much pressure as if it were pushed by a pound weight (without counting the weight of the water, of which I do not here speak, for I am speaking only of the weight of the piston), because the weight of a pound presses the piston which is in the opening, and each portion of the vessel more or less great is subjected to pressure precisely more or less in proportion to its size, whether this portion is opposite the opening or at the side, far or near; for the continuity and the fluidity of the water make all these things equal and indifferent; so that it is necessary that the material of which the vessel is made should have sufficient resistance in all its parts to sustain all these forces; if the resistance is less in any one of these places, it yields; if it is greater, it provides all the necessary force and the rest remains useless in the circumstances; so that if a new opening is made in the vessel there would be needed to check the water which would gush out, a force equal to the resistance which this part ought to present, that is to say, a force which should be to the force of a pound as this last opening is to the first.

There follows a proof that the displacements which occur are consistent with the principle that a body is never moved by its own weight unless the center of gravity can descend.

Chapter III

Examples and Reasons for the Equilibrium

If a vessel full of water has two openings, to each one of which is fastened a vertical tube; if water is poured into the one and into the other to the same height, the columns of water will be in equilibrium.   (Fig. 25.)

For since their heights are equal, they will be in proportion to their magnitudes, that is to say, to their openings, therefore the two masses of water in these tubes are properly two pistons whose weights are proportional to their openings; therefore they will be in equilibrium, by the preceding demonstration.

Hence it follows that, if we pour water into one of these tubes only, it will make the water rise in the other one until it has come to the same height, and then the water will remain in equilibrium, for then the water columns will be two pistons, the weights of which are in proportion to their openings.

This is the reason why water rises as high as its source.

If different liquids are placed in the tubes, such as water in the one and quicksilver in the other, these two liquids will be in equilibrium when their heights are inversely proportional to their weights; that is to say, when the height of the water is fourteen times as great as the height of the quicksilver, because quicksilver weighs fourteen times as much as water does; for then there will be two pistons, one of water and one of quicksilver, of which the weights are proportional to the openings.

And even when the tube full of water is one hundred times narrower than that in which the quicksilver is, this little thread of water will hold in equilibrium all the great mass of quicksilver, provided that it is fourteen times as high.

All that we have said up to this point about tubes can be applied to any vessel whatever, regular or not; for the same equilibrium occurs in any case: so that, if, in place of these two tubes that we have represented at the two openings, there are two vessels used; which are applied also to these openings but which are large in some places, narrow in others, and indeed entirely irregular in all their extent; when liquids have been poured in to the heights which we have described, these liquids will also be in equilibrium in these irregular tubes, just as they were in the uniform tubes, because the weight of liquids is measured only by their heights and not by their sizes.

Chapter IV deals with equilibrium between a column of water and a block of copper which is placed as a piston in the lower end of a vertical tube that projects above the surface of the water. The experiment shows that the copper is in equilibrium when it is so far below the surface of the water that the water column of the same cross-section weighs as much as the copper block does. Another experiment shows that the pressure of the water may also be exerted downward on a block of wood used as a piston in a recurved tube.

Chapter V

On Bodies Which Are Immersed in Water

We see by the former chapter that water pushes upward those bodies which it touches from below; that it pushes downward those bodies which it touches from above; and that it pushes toward one side those which it touches on the opposite side: from which it is easy to conclude that when a body is entirely immersed in water, since the water touches it above, below and on all its sides, it strives to push the body upward, downward and in every direction: but since its height is the measure of its force in all these actions, it is easy to see which of all these forces ought to prevail.

For it appears, first, that since it has the same height of water on all the faces of its sides, they will be pushed equally; and therefore the body will not be moved sideways in any direction, any more than a weather-vane between two equal winds. But since the water has more height on the lower face than on the upper face, it is clear that it will push the body more upward than downward: since the difference of these heights of the water is the height of the body itself, it is easy to understand that the water pushes it more upward than downward with a force equal to the weight of the volume of water equal to that of the body.

So that a body which is in water is borne up, in the same way as if it were in one scale of a balance, and the other were loaded with a volume of water equal to its own, from which it appears that a body of copper or of any other matter which weighs more than an equal volume of water, will sink; for its weight exceeds that which counterbalances it.

If a body is of wood, or of some other matter lighter than an equal volume of water, it rises with the force by which the weight of the water is greater than its weight.

And if it is of equal weight, it neither sinks nor rises; as wax does, which stays almost in the same place in water when it is once put there.

VON GUERICKE

Otto von Guericke was born in Magdeburg on November 20, 1602. Members of his family had for many years been prominent in the life of the city. He studied law in several of the German universities and in Leyden he occupied himself with mathematics and mechanics. On his return to Magdeburg he devoted himself to the interests of the city and was chosen one of its magistrates in 1627. He was in the city when it was sacked in 1631. He and his family escaped and his house was protected by a special order. In the various difficult situations which arose in the remaining years of the Thirty Years' War, von Guericke was actively engaged in the negotiations by which the city endeavored to avoid further calamities. After the cessation of the troubles he lived for many years in peace and with the highest respect of his fellow citizens. When nearly eighty years of age he was driven from his home by the plague and spent the remaining years of his life in the house of his son in Hamburg, where he died in May, 1686.

Von Guericke is best known for his invention of the air pump and for the experiments which he tried with the vacuum made by it. The following extract gives his account of his early experiments. He also made the observation of electrical repulsion which will be given in its proper place. These extracts are taken from a book published in 1672, entitled *Ottonis de Guericke Experimenta nova (ut vocantur) magdeburgica*.

THE AIR PUMP

First Attempt to Produce a Vacuum by Drawing Out Water

While I was engaged in the consideration of the immeasurability of space and considered that it must be everywhere present, I thought of the following investigation.

I thought of filling a wine or beer cask with water and caulking it everywhere so that the outside air could not enter. In the lower part of the cask a metal tube was to be introduced by means of which the water could be drawn out; the water then, in consequence of its weight would sink, and leave behind it in the cask a space empty of air and therefore of any body.

That the result should correspond to this plan, I arranged a brass force pump, *abc* (Fig. 26), like those used for fires, with a piston *c* or *f* and a plug *g* which was worked very accurately so that the air could not go in or out past its sides.   In the pump were further introduced two leather valves, of which the inner one *a* or *d* in the end of the pump permitted the entrance of the water, the outer one, *b*, its outflow.   After fastening the pump by means of the ring *e*, furnished with four lugs, to the lower part of the cask I undertook to draw out the water.   Before the water followed the piston, the lugs were broken and the iron screws by means of which the pump was fastened to the cask were drawn out.

The trial, however, was not without value.   After providing stronger screws, it came about at last that three strong men pulling on the piston of the pump drew out the water through the upper valve *b*.   But when this happened a rustling noise was heard in all parts of the cask as if the water was boiling vigorously, and this lasted until the cask was filled with air in place of the water that had been drawn out.

Some way had to be found to avoid this unfortunate result. I therefore prepared a smaller cask which I placed in a larger one. After a longer pipe attached to the pump had been carried up through the bottoms of both the casks, I filled the smaller cask with water, sealed up its opening and, after the larger cask was also filled with water, began the task again.   This time we were able to draw the water out of the smaller cask, in place of which, without doubt, a vacuum remained.

However, as the day went by and the work was stopped and everything around became quiet, we perceived a varying tone, interrupted from time to time, like that of a twittering song-bird. This lasted for three whole days.

When, after that, the mouth of the smaller cask was opened, it was found to be filled for the greater part with air and water. Notwithstanding, a part of it was empty, since, while it was being opened, some air entered.

All of us were astonished that the water entered a cask which had been so carefully pitched and closed everywhere.   I perceived finally, after many repeated researches, that the water under high pressure passed through the wood, and on account of the pressure and the friction occasioned by passing through the wood, a little air developed itself from the water in the cask (which in the future must be kept in mind).   The cask, however, could not be filled entirely with air on account of the resistance which the wood offered to its passage.   When the pressure ceased, the

entrance of water and air came to an end; therefore, we obtained
an only half-emptied cask.

Second Attempt to Obtain Vacuum by Drawing Out Air

After the porosity of wood was proved by inspection, as well
as by the investigation, it appeared to me that for my purposes a

FIG. 26.

FIG. 27.

copper sphere (which the Reverend Father Schott, in his book
on the Magdeburgian investigation, calls "Cacabus") would be
more suitable. This sphere A (Fig. 27) contained from 60 to

70 Magdeburgian quarts and was furnished with a brass stopcock *B* at the top; at the bottom the pump was introduced and joined closely to it. Then I again undertook, as before, to draw out water and air.

At first the piston moved easily, but soon it became more difficult to move it, so that two strong men were hardly able to pull the piston out. While they were still occupied with pulling it in and out, and already believed that nearly all the air was drawn out, suddenly with a loud noise and to the astonishment of all the metal sphere was crushed in as a cloth can be rolled up between the fingers, or as if the sphere had been thrown down from the summit of a tower with a violent shock.

I believe that the cause of this was the inexpertness of the workmen, who perhaps had not made this sphere exactly spherical. The flat part, wherever it was, could not sustain the pressure of the surrounding air, whereas on the other hand an exactly made sphere could easily have sustained it on account of the mutual support of its parts which sustain each other in overcoming resistance.

It was therefore necessary that the workmen should make a perfectly round sphere. From this the air was pumped out, at first easily but with great trouble toward the end.

As a proof that the sphere was completely evacuated, we have the circumstance that at last no more air passed out of the upper valve of the pump.

Thus for the second time a vacuum was obtained.

On opening the stop-cock *B*, the air rushed with such force into the copper sphere that it seemed as if it would draw in a man standing before it. If the face was brought fairly near, the breath was taken away and one could not hold one's hand above the stop-cock without the risk that it would be violently drawn to it.

Although the sphere seemed to be completely emptied, yet experience showed that when it was left for a day or two, it filled up again with air, which passed in along the plug of the pump and also through the valve and the stop-cock. It was therefore necessary to avoid this defect, as will be shown later.

The remaining chapters are largely concerned with the perfection of the air pump, of which the instrument described is an imperfect model. Among the experiments described is the famous one of the Magdeburg hemispheres. In this, two hemispheres nearly three-quarters of a Magdeburgian ell in diameter were so strongly pressed together, when the space within them was evacuated, that sixteen horses pulling in teams of eight against each other could not separate them.   Similar hemispheres were arranged which sustained heavy weights.

BOYLE

Robert Boyle was born in Ireland on January 25, 1627.   His father was the Earl of Cork.   At an early age he went to Eton and after a few years of study there traveled extensively on the Continent with a tutor.   He returned home in 1644 and by the death of his father came into possession of an estate on which for some years he remained quietly at work.   Ten years later he removed to Oxford, where he lived for fourteen years.   He was one of the first members of the Royal Society and after moving to London in 1668 he became one of its most active supporters.   He died on December 30, 1691.

Boyle's experimental researches covered a wide range in chemistry and physics.   His most important contribution was his discovery of the law connecting the pressure and volume of a gas.   The extract which follows describes his experiments and their results.   It is taken from his book entitled *A Defense of the Doctrine Touching the Spring and Weight of the Air*, third edition, 1682. The first edition was published in 1660, the second, containing the account here given, in 1662.

Relations of Pressure and Volume of Air

Two New Experiments Touching the Measure of the Force of the Spring of Air Compressed and Dilated.

We took then a long glass-tube, which, by a dexterous hand and the help of a lamp, was in such a manner crooked at the bottom, that the part turned up was almost parallel to the rest of the tube, and the orifice of this shorter leg of the siphon (if I may so call the whole instrument) being hermetically sealed, the length of it was divided into inches (each of which was subdivided into eight parts) by a streight list of paper, which containing those divisions, was carefully pasted all along it.   Then putting in as much quicksilver as served to fill the arch or bended part of the siphon, that the mercury standing in a level might reach in the one leg to the bottom of the divided paper, and just to the same height or horizontal line in the other; we took care, by frequently inclining the tube, so that the air might freely pass from one leg into the other by the sides of the mercury (we took, I say, care) that the air at last included in the shorter cylinder should be of the same laxity with the rest of the air about it.   This done, we began to pour quicksilver into the longer leg of the siphon, which by its weight pressing up that in the shorter leg, did by degrees streighten the included air: and continuing this pouring in of quicksilver till the air in the shorter leg was by condensation reduced to take up by half the space it possessed (I say, possessed, not filled) before; we cast our eyes upon the longer leg of the glass, on which was likewise pasted a list of paper carefully divided into inches and parts, and we observed, not without delight and satisfaction, that the quicksilver in that longer part of the tube was 29 inches higher than the other. Now that this observation does both very well agree with and confirm our hypothesis, will be easily discerned by him that takes notice what we teach; and Monsieur Paschal and our English friend's experiments prove, that the greater the weight is that leans upon the air, the more forcible is its endeavour of dilatation, and consequently its power of resistance (as other springs are stronger when bent by greater weights). For this being considered, it will appear to agree rarely-well with the hypothesis, that as according to it the air in that degree of density and correspondent measure of resistance, to which the weight of the incumbent atmosphere had brought it, was able to counter-balance and resist the pressure of a mercurial cylinder of about 29 inches, as we are taught by the Torricellian experiment; so here the same air being brought to a degree of density about twice as great as that it had before, obtains a spring twice as strong as formerly. As may appear by its being able to sustain or resist a cylinder of 29 inches in the longer tube, together with the weight of the atmospherical cylinder, that leaned upon those 29 inches of mercury; and, as we just now inferred from the Torricellian experiment, was equivalent to them.

We were hindered from prosecuting the trial at that time by the casual breaking of the tube. But because an accurate experiment of this nature would be of great importance to the doctrine of the spring of the air, and has not yet been made (that I know) by any man; and because also it is more uneasy to be made than one would think, in regard of the difficulty as well of procuring crooked tubes fit for the purpose, as of making a just estimate of the true place of the protuberant mercury's surface; I suppose it will not be unwelcome to the reader to be informed, that after some other trials, one of which we made in a tube whose longer leg was perpendicular, and the other, that contained the air, parallel to the horizon, we at last procured a tube of the figure expressed in the scheme; which tube, though of a pretty bigness, was so long, that the cylinder, whereof the shorter leg of it consisted, admitted a list of paper, which had before been divided into 12 inches and their quarters, and the longer leg admitted another list of paper of divers feet in length, and divided after the same manner. Then quicksilver being poured in to fill up the bended part of the glass, that the surface of it in either leg might rest in the same horizontal line, as we lately taught, there was more and more quicksilver poured into the longer tube; and notice being watchfully taken how far the mercury was risen in that longer tube, when it appeared to have ascended to any of the divisions in the shorter tube, the several observations that were thus successively made, and as they were made set down, afforded us the ensuing table:

A TABLE OF THE CONDENSATION OF THE AIR

A	A	B	C	D	E	
48	12	00		29 2/16	29 2/16	AA. The number of equal spaces in the shorter leg,
46	11½	01 7/16		30 9/16	30 6/16	that contained the same
44	11	02 13/16		31 15/16	31 12/16	parcel of air diversely
42	10½	04 6/16		33 8/16	33¼	extended.
40	10	06 3/16		35 5/16	35	
38	9½	07 14/16		37	36 15/19	B. The height of the mer-
36	9	10 2/16		39 5/16	38⅞	curial cylinder in the longer
34	8½	12 8/16		41 10/16	41 2/17	leg, that compressed the
32	8	15 1/16	Added to 29⅞ makes	44 3/16	43 11/16	air into those dimensions.
30	7½	17 15/16		47 1/16	46⅜	
28	7	21 3/16		50 5/16	50	C. The height of the mer-
26	6½	25 3/16		54 5/16	53 10/13	curial cylinder, that coun-
24	6	29 11/16		58 13/16	58⅜	terbalanced the pressure
23	5¾	32 3/16		61 5/16	60 18/23	of the atmosphere.
22	5½	34 15/16		64 1/16	63 6/11	
21	5¼	37 15/16		67 1/16	66 4/7	D. The aggregate of the
20	5	41 9/16		70 11/16	70	two last columns, B and
19	4¾	45		74 2/16	73 11/19	C, exhibiting the pressure
18	4½	48 12/16		77 14/16	77⅔	sustained by the included
17	4¼	53 11/16		82 12/16	82 4/7	air.
16	4	58 2/16		87 14/16	87⅜	E. What that pressure should
15	3¾	63 15/16		93 1/16	93⅕	be according to the hypo-
14	3½	71 5/16		100 7/16	99 6/7	thesis, that supposes the
13	3¼	78 11/16		107 13/16	107 7/13	pressures and expansions
12	3	88 7/16		117 9/16	116⅝	to be in reciprocal proportion.

· · · · · · · · · · · · · · · · · ·

And to let you see, that we did not (as a little above) inconsiderately mention the weight of the incumbent atmospherical cylinder as a part of the weight resisted by the imprisoned air, we will here annex, that we took care, when the mercurial cylinder in the longer leg of the pipe was about an hundred inches high, to cause one to suck at the open orifice; whereupon (as we expected) the mercury in the tube did notably ascend. . . . And therefore we shall render this reason of it that the pressure of the incumbent air being in part taken off by its expanding itself into the sucker's dilated chest, the imprisoned air was thereby enabled to dilate itself manifestly, and repel the mercury, that comprest it, till there was an equality of force betwixt the strong spring of the comprest air on the one part, and the tall mercurial cylinder, together with the contiguous dilated air, on the other part.

Now, if to what we have thus delivered concerning the compression of the air, we add some observations concerning its spontaneous expansion, it will the better appear, how much the phaenomena of these mercurial experiments depend upon the differing measures of strength to be met with in the air's spring, according to its various degrees of compression and laxity.

A TABLE OF THE RAREFACTION OF THE AIR

	A	B	C	D	E
A. The number of equal spaces at the top of the tube, that contained the same parcel of air.	1	$00\frac{6}{8}$		$29\frac{3}{4}$	$29\frac{3}{4}$
	$1\frac{1}{2}$	$10\frac{5}{8}$		$19\frac{1}{8}$	$19\frac{5}{8}$
	2	$15\frac{3}{8}$		$14\frac{3}{8}$	$14\frac{7}{8}$
	3	$20\frac{2}{8}$		$9\frac{4}{8}$	$9\frac{15}{12}$
B. The height of the mercurial cylinder, that together with the spring of the included air counterbalanced the pressure of the atmosphere.	4	$22\frac{5}{8}$		$7\frac{1}{8}$	$7\frac{7}{16}$
	5	$24\frac{1}{8}$		$5\frac{5}{8}$	$5\frac{19}{25}$
	6	$24\frac{7}{8}$	Subtracted from $29\frac{3}{4}$ leaves	$4\frac{7}{8}$	$4\frac{23}{24}$
	7	$25\frac{4}{8}$		$4\frac{2}{8}$	$4\frac{1}{4}$
	8	$26\frac{0}{8}$		$3\frac{6}{8}$	$3\frac{23}{32}$
	9	$26\frac{3}{8}$		$3\frac{3}{8}$	$3\frac{11}{36}$
C. The pressure of the atmosphere.	10	$26\frac{6}{8}$		$3\frac{0}{8}$	$2\frac{39}{40}$
	12	$27\frac{1}{8}$		$2\frac{5}{8}$	$2\frac{23}{48}$
	14	$27\frac{4}{8}$		$2\frac{2}{8}$	$2\frac{1}{8}$
D. The complement of B to C, exhibiting the pressure sustained by the included air.	16	$27\frac{6}{8}$		$2\frac{0}{8}$	$1\frac{55}{64}$
	18	$27\frac{7}{8}$		$1\frac{7}{8}$	$1\frac{47}{72}$
	20	$28\frac{0}{8}$		$1\frac{6}{8}$	$1\frac{9}{80}$
	24	$28\frac{2}{8}$		$1\frac{4}{8}$	$1\frac{23}{96}$
E. What that pressure should be according to the hypothesis.	28	$28\frac{3}{8}$		$1\frac{3}{8}$	$1\frac{1}{16}$
	32	$28\frac{4}{8}$		$1\frac{2}{8}$	$0\frac{119}{128}$

MARIOTTE

Edme Mariotte was born in Bourgogne in 1620. He joined a religious order and rose to be the Prior of a convent near Dijon. He was one of the first members of the French Academy of Sciences. He died on May 12, 1684.

His investigations in physics cover a wide range, though they were principally restricted to the field of mechanics. He independently discovered the law connecting the pressure and volume of a gas, of which he published an account in 1676. Boyle's previous work had been overlooked and the law became known and is still often known by his name, especially on the European Continent. The following extract describing the experiments to prove the law is taken from his paper on the "Nature of Air" (1676 and 1679) as it appears in his *Collected Works*, p. 150, published at Leyden in 1717.

RELATIONS OF PRESSURE AND VOLUME OF AIR

The second property of air is that of being capable of extreme condensation and dilatation and of always retaining the power of a spring, by which it pushes back or endeavors to push back the bodies which press on it, until it has recovered its natural extension. For the most part other springs get weaker and weaker but one does not notice that the spring of the air gets weaker; and some people have told me that they have seen air guns, which had been charged for more than a year, have the same effect as if they had just been charged. The air also dilates very easily by heat and condenses by cold, as we can notice any day by experiment.

It should not be thought that the air near the surface of the earth, which we breathe, has its natural extension: for since that which is above is heavy and has the power of a spring, that which is here below, being loaded with the weight of the whole atmosphere, ought to be much more condensed than that which is higher up, which is free to dilate; and that which is between the two extremes ought to be less condensed than that which is on the earth and less dilated than that which is the most remote.

We can understand in a way this difference of condensation of air by using the example of many sponges, which have been piled one on the other. For it is evident that those which are at the top of the pile will have their natural extent; that those which are just below them will be a little less dilated; and that those which are at the bottom of the whole pile will be very closely compressed and condensed. It is further plain that, if we take away all those at the top, those below will recover their natural extent by the power of the spring which they have, and that if we take away only a part they will recover only a part of their dilatation.

The first question that we can ask is to know if the air condenses itself precisely in the proportion of the weights with which it is loaded, or if this condensation follows other laws and other proportions. I give here the steps of the argument which I used to determine whether the condensation of air is in proportion to the weights which compress it.

If we suppose, as experiment shows, that air condenses more when it is compressed by a greater weight, it necessarily follows that if the air, which extends from the surface of the earth to the greatest height at which it terminates, becomes lighter, the lowest part of it would be more dilated than it is at present, and if it becomes heavier this same part would be more condensed. We must therefore conclude that the condensation which it has near the earth is in a certain proportion to the weight of the air above it by which it is compressed, and that in this condition it is precisely in equilibrium by its spring with the whole weight of the air which it sustains.

From which it follows that if we introduce some air along with the mercury of a barometer and perform the experiment of the vacuum, the mercury will not remain in the tube at its ordinary height: for the air which is enclosed in it before the experiment makes equilibrium by its spring with the weight of the whole atmosphere, that is to say, with the column of air of the same section which extends from the surface of the mercury of the vessel to the limits of the atmosphere, and consequently, since the mercury which is in the tube encounters nothing which makes equilibrium with it, it will descend, but it will not descend altogether; for when it descends the air enclosed in the tube dilates, and consequently its spring is no longer sufficient to make equilibrium with all the weight of the air above it. Therefore a part of the mercury must remain in the tube at such a height that, since the air which is enclosed is in a condition of condensation which gives it a force of spring that can sustain only a part of the weight of the atmosphere, the mercury which remains in the tube makes equilibrium with the rest, and then there will be equilibrium between the weight of the whole column of air and the weight of the remaining mercury joined to the force of the spring of the enclosed air. Now if the air is condensed in proportion to the weight with which it is loaded, it follows necessarily that if we make an experiment in which the mercury remains in the tube at the height of fourteen inches, the air which is enclosed in the rest of the tube will be then dilated twice as much as it was before the experiment; provided that at the same time in the barometers without air the mercury is raised precisely to twenty-eight inches.

To determine if this conclusion was the true one I carried out experiments with M. Hubin, who is very expert in constructing barometers and thermometers of many sorts. We used a tube forty inches long, which I had filled with mercury up to twenty-seven inches and a half, so that there were twelve and a half inches of air, and when it was immersed by an inch in the mercury of the vessel, there were thirty-nine inches remaining to contain fourteen inches of mercury and twenty-five inches of air expanded to twice its original volume.   I was not deceived in my expectations: for when the end of the inverted tube was plunged in the mercury of the vessel, the mercury of the tube descended and after some oscillations stood quiet at a height of fourteen inches; and consequently the enclosed air, which then occupied twenty-five inches, was extended to double the volume of that which had been enclosed in the tube when it occupied only twelve and a half inches.

I had him make another experiment in which he left twenty-four inches of air above the mercury and the mercury descended to seven inches, in agreement with this hypothesis; for seven inches of mercury made equilibrium with a quarter of the weight of the whole atmosphere and the three quarters which remained were sustained by the spring of the enclosed air, which then being extended to thirty-two inches was in the same ratio with its original extent of twenty-four inches as the entire weight of the air is to three quarters of the same weight.

I had him try some other similar experiments, in which more or less air was left in the same tube or in others of different sizes; and I always found that when the experiment was tried the proportion of the dilated air to the extent of that which had been left above the mercury before the experiment was the same as that of twenty-eight inches of mercury, which is the whole weight of the atmosphere, to the difference between twenty-eight inches and the height at which the mercury stood after the experiment: and thus it is made sufficiently clear that we can take as a fixed rule or law of nature that air is condensed in proportion to the weight with which it is loaded.

If we wish to make more delicate experiments we must have a recurved tube, of which the two branches are parallel, one of which is about eight feet long and the other twelve inches; the long tube ought to be open at the top and the other carefully sealed.

We begin by pouring in a little mercury to fill up the bottom where the two branches communicate, and we arrange it so that the mercury stands no higher in one than in the other, so as to be assured that the enclosed air is neither more condensed nor more dilated than the free air.

We then pour mercury into the tube little by little, taking care that the shock does not introduce any additional air into that which is enclosed, and we shall see, as I have seen several times, that when the mercury has risen four inches in the small branch the mercury in the other will be fourteen inches higher, that is to say eighteen inches above the communicating tube; which is what ought to happen if air condenses in proportion to the weight with which it is loaded; since the enclosed air is then loaded with the weight of the atmosphere, which is equal to the weight of twenty-eight inches of mercury, and also with that of fourteen inches, of which the sum 42 inches is to 28 inches, the first weight which kept the air at twelve inches in the small branch, reciprocally as this extent of twelve inches is to the remaining extent of eight inches.

If we again pour in mercury until it rises to 6 inches in the small branch, so that there remains only 6 inches of air, the mercury in the other branch will be higher by 28 inches than the height of these six inches; which is what ought to happen according to the same hypothesis: for then the enclosed air will be loaded with 28 inches of mercury and with the weight of the atmosphere, which is also 28, the sum of which 56 is twice 28 as the first extent of 12 inches of air is twice that of the 6 inches which remain; and when as we continue to pour mercury into the long branch it rises in the small branch to a height of 8 inches, there will be 56 inches of mercury above it in the long branch, which makes again the same proportion.

If we wish to carry the experiment further we may pour in still more mercury, until the air in the small branch is reduced to 3 inches; and we shall see that in the other branch the mercury will be raised to 84 inches above it, which with the 28 of the weight of the atmosphere makes 112, a number four times 28, just as the first extent of 12 inches is four times the last extent of 3 inches.

To carry out these experiments successfully the small branch should have a section which is uniformly the same; but it is not necessary that the section of the long branch should be precisely the same in all its length.

NEWTON

The demonstration of the law of gravitation is the theme of Newton's *Principia* (p. 30). In the Third Book he applies the mathematical theorems of the First Book especially to prove the two elements which go to make up the law, namely, the attraction of any two bodies to each other with a force proportional to their masses and inversely proportional to the square of their distances apart. The proof is left to work its way into the mind by a series of successful applications of the law, and there is nowhere to be found an epitome of the argument and a succinct statement of the law, which would be suitable for quotation in this book. It is greatly to be regretted that no satisfactory presentation of this, the most renowned discovery in the history of physics, can be made.

The short extract which follows is taken from the *General Scholium* which Newton added to the Third Book in the second edition. It contains a statement of the law of gravitation and also an outline of Newton's philosophical method, including the famous statement *hypotheses non fingo*.

Law of Gravitation

Hitherto we have explained the phaenomena of the heavens and of our sea by the power of gravity, but have not yet assigned the cause of this power. This is certain, that it must proceed from a cause that penetrates to the very centres of the sun and planets, without suffering the least diminution of its force; that operates not according to the quantity of the surfaces of the particles upon which it acts (as mechanical causes use to do), but according to the quantity of the solid matter which they contain, and propagates its virtue on all sides to immense distances, decreasing always in the duplicate proportion of the distances. Gravitation towards the sun is made up out of the gravitations towards the several particles of which the body of the sun is composed; and in receding from the sun decreases accurately in the duplicate proportion of the distances as far as the orb of Saturn, as evidently appears from the quiescence of the aphelions of the planets; nay, and even to the remotest aphelions of the

comets, if those aphelions are also quiescent. But hitherto I
have not been able to discover the cause of those properties of
gravity from phaenomena, and I frame no hypotheses; for whatever is not deduced from the phaenomena is to be called an hypothesis; and hypotheses, whether metaphysical or physical, whether
of occult qualities or mechanical, have no place in experimental
philosophy. In this philosophy particular propositions are
inferred from the phaenomena, and afterwards rendered general
by induction. Thus it was that the impenetrability, the mobility,
and the impulsive force of bodies, and the laws of motion and of
gravitation, were discovered. And to us it is enough that gravity
does really exist, and acts according to the laws which we have
explained, and abundantly serves to account for all the motions
of the celestial bodies, and of our sea.

HOOKE

Robert Hooke was born on the Isle of Wight on July 18, 1635. He studied
at Oxford and became assistant to Robert Boyle. In 1662 he became curator
of experiments to the Royal Society and soon after professor of geometry at
Gresham College. He died in London, March 3, 1703.

Hooke was a man of extraordinary industry and ability and yet he contributed but little to the history of physics. His mind was full of ideas which
he did not seem able to carry to any final conclusion. He was consequently
involved in disputes about priority. He was one of the earliest exponents
of the undulatory theory of light and was recognized by Newton along with
Wren and Halley as having suggested the law of gravitation.

The extract which follows, taken from his *De Potentia Restitutiva*, 1678, contains an account of his demonstration of the law which is known by his name.

Law of Elastic Force

The theory of springs, though attempted by divers eminent
mathematicians of this age has hitherto not been published by
any. It is now about eighteen years since I first found it out,
but designing to apply it to some particular use, I omitted the
publishing thereof.

About three years since His Majesty was pleased to see the
experiment that made out this theory tried at *White-Hall*, as
also my spring watch.

About two years since I printed this theory in an anagram
at the end of my book of the descriptions of helioscopes, *viz.*

c e i i i n o s s s t t u u, id est, ut tensio sic vis; that is, the power
of any spring is in the same proportion with the tension thereof:
that is, if one power stretch or bend it one space, two will bend
it two, and three will bend it three, and so forward. Now as
the theory is very short, so the way of trying it is very easie.

Take then a quantity of even-drawn wire, either steel, iron, or
brass, and coyl it on an even cylinder into a helix of what length
or number of turns you please, then turn the ends of the wire into
loops, by one of which suspend this coyl upon a nail, and the other
sustain the weight that you would have to extend it, and hanging
on several weights observe exactly to what length each of the
weights do extend it beyond the length that its own weight doth
stretch it to, and you shall find that if one ounce, or one pound,
or one certain weight doth lengthen it one line, or one inch, or one
certain length, then two ounces, two pounds, or two weights will
extend it two lines, two inches, or two lengths; and three ounces,
pounds, or weights, three lines, inches, or lengths; and so forwards.
And this is the rule or law of nature, upon which all manner of
restituent or springing motion doth proceed, whether it be of
rarefaction, or extension, or condensation and compression.

Or take a watch spring, and coyl it into a spiral, so as no part
thereof may touch another, then provide a very light wheel of
brass, or the like, and fix it on an arbor that hath two small pivots
of steel, upon which pivot turn the edge of the said wheel very
even and smooth, so that a small silk may be coyled upon it, then
put this wheel into a frame, so that the wheel may move very
freely on its pivots; fasten the central end of the aforesaid spring,
close to the pivot hole or center of the frame in which the arbor
of the wheel doth move, and the other end thereof to the rim of
the wheel, then coyling a fine limber thread of silk upon the edge
of the wheel hang a small light scale at the end thereof fit to
receive the weight that shall be put thereinto; then suffering the
wheel to stand in its own position by a little index fastened to the
frame, and pointing to the rim of the wheel, make a mark with
ink, or the like, on that part of the rim that the index pointeth at;
then put in a drachm weight into the scale, and suffer the wheel
to settle, and make another mark on the rim where the index
doth point; then add a drachm more, and let the wheel settle
again, and note with ink, as before, the place of the rim pointed
at by the index; then add a third drachm, and do as before, and so
a fourth, fifth, sixth, seventh, eighth, etc., suffering the wheel

to settle, and marking the several places pointed at by the index, then examine the distances of all those marks, and comparing them together you shall find that they will all be equal the one to the other, so that if a drachm doth move the wheel ten degrees, two drachms will move it twenty, and three thirty, and four forty, and five fifty, and so forwards.

Or take a wire string of twenty, or thirty, or forty foot long, and fasten the upper part thereof to a nail, and to the other end fasten a scale to receive the weights: then with a pair of compasses take the distance of the bottom of the scale from the ground or floor underneath, and set down the said distance, then put in weights into the said scale in the same manner as in the former trials, and measure the several stretchings of the said string, and set them down. Then compare the several stretchings of the said string, and you will find that they will always bear the same proportions one to the other that the weights do that made them.

The same will be found, if trial be made, with a piece of dry wood that will bend and return, if one end thereof be fixed in a horizontal posture, and to the other end be hanged weights to make it bend downwards.

The manner of trying the same thing upon a body of air, whether it be for the rarefaction or for the compression thereof I did about fourteen years since publish in my *Micrographia*, and therefore I shall not need to add any further description thereof.

. . . .

YOUNG

The following extract contains the contribution of Thomas Young (p. 59) to the subject of elasticity. In it he introduces the name *modulus* and the definition of it which is now generally used. The extract is taken from Lecture XIII of his *Lectures on Natural Philosophy*, published in 1807.

Modulus of Elasticity

. . . .

Extension and compression follow so nearly the same laws, that they may be best understood by comparison with each other. The cohesive and repulsive forces which resist these effects, depend almost as much on the solidity or rigidity of the substances, as on the attractions and repulsions which are their immediate causes:

for a substance perfectly liquid, although its particles are in full possession of their attractive and repulsive powers, may be extended or compressed by the smallest force that can be applied to it. It is not indeed certain that the actual distances of the particles of all bodies are increased when they are extended, or diminished when they are compressed: for these changes are generally accompanied by contrary changes in other parts of the same substance, although probably in a smaller degree. We may easily observe that if we compress a piece of elastic gum in any direction, it extends itself in other directions; and if we extend it in length, its breadth and thickness are diminished.

If the rigidity of a body were infinite, and all lateral motions of its particles were prevented, the direct cohesion alone would be the measure of the force required to produce extension, and the direct repulsion, of the force required to produce compression; in this respect indeed, the actual rigidity of some substances may be considered as infinite, wherever the extension or compression is moderate, and no permanent alteration of form is produced; and within these limits these substances may be called perfectly elastic. If the cohesion and repulsion were infinite, and the rigidity limited, the only effect of force would be to produce alteration of form: and such bodies would be perfectly inelastic, but they would be harder or softer according to the degree of rigidity.

It is found by experiment, that the measure of the extension and compression of uniform elastic bodies is simply proportional to the force which occasions it; at least when the forces are comparatively small. Thus if a weight of 100 pounds lengthened a rod of steel one hundredth of an inch, a weight of 200 would lengthen it very nearly two hundredths, and a weight of 300 pounds three hundredths. The same weights acting in a contrary direction would also shorten it one, two, or three hundredths respectively. The former part of this law was discovered by Dr. Hooke, and the effects appear to be perfectly analogous to those which are more easily observable in elastic fluids.

According to this analogy, we may express the elasticity of any substance by the weight of a certain column of the same substance, which may be denominated the modulus of its elasticity, and of which the weight is such, that any addition to it would increase it in the same proportion as the weight added would shorten, by its pressure, a portion of the substance of equal diameter. Thus if a rod of any kind, 100 inches long, were compressed 1 inch by a weight of 1000 pounds, the weight of the modulus of its elasticity would be 100 thousand pounds, or more accurately 99,000, which is to 100,000 in the same proportion as 99 to 100. In the same manner, we must suppose that the subtraction of any weight from that of the modulus will also diminish it, in the same ratio that the equivalent force would extend any portion of the substance. The height of the modulus is the same for the same substance, whatever its breadth and thickness may be: for atmospheric air, it is about 5 miles, and for steel nearly 1500. This supposition is sufficiently confirmed by experiments to be considered at least as a good approximation: it follows that the weight of the modulus must always exceed the utmost cohesive strength of the substance, and that the compression produced by such a weight must reduce its dimensions to one half: and I have found that a force capable of compressing a piece of elastic gum to half its length will usually extend it to many times that length, and then break or tear it; and also that a force capable of extending it to twice its length will only compress it two thirds. In this substance, and others of a similar nature, the resistance appears to be much diminished by the facility by which a contrary change is produced in a different direction; so that the cohesion and repulsion thus estimated appears to be very weak, unless when the rigidity is increased by a great degree of cold. It would be easy to ascertain the specific gravity of such a substance in different states of tension and compression, and some light might be thrown by the comparison, on the nature and operation of the forces which are concerned. It has indeed been asserted that the specific gravity of elastic gum is even diminished by tension, so that the actual distances of the particles cannot, in this case, be supposed to be materially increased.

COULOMB

Charles-Augustin Coulomb was born in Angoulême on June 14, 1736. After studying in Paris he entered the army as an officer of engineers. He spent several years in the West Indies until failing health made necessary his return home. He then had some leisure to devote to science. He won a prize from the Academy for a memoir entitled, *Théorie des Machines Simples*, in which he announced the laws of friction and the law of torsion which are known by his name. He became a member of the Academy in consequence of the reputation which this work gave him. On the outbreak of the Revolution he resigned his various positions and withdrew to a small property in the country near Blois. After the Consulate was established he returned to Paris and became a member of one of the newly established Institutes. He died there on August 23, 1806.

Besides the researches on torsion and on friction which are described in the following extracts, Coulomb is known for his discovery of the laws of electric and magnetic force. His work on these subjects will be given in the proper place.

The Force of Torsion

In these experiments a wire of known length and weight was stretched by a cylindrical lead weight. It was then slightly twisted and the oscillations of the suspended weight were observed. The diameters of the cylindrical weights were all alike. In the formula which appears in the following sections T represents the time of one complete oscillation, M the mass of the suspended weight, a the radius of the cylindrical weight, and n a constant for each experiment, the value of which depends on the nature of the metal, on the length and on the thickness of the wire. After giving the theory of the experiment and the details of the individual experiments the results are examined in the following paragraphs.

The force or reaction of the torsion of metallic wires ought to depend on their length, on their thickness and on their tension. Thus to be able to determine generally the law of this reaction we have been compelled in the preceding experiments to suspend different weights by iron and brass wires of different lengths and thicknesses: the results given by these experiments are as follows:

If we turn the cylinder about its axis without moving the axis from the vertical line the wire will be twisted; when the cylinder is released, the wire by its force of reaction will strive to regain its natural condition; this effort will make the cylinder oscillate about its axis for a longer or a shorter time according as the elastic force is more or less perfect.

But we have found in all the preceding experiments that when the angle of torsion is not very great the time of the oscillations is sensibly isochronous; thus we can accept as a first law that for all metallic wires, when the angles of torsion are not very great, the force of torsion is sensibly proportional to the angle of torsion.

Since it has been found by experiment that the force of reaction of torsion is proportional to the angle of torsion it follows that all the formulas of oscillation which we have given, on the supposition that the force of torsion is proportional to the angle of torsion or is modified only by a very small term, can be applied to these experiments.

As we have obtained by means of these formulas

$$T = \left(\frac{Ma^2}{2n}\right)^{\frac{1}{2}} \cdot 180 \text{ degrees},$$

and as in all the preceding experiments the half pound cylinder and the 2 pound cylinder had the same diameter, it follows that n ought always to be proportional to $\left(\dfrac{M}{T^2}\right)$.

Thus if the greater or less tension of the wire has no influence on the force of torsion, the quantity n for the same wire will be the same for tensions of half a pound and of 2 pounds, and consequently we shall have T proportional to $M^{\frac{1}{2}}$. Compare our experiments made with two weights, the one of half a pound, the other of 2 pounds, of which the square roots are as 1 to 2.

FIRST EXPERIMENT. An iron wire no. 12, stretched by the half pound weight makes 20 oscillations in. . . . . .120″

SECOND EXPERIMENT. The same wire stretched by the 2 pound weight makes 20 oscillations in. . . . . .242″

THIRD EXPERIMENT. An iron wire no. 7 stretched by the half pound weight makes 20 oscillations in. . . . . 43″

FOURTH EXPERIMENT. An iron wire no. 7 stretched by the 2 pound weight makes 20 oscillations in. . . . . 85″

The FIFTH EXPERIMENT cannot be compared with the SIXTH.

SEVENTH EXPERIMENT. A brass wire no. 12 stretched by the half pound weight makes 20 oscillations in. . . . . . .220″

EIGHTH EXPERIMENT. A brass wire no. 12 stretched by the 2 pound weight makes 20 oscillations in . . . . .442″

NINTH EXPERIMENT. A brass wire no. 7 stretched by the half pound weight makes 20 oscillations in. . . . . 57″

TENTH EXPERIMENT. A brass wire no. 7 stretched by the 2 pound weight makes 20 oscillations in. . . . . .110″

The ELEVENTH and TWELFTH EXPERIMENTS cannot be compared with each other.

It results from all these experiments that with the same wire a 2 pound weight makes its oscillations sensibly in a time twice that in which a half pound weight makes its oscillations; that consequently the periods of the oscillations are as the square roots of the weights; and therefore, that the greater or less tension has no sensible effect on the reaction of the force of torsion.

Nevertheless, by a number of experiments made with tensions which are very great compared with the force of the wire, it turns out that great tensions diminish or change the force of torsion a little. In fact we can easily see that as the tension increases the wire lengthens and its diameter diminishes, and both of these changes ought to increase the periods of the oscillations.

. . . .

On the Force of Torsion as It Depends on the Lengths of the Wires

We have found in the preceding section that the greater or less tension of the wire has only a trifling effect on the force of torsion. We proceed to investigate from the same experiments by how much, with the same angle of torsion, the length of the wire increases or diminishes this force. Now it is plain that as we increase the length of the wire we can make the cylinder describe a greater number of revolutions, in the same proportion, without changing the amount of the torsion; and so the force of reaction of torsion ought to be in the inverse ratio of the length of the wire for the same number of revolutions. Let us see if this reasoning agrees with experiments.

The formula of a preceding article gives us

$$T = \left(\frac{Ma^2}{2n}\right)^{\frac{1}{2}} \cdot 180 \text{ degrees,}$$

or for the same weight T is proportional to $1/(\sqrt{n})$. Thus if n is in the inverse ratio of the lengths, as theory suggests, T will be as the square roots of the lengths of the suspended wires. Let us compare this with experiment.

We find in the tenth experiment that the brass wire no. 7, 9 inches long, when stretched with a 2 pound weight makes 20 oscillations in . . . . .110″

We find in the thirteenth experiment that the same brass wire, 36 inches long, stretched by a 2 pound weight makes 20 oscillations in . . . . .220″

Thus the lengths of the wires are to each other as 1 to 4 while the times of the oscillations of the wires are as 1 to 2; thus experiment proves that the times of the same number of oscillations are, for the same wires, stretched with the same weights, as the square roots of the lengths of these wires, as the theory suggested.

We have made many experiments of the same sort, which have all confirmed this law with great precision. We have not thought it necessary to report them here.

On the Force of Torsion as It Depends on the Thickness of the Wires

We have determined the laws of the force of torsion as they depend on the tension and the length of the wires; it only remains to determine them as they depend on the thickness of the same wires.

In the first six experiments we used three iron wires of different thicknesses and of the same length; and in the six experiments following, three brass wires of the same length and of different thicknesses; but as we know the weights of a length of 6 feet of each of these wires it is easy to determine from that the ratio of their diameters. Let us see what reasoning will predict; the *moment* of the reaction of torsion ought to increase with the thickness of the wires in three ways. Let us take, for example, two wires of the same material and of the same length, of which the diameter of one is twice that of the other. It is plain that in that wire which has the double diameter there are four times more parts stretched by the torsion than in the wire with the single diameter.; and that the mean extension of all these parts will be proportional to the diameter of the wire, like the mean lever arms relative to the axis of rotation. Thus we are led to believe from theory, that the forces of torsion of two metallic wires of the same material and of the same length but of different thicknesses are proportional to the fourth powers of their diameters, or for the same length to the squares of their weights. Let us compare this with experiment.

We shall consider only the experiments in which the tension was 2 pounds, so as to be able to compare all the numbers, the wires of no. 1 not being completely stretched by the half pound weight: we have

Iron Wire
SECOND EXPERIMENT. The iron wire no. 12, 6 feet of which weigh 5 grains, makes 20 oscillations in. . . . . . .242."
FOURTH EXPERIMENT. The iron wire no. 7, 6 feet of which weigh 14 grains, makes 20 oscillations in . . . . . . 85"
SIXTH EXPERIMENT. The iron wire no. 1, 6 feet of which weigh 56 grains, makes 20 oscillations in . . . . . . 23"

Brass Wire

> EIGHTH EXPERIMENT. The brass wire no. 12, 6 feet of which weigh 5 grains, makes 20 oscillations in. . . .442″
> TENTH EXPERIMENT. The brass wire no. 7, 6 feet of which weigh 18½ grains, makes 20 oscillations in. . . . .110″
> TWELFTH EXPERIMENT. The brass wire no. 1, 6 feet of which weigh 66 grains, makes 20 oscillations in. . . 32″

To determine from these experiments the law of the reaction of the force of torsion as it depends on the diameter of the suspended wire, let us suppose that

$$T:T'::D^m:D'^m::\varphi^{\frac{m}{2}}:\varphi'^{\frac{m}{2}},$$

in which we represent by T and T' the time of a certain number of oscillations for metallic wires of which the diameters are D and D' and the weights for the same length are φ and φ'; m is the power that we are trying to determine. From this proportion we obtain

$$m = \frac{2(\log \cdot T - \log \cdot T')}{\log \cdot \varphi - \log \cdot \varphi'},$$

and this formula is to be compared with experiment.
The second experiment compared with the fourth gives $m = -1.82$
The second experiment compared with the sixth. . .$m = -1.95$
The eighth experiment compared with the tenth. . .$m = -2.04$
The eighth experiment compared with the twelfth . .$m = -2.02$
From which it follows that

$$T:T'::\frac{1}{D^2}:\frac{1}{D'^2}::\frac{1}{\varphi}:\frac{1}{\varphi'}.$$

But the formula of the oscillatory motion

$$T = \left(\frac{Ma^2}{2n}\right)^{\frac{1}{2}} \cdot 180 \text{ degrees,}$$

gives from the preceding experiments, because of the equality of the weights, n proportional to $1/T^2$; thus the force of torsion for wires of the same material, of the same length, but of different thickness, is as the fourth power of the diameter, as the theory suggested.

General Result

It results from all the preceding experiments that the *moment* of the force of torsion for wires of the same metal is in a ratio compounded of the angle of torsion, and of the fourth power of the diameter directly, and of the length of the wire inversely; so that if we represent by l the length of the wire, by D its diameter, by B the angle of torsion, we shall have for the expression which represents the force of torsion $\mu BD^4/l$ where μ is a constant coefficient which depends on the natural rigidity of each metal: this quantity μ, which is constant for wires of the same metal, can easily be determined by experiment, as we shall see in the following section.

. . . .

SLIDING FRICTION

. . . .

Friction in this sort of movement [sliding] can be looked at from two points of view, either when the planes have been placed one on the other for a certain time and we desire to start up motion by a pull in the direction of the plane of contact, or when these planes have already a certain amount of uniform velocity and we investigate the friction for this particular velocity.

In the first case, in which we wish to make one surface slide over another when it starts from rest, the friction may depend on four causes.

1. On the nature of the materials in contact and of their lubrications.

2. On the extent of the surfaces.

3. On the pressure which these surfaces experience.

4. On the length of time which has elapsed since the surfaces were put in contact.

To these four causes we might perhaps add a fifth, that is, the dampness or dryness of the atmosphere. One might imagine that the particles of moisture contained in the air might attach themselves to the surfaces in contact and there act as a lubricant which would change their nature. But since it seems unlikely that this last cause can have any sensible effect on the results, we have not taken it into consideration in our experiments.

When the surfaces are sliding over each other with a certain velocity, even then the friction may still depend on the first three causes mentioned in the preceding paragraphs and also on the greater or less velocity of the planes in contact.

The physical cause of the resistance offered by friction to the motion of surfaces which slide on each other can be explained either by supposing an interlocking of the roughnesses of the surfaces, which cannot be separated unless they yield or break or are lifted over the tops of one another; or by supposing that the molecules of the surfaces of the two planes in contact are so close together that they develop a cohesion which must be overcome to produce the motion: experiment alone can enable us to determine the reality of these different causes.

The apparatus used in these experiments was of the familiar type and constructed on a large scale, since the experiments were designed to give information which could be immediately applied, particularly in construction and for the use of the army and navy. The first experiments were made with oak wood resting on oak. The weights pressing the blocks together were several hundred pounds. The extent of the surface in contact was made very different in some of the experiments. From these experiments Coulomb draws the following conclusions.

If we wish to determine from the last three experiments the ratio of the pressure to the friction we shall observe, first, that when the points of contact are reduced to the smallest possible dimensions, as they are in these experiments, the friction attains its maximum value in a very short time: for I have never found it possible in these experiments, however short a time of rest was allowed, to make the friction vary and to find it less than the quantity which in these cases represents its limit. The ratio of the pressure to the friction given by the last three experiments is

IV experiment.................. $\frac{250}{106}$ 2.36

V experiment.................. $\frac{450}{186}$ 2.42

VI experiment.................. $\frac{856}{356}$ 2.40

We find that when the surfaces of contact are reduced to the smallest possible dimensions, as they are in these cases, since the carriage is supported only on blunted angles, the ratio of the pressure to the friction is given by a constant quantity. We find also that this ratio is very little different from that which we have found in the first three experiments, since the mean ratio of the pressure to the friction given by the first three experiments is 2.28 and by the last three 2.39; quantities which do not differ by more than a twenty-third part, although the extent of the surfaces are in a ratio which is almost infinite. The definite result of the preceding experiments is that when surfaces of oak wood slip over each other without any lubricant, the ratio of the pressure to the friction is always a constant quantity, and the extent of the surfaces in contact has only a negligible effect.

A great many additional experiments are described, in which different kinds of wood and of metal are used without, as well as with, a lubricant. The general results of these experiments confirm the statement already made.

The next chapter deals with the friction of surfaces in motion. It is not necessary to describe the experiments in detail but the following general conclusion may be given.

In the nine preceding experiments, after the carriage was started in motion with a very small velocity, we were always careful to observe the motion through a distance of 4 feet divided into two equal parts of 2 feet each: in this motion it appeared that in general the first 2 feet were traversed in a time a little more than twice that in which the last 2 feet were traversed. Now when a body is set in motion by a constant force and its motion is uniformly accelerated, two equal distances are traversed consecutively in times which are to each other nearly as 100 to 42; and thus our carriage traversed its course of 4 feet with a motion almost uniformly accelerated and so, since it was kept in motion by a constant weight, the retarding force of friction must also have been a constant quantity: consequently it must have been nearly the same for all velocities.

. . . .

CAVENDISH

Henry Cavendish was born on October 10, 1731. He was a younger son of Lord Charles Cavendish. He studied at Cambridge. Through the death of his uncle he became the possessor of a great fortune, and withdrew from all public life and even from society to devote himself to scientific pursuits. He was best known during his lifetime as a chemist. He is one of the claimants for the discovery of the composition of water and of the existence and nature of hydrogen gas. He published but few papers on physical subjects and it was not until his manuscripts were inspected and edited by Maxwell that it was known that he had made many important discoveries in electricity which anticipated the results of Coulomb and Faraday. He died in London on February 24, 1810.

In the following extract, taken from the *Philosophical Transactions*, Vol. 17, 1798, Cavendish describes his experiment to determine the density of the earth, or to prove by direct observation the law of gravitation. The paper in which this matter was first studied by Nevil Maskelyne in 1774 appears in the *Source Book in Astronomy.*

THE DENSITY OF THE EARTH

Many years ago, the late Rev. John Michell, of this Society, contrived a method of determining the density of the earth, by rendering sensible the attraction of small quantities of matter; but, as he was engaged in other pursuits, he did not complete the apparatus till a short time before his death, and did not live to make any experiments with it. After his death, the apparatus came to the Rev. Francis John Hyde Wollaston, Jacksonian Professor at Cambridge, who, not having conveniences for making experiments with it, in the manner he could wish, was so good as to give it to me. The apparatus is very simple; it consists of a wooden arm, 6 feet long, made so as to unite great strength with little weight. This arm is suspended in an horizontal position, by a slender wire 40 inches long, and to each extremity is hung a leaden ball, about 2 inches in diameter; and the whole is inclosed in a narrow wooden case, to defend it from the wind.

As no more force is required to make this arm turn round on its centre, than what is necessary to twist the suspending wire, it is plain, that if the wire is sufficiently slender, the most minute force, such as the attraction of a leaden weight a few inches in diameter, will be sufficient to draw the arm sensibly aside. The weights which Mr. Michell intended to use, were 8 inches diameter. One of these was to be placed on one side the case, opposite to one of the balls, and as near it as could conveniently be done, and the other on the other side, opposite to the other ball, so that the attraction of both these weights would conspire in drawing the arm aside; and when its position, as affected by these weights, was ascertained, the weights were to be removed to the other side of the case, so as to draw the arm the contrary way, and the position of the arm was to be again determined; consequently, half the difference of these positions would show how much the arm was drawn aside by the attraction of the weights. In order to determine from hence the density of the earth, it is necessary to ascertain what force is required to draw the arm aside through a given space. This Mr. Michell intended to do, by putting the arm in motion, and observing the time of its vibrations, from which it may easily be computed.

Mr. Michell had prepared 2 wooden stands, on which the leaden weights were to be supported, and pushed forwards, till they came almost in contact with the case; but he seems to have intended to move them by hand. As the force with which the balls are attracted by these weights is excessively minute, not more than $\frac{1}{50000000}$ of their weight, it is plain that a very minute disturbing force will be sufficient to destroy the success of the experiment; and from the following experiments it will appear, that the disturbing force most difficult to guard against, is that arising from the variations of heat and cold; for if one side of the

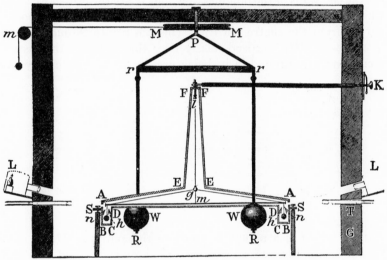

FIG. 28.

case be warmer than the other, the air in contact with it will be rarefied, and in consequence will ascend, while that on the other side will descend, and produce a current which will draw the arm sensibly aside. As I was convinced of the necessity of guarding against this source of error, I resolved to place the apparatus in a room which should remain constantly shut, and to observe the motion of the arm from without, by means of a telescope; and to suspend the leaden weights in such manner, that I could move them without entering into the room. This difference in the manner of observing, rendered it necessary to make some alteration in Mr. Michell's apparatus; and as there were some parts of it which I thought not so convenient as could be wished, I chose to make the greatest part of it anew.

Fig. 28 is a longitudinal vertical section through the instrument, and the building in which it is placed: *ABCDDCBAEFFE* is the case; *x* and *x* are the two balls, which are suspended by the wires *bx* from the arm *ghmh*, which is itself suspended by the slender wire *gl*. This arm consists of a slender deal rod *hmh*, strengthened by a silver wire *hgh*; by which means it is made strong enough to support the balls, though very light. The case is supported, and set horizontal, by 4 screws, resting on posts fixed firmly into the ground: 2 of them are represented in the figure, by *S* and *S*; the other 2 are not represented, to avoid confusion. *GG* and *GG* are the end walls of the building. *W* and *W* are the leaden weights; which are suspended by the copper rods *RrPrR*, and the wooden bar *rr*, from the centre pin *Pp*. This pin passes through a hole in the beam *HH*, perpendicularly over the centre of the instrument, and turns round in it, being prevented from falling by the

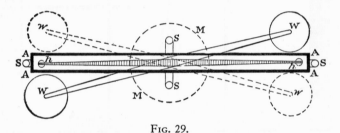

FIG. 29.

plate *p*. *MM* is a pulley, fastened to this pin; and *Mm* a cord wound round the pulley, and passing through the end wall; by which the observer may turn it round, and so move the weights from one situation to the other.

Fig. 29 is a plan of the instrument. *AAAA* is the case; *SSSS* the 4 screws for supporting it; *bb* the arm and balls, *W* and *W* the weights; *MM*, the pulley for moving them. When the weights are in this position, both conspire in drawing the arm in the direction *bW*; but, when they are removed to the situation *w* and *w*, represented by the dotted lines, both conspire in drawing the arm in the contrary direction *bw*. These weights are prevented from striking the instrument, by pieces of wood, which stop them as soon as they come within ⅕ of an inch of the case. The pieces of wood are fastened to the wall of the building; and I find that the weights may strike against them with considerable force, without sensibly shaking the instrument.

In order to determine the situation of the arm, slips of ivory are placed within the case, as near to each end of the arm as can be done without danger of touching it, and are divided to 20ths of an inch. Another small slip of ivory is placed at each end of the arm, serving as a vernier, and subdividing these divisions into 5 parts; so that the position of the arm may be observed with ease to 100ths of an inch, and may be estimated to less. These divisions are viewed, by means of the short telescopes T and T, Fig. 28, through slits cut in the end of the case, and stopped with glass; they are enlightened by the lamps L and L, with convex glasses, placed so as to throw the light on the divisions; no other light being admitted to the room. The divisions on the slips of ivory run in the direction Ww, Fig. 29, so that, when the weights are placed in the positions w and w, represented by the dotted circles, the arm is drawn aside, in such direction as to make the index point to a higher number on the slips of ivory; for which reason, I call this the positive position of the weights.

FK, Fig. 28, is a wooden rod, which, by means of an endless screw, turns round the support to which the wire gl is fastened, and so enables the observer to turn round the wire, till the arm settles in the middle of the case, without danger of touching either side. The wire gl is fastened to its support at top, and to the centre of the arm at bottom, by brass clips, in which it is pinched by screws. In these 2 figures, the different parts are drawn nearly in the proper proportion to each other.

. . . .

The account of the method of observation and the study of the corrections which were applied is too long for insertion. Seventeen experiments are recorded in the final Table. Of these the first three were tried with a wire so thin that the time of vibration was nearly fifteen minutes. The time of vibration in the remaining experiments was usually over seven minutes. It is not necessary to give this long table, but Cavendish's conclusions may be stated.

From this table it appears, that though the experiments agree pretty well together, yet the difference between them, both in the quantity of motion of the arm and in the time of vibration, is greater than can proceed merely from the error of observation. As to the difference in the motion of the arm, it may very well be accounted for, from the current of air produced by the difference of temperature; but whether this can account for the difference in the time of vibration, is doubtful. If the current of air was regular, and of the same swiftness in all parts of the vibration of the ball, I think it could not; but as there will most likely be much irregularity in the current, it may very likely be sufficient to account for the difference.

By a mean of the experiments made with the wire first used, the density of the earth comes out 5.48 times greater than that of water; and by a mean of those made with the latter wire, it comes out the same; and the extreme difference of the results of the 23 observations made with this wire, is only .75; so that the extreme results do not differ from the mean by more than .38, or $\frac{1}{14}$ of the whole, and therefore the density should seem thus to be determined, to great exactness. It may indeed be objected, that as the result appears to be influenced by the current of air, or some other cause, the laws of which we are not well acquainted with, this cause may perhaps act always, or commonly, in the same direction, and so make a considerable error in the result. But yet, as the experiments were tried in various weathers, and with considerable variety in the difference of temperature of the weights and air, and with the arm resting at different distances from the sides of the case, it seems very unlikely that this cause should act so uniformly in the same way, as to make the error of the mean result nearly equal to the difference between this and the extreme; and therefore it seems very unlikely that the density of the earth should differ from 5.48 by so much as $\frac{1}{14}$ of the whole.

Another objection perhaps may be made to these experiments, namely, that it is uncertain whether, in these small distances, the force of gravity follows exactly the same law as in greater distances. There is no reason however to think that any irregularity of this kind takes place, until the bodies come within the action of what is called the attraction of cohesion, and which seems to extend only to very minute distances. With a view to see whether the result could be affected by this attraction, I made the 9th, 10th, 11th, and 15th experiments, in which the balls were made to rest as close to the sides of the case as they could; but there is no difference to be depended on, between the results under that circumstance, and when the balls are placed in any other part of the case.

According to the experiments made by Dr. Maskelyne, on the attraction of the hill Schehallian, (*See* A Source Book in

Astronomy, p. 133.) the density of the earth is 4½ times that of water; which differs rather more from the preceding determination than I should have expected. But I forbear entering into any consideration of which determination is most to be depended on, till I have examined more carefully how much of the preceding determination is affected by irregularities whose quantity I cannot measure.

TORRICELLI

Torricelli (p. 70) published in 1641 a book entitled *Trattato del Moto dei Gravi*, in which he investigated the motions of projectiles and the flow of liquids from small openings. The extract which follows gives an account of the discovery of the theorem of fluid motion which is known by his name. It may be found in Torricelli's *Collected Works*, published in 1919, Vol. II, p. 185.

EFFLUX OF LIQUIDS

It will not be out of place to introduce in this book some consideration of liquids, for liquids are distinguished from other sublunary bodies by a motion which is peculiar to them, so that they are almost never at rest. I do not deal with the great motions of oceanic waves; I pass over also the measurement of the currents of rivers and of running water in general and their practical applications, the theory of which was first investigated by Abbot Benedictus Castellius, my teacher. He wrote what he himself knew and confirmed it not only by demonstration but by practical tests, of the greatest use to princes and people, and to the great admiration of philosophers. His book, truly a golden book, exists for all. We shall investigate certain minor questions, almost useless and yet not altogether uninteresting, about this matter.

Hypothesis

Liquids which issue with violence [*from an opening in a vessel*] have at the point of issue the same velocity which any heavy body would have, or any drop of the same liquid, if it were to fall from the upper surface of the liquid to the orifice from which it issues.

For example, if the tube *AB* (Fig. 30) of some convenient capacity, which is not definitely fixed, is assumed to be always full of water up to the level A, and if it is pierced at B with a small orifice, we assume that the water which issues at B will have the same velocity which any heavy body would have if it were to fall from A to B.

It seems as if we could confirm this statement by considering the fact that if another tube is connected with the opening B (Fig. 31) and accurately fitted to it, the water flowing from B into the tube BC has such a force that it lifts itself to the surface A.

Thus it seems very probable that even when the water issues freely from B, it will have a force which will carry it up to the

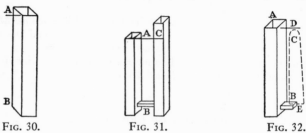

FIG. 30.   FIG. 31.   FIG. 32.

horizontal line drawn through A; or what is the same thing, that it will have the same velocity as that of any heavy body or of one drop of liquid, falling freely from A to B.

An experiment also proves our principle in another way, although it may seem open to criticism in some respects. For, if the orifice B (Fig. 32) is directed upward and is suitably rounded and smoothed and if the width of the tube is much greater than that of the orifice B, we shall see the water leap up in the line BC as if to ascend to the level AD. We may assign as the reason for the difference in height CD, partly the resistance of the air, which always acts against the motion of any body; and partly also the resistance of the water itself, which when it falls back again from the point C interferes with the stream that is rising and retards it, and does not allow the succeeding drops to rise to that mark to which they would have come by their velocity. This will appear very clearly if the opening B is closed with the hand and is then opened again when the hand is suddenly withdrawn, for the first and leading drops will be seen to rise higher than the highest point C of the water after it begins to fall again; for those first drops do not have any water ahead of them, which by flowing back against them checks their motion at the end of the rise, at least if BC is perpendicular.

It may be added that if we observe the air that is around the water column *BC*, it will be found to be in agitation and to be moving upward. Now this motion of the air cannot be effected without force, and hence the motion of the ascending water is interfered with. For this reason, if one wishes to try an experiment to prove this principle, one should take quicksilver, which on account of its greater weight, is better adapted both to retain the velocity which it assumes at the start and to overcome the resistance of the air.

But water, on account of its levity, will be found to deviate considerably from the principle, especially if the tube is of great height; for then, on account of its very great velocity, it is broken up into minute drops like a mist, and it will not ascend one-half, perhaps not even a third or a fourth part of the distance through which, speaking theoretically, and assuming all impediments to the motion removed, we might expect it to rise by its original velocity. But, if anyone does not agree with the reasons which have already been presented, let him see whether he approves of any of the following propositions: because if he does so we may easily, from an argument based on the approved proposition, demonstrate our first hypothesis. But if he skips this whole appendix on the motion or flow of liquids or leaves it out of the book altogether, which I freely concede he may do, nevertheless the experiments which have been made with all due diligence have most exactly confirmed a great many of the following propositions.

After these statements let us consider water falling back to the point *E* in a horizontal plane drawn through the level of the orifice *B*. From the theorem of Galileo we know that the velocity of water falling from *C* to *E* is just that which can carry the same water from *E* to *C*. Therefore the velocity in *E* is the same as that in *B*, but in *E* the velocity is that of a heavy body falling from *C* to *E*, (for we assume that the point *E* ought to be in the level *AD* if the obstacles which interfere with the free motion of the water are removed) therefore the velocity in *B* is that of a heavy body falling naturally from *A* to *B*.

These things being supposed we shall demonstrate certain theorems about liquids issuing from orifices which seem in a wonderful way to fit in with the theory of projectiles.

AMPÈRE

The short extract from the *Bibliothèque Universelle, Sciences et Arts,* Vol. 69, p. 225, is given because it contains a clear statement (Ampére, p. 446) for the first time of the distinction between particles, molecules, and atoms, which has been so influential in all subsequent speculation.

MOLECULES AND ATOMS

The principle upon which these ideas are founded is the distinction that M. Ampère has for some time maintained between particles, molecules, and atoms. He gives the name *particle* to an infinitely small portion of a body of the same nature as the body, so that a particle of a solid body is solid, that of a liquid, liquid and that of a gas, is in the gaseous condition.

The particles are made up of molecules held apart from each other, 1. by that which remains at this distance of the attractive and repulsive forces which belong to the atoms, 2. by the repulsion which is set up between them by the vibratory motion of the interposed ether, 3. by the attraction in the direct ratio of the masses and in the inverse ratio of the squares of their distances. He gives the name *molecule* to an assemblage of atoms kept apart by attractive and repulsive forces belonging to each atom, forces which he conceives so much greater than the former ones that these may be considered relatively completely insensible. The name *atom* is given to the material points from which these attractive and repulsive forces proceed.

SOUND

MERSENNE

Marin Mersenne was born in 1588 in Soultière, and died in Paris in 1648. He was educated in the Jesuit College of LaFlèche, where he was a fellow student of Descartes. He joined the Minorite Order.

Mersenne was actively interested in physical problems and corresponded with many of the learned men of the time. By acting as the intermediary through whom their work could be communicated to others, he served as a substitute for the scientific journals which had not yet been started. His most important work was his study of acoustics, the results of which he published in his *Harmonie Universelle*, in twelve books, dealing with the whole subject of music as it was then understood and with its physical basis. The extract which follows contains his account of experiments with the monochord, in which he gives the relations between the dimensions of a string and the tones which it produces. The part of the whole work from which it is taken is the *Traité des Instrumens a Chordes, Livre Troisiesme*, p. 123, 1636.

MUSICAL TONES PRODUCED BY STRINGS

A deaf man may tune a lute, a viol, or a spinet and other stringed instruments and may find such sounds as he wishes if he knows the length and the mass of the strings: the rules for the deaf are given in a table.

We may have several sorts of strings; those which are equal in length and mass, like those of monochords; or unequal in length and equal in mass; or unequal in length and mass, like those of harps and of the spinet; or equal in length and unequal in mass, like those of the viol or of the lute. Now, however different they are, a deaf man can bring them to any accord that he wishes provided that he knows their differences in material, in length and in mass. I shall first demonstrate this in the case of strings which are similar in all respects, so as to begin with the simplest case, because when they are stretched with equal forces they are in unison, since equal things added to equal things leave them equal.

I shall now give the general rules which must be used to make all sorts of accords, which will serve for a proof and a demonstration, as we have elsewhere shown, that they are valid and infallible.

115

First Rule

If the strings are equal in length and mass and one of them gives the low tone C, when it is stretched with a one pound weight, the other must be stretched with four pounds to make it give the octave, in as much as the weights are as the squares of the harmonic intervals which the strings are to give; now the interval of the octave is as 2 to 1, of which the ratio squared is 4 to 1.

Second Rule

It is necessary to add to the weight mentioned a sixteenth part of the larger weight or one fourth of the smaller weight, so that they may be in exact accord: for example, we must add four ounces to the four pounds in the preceding example to give the exact octave: so that $4\frac{1}{4}$ pounds and 1 pound hung on two equal strings will give the perfect octave.

Third Rule

When the strings are equal in mass (*of unit length*) and unequal in length, and we wish to bring them to unison, the forces which stretch the strings should be in the squared ratio of the lengths of the strings: for example, if one is two feet long and the other one foot and the latter is stretched by a force (*of one pound*) we must stretch the former by four pounds and add a quarter of a pound, as I have said in the other rule, to make it pass from the lower octave which it gives to unison with the shorter string.

Fourth Rule

When the strings are unequal in mass and equal in length, the forces which are in the ratio of the masses put them in unison; for example, if one has a mass of 2 and the other of 3 and the first is stretched with 2 forces and the second with 3 they will be in unison: and if the string was a hundred times greater, a force a hundred times greater would put it in unison.

The remaining rules contain developments from these fundamental laws of strings.

WALLIS

John Wallis was born in Ashford on November 23, 1616. He studied at Cambridge and took orders in the church. His bent for mathematics showed itself early and in 1649 he became professor of mathematics at Oxford. He was distinguished for his contributions to mathematics. He died on October 28, 1703.

Wallis was one of those who successfully investigated the laws of collision. The extract which follows deals with the vibrations of strings and is given because it contains an account of the discovery of the vibrations which produce partial tones. The short paper entitled "Of the Trembling of Consonant Strings" appeared in the *Philosophical Transactions, Abridged*, Vol. I, p. 606, 2d ed., 1716.

VIBRATIONS OF PARTS OF STRINGS

It hath been long since observed, that if a viol-string, or lute-string, be touched with the bow or hand, another string on the same or another instrument not far from it, (if an unison to it, or an octave, or the like) will at the same time tremble of its own accord. But I can now add, that not the whole of that other string doth thus tremble, but the several parts severally, according as they are unisons to the whole, or the parts of that string so struck. For instance, supposing AC (Fig. 33a) to be an upper octave to $\alpha\gamma$, and therefore an unison to each half of it, stopped at β. If, while $\alpha\gamma$, is open, AC, be struck; the two halves of this other, that is $\alpha\beta$, and $\beta\gamma$, will both tremble; but not the middle point at β. Which will easily be observed, if a little bit of paper be lightly wrapt about the string $\alpha\gamma$, and removed successively from one end of the string to the other.

In like manner, if AD (Fig. 33b), be an upper twelfth to $\alpha\delta$, and consequently an unison to its three parts equally divided in β, γ; if $\alpha\delta$, being open, AD, be struck, its three parts $\alpha\beta$, $\beta\gamma$, $\gamma\delta$, will severally tremble, but not the points, β, γ. In the like manner, if AE (Fig. 33c), be a double octave to $\alpha\epsilon$; the four quarters of this will tremble, when that is struck, but not the points β, γ, δ. So if AG (Fig. 33d), be a fifth to $\alpha\eta$; and consequently each half of that stopped in D, an unison to each third part of this stopped in β, γ; while that is struck, each part of this will tremble severally, but not the points β, γ; and while this is struck, each of that will tremble, but not the point D. The like will hold in lesser concords; but the less remarkably, as the number of divisions encreases.

This was first of all (that I know of) discovered by Mr. Will. Noble M.A. of Merton College; and by him shewed to some of our musicians about three years since; and after him by Mr. Tho. Pigot A.B. of Wadham College, without knowing that Mr. Noble had discover'd it before.   I add this further (which I took notice of upon occasion of making tryal of the other,) that the same string, as $\alpha\gamma$, being struck in the midst at β, each part being unison to the other, will give no clear sound at all, but very confused.   And not only so (which others have observed, that a string doth not

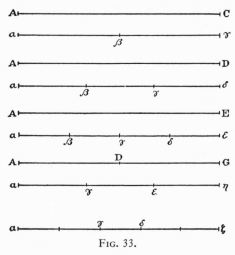

FIG. 33.

sound clear if struck in the midst;) but also if $\alpha\delta$, be struck at β, or γ, where one part is an octave to the other; and in like manner, if $\alpha\epsilon$, be struck at β, or δ; the one part being a double octave to the other.   And so if $\alpha\zeta$ (Fig. 33e), be struck in γ, or δ; the one part being a fifth to the other; and so in other like consonant divisions; but still the less remarkable as the number of divisions encreaseth.   This and the former I judge to depend upon one and the same cause; viz. the contemporary vibrations of the several unison parts, which make the one tremble at the motion of the other: but when struck at the respective points of divisions, the sound is incongruous, by reason the point is disturbed which should be at rest.

SAUVEUR

Joseph Sauveur was born in LaFlèche on March 24, 1653. He studied and taught mathematics and became professor of mathematics at the Collège Royal. He was a member of the Academy. He died July 9, 1716.

The extract which follows is from a memoir published in the *Mémoires de l'Académie Royale des Sciences*, p. 297, Paris, 1701, entitled "Système Général des Intervalles des Sons, et son Applications à tous les Systèmes et à tous les Instrumens de Musique." This memoir deals with the relations of the tones of the musical scale. It proposes to establish the practice of music upon a science superior to it which the author calls "acoustics," the subject of which is sound in general, while music deals with the subject of sound so far as it is agreeable to the sense of hearing. The extract presents the author's view of harmonics or overtones.

Harmonics or Overtones

I call a tone the harmonic of a fundamental tone if it makes several vibrations while the fundamental tone makes only one. Thus the tone which is the twelfth of the fundamental tone is harmonic, because it makes 3 vibrations while the fundamental tone makes only 1.

A table giving the ratios of the vibrations to those of the fundamental tone and assigning to all the tones their position in the musical scale is then given and explained.

After having defined and determined the harmonic tones, it remains to make them perceptible to the ear and even to the eye and to indicate their properties.

Divide the string of the monochord into equal parts, for example into 5 parts, (we may divide a rule of the same length and lay it along the string) pluck this string as it stands and it will give a tone which I call its fundamental; as soon as possible bring a light obstacle *C* on one of these divisions at *D*, like the end of a feather if the string is thin; in such a way that the movement of the string can be communicated on one side and the other of the obstacle; the string will give the 5th. harmonic tone, that is to say the XVIIth.

To understand the reason for this result, notice that when we pluck the string *AB* (Fig. 34) as a whole, it makes oscillations through all its length; but when we put an obstacle *C* on the first division mark *D* of the string, which I suppose to be divided into

5 equal parts, the total oscillation *AB* is divided first into the
2 oscillations *AD, DB,*

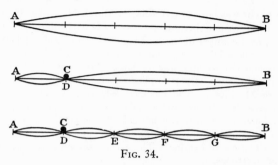

FIG. 34.

and since *AD* is ⅕ of *AB* or ¼ of *DB*, it makes its oscillations
5 times as fast as the whole cord *AB*, or 4 times as fast as the other
part *DB*; so that the part *AB* carries with it its neighboring part
DB and compels it to follow its motion. This part *DE* should
consequently be equal to it; for a greater part will move more
slowly and a smaller part more quickly. Then the part *DE* will
compel the next part *EF* to take up the same motion and so on
to the last; so that all the parts will make oscillations which will

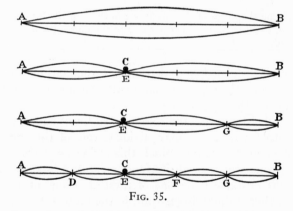

FIG. 35.

cross at the points of division *D, E, F, G;* and consequently the
string will give the 5th. harmonic tone or a XVIIth.

I shall call these points *A, D, E, F, G, B* (Fig. 35), the nodes
of these oscillations and the middle points of these oscillations
will be called the ventres.

If the obstacle *C* is placed at the second dividing point *E* of
the string it will give the same harmonic tone; for (1) the obstacle

C will at first compel the string to make the two oscillations *AE*, *EB*; (2) since the oscillation of *AE* is more rapid than the other, it will compel the part *EG* which is equal to it to follow its motion; (3) the remaining part *GB*, which is the half of one of these, going twice as fast, will force the equal part *GF* to follow its motion, and that will compel the next part *FE* to take up the same motion, and so on to the last; so that all the strings will be divided by oscillating parts into parts equal to the greatest common measure of the parts *AC, CB* which are divided by the light obstacle *C*.

We may detect these oscillations, (1) by the ear, for those who have a fine ear will perceive a harmonic tone proportional to the parts which form these oscillations, or we may make certain of them by setting the monochord in unison with this harmonic tone: (2) by the eye; for if we divide the string into equal parts, for example into 5, and if we put a movable stop *C* at the points *D* or *E*, and place small pieces of black paper at the points of division *D, F*, and pieces of white paper half way between these points, when we strike the part *AC* we shall notice that the pieces of white paper which are on the ventral segments will be thrown off and that the pieces of black paper which are at the nodes will remain at rest.

From this which we have established we shall draw the following consequences and remarks.

I. If we have produced a harmonic tone by placing a light obstacle at *B* it will persist if the obstacle is removed or if another one is put at any other node or at all the nodes.

II. After a harmonic tone has been formed, for example the 5th., if we touch the light obstacle to the ventral segment of an oscillation so as to divide it, for example, into 3 parts, there will be produced a 3d. harmonic tone of the 1st. harmonic tone, that is to say the 15th. harmonic tone of the fundamental.

III. Without touching the string with the light obstacle we may give it a harmonic tone, (1) if another string in unison with some one of its harmonic tones is sounded alongside of it; (2) if the first string is not in unison with one of the harmonic tones of the other they are broken up by their oscillations into harmonic tones which will be the greatest common measures of the fundamental tones of the two strings, so that, for example, if the two strings are in the ratio of 3 to 4 the shorter one will give the 3rd. harmonic tone and the longer one its 4th. which are in unison.

. . . .

HELMHOLTZ

Hermann Ludwig Ferdinand von Helmholtz was born in Potsdam on August 31, 1821. His father was a teacher in the Gymnasium: his mother was a lineal descendant of William Penn. He studied medicine and surgery and became a surgeon in the Prussian army. His first paper, published when he was twenty-one, was on a physiological subject. He lived in Berlin from 1842 to 1849, when he was called to be professor of physiology in Königsberg. He subsequently occupied similar positions in Bonn and in Heidelberg. In 1871 he became professor of physics at the University of Berlin, where he remained until his death, which occurred on September 8, 1894.

Helmholtz' contributions to physics cover a very wide range. His work in physiological optics, which he embodied in a great book, is of fundamental value. His researches in acoustics, also embodied in a book, are equally important. His paper of 1847 on the conservation of energy is the first one in which that principle is presented with convincing arguments. His study of vortex motions opened a new field in hydrodynamics, and his electrical researches and his applications of thermodynamics to chemistry are also important.

The extract which follows is from Helmholtz' book entitled *Tonempfindungen*, or the *Sensations of Tone*, as translated by Alexander J. Ellis, 1885. The preface to the first edition of this great work is dated October, 1862. The translation from which the extracts have been made is a revision published as a second English edition in 1885.

After showing how the vibrations of tuning forks can be represented by making the forks draw a wavy line, and after showing further that all vibrations or periodic motions in general may be represented by curves, the author proceeds as follows:

The Quality of Musical Tones

Physicists, then, having in their mind such curvilinear forms, representing the law of motion of sounding bodies, speak briefly of the *form of vibration* of a sounding body, and assert that *the quality of tone depends on the form of vibration*. This assertion, which has hitherto been based simply on the fact of our knowing that the quality of the tone could not possibly depend on the periodic time of a vibration, or on its amplitude, will be strictly examined hereafter. It will be shewn to be in so far correct that every different quality of tone requires a different form of vibration, but on the other hand it will also appear that different forms of vibration may correspond to the same quality of tone.

On exactly and carefully examining the effect produced on the ear by different forms of vibration we meet with a strange and unexpected phenomenon, long known indeed to individual musicians and physicists, but commonly regarded as a mere curiosity, its generality and its great significance for all matters relating to musical tones not having been recognized. The ear when its attention has been properly directed to the effect of the vibrations which strike it, does not hear merely that one musical tone whose pitch is determined by the period of vibrations in the manner already explained, but in addition to this it becomes aware of a whole series of higher musical tones, which we will call the *harmonic upper partial tones,* and sometimes simply the *upper partials* of the whole musical tone or note, in contradistinction to the *fundamental* or *prime partial tone* or simply the *prime,* as it may be called, which is the lowest and generally the loudest of all the partial tones, and by the pitch of which we judge of the pitch of the whole *compound musical tone* itself. The series of these upper partial tones is precisely the same for all compound musical tones which correspond to a uniformly periodical motion of the air. It is as follows:—

The first upper partial tone (or second partial tone) is the upper Octave of the prime tone, and makes double the number of vibrations in the same time. If we call the prime *C,* this upper Octave will be *c*.

The second upper partial tone (or third partial tone) is the Fifth of this Octave, or *g,* making three times as many vibrations in the same time as the prime.

The third upper partial tone (or fourth partial tone) is the second higher Octave, or *c',* making four times as many vibrations as the prime in the same time.

The fourth upper partial tone (or fifth partial tone) is the major third of this second higher Octave, or *e',* with five times as many vibrations as the prime in the same time.

The fifth upper partial tone (or sixth partial tone) is the Fifth of the second higher Octave, or *g',* making six times as many vibrations as the prime in the same time.

And thus they go on, becoming continually fainter, to tones making 7, 8, 9, etc., times as many vibrations in the same time, as the prime tone.

The rest of the paragraph containing the representation of the partial tones in musical notation is omitted.

The whole sensation excited in the ear by a periodic vibration of the air we have called a *musical tone*. We now find that this is *compound*, containing a series of different tones, which we distinguish as the *constituents* or *partial tones* of the *compound*. The first of these constituents is the *prime partial tone* of the compound, and the rest its *harmonic upper partial tones*. The *number* which shews the *order* of any partial tone in the series shews how many times its vibrational number exceeds that of the prime tone. Thus, the second partial tone makes twice as many, the third, three times as many vibrations in the same time as the prime tone, and so on.

HEAT

NEWTON

The short paper which follows was published anonymously in the *Philosophical Transactions, Abridged,* Vol. II. Its author was known to be Newton (p. 30) and it is given as his in his *Opuscula.* It is dated 1701. Newton's thermometer was graduated by the use of two fixed points and may have been the first one of the sort with which useful observations were made, though others had suggested the same method of graduation, particularly Dalencé in 1688 in his *Traittez des baromètres, thermomètres et notiomètres.* The paper contains also Newton's law of cooling.

A Scale of the Degrees of Heat

0		The heat of the air in winter at which water begins to freeze. This heat is determined by placing a thermometer in packed snow, while it is melting.
0, 1, 2		Heats of the air in winter.
2, 3, 4		Heats of the air in spring and in autumn.
4, 5, 6		Heats of the air in summer.
6		Heat of the air in the middle of the day in the month of July.
12	1	The greatest heat which a thermometer takes up when in contact with the human body. This is about the heat of a bird hatching its eggs.
$14\frac{3}{11}$	$1\frac{1}{4}$	About the greatest heat of a bath which one can endure for some time when the hand is dipped in it and constantly moved. The same heat is almost the heat of recently shed blood.
17	$1\frac{1}{2}$	The greatest heat of a bath which one can endure for some time when the hand is dipped in it and is kept still.
$20\frac{3}{11}$	$1\frac{3}{4}$	The heat of a bath in which wax floating around and liquified hardens and loses its transparency as the bath cools.
24	2	The heat of a bath in which wax floating around melts as the bath gets warmer and is kept in continual movement without boiling.
$28\frac{3}{11}$	$2\frac{1}{4}$	A mean heat between the heats at which wax melts and water boils.
34	$2\frac{1}{2}$	The heat at which water boils violently, and at which a mixture of two parts of lead, three parts of tin and five parts of bismuth hardens as it cools. Water begins to boil at a heat of 33 parts and scarcely reaches a heat of more than $34\frac{1}{2}$ when boiling. Iron, while cooling, at a heat of 35 or 36 parts

		if warm water, and of 37 parts if cold water is poured on it drop by drop, ceases to cause boiling.
$40\frac{4}{11}$	$2\frac{3}{4}$	The lowest heat at which a mixture of one part of lead, four parts of tin and five parts of bismuth liquifies as it is heated and is kept in a continual flux.
48	3	The lowest heat at which a mixture of equal parts of tin and bismuth liquifies. This mixture hardens as it cools at a heat of 47 parts.
57	$3\frac{11}{14}$	The heat at which a mixture of two parts of tin and one part of bismuth is melted, as also a mixture of three parts of tin and two of lead; but a mixture of five parts of tin and two of bismuth hardens at this temperature when cooling. The same thing happens to a mixture of equal parts of lead and bismuth.
68	$3\frac{1}{2}$	The lowest heat at which a mixture of one part of bismuth and eight parts of tin is melted. Tin itself is melted at a heat of 72 parts and when it cools hardens at a heat of 70 parts.
81	$3\frac{3}{4}$	The heat at which bismuth melts as also a mixture of four parts of lead and one part of tin. But a mixture of five parts of lead and one part of tin, when it is melted and cools, hardens at this heat.
96	4	The lowest heat at which lead melts. When it is heated lead melts at a heat of 96 or 97 parts, and as it cools it hardens at a heat of 95 parts.
114	$4\frac{1}{4}$	The heat at which glowing bodies, as they cool, cease to shine in the darkness of the night, and in turn as they are warmed begin to shine in the same darkness, but with a very feeble light, which can hardly be seen. At this heat there melts the mixture of equal parts of tin and of antimony, and a mixture of seven parts of bismuth and four parts of antimony hardens on cooling.
136	$4\frac{1}{2}$	The heat at which glowing bodies shine at night though hardly in the twilight. At this heat not only a mixture of two parts of antimony and one part of bismuth but also a mixture of five parts of antimony and one part of tin hardens when it cools. Antimony itself hardens at a heat of 146 parts.
161	$4\frac{3}{4}$	The heat at which glowing bodies in the twilight, just before the rising of the sun or after its setting, plainly shine, but in the clear light of day not at all or only very slightly.
192	5	The heat of coals in a little kitchen fire made from bituminous coal and excited by the use of a bellows. The same is the heat of iron in such a fire which is shining as much as it can. The heat of a little kitchen fire which is made of wood is a little greater, perhaps 200 or 210 parts. And the heat of a great fire is greater still, especially if it is excited by a bellows.

In the first column are the degrees of heat in arithmetical progression, computed by beginning with the heat at which water begins to freeze, as from the lowest degree of heat or from the common boundary of heat and cold, and by setting the external heat of the human body at 12 parts. In the second column are the degrees of heat in geometrical progression, so that the second degree is twice as great as the first, the third twice as great as the second, and the fourth twice as great as the third; and the first is the external heat of the human body, which is sensibly constant. It appears by this table that the heat of boiling water is almost three times as great as the heat of the human body, that the heat of melting tin is six times as great, the heat of melting lead eight times as great, the heat of melting antimony twelve times as great and the ordinary heat of a kitchen fire sixteen or seventeen times as great as the heat of the human body.

This table was constructed by the help of a thermometer and of heated iron. With the thermometer I found the measure of all the heats up to that at which lead melts and by the hot iron I found the measure of the other heats. For the heat which the hot iron communicates in a given time to cold bodies which are near it, that is, the heat which the iron loses in a given time, is proportional to the whole heat of the iron. And so, if the times of cooling are taken equal, the heats will be in a geometrical progression and consequently can easily be found with a table of logarithms. First therefore, I found with a thermometer made with linseed oil that if the oil, when the thermometer was placed in melting snow, occupied a space taken as 10000, the same oil expanded by the heat of the first degree, or of the human body, occupied the space 10256, and by the heat of water just beginning to boil the space 10705, and by the heat of water boiling violently the space 10725, and by the heat of melted tin when it begins to harden on cooling and to take on the consistency of an amalgam the space 11516, and when it is completely hardened the space 11496. Therefore the oil was expanded in the ratio of 40 to 39 by the heat of the human body, in the ratio of 15 to 14 by the heat of boiling water, in the ratio of 15 to 13 by the heat of tin which is cooling, when it begins to coagulate and stiffen, and in the ratio of 23 to 20 by the heat at which cooling tin is completely hardened. The expansion of air by an equal heat is ten times as great as the expansion of oil and the expansion of oil is about fifteen steps greater than the expansion of alcohol. And when these were found, by setting the heats proportional to the expansion of the oil and writing 12 parts for the heat of the human body, we obtained for the heat of water when it begins to boil 33 parts, and when it boils more violently 34 parts; and for the heat of tin, either when it melts or when on cooling it begins to stiffen and to take on the consistency of an amalgam, we found 72 parts and when on cooling it hardens completely we found 70 parts. When I had these points determined, in order to investigate the others, I heated a large enough block of iron until it was glowing, and taking it from the fire with a forceps while it was glowing I placed it at once in a cold place where the wind was constantly blowing; and placing on it little pieces of various metals and other liquefiable bodies, I noted the times of cooling until all these bodies lost their fluidity and hardened, and until the heat of the iron became equal to the heat of the human body. Then by assuming that the excess of the heat of the iron and of the hardening bodies above the heat of the atmosphere, found by the thermometer, were in geometrical progression when the times were in arithmetical progression, all the heats were determined. I placed the iron not in quiet air but in a uniformly blowing wind, so that the air warmed by the iron would continually be taken away by the wind, and cold air would come in its place with a uniform motion. For thus equal parts of the air are warmed in equal times and carry away a heat proportional to the heat of the iron. The heats so found had the same ratio to one another as those found by the thermometer and therefore we have correctly assumed that the expansions of the oil were proportional to its heats.

AMONTONS

Guillaume Amontons was born in Paris on August 31, 1663, and died on October 11, 1705. Having lost his hearing in early life he could not engage in a profession and he devoted himself entirely to his studies.

The paper which follows, entitled "Discours sur quelques proprietez de l'air, et la moyen d'en connoître la temperature dans tous les climats de la terre," was published in Paris under date of June 18, 1702, in the *Mémoires de l'Académie Royal e des Sciences*. It contains Amontons' account of the first thermometer with which temperature was measured by the pressure of air.

An Air Pressure Thermometer

Any experiments which can lead to a knowledge of the nature of the air in which we live are sufficiently important to deserve special attention. Those which I made three years ago on the expansion of air by the heat of boiling water showed that unequal masses of air under the same weight or equal weights increase equally the force of their spring for equal degrees of heat; and as my principal aim in these experiments was to determine by how much the heat of boiling water would increase the spring of the air above that which it had in water which we call cold, these experiments led me at that time to believe that this increase was only by an amount capable of sustaining a mercury column ten inches high besides the weight of the atmosphere: but having since carried these experiments further, I have found that the spring of the air increased by the heat of boiling water does not always sustain ten inches of mercury more than the weight of the atmosphere; but that it will sustain more or less in proportion to the weight with which it is charged, and that this increase is always about a third of these weights, when the air is at first in the condition which we call temperate, and less than a third when the air is in a condition hotter than the temperate, and on the contrary more than a third when the air is colder than the temperate. For example, if in the temperate state a mass of air is loaded with 30 inches of mercury, including in that the load of the atmosphere, and has increased its spring by the heat of boiling water so as to sustain ten inches of mercury in addition to the load of thirty inches of mercury; then when this same mass is loaded with sixty inches it will increase its spring by twenty inches, and by thirty inches when it is loaded with ninety and similarly in other cases. From which it appears that we can deduce this consequence, that the same degree of heat, however small it may be, can always increase more and more the force of the spring of the air, if this air is loaded with a greater and greater weight. And as we have already noticed that unequal masses of air increase equally the force of their spring by equal degrees of heat, we can deduce this other consequence, that a very small portion of air, however small it may be, can acquire a force of spring greater and greater for a very small degree of heat if this small portion is always loaded more and more. These properties of air can perhaps in the future help us to explain several effects of which at present we do not know the causes.

I have just said that experiment led me to perceive that unequal masses of air loaded with equal weights would equally increase the force of their spring by equal degrees of heat, and that the forces of spring that they would acquire would be so much more considerable as the weights by which they were compressed were greater, of which the reason is that, since these masses of air are either in the same surroundings, or are considered as being so, and are loaded with equal weights, there is no reason why one of them should acquire a greater force of spring than the other. For although it may be true, that if these masses of air were free to expand, the larger ones would expand more than the smaller ones; yet this ought not to occur for the increase of their spring, since, according to the rule of M. Mariotte, unequal masses of air loaded equally ought to reduce their volumes in proportion to their original masses to acquire new equal degrees of force of spring; and by the inverse of this same rule, if equal masses of air loaded unequally are free to expand they will occupy spaces in proportion to the weights by which they are loaded; but if they cannot expand they ought necessarily to acquire forces of spring proportional to these same weights.

After I had recognized these truths, I tried to make an application of them; and I believed that I could use them advantageously in perfecting those instruments which serve to measure degrees of heat, and which we call for this reason thermometers.

The author discusses some forms of thermometers which had already been invented and points out the reasons why they did not give correct indications of the temperature. He then proceeds to describe a thermometer which he had invented, from which at least some of these defects had been eliminated.

The degree of heat which is necessary to establish uniformity in the construction of thermometers may be that of boiling water, experiment having shown me that water cannot acquire a greater degree of heat, however long it is on the fire, and however great the fire is.

ABCD (Fig. 36) is one of the glass tubes which I used in the experiments already reported in the Mémoires of 1699 to determine the increase of the spring of the air by the heat of boiling water, open at *A*, recurved at *C*, and ending in a globe *D*. The size of the tube is about half a line interior measurement, that of the globe is 3 inches and a quarter more or less and in this these thermometers have a great advantage over others by the equality of their motion, so easy to find in these new thermometers and so difficult to find in the old ones; the length of this tube from *A* to *B* is 46 inches, so that the whole length *AC* is about 48. There is mercury in the tube from the opening *E* of the globe and in all the rest of the tube nearly up to the opening *A*, so that when the globe *D* is in boiling water the air which it contains sustains by its spring 73 inches of mercury, including the weight of the atmosphere, which we shall suppose always equal to 28 inches, and only 45 inches without including that, measuring from the level of the mercury which is at *E*; then the surface of the mercury in the tube *AB* near the opening *A* will be the point from which we can begin to count all the other degrees of heat, which will be less than that of boiling water; for it is unheard of that in any climate the heat is equal to that of boiling water, and there is no place on the earth where one cannot easily obtain it; consequently we shall have a degree of heat known in all countries which will take in all those below it and from which we can begin to count them.

. . . .

FAHRENHEIT

Daniel Gabriel Fahrenheit was born in Danzig on May 14, 1686, and died in Holland on September 16, 1736. His father was a merchant. He was the first who succeeded in making thermometers which were comparable with each other through the whole length of the scale. He also introduced mercury as the thermometric substance.

In this short extract, taken from a paper entitled "Experimenta circa gradum caloris liquorum nonnullorum ebullientium instituta" published in the *Philosophical Transactions*, Vol. 33, 1724, p. 1, he describes the scale which he used.

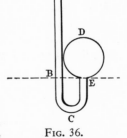

Fig. 36.

The Fahrenheit Scale

About ten years ago I read in the History of the Sciences issued by the Royal Academy of Paris, that the celebrated Amontons, using a thermometer of his own invention, had discovered that

water boils at a fixed degree of heat.   I was at once inflamed with
a great desire to make for myself a thermometer of the same sort,
so that I might with my own eyes perceive this beautiful phenomenon of nature, and be convinced of the truth of the experiment.

I therefore attempted to construct a thermometer, but because
of my lack of experience in its construction, my efforts were in
vain, though they were often repeated; and since other matters
prevented my going on with the development of the thermometer,
I postponed any further repetition of my attempts to some more
fitting time.   Though my powers and my time failed me, yet my
zeal did not slacken, and I was always desirous of seeing the outcome of the experiment.   It then came into my mind what that
most careful observer of natural phenomena had written about
the correction of the barometer; for he had observed that the
height of the column of mercury in the barometer was a little
(though sensibly enough) altered by the varying temperature of
the mercury.   From this I gathered that a thermometer might
perhaps be constructed with mercury, which would not be so hard
to construct, and by the use of which it might be possible to carry
out the experiment which I so greatly desired to try.

When a thermometer of that sort was made (perhaps imperfect
in many ways) the result answered to my prayer; and with great
pleasure of mind I observed the truth of the thing.

Three years then passed, in which I was occupied with optical
and other work, when I became anxious to try by experiment
whether other liquids boiled at fixed degrees of heat.

The results of my experiments are contained in the following
table, of which the first column contains the liquids used, the
second, their specific gravity, the third, the degree of heat which
each liquid attains when boiling.

Liquids	Specific Gravity of Liquids at 48° of heat	Degree Attained by Boiling
Spirits of Wine or Alcohol	8260	176
Rain Water	10000	212
Spirits of Niter	12935	242
Lye prepared from wine lees	15634	240
Oil of Vitriol	18775	546

I thought it best to give the specific gravity of each liquid,
so that, if the experiments of others already tried, or which may be

tried, give different results, it might be determined whether the difference should be looked for as resulting from differences in the specific gravities or from other causes. The experiments were not made at the same time, and hence the liquids were affected by different degrees of temperature or heat, but since their gravity is altered in a different way and unequally, I reduced it by calculation to the degree 48, which in my thermometers holds the middle place between the limit of the most intense cold obtained artificially in a mixture of water, of ice and of sal-ammoniac or even of sea-salt, and the limit of the heat which is found in the blood of a healthy man.

· · · · · · · · · · · · · · · · · · · ·

TAYLOR

Brook Taylor was born on August 18, 1685, and died in London on December 29, 1731. He was a man of means and without taking up a profession was able to devote himself to science. He was the discoverer of Taylor's theorem.

The experiment described in the following extract, taken from the *Philosophical Transactions*, Vol. 6, Part II, p. 17, is important because it furnished a basis for the belief in the indestructibility of heat. A similar rule for the temperature of mixtures was discovered by Richmann. The formula deduced from these observations for the temperature of mixtures is known as the Taylor-Richmann rule.

THE TEMPERATURES OF MIXTURES

I provided a good linseed oil thermometer, which I marked with small divisions, not equal in length, but equal according to the capacity of the tube in the several parts of it, as all thermometers ought to be graduated. I likewise provided two vessels of thin tin, of the same shape, and equal in capacity, containing each about a gallon. Then (observing in every trial, that the vessels were cold, before the water was put in them, as also that the vessel I measured the hot water with, was well heated with it) I successively filled the vessels with one, two, three, etc. parts of hot boiling water, and the rest cold; and at last with all the water boiling hot; and in every case I immersed the thermometer into the water, and observed to what mark it rose, making each trial in both vessels for the greater accuracy. And having first observed where the thermometer stood in cold water, I found that its rising from that mark, or the expansion of the oil, was accurately proportional to the quantity of hot water in the mixture, that is to the degree of heat.

BLACK

Joseph Black was born in Bordeaux in 1728, and died in Edinburgh on November 26, 1799. He was a physician and professor of chemistry at Glasgow and Edinburgh. He was one of the discoverers of carbon dioxide. He published very little during his life, but presented his views and discoveries in his academic lectures.

The following extracts, taken from Black's *Lectures on the Elements of Chemistry*, published after his death in 1803, contain the account of his discoveries of specific and latent heat.

SPECIFIC HEAT

. . . .

A second improvement in our knowledge of heat, which has been attained by the use of thermometers, is the more distinct notion we have now than formerly, of the *Distribution* of heat among different bodies.

I remarked formerly, that, even without the help of thermometers, we can perceive a tendency of heat to diffuse itself from any hotter body to the cooler around, until it be distributed among them, in such a manner that none of them are disposed to take any more heat from the rest. The heat is thus brought into a state of equilibrium. This equilibrium is somewhat curious. We find that when all mutual action is ended, a thermometer, applied to any one of the bodies, acquires the same degree of expansion: Therefore the temperature of them all is the same, and the equilibrium is universal. No previous acquaintance with the peculiar relation of each to heat could have assured us of this, and we owe the discovery entirely to the thermometer. We must therefore adopt, as one of the most general laws of heat, that "all bodies communicating freely with each other, and exposed to no inequality of external action, acquire the same temperature, as indicated by a thermometer." All acquire the temperature of the surrounding medium.

By the use of these instruments we have learned, that if we take 1000, or more, different kinds of matter, such as metals, stones, salts, woods, cork, feathers, wool, water and a variety of other fluids, although they be all at first of different heats, let them be placed together in the same room without a fire, and into which the sun does not shine, the heat will be communicated from the hotter of these bodies to the colder, during some hours perhaps, or the course of a day, at the end of which time, if we apply a thermometer to them all in succession, it will point precisely to the same degree. The heat, therefore, distributes itself upon this occasion, until none of these bodies has a greater demand or attraction for heat than every other of them has; in consequence of which, when we apply a thermometer to them all in succession, after the first to which it is applied has reduced the instrument to its own temperature, none of the rest are disposed to increase or diminish the quantity of heat which that first one left in it. This is what has been commonly called an equal heat, or the equality of heat among different bodies; I call it the *equilibrium of heat*. The nature of this equilibrium was not well understood, until I pointed out a method of investigating it. Dr. Boerhaave imagined, that when it obtains, there is an equal quantity of heat in every equal measure of space, however filled up with different bodies; and Professor Muschenbroeck expresses his opinion to the same purpose: "Est enim ignis aequaliter per omnia, non admodum magna, distributus, ita ut in pede cubico auri et aëris et plumarum, par ignis fit quantitas." The reason they give for this opinion is, that to whichever of those bodies the thermometer be applied, it points to the same degree.

But this is taking a very hasty view of the subject. It is confounding the quantity of heat in different bodies with its general strength or intensity, though it is plain that these are two different things, and should always be distinguished, when we are thinking of the distribution of heat.

. . . .

It was formerly a common supposition, that the quantities of heat required to increase the heat of different bodies by the same number of degrees, were directly in proportion to the quantity of matter in each; and therefore, when the bodies were of equal size, the quantities of heat were in proportion to their density. But very soon after I began to think on this subject, (anno 1760) I perceived that this opinion was a mistake, and that the quantities of heat which different kinds of matter must receive, to reduce them to an equilibrium with one another, or to raise their temperature by an equal number of degrees, are not in proportion to the quantity of matter in each, but in proportions widely different from this, and for which no general principle or reason can yet be assigned. It will be proper to consult, on this subject, the Comment. de Rebus in Medicina Gestis, vol. 21, and vol. 26, containing the valuable experiments of Jo. Carl. Wilcke, extracted from the Swedish Transactions. Also experiments of Professor Godolin, in the Nova Acta Reg. Societ. Upsalensis, tom. 5. This opinion was first suggested to me by an experiment described by Dr. Boerhaave (Boerhaave Elementa Chemiae, exp. 20, cor. 11.) After relating the experiment which Fahrenheit made at his desire, by mixing hot and cold water, he also tells us, that Fahrenheit agitated together quicksilver and water unequally heated. From the Doctor's account, it is quite plain, that quicksilver, though it has more than 13 times the density of water, produced less effect in heating or cooling water to which it was applied, than an equal measure of water would have produced. He says expressly, that the quicksilver, whether it was applied hot to cold water, or cold to hot water, never produced more effect in heating or cooling an equal measure of the water than would have been produced by the water equally hot or cold with the quicksilver, and only two-thirds of its bulk. He adds, that it was necessary to take three measures of quicksilver to two of water, in order to produce the same middle temperature that is produced by mixing equal measures of hot and cold water.

To make this plainer by an example in numbers, let us suppose the water to be at the 100th degree of heat, and that an equal measure of warm quicksilver at the 150th degree, is suddenly mixed and agitated with it. We know that the middle temperature between 100 and 150 is 125, and we know that this middle temperature would be produced by mixing the cold water at 100 with an equal measure of warm water at 150; the heat of the warm water being lowered by 25 degrees, while that of the cold is raised just as much. But when warm quicksilver is used in place of warm water, the temperature of the mixture turns out 120 degrees only instead of 125. The quicksilver, therefore, is become less warm by 30 degrees, while the water has become warmer by twenty degrees only; and yet the quantity of heat which the water has gained is the very same quantity which the quicksilver has lost. This shews that the same quantity of the matter of heat has more effect in heating quicksilver than in heating an equal measure of water, and therefore that a smaller *quantity* of it is sufficient for increasing the sensible heat of quicksilver by the same number of degrees. The same thing appears, whatever way we vary the experiment; for, if the water is the warmer mass, and quicksilver the less warm one, by the above difference, the temperature produced is 130. The water, in this case, is become less warm by 20 degrees, while the heat it has lost, being given to the quicksilver, has made this warmer by 30 degrees. And lastly, if we take three measures of quicksilver to two of water, it is no matter which of them be the hotter. The temperature produced is always the middle temperature between the two, or 125 degrees, in the temperatures already mentioned. Here it is manifest that the same quantity of the matter of heat which makes *two* measures of water warmer by 25 degrees, is sufficient for making *three* measures of quicksilver warmer by the same number of degrees. Quicksilver, therefore, has less *capacity* for the matter of heat than water (if I may be allowed to use this expression) has; it requires a smaller quantity of it to raise its temperature by the same number of degrees.

The inference which Dr. Boerhaave drew from this experiment is very surprising. Observing that heat is not distributed among different bodies in proportion to the quantity of matter in each, he concludes that it is distributed in proportion to the space occupied by each body; a conclusion contradicted by this very experiment. Yet Muschenbroeck has followed him in this opinion.

As soon as I understood this experiment in the manner I have now explained it, I found a remarkable agreement between it and some experiments made by Dr. Martin (Essay on the Heating and Cooling of Bodies) which appeared at first very surprising and unaccountable; but, being compared with this one, may be explained by the same principle. Dr. Martin placed before a good fire, and at an equal distance from it, a quantity of water, and an equal bulk or measure of quicksilver, each of them contained in equal and similar glass vessels, and each having a delicate thermometer immersed into it. He then carefully observed the progress, or celerity, with which each of these fluids was heated by the fire, and raised the thermometers. He found, by repeated experiments, that the quicksilver was warmed by the fire much faster than the water, almost twice as fast; and after each experiment, having heated these two fluids to the same degree, he placed them in a stream of cold air, and found that the quicksilver was always cooled much faster than the water. Before these experiments were made, it was supposed that the quicksilver would require to heat or cool it a longer time than an equal bulk of water, in the proportion of 13 or 14 to one.

But, from the view I have given of Fahrenheit's, or Boerhaave's experiment with quicksilver and water, the above of Dr. Martin's is easily explained. We need only to suppose that the matter of heat, communicated by the fire, was communicated equally to the quicksilver and to the water, but that, as less of it was required for heating the quicksilver, than for heating the water, the quicksilver necessarily was warmed fastest of the two. And when both, being equally heated, were exposed to the cold air to cool, the air at first took their heat from them equally fast, but the quicksilver, by losing the same quantity of the matter of heat that the water lost, was necessarily cooled to a greater degree; it therefore became cold much faster than the water. These experiments of Dr. Martin, therefore, agreeing so well with Fahrenheit's experiment, plainly shew that quicksilver, notwithstanding its great density and weight, requires less heat to heat it, than that which is necessary to heat, by the same number of degrees, an equal measure of equally cold water. The quicksilver, therefore, may be said to have less capacity for the matter of heat. And we are thus taught, that, in cases in which we may have occasion to investigate the capacity of different bodies for heat, we can learn it only by making experiments. Some have accordingly been made, both by myself and others. Dr. Crawford has made a great number of very curious ones, and his Theory of the Heat of Animals is founded partly on some experiments made in this manner, the result of which is given in his book on that subject.

It appears, therefore, from the general result of such experiments, that if we had a thousand masses of matter, of the same size and form, but of different materials, and were to place them all in the same room, until they assumed the same temperature; were we then to introduce into that room a great mass of red hot iron, the heat of which, when communicated with all these different bodies at the same time, might be sufficient for raising the temperature of them all, by 20 degrees; the heat thus communicated from the iron, although it produced an equal effect on each of these bodies, in raising its temperature by 20 degrees, would not however be equally divided or distributed among them. Some of them would attract and retain a much greater quantity of this heat, or matter of heat, than others; and the quantity received by each would not be in proportion to their densities, but in proportions totally unconnected with it; and perhaps not any two of them would receive precisely the same quantity, but each, according to its particular capacity, or its particular force of attraction for this matter, would attract and require its own peculiar quantity to raise its temperature by the 20 degrees, or to reduce it to an equilibrium or equality of saturation with the surrounding bodies. We must, therefore, conclude that different bodies, although they be of the same size, or even of the same weight, when they are reduced to the same temperature or degree of heat, whatever that be, may contain very different quantities of the matter of heat; which different quantities are necessary to bring them to this level, or equilibrium, with one another.

LATENT HEAT

. . . .

Fluidity was universally considered as produced by a small addition to the quantity of heat which a body contains, when it is once heated up to its melting point; and the return of such a body to a solid state, as depending on a very small diminution of the quantity of its heat, after it is cooled to the same degree; that a solid body, when it is changed into a fluid, receives no greater addition to the heat within it than what is measured by the elevation of temperature indicated after fusion by the thermometer; and that, when the melted body is again made to congeal, by a diminution of its heat, it suffers no greater loss of heat than what is indicated also by the simple application to it of the same instrument.

This was the universal opinion on this subject, so far as I know, when I began to read my lectures in the University of Glasgow, in the year 1757. But I soon found reason to object to it, as inconsistent with many remarkable facts, when attentively considered; and I endeavoured to shew, that these facts are convincing proofs that fluidity is produced by heat in a very different manner.

I shall now describe the manner in which fluidity appeared to me to be produced by heat, and we shall then compare the former and my view of the subject with the phenomena.

The opinion I formed from attentive observation of the facts and phenomena, is as follows. When ice, for example, or any other solid substance, is changing into a fluid by heat, I am of opinion that it receives a much greater quantity of heat than what is perceptible in it immediately after by the thermometer. A greater quantity of heat enters into it, on this occasion, without making it apparently warmer, when tried by that instrument. This heat, however, must be thrown into it, in order to give it the form of a fluid; and I affirm, that this great addition of heat is the principal, and most immediate cause of the fluidity induced.

And, on the other hand, when we deprive such a body of its fluidity again, by a diminution of its heat, a very great quantity of heat comes out of it, while it is assuming a solid form, the loss of which heat is not to be perceived by the common manner of using the thermometer. The apparent heat of the body, as measured by that instrument, is not diminished, or not in proportion to the loss of heat which the body actually gives out on this occasion; and it appears from a number of facts, that the state of solidity cannot be induced without the abstraction of this great quantity of heat. And this confirms the opinion, that this quantity of heat, absorbed, and, as it were, concealed in the composition of fluids, is the most necessary and immediate cause of their fluidity.

To perceive the foundation of this opinion, and the inconsistency of the former with many obvious facts, we must consider, in the first place, the appearances observable in the melting of ice, and the freezing of water.

If we attend to the manner in which ice and snow melt, when exposed to the air of a warm room, or when a thaw succeeds to frost, we can easily perceive, that however cold they might be at the first, they are soon heated up to their melting point, or begin soon at their surface to be changed into water. And if the common opinion had been well founded, if the complete change of them into water required only the further addition of a very small quantity of heat, the mass, though of considerable size, ought all to be melted in a very few minutes or seconds more, the heat continuing incessantly to be communicated from the air around. Were this really the case, the consequences of it would be dreadful in many cases; for, even as things are at present, the melting of great quantities of snow and ice occasions violent torrents, and great inundations in the cold countries, or in the rivers that come from them. But, were the ice and snow to melt as suddenly as they must necessarily do, were the former opinion of the action of heat in melting them well founded, the torrents and inundations would be incomparably more irresistible and dreadful. They would tear up and sweep away every thing, and that so suddenly, that mankind should have great difficulty to escape from their ravages. This sudden liquefaction does not actually happen; the masses of ice or snow melt with a very slow progress, and require a long time, especially if they be of a large size, such as are the collections of ice, and wreaths of snow, formed in some places during the winter. These, after they begin to melt, often require many weeks of warm weather, before they are totally dissolved into water. This remarkable slowness with which ice is melted, enables us to preserve it easily during the summer, in the structures called Ice-houses. It begins to melt in these, as soon as it is put into them; but, as the building exposes only a small surface to the air, and has a very thick covering of thatch, and the access of the external air to the inside of it is prevented as much as possible, the heat penetrates the ice-house with a slow progress, and this, added to the slowness with which the ice itself is *disposed* to melt, protracts the total liquefaction of it so long, that some of it remains to the end of summer. In the same manner does snow continue on many mountains during the whole summer, in a melting state, but melting so slowly, that the whole of that season is not a sufficient time for its complete liquefaction.

This remarkable slowness with which ice and snow melt, struck me as quite inconsistent with the common opinion of the modification of heat, in the liquefaction of bodies.

And this very phenomenon is partly the foundation of the opinion I have proposed; for if we examine what happens, we may perceive that a great quantity of heat enters the melting ice, to form the water into which it is changed, and that the length of time necessary for the collection of so much heat from the surrounding bodies, is the reason of the slowness with which the ice is liquefied. If any person entertain doubts of the entrance and absorption of heat in the melting ice, he needs only to touch it; he will instantly feel that it rapidly draws heat from his warm hand. He may also examine the bodies that surround it, or are in contact with it, all of which he will find deprived by it of a great part of their heat; or if he suspend it by a thread, in the air of a warm room, he may perceive with his hand, or by a thermometer, a stream of cold air descending constantly from the ice; for the air in contact is deprived of a part of its heat, and thereby condensed and made heavier than the warmer air of the rest of the room; it therefore falls downwards, and its place round the ice is immediately supplied by some of the warmer air; but this, in turn, is soon deprived of some heat, and prepared to descend in like manner; and thus there is a constant flow of warm air from around, to the sides of the ice, and a descent of the same in a cold state, from the lower part of the mass, during which operation the ice must necessarily receive a great quantity of heat.

It is, therefore, evident, that the melting ice receives heat very fast, but the only effect of this heat is to change it into water, which is not in the least sensibly warmer than the ice was before. A thermometer, applied to the drops or small streams of water, immediately as it comes from the melting ice, will point to the same degree as when it is applied to the ice itself, or if there is any difference, it is too small to deserve notice. A great quantity, therefore, of the heat, or of the matter of heat, which enters into the melting ice, produces no other effect but to give it fluidity, without augmenting its sensible heat; it appears to be absorbed and concealed within the water, so as not to be discoverable by the application of a thermometer.

In order to understand this absorption of heat into the melting ice, and concealment of it in the water, more distinctly, I made the following experiments.

. . . .

Black describes an experiment, in which, when equal masses of water and ice were exposed to similar sources of heat, it was found that the temperature of the water rose regularly, while that of the water formed from the melting ice did not rise. He also describes the results which he obtained by mixing water and ice. The account of one of these experiments follows.

I have, in the same manner, put a lump of ice into an equal quantity of water, heated to the temperature 176, and the result was, that the fluid was no hotter than water just ready to freeze. Nay, if a little sea salt be added to the water, and it be heated only to 166 or 170, we shall produce a fluid sensibly colder than the ice was in the beginning, which has appeared a curious and puzzling thing to those unacquainted with the general fact.

It is, therefore, proved that the phenomena which attended the melting of ice in different circumstances, are inconsistent with the common opinion which was established upon this subject, and that they support the one which I have proposed.

. . . .

In the above described common process of freezing water, the extrication and emergence of the latent heat, if I may be allowed to use these terms, is performed by such minute steps, or rather with such a smooth progress, that many may find difficulty in apprehending it; but I shall now mention another example, in which this extrication of the concealed heat becomes manifest and striking.

This example is an experiment, first made by Fahrenheit, but since repeated and confirmed by many others.

He wished to freeze water from which the air had been carefully extracted. This water was contained in small glass globes, about one-third filled, and accurately closed, to prevent the return of the air into them. These globes were exposed to the air in frosty weather, and remained so long exposed, that he had reason to be satisfied that they were cooled down to the degree of the air, which was six or seven degrees below the freezing point. The water, however, still remained fluid, so long as the glasses were left undisturbed, but, on being taken up and shaken a little, a sudden congelation was instantly seen.

It has since been found, by the trials of others, that the experiment will succeed, although the water be not deprived of its air, and that the circumstances the most essentially necessary are, that it be contained in vessels of small size, and preserved carefully from the least disturbance. The vessels, therefore, ought to be covered with paper, or otherwise, to prevent slight motions of the air from affecting the surface of the water. In these circumstances, it may be cooled to six, or seven, or eight degrees below the freezing point, without being frozen; but, if it be then disturbed, there is a sudden congelation, not of the whole, but of a small part only, which is formed into feathers of ice, traversing the water, in every direction, and forming a spongy contexture of ice, which contains the water in its vacuities, so as to give to the whole the appearance of being frozen. But the most remarkable fact is, that while this happens, (and it happens in a moment of time) this mixture of ice and water suddenly becomes warmer, and makes a thermometer, immersed in it, rise to the freezing point.

Nothing can be more inconsistent with the old opinion concerning the cause of congelation than the phenomena of this experiment. It shews that the loss of a little more heat, after the water is cooled down to the freezing point, is not the most necessary and inseparable cause of its congelation, since the water is cooled 6, 7, or 8 degrees below that point, without being congealed.

· · · · · · · · · · · · · · · · · · · ·

OF VAPOUR AND VAPORISATION

· · · · · · · · · · · · · · · · · · · ·

A more just explanation will occur to any person, who will take the trouble to consider this subject with patience and attention. In the ordinary manner of heating water, the heating cause is applied to the lower parts of the fluid. If the pressure on the surface be not increased, the water soon acquires the greatest heat which it can bear, without assuming the form of vapour. Subsequent additions of heat, therefore, in the same instant in which they enter the water, must convert into vapour that part which they thus affect. As these additions of heat all enter at the bottom of the fluid, there is a constant production of elastic vapour there, which, on account of its weighing almost nothing, must rise through the surrounding water, and appear to be thrown up to the surface with violence, and from thence it is diffused through the air. The water is thus gradually wasted, as the boiling continues, but its temperature is never increased, at least in that part which remains after long continued and violent boiling. The parts, indeed, in contact with the bottom of the vessel may be supposed to have received a little more heat, but this is instantly communicated to the surrounding water through which the elastic vapour rises.

This has the appearance of being a simple, plain, and complete account of the production of vapour, and of the boiling of fluids; and it is the only account that was given of this subject before I began to deliver these lectures: But I am persuaded that it is by no means a full accunt of the matter. According to this account, and the notion that was conceived of the formation of vapour, it was taken for granted that, after a body is heated up to its vaporific point, nothing further is necessary but the addition of a little more heat to change it into vapour. It was also supposed, on the other hand, that when the vapour of water is so far cooled as to be ready for condensation, this condensation, or return into the state of water, will happen at once, or in consequence of its losing only a very small quantity of heat.

But I can easily shew, in the same manner as in the case of fluidity, that a very great quantity of heat is necessary to the production of vapour, although the body be already heated to that temperature which it cannot pass, by the smallest possible degree, without being so converted. The undeniable consequence of this should be, an explosion of the whole water, with a violence equal to that of gunpowder. But I can shew, that this great quantity of heat enters into the vapour gradually, while it is forming, without making it perceptibly hotter to the thermometer. The vapour, if examined with a thermometer, is found to be exactly of the same temperature as the boiling water from which it arose. The water must be raised to a certain temperature, because, at that temperature only, is it disposed to absorb heat; and it is not instantly exploded, because, in that instant, there cannot be had a sufficient supply of heat through the whole mass. On the other hand, I can shew that when the vapour of water is condensed into a liquid, the very same great quantity of heat comes out of it into the colder matter by which it is condensed; and the matter of the vapour, or the water into which it is changed, does not become sensibly colder by the loss of this great quantity of heat. It does not become colder in proportion to the quantity of heat obtainable from it during its condensation.

All this will become evident, when we consider with attention the gradual formation of vapour, in consequence of the continued application of a heating cause, and the like gradual condensation of this vapour, when we continue to apply to it a body that is colder.

. . . .

I, therefore, set seriously about making experiments, conformable to the suspicion that I entertained concerning the boiling of fluids. My conjecture, when put into form, was to this purpose. I imagined that, during the boiling, heat is absorbed by the water, and enters into the composition of the vapour produced from it, in the same manner as it is absorbed by ice in melting, and enters into the composition of the produced water. And, as the ostensible effect of the heat, in this last case, consists, not in warming the surrounding bodies, but in rendering the ice fluid; so, in the case of boiling, the heat absorbed does not warm surrounding bodies, but converts the water into vapour. In both cases, considered as the cause of warmth, we do not perceive its presence: it is concealed, or latent, and I gave it the name of LATENT HEAT.

RUMFORD

Benjamin Thompson, Count Rumford, was born in Rumford, now Concord, New Hampshire on March 26, 1753. The circumstances of his early life were such that he had very little systematic education. At the outbreak of the American Revolution he served for awhile in the American army but was offended by some slight, and perhaps was influenced by political principle, so that he left the service and sailed for England. There he formed the acquaintance of Lord George Sackville, who patronized him and gave him opportunity for scientific research. He returned to America for a short time in the British service and after peace was concluded he went to Germany, intending to serve in the war against the Turks. There he was received by the reigning family of Bavaria and was employed by them as minister of war and in various offices. He was given the title of Count Rumford. He remained in the Bavarian service, except for a short interval, until 1799. He then went to Paris, where he lived in retirement in Auteuil until his death on August 21, 1814.

The extracts which follow are taken from Rumford's *Collected Works*, of which several editions have been published. The first one, which deals with the propagation of heat in fluids, in which Rumford describes his discovery of the convection of heat, is taken from Vol. II, Essay II (VII). The second one, "An Inquiry Concerning the Source of the Heat Which Is Excited by Friction," Vol. II, Essay IX, was read before the Royal Society on January 25, 1798 and appeared also in the *Philosophical Transactions*. In it is described the experiment by which heat was produced in apparently unlimited quantity by friction.

Convection of Heat

It is certain, that there is nothing more dangerous, in philosophical investigations, than to take any thing for granted, however unquestionable it may appear, till it has been proved by direct and decisive experiment.

I have very often, in the course of my philosophical researches, had occasion to lament the consequences of my inattention to this most necessary precaution.

There is not, perhaps, any phenomenon that more frequently falls under our observation, than the Propagation of Heat. The changes of the temperature of sensible bodies—of solids—liquids—and elastic fluids, are going on perpetually under our eyes; and there is no fact which one would not as soon think of calling in question, as to doubt of the free passage of Heat, in all directions, through all kinds of bodies. But, however obviously this conclusion appears to flow, from all that we observe and experience in the common course of life, yet it is certainly not true;—and to the erroneous opinion respecting this matter, which has been universally entertained—by the *learned*, and by the *unlearned*—and which has, I believe, never even been called in question, may be attributed the little progress that has been made in the investigation of the science of Heat:—a science, assuredly, of the utmost importance to mankind!

Under the influence of this opinion, I, many years ago, began my experiments on Heat; and had not an accidental discovery drawn my attention with irresistible force, and fixed it on the subject, I probably never should have entertained a doubt of the free passage of Heat *through air;* and even after I had found reason to conclude, from the results of experiments which to me appeared to be perfectly decisive, that air is a *non-conductor* of Heat; or that Heat cannot pass through it, without being transported by its particles; which, in this process, act individually, or independently of each other; yet, so far from pursuing the subject, and contriving experiments to ascertain the manner in which Heat is communicated in other bodies, I was not sufficiently awakened to suspect it to be even possible, that this quality could extend farther than to elastic Fluids.

With regard to liquids, so entirely persuaded was I, that Heat could pass freely, *in them*, in all directions, that I was perfectly blinded by this prepossession, and rendered incapable of seeing the most striking and most evident proofs of the fallacy of this opinion.

I have already given an account, in one of my late publications, of the manner in which I was led to discover, that *steam* and *flame* are *non-conductors* of Heat: I shall now lay before the Public an account of a number of experiments I have lately made, which seem to show that *water*,—and probably all other liquids,—and Fluids of every kind, possess the same property. That is to say, that although the particles of any Fluid, *individually*, can receive heat from other bodies, or communicate it to them; yet, among these particles themselves all *interchange* and *communication* of Heat is absolutely impossible.

It may, perhaps, be thought not altogether uninteresting, to be acquainted with the various steps by which I was led to an experimental investigation of this curious subject of inquiry.

When dining, I had often observed that some particular dishes retained their Heat much longer than others; and that apple pies, and apples and almonds mixed, (a dish in great repute in England) remained hot a surprising length of time. Much struck with this extraordinary quality of retaining Heat, which apples appeared to possess, it frequently occurred to my recollection; and I never burnt my mouth with them, or saw others meet with the same misfortune, without endeavouring, but in vain, to find out some way of accounting, in a satisfactory manner, for this surprising phenomenon.

About four years ago, a similar accident awakened my attention, and excited my curiosity still more; being engaged in an experiment which I could not leave, in a room heated by an iron stove, my dinner, which consisted of a bowl of thick rice soup, was brought into the room; and as I happened to be too much engaged at the time to eat it, in order that it might not grow cold, I ordered it to be set down on the top of the stove; about an hour afterwards, as near as I can remember, beginning to grow hungry, and seeing my dinner standing on the stove, I went up to it, and took a spoonful of the soup, which I found almost cold, and quite thick. Going, by accident, deeper with the spoon the second time, this second spoonful burnt my mouth. (It is probable that the stove happened to be nearly cold when the bowl was set down upon it, and that the soup had grown almost cold; when a fresh quantity of fuel being put into the stove, the Heat had been suddenly increased.) This accident recalled very forcibly to my mind the recollection of the hot apples and almonds with which I had so often burned my mouth, a dozen years before, in England; but even this, though it surprised me very much, was not sufficient to open my eyes, and to remove my prejudices respecting the conducting power of water.

Being at Naples, in the beginning of the year 1794, among the many natural curiosities which attracted my attention, I was much struck with several very interesting phenomena which the hot baths of BAIA presented to my observation; and among them there was one, which quite astonished me: standing on the sea-shore, near the baths, where the hot steam was issuing out of every crevice of the rocks, and even rising up out of the ground, I had the curiosity to put my hand into the water. As the waves which came in from the sea followed each other without intermission, and broke over the even surface of the beach, I was not surprised to find the water cold; but I was more than surprised, when, on running the ends of my fingers through the cold water into the sand, I found the heat so intolerable, that I was obliged instantly to remove my hand. The sand was perfectly wet; and yet, the temperature was so very different at the small distance of two or three inches! I could not reconcile this with the supposed great conducting power of water. I even found that the top of the sand was, to all appearance, quite as cold as the water which flowed over it; and this increased my astonishment still more. I then, for the first time, began to doubt of the conducting power of water, and resolved to set about making experiments to ascertain the fact. I did not however put this resolution into execution till about a month ago; and should perhaps never have done it, had not another unexpected appearance again called my attention to it, and excited afresh all my curiosity.

In the course of a set of experiments on the communication of Heat, in which I had occasion to use thermometers of an uncommon size, (their globular bulbs being above four inches in diameter) filled with various kinds of liquids, having exposed one of them, which was filled with spirits of wine, in as great a heat as it was capable of supporting, I placed it in a window, where the sun happened to be shining, to cool; when, casting my eye on its tube, which was quite naked, (the divisions of its scale being marked in the glass with a diamond) I observed an appearance which surprised me, and at the same time interested me very much indeed. I saw the whole mass of the liquid in the tube in a most rapid motion, running swiftly in two opposite directions, *up*, and *down*, at the same time. The bulb of the thermometer, which is of copper, had been made two years before I found leisure to begin my experiments; and having been left unfilled, without being closed with a stopple, some fine particles of dust had found their way into it, and these particles, which were intimately mixed with the spirits of wine, on their being illuminated by the sun's beams, became perfectly visible, (as the dust in the air of a darkened room is illuminated and rendered visible by the sun-beams which come in through a hole) and by their motion discovered the violent motions by which the spirits of wine in the tube of the thermometer was agitated.

This tube, which is $\frac{43}{100}$ of an inch in diameter internally, and very thin, is composed of very transparent, colourless glass, which rendered the appearance clear and distinct, and exceedingly beautiful. On examining the motion of the spirits of wine with a lens, I found that the ascending current occupied the *axis of the tube*, and that it descended by the *sides of the tube*.

On inclining the tube a little, the rising current moved out of the axis, and occupied the side of the tube which was uppermost, while the *descending* current occupied the whole of the lower side of it.

When the cooling of the spirits of wine in the tube was hastened by wetting the tube with ice cold water, the velocities of both the ascending and the descending currents were sensibly accelerated.

The velocity of these currents was gradually lessened, as the thermometer was cooled; and when it had acquired nearly the temperature of the air of the room, the motion ceased entirely.

By wrapping up the bulb of the thermometer in furs, or any other warm covering, the motion might be greatly prolonged.

I repeated the experiment with a similar thermometer of equal dimensions, filled with linseed-oil, and the appearances, on setting it in the window to cool, were just the same. The directions of the currents, and the parts they occupied in the tube, were the same; and their motions were to all appearance quite as rapid as those in the thermometer which was filled with spirits of wine.

Having now no longer any doubt with respect to the cause of these appearances, being persuaded that the motion in these liquids was occasioned by their particles *going individually*, and *in succession*, to give off their Heat to the cold side of the tube, in the same manner as I have shown in another place, that the particles of air give off *their* Heat to other bodies, I was led to conclude that these, and probably all other liquids, are in fact *non-conductors* of Heat; and I went to work immediately to contrive experiments to put the matter out of all doubt.

On considering the subject attentively, it appeared to me, that if liquids were in fact *non-conductors* of Heat, or if it be propagated in them *only* in consequence of the internal motions of their particles; in that case, every thing which tends to obstruct those motions, ought certainly to retard the operation, and render the propagation of the Heat slower, and more difficult. I had found that this is actually the case in respect to air; and though (under the influence of a strong and deep-rooted prejudice) I had, from the result of one imperfect experiment, too hastily concluded, that it did not take place in regard to water; yet I now found strong reasons to call in question the result of that experiment, and to give the subject a careful and thorough investigation.

Rumford then describes experiments on the rate of cooling of thermometers whose bulbs were immersed in pure water, and in water thickened with starch or containing eiderdown or stewed apples. The cooling was invariably slower when the free motion of the water was restricted by foreign bodies.

He also describes an experiment in which powdered amber, suspended in water of which the specific gravity had been made the same as that of the amber by adding to it a proper amount of vegetable alkali, moved with the currents in the water when it was either receiving or giving out heat, so as to show the way in which the heat was carried through the water by convection.

. . . .

HEAT PRODUCED BY FRICTION

It frequently happens, that in the ordinary affairs and occupations of life, opportunities present themselves of contemplating some of the most curious operations of Nature; and very interesting philosophical experiments might often be made, almost without trouble or expense, by means of machinery contrived for the mere mechanical purposes of the arts and manufactures.

I have frequently had occasion to make this observation; and am persuaded, that a habit of keeping the eyes open to every thing that is going on in the ordinary course of the business of life has oftener led, as it were by accident, or in the playful excursions of the imagination, put into action by contemplating the most common appearances, to useful doubts, and sensible schemes for investigation and improvement, than all the more intense meditations of philosophers, in the hours expressly set apart for study.

It was by accident that I was led to make the Experiments of which I am about to give an account; and, though they are not perhaps of sufficient importance to merit so formal an introduction, I cannot help flattering myself that they will be thought curious in several respects, and worthy of the honour of being made known to the Royal Society.

Being engaged, lately, in superintending the boring of cannon, in the workshops of the military arsenal at Munich, I was struck with the very considerable degree of Heat which a brass gun acquires, in a short time, in being bored; and with the still more intense Heat (much greater than that of boiling water, as I found by experiment) of the metallic chips separated from it by the borer.

The more I meditated on these phaenomena, the more they appeared to me to be curious and interesting. A thorough investigation of them seemed even to bid fair to give a farther insight into the hidden nature of Heat; and to enable us to form some reasonable conjectures respecting the existence, or non-existence, of an igneous fluid: a subject on which the opinions of philosophers have, in all ages, been much divided.

In order that the Society may have clear and distinct ideas of the speculations and reasonings to which these appearances gave rise in my mind, and also of the specific objects of philosophical investigation, they suggested to me, I must beg leave to state them at some length, and in such manner as I shall think best suited to answer this purpose.

From whence comes the Heat actually produced in the mechanical operation above mentioned?

Is it furnished by the metallic chips which are separated by the borer from the solid mass of metal?

If this were the case, then, according to the modern doctrines of latent Heat, and of caloric, the capacity for Heat of the parts of the metal, so reduced to chips, ought not only to be changed, but the change undergone by them should be sufficiently great to account for all the Heat produced.

But no such change had taken place; for I found, upon taking equal quantities, by weight, of these chips, and of thin slips of the same block of metal separated by means of a fine saw, and putting them, at the same temperature, (that of boiling water) into equal quantities of cold water, (that is to say, at the temperature of 59°½F.) the portion of water into which the chips were put was not, to all appearance, heated either less or more than the other portion, in which the slips of metal were put.

This Experiment being repeated several times, the results were always so nearly the same, that I could not determine whether any, or what change, had been produced in the metal, in regard to its capacity for Heat, by being reduced to chips by the borer.

From hence it is evident, that the Heat produced could not possibly have been furnished at the expense of the latent Heat of the metallic chips. But, not being willing to rest satisfied with these trials, however conclusive they appeared to me to be, I had recourse to the following still more decisive Experiment:

Taking a cannon, (a brass six-pounder) cast solid, and rough as it came from the foundry, (see *Fig. 37, 1*) and fixing it (horizontally) in the machine used for boring, and at the same time finishing the outside of the cannon by turning, (see *Fig. 37, 2.*) I caused its extremity to be cut off; and, by turning down the metal

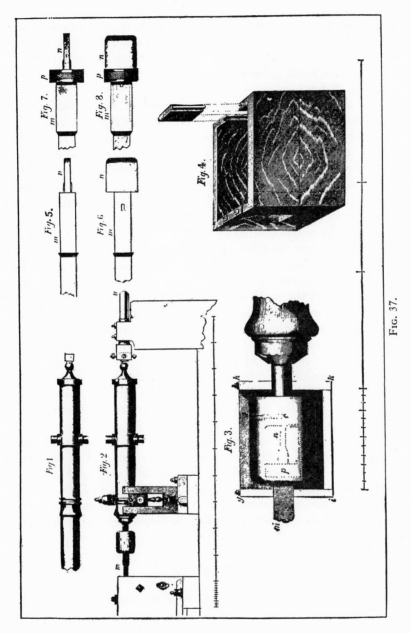

FIG. 37.

in that part, a solid cylinder was formed, 7¾ in. in diameter, and 9⅞₀ inches long; which, when finished, remained joined to the rest of the metal (that which, properly speaking, constituted the cannon) by a small cylindrical neck, only 2⅕ inches in diameter, and 3⅞₀ inches long.

This short cylinder, which was supported in its horizontal position, and turned round its axis, by means of the neck by which it remained united to the cannon, was now bored with the horizontal borer used in boring cannon; but its bore, which was 3.7 inches in diameter, instead of being continued through its whole length (9.8 inches) was only 7.2 inches in length; so that a solid bottom was left to this hollow cylinder, which bottom was 2.6 inches in thickness.

This cavity is represented by dotted lines in *Fig. 37, 2*; as also in *Fig. 37, 3*, where the cylinder is represented on an enlarged scale.

This cylinder being designed for the express purpose of generating Heat by friction, by having a blunt borer forced against its solid bottom at the same time that it should be turned round its axis by the force of horses, in order that the Heat accumulated in the cylinder might from time to time be measured, a small round hole, (see *d, e, Fig. 37, 3*) 0.37 of an inch only in diameter, and 4.2 inches in depth, for the purpose of introducing a small cylindrical mercurial thermometer, was made in it, on one side, in a direction perpendicular to the axis of the cylinder, and ending in the middle of the solid part of the metal which formed the bottom of its bore.

The solid contents of this hollow cylinder, exclusive of the cylindrical neck by which it remained united to the cannon, were 383¾ cubic inches, English measure; and it weighed 113.13 lb. Avoirdupois: as I found, on weighing it at the end of the course of Experiments made with it, and after it had been separated from the cannon with which, during the Experiments, it remained connected.

Experiment, No. 1

This Experiment was made in order to ascertain how much Heat was actually generated by friction, when a blunt steel borer being so forcibly shoved (by means of a strong screw) against the bottom of the bore of the cylinder, that the pressure against it was equal to the weight of about 10000 lb. Avoirdupois, the cylinder was turned round on its axis (by the force of horses) at the rate of about 32 times in a minute.

This machinery, as it was put together for the Experiment, is represented by *Fig.* 37, 2. *W* is a strong horizontal iron bar, connected with proper machinery carried round by horses, by means of which the cannon was made to turn round its axis.

To prevent, as far as possible, the loss of any part of the Heat that was generated in the Experiment, the cylinder was well covered up with a fit coating of thick and warm flannel, which was carefully wrapped round it, and defended it on every side from the cold air of the atmosphere. This covering is not represented in the drawing of the apparatus, *Fig.* 37, 2.

I ought to mention, that the borer was a flat piece of hardened steel, 0.63 of an inch thick, 4 inches long, and nearly as wide as the cavity of the bore of the cylinder, namely, $3\frac{1}{2}$ in. Its corners were rounded off at its end, so as to make it fit the hollow bottom of the bore; and it was firmly fastened to the iron bar (*m*) which kept it in its place. The area of the surface by which its end was in contact with the bottom of the bore of the cylinder was nearly $2\frac{1}{3}$ inches. This borer, which is distinguished by the letter *n*, is represented in most of the figures.

At the beginning of the Experiment, the temperature of the air in the shade, as also that of the cylinder, was just 60°F.

At the end of 30 minutes, when the cylinder had made 960 revolutions about its axis, the horses being stopped, a cylindrical mercurial thermometer, whose bulb was $\frac{32}{100}$ of an inch in diameter, and $3\frac{1}{4}$ inches in length, was introduced into the hole made to receive it, in the side of the cylinder, when the mercury rose almost instantly to 130°.

Though the Heat could not be supposed to be quite equally distributed in every part of the cylinder, yet, as the length of the bulb of the thermometer was such that it extended from the axis of the cylinder to near its surface, the Heat indicated by it could not be very different from that of the mean temperature of the cylinder; and it was on this account that a thermometer of that particular form was chosen for this Experiment.

To see how fast the Heat escaped out of the cylinder, (in order to be able to make a probable conjecture respecting the quantity given off by it, during the time the Heat generated by the friction was accumulating) the machinery standing still, I suffered the thermometer to remain in its place near three quarters of an

hour, observing and noting down, at small intervals of time, the
height of the temperature indicated by it.

Thus at the end of

The Heat as
shown by the
thermometer, was

	4 minutes.............................	126°
after	5 minutes, always reckoning from the first observation.........................	125°
at the end of	7 minutes.............................	123°
	12 minutes.............................	120°
	14 minutes.............................	119°
	16 minutes.............................	118°
	20 minutes.............................	116°
	24 minutes.............................	115°
	28 minutes.............................	114°
	31 minutes.............................	113°
	34 minutes.............................	112°
	37½ minutes..........................	111°
and when	41 minutes had elapsed...................	110°

Having taken away the borer, I now removed the metallic
dust, or rather scaly matter, which had been detached from the
bottom of the cylinder by the blunt steel borer, in this Experiment;
and, having carefully weighed it, I found its weight to be 837 grains
Troy.

Is it possible that the very considerable quantity of Heat that
was produced in this Experiment (a quantity which actually raised
the temperature of above 113 lb. of gun-metal at least 70 degrees
of Fahrenheit's thermometer, and which, of course, would have
been capable of melting 6½ lb. of ice, or of causing near 5 lb. of
ice-cold water to boil) could have been furnished by so inconsiderable a quantity of metallic dust? and this merely in consequence
of a change of its capacity for Heat?

As the weight of this dust (837 grains Troy) amounted to
no more than $\frac{1}{948}$th part of that of the cylinder, it must have lost
no less than 948 degrees of Heat, to have been able to have raised
the temperature of the cylinder 1 degree; and consequently it
must have given off 66360 degrees of Heat, to have produced
the effects which were actually found to have been produced in
the Experiment!

But, without insisting on the improbability of this supposition,
we have only to recollect, that from the results of actual and
decisive Experiments, made for the express purpose of ascertaining
that fact, the capacity for Heat, of the metal of which great guns

are cast, it not sensibly changed by being reduced to the form of metallic chips, in the operation of boring cannon; and there does not seem to be any reason to think that it can be much changed, if it be changed at all, in being reduced to much smaller pieces, by means of a borer that is less sharp.

If the Heat, or any considerable part of it, were produced in consequence of a change in the capacity for Heat of a part of the metal of the cylinder, as such change could only be superficial, the cylinder would by degrees be exhausted; or the quantities of Heat produced, in any given short space of time, would be found to diminish gradually in successive Experiments. To find out if this really happened or not, I repeated the last mentioned Experiment several times, with the utmost care; but I did not discover the smallest sign of exhaustion in the metal, notwith- standing the large quantities of Heat actually given off.

Finding so much reason to conclude, that the Heat generated in these Experiments, or excited, as I would rather choose to express it, was not furnished at the expense of the latent Heat or combined caloric of the metal, I pushed my inquiries a step farther, and endeavoured to find out whether the air did, or did not, contribute any thing in the generation of it.

Experiment 2 was devised to prove, by preventing the access of the external air, that the air did not produce the heat that was developed when the cannon was turned. It has been omitted as of relatively little importance.

Experiment, No. 3.

A quadrangular oblong deal box (*Fig.* 37, 4) water-tight, 11½ English inches long, 9$\frac{2}{10}$ inches wide, and 9$\frac{6}{10}$ inches deep, (measured in the clear) being provided, with holes or slits in the middle of each of its ends, just large enough to receive, the one, the square iron rod to the end of which the blunt steel borer was fastened, the other, the small cylindrical neck which joined the hollow cylinder to the cannon; when this box (which was occa- sionally closed above, by a wooden cover or lid moving on hinges) was put into its place; that is to say, when, by means of the two vertical openings or slits in its two ends, (the upper parts of which openings were occasionally closed, by means of narrow pieces of wood sliding in vertical grooves) the box (g, b, i, k, Fig. 3.) was fixed to the machinery, in such a manner that its bottom (i, k) being in the plane of the horizon, its axis coincided with the axis of the hollow metallic cylinder; it is evident from the description, that the hollow metallic cylinder would occupy the middle of the box, without touching it on either side (as it is represented in *Fig. 3;*) and that, on pouring water into the box, and filling it to the brim, the cylinder would be completely covered, and surrounded on every side, by that fluid. And farther, as the box was held fast by the strong square iron rod (*m*) which passed, in a square hole, in the center of one of its ends (*a, Fig. 4.*) while the round or cylindrical neck, which joined the hollow cylinder to the end of the cannon, could turn round freely on its axis in the round hole in the centre of the other end of it, it is evident that the machinery could be put in motion, without the least danger of forcing the box out of its place, throwing the water out of it, or deranging any part of the apparatus.

Every thing being ready, I proceeded to make the Experiment I had projected, in the following manner:

The hollow cylinder having been previously cleaned out, and the inside of its bore wiped with a clean towel till it was quite dry, the square iron bar, with the blunt steel borer fixed to the end of it, was put into its place; the mouth of the bore of the cylinder being closed at the same time, by means of the circular piston, through the centre of which the iron bar passed.

This being done, the box was put in its place, and the joinings of the iron rod, and of the neck of the cylinder, with the two ends of the box, having been made water-tight, by means of collars of oiled leather, the box was filled with cold water, (viz. at the temperature of 60°) and the machine was put in motion.

The result of this beautiful Experiment was very striking, and the pleasure it afforded me amply repaid me for all the trouble I had had, in contriving and arranging the complicated machinery used in making it.

The cylinder, revolving at the rate of about 32 times in a minute, had been in motion but a short time, when I perceived, by putting my hand into the water, and touching the outside of the cylinder, that Heat was generated; and it was not long before the water which surrounded the cylinder began to be sensibly warm.

At the end of 1 hour I found, by plunging a thermometer into the water in the box, (the quantity of which fluid amounted to 18.77 lb. Avoirdupois, or 2¼ wine gallons) that its temperature had been raised no less than 47 degrees; being now 107° of Fahrenheit's scale.

When 30 minutes more had elapsed, or 1 hour and 30 minutes after the machinery had been put in motion, the Heat of the water in the box was 142°.

At the end of 2 hours, reckoning from the beginning of the Experiment, the temperature of the water was found to be raised to 178°.

At 2 hours 20 minutes it was 200°; and at 2 hours 30 minutes it ACTUALLY BOILED!

It would be difficult to describe the surprise and astonishment expressed in the countenances of the by-standers, on seeing so large a quantity of cold water heated, and actually made to boil without any fire.

Though there was, in fact, nothing that could justly be considered as surprising in this event, yet I acknowledge fairly that it afforded me a degree of childish pleasure, which, were I ambitious of the reputation of a grave philosopher, I ought most certainly rather to hide than to discover.

The quantity of Heat excited and accumulated in this Experiment was very considerable; for, not only the water in the box, but also the box itself, (which weighed 15¼ lb.) and the hollow metallic cylinder, and that part of the iron bar which, being situated within the cavity of the box, was immersed in the water, were heated 150 degrees of Fahrenheit's scale; viz. from 60° (which was the temperature of the water, and of the machinery, at the beginning of the Experiment) to 210°, the Heat of boiling water at Munich.

The total quantity of Heat generated may be estimated with some considerable degree of precision, as follows:

The calculation has been omitted.

From the result of these computations it appears, that the quantity of Heat produced equably, or in a continual stream, (if I may use that expression) by the friction of the blunt steel borer against the bottom of the hollow metallic cylinder, in the Experiment under consideration, was greater than that produced equably in the combustion of nine wax-candles, each ¾ of an inch in diameter, all burning together, or at the same time, with clear bright flames.

As the machinery used in this Experiment could easily be carried round by the force of one horse, (though, to render the work lighter, two horses were actually employed in doing it) these computations show further how large a quantity of Heat might be produced, by proper mechanical contrivance, merely by the strength of a horse, without either fire, light, combustion, or chemical decomposition; and, in a case of necessity, the Heat thus produced might be used in cooking victuals.

But no circumstances can be imagined, in which this method of procuring Heat would not be disadvantageous; for, more Heat might be obtained by using the fodder necessary for the support of a horse, as fuel.

By meditating on the results of all these Experiments, we are naturally brought to that great question which has so often been the subject of speculation among philosophers; namely:

What is Heat?—Is there any such thing as an *igneous fluid?*—Is there any thing that can with propriety be called *caloric?*

We have seen that a very considerable quantity of Heat may be excited in the Friction of two metallic surfaces, and given off in a constant stream or flux, *in all directions*, without interruption or intermission, and without any signs of diminution or exhaustion.

From whence came the Heat which was continually given off in this manner, in the foregoing Experiments? Was it furnished by the small particles of metal, detached from the larger solid masses, on their being rubbed together? This, as we have already seen, could not possibly have been the case.

Was it furnished by the air? This could not have been the case; for, in three of the Experiments, the machinery being kept immersed in water, the access of the air of the atmosphere was completely prevented.

Was it furnished by the water which surrounded the machinery? That this could not have been the case is evident: *first*, because this water was continually *receiving Heat* from the machinery, and could not, at the same time, be *giving to*, and *receiving Heat from*, the same body; and secondly, because there was no chemical decomposition of any part of this water. Had any such decomposition taken place, (which indeed could not reasonably have been expected) one of its component elastic fluids (most probably inflammable air) must, at the same time, have been set at liberty, and in making its escape into the atmosphere, would have been detected; but though I frequently examined the water to see if any air bubbles rose up through it, and had even made preparations for catching them, in order to examine them, if any should appear,

I could perceive none; nor was there any sign of decomposition of any kind whatever, or other chemical process, going on in the water.

Is it possible that the Heat could have been supplied by means of the iron bar to the end of which the blunt steel borer was fixed? or by the small neck of gun-metal by which the hollow cylinder was united to the cannon? These suppositions appear more improbable even than either of those before mentioned; for Heat was continually going off, or *out of the machinery,* by both these passages, during the whole time the Experiment lasted.

And, in reasoning on this subject, we must not forget to consider that most remarkable circumstance, that the source of the Heat generated by friction, in these Experiments, appeared evidently to be *inexhaustible.*

It is hardly necessary to add, that any thing which any *insulated* body, or system of bodies, can continue to furnish *without limitation,* cannot possibly be *a material substance:* and it appears to me to be extremely difficult, if not quite impossible, to form any distinct idea of any thing, capable of being excited and communicated, in the manner the Heat was excited and communicated in these Experiments, except it be MOTION.

DAVY

Sir Humphry Davy was born in Penzance on Decenber 17, 1778. He worked as assistant and student with a surgeon who was also a druggist. He became interested in chemistry and was early distinguished for his discoveries in chemistry and electricity. He was for many years professor of chemistry at the Royal Institution and for seven years was president of the Royal Society. He isolated sodium and potassium by means of the electric current and was the first to observe the electric arc. He died in Geneva on May 29, 1829.

The paper which follows, entitled "An Essay on Heat, Light, and the Combinations of Light," was published in a book entitled *Contributions to Physical and Medical Knowledge, Principally from the West of England,* Collected by Thomas Beddoes, M.D., 1799. It also may be found in Davy's Collected Works. In it Davy describes his experiments on the production of heat by friction.

Heat Produced by Friction

Matter is possessed of the power of attraction. By this power the particles of bodies tend to approximate, and to exist in a state of contiguity. The particles of all bodies with which we are acquainted, can be made to approach nearer to each other, by peculiar means, that is, the specific gravity of all bodies can be increased by diminishing their temperatures. Consequently (on the supposition of the impenetrability of matter) the particles of bodies are not in actual contact. There must then act on the corpuscles of bodies some other power, which prevents their actual contact; this may be called repulsion. The phaenomena of repulsion have been supposed, by the greater part of chemical philosophers, to depend on a peculiar elastic fluid; to which the names of latent heat, and caloric, have been given. The peculiar modes of existence of bodies, solidity, fluidity, and gazity, depend (according to the calorists) on the quantity of the fluid of heat entering into their composition; this substance insinuating itself between their corpuscles, separating them from each other, and preventing their actual contact, is, by them, supposed to be the cause of repulsion.

Other philosophers, dissatisfied with the evidences produced in favour of the existence of this fluid, and perceiving the generation of heat by friction and percussion, have supposed it to be motion.

Considering the discovery of the true cause of the repulsive power as highly important to philosophy, I have endeavoured to investigate this part of chemical science by experiments: from these experiments (of which I am now about to give a detail), I conclude that heat, or the power of repulsion, is not matter.

. . . .

The Phaenomena of Repulsion Are Not Dependent on a Peculiar Elastic Fluid for Their Existence, or Caloric Does Not Exist

Without considering the effects of the repulsive power on bodies, or endeavouring to prove from these effects that it is motion, I shall attempt to demonstrate by experiments that it is not matter; and in doing this, I shall use the method called by mathematicians, *reductio ad absurdum.*

Let heat be considered as matter, and let it be granted that the temperature of bodies cannot be increased, unless their capacities are diminished from some cause, or heat added to them from some bodies in contact.

Now the temperatures of bodies are uniformly raised by friction and percussion. And since an increase of temperature is consequent on friction and percussion, it must consequently be generated in one of these modes. First, either from a diminution of the capacities of the acting bodies from some change induced in them by friction, a change producing in them an increase of temperature.

Secondly, or from heat communicated, from the decomposition of the oxygen gas in contact by one or both of the bodies, and then friction must effect some change in them (similar to an increase of temperature), enabling them to decompose oxygen gas, and they must be found after friction, partially or wholly oxydated.

Thirdly, or from a communication of caloric from the bodies in contact, produced by a change induced by friction in the acting bodies, enabling them to attract caloric from the surrounding bodies.

Now first let the increase of temperature produced by friction and percussion be supposed to arise from a diminution of the capacities of the acting bodies. In this case it is evident some change must be induced in the bodies by the action, which lessens their capacities and increases their temperatures.

.   .   .   .

Experiment II

I procured two parallelopipedons of ice, of the temperature of 29°, six inches long, two wide, and two-thirds of an inch thick: they were fastened by wires to two bars of iron. By a peculiar mechanism, their surfaces were placed in contact, and kept in a continued and violent friction for some minutes. They were almost entirely converted into water, which water was collected, and its temperature ascertained to be 35°, after remaining in an atmosphere of a lower temperature for some minutes. The fusion took place only at the plane of contact of the two pieces of ice, and no bodies were in friction but ice. From this experiment it is evident that ice by friction is converted into water, and according to the supposition its capacity is diminished; but it is a well-known fact, that the capacity of water for heat is much greater than that of ice; and ice must have an absolute quantity of heat added to it, before it can be converted into water. Friction consequently does not diminish the capacities of bodies for heat.

From this experiment it is likewise evident, that the increase of temperature consequent on friction cannot arise from the decomposition of the oxygen gas in contact, for ice has no attraction for oxygen. Since the increase of temperature consequent on friction cannot arise from the diminution of capacity, or oxydation of the acting bodies, the only remaining supposition is, that it arises from an absolute quantity of heat added to them, which heat must be attracted from the bodies in contact. Then friction must induce some change in bodies, enabling them to attract heat from the bodies in contact.

Experiment III

I procured a piece of clock-work so constructed as to be set to work in the exhausted receiver; one of the external wheels of this machine came in contact with a thin metallic plate. A considerable degree of sensible heat was produced by friction between the wheel and plate when the machine worked uninsulated from bodies capable of communicating heat. I next procured a small piece of ice; round the superior edge of this a small canal was made and filled with water. The machine was placed on the ice, but not in contact with the water. Thus disposed, the whole was placed under the receiver, (which had been previously filled with carbonic acid,) a quantity of potash (i.e. caustic vegetable alkali) being at the same time introduced.

The receiver was now exhausted. From the exhaustion, and from the attraction of the carbonic acid gas by the potash, a vacuum nearly perfect was, I believe, made.

The machine was now set to work. The wax rapidly melting, proved the increase of temperature.

Caloric then was collected by friction; which caloric, on the supposition, was communicated by the bodies in contact with the machine. In this experiment, ice was the only body in contact with the machine. Had this ice given out caloric, the water on the top of it must have been frozen. The water on the top of it was not frozen, consequently the ice did not give out caloric. The caloric could not come from the bodies in contact with the ice; for it must have passed through the ice to penetrate the machine, and an addition of caloric to the ice would have converted it into water.

Heat, when produced by friction, cannot be collected from the bodies in contact, and it was proved by the second experiment, that the increase of temperature consequent on friction cannot arise from diminution of capacity, or from oxydation. But if it be considered as matter, it must be produced in one of these modes. Since (as is demonstrated by these experiments) it is produced in neither of these modes, it cannot be considered as matter. It has then been experimentally demonstrated that caloric, or the matter of heat, does not exist.

Solids, by long and violent friction, become expanded, and if of a higher temperature than our bodies, affect the sensory organs with the peculiar sensation known by the common name of heat.

Since bodies become expanded by friction, it is evident that their corpuscles must move or separate from each other. Now a motion or vibration of the corpuscles of bodies must be necessarily generated by friction and percussion. Therefore we may reasonably conclude that this motion or vibration is heat, or the repulsive power.

Heat, then or that power which prevents the actual contact of the corpuscles of bodies, and which is the cause of our peculiar sensations of heat and cold, may be defined as a peculiar motion, probably a vibration, of the corpuscles of bodies, tending to separate them. It may with propriety be called the repulsive motion.

GAY-LUSSAC

Louis-Joseph Gay-Lussac was born on December 6, 1778, in St. Leonard. His father was a judge who was imprisoned during the Revolution as a suspect. After his release in 1795 he sent his son to Paris where he studied in the École Polytechnique and the École des Ponts et Chaussées. In 1802 Gay-Lussac was appointed a demonstrator in the École Polytechnique and in 1809 he became professor of chemistry in that institution. He was an ardent and successful investigator in chemistry. On the basis of his own researches he announced the law of combining volumes, which has had so great an influence in that science. He was also successful in the study of those physical subjects which lie near to chemistry. Gay-Lussac died on May 9, 1850, in Paris.

The following extract deals with the expansion of gases and vapors and contains the first statement of what is now known as Gay-Lussac's law. It is taken from the *Annales de Chimie*, Vol. 43, p. 137, 1802. The next one, by the same author, is taken from the *Mémoires de la société d'Arcueil*, Vol. I, p. 180, 1807. The translation, by J. S. Ames, appeared in Harper's *Scientific Memoirs*. In this paper is described Gay-Lussac's experiment on the expansion of a gas into a vacuum, by which he was led to the same conclusion as that later reached by Joule (p. 203). This result is interesting because it was appealed to by J. R. Mayer to justify the assumption made by him in his calculation of the mechanical equivalent of heat.

THE EXPANSION OF GASES BY HEAT

First Section

Object of this Memoir

For many years physicists have studied the expansion of gases; but the results which they obtained showed such great differences that instead of reaching a definite result they render further examination desirable.

The expansion of vapors has been given less consideration, although we have known for many years the prodigious effects of water vapor; and although we have made most fortunate applications of these effects, Ziegler and Bettancourt are the only ones, so far as I know, who have tried to measure them. Their experiments cannot give the true expansion of this vapor; for since they always had water in their apparatus, for each new degree of heat they had the expansion of the vapor formed by the preceding degrees of heat, and an increase of volume by the formation of new vapor; two causes which manifestly conspire to raise the mercury in their manometer.

The thermometer, such as we have it to-day, cannot be used to indicate exact ratios of heat, because we still do not know what ratio there is between the degrees of the thermometer and the quantities of heat which they can indicate. It is generally believed, it is true, that equal divisions of its scale represent equal tensions of caloric; but this opinion is not based on any definite fact.

It must then be admitted that we are still very far from having certain knowledge of the expansion of gases and vapors, and of the progress of a thermometer; and nevertheless we need every day, in physics and in chemistry, to bring a given volume of a gas from one temperature to another, to measure the heat disengaged or absorbed when a body changes its state, or that disengaged or absorbed by the same body in passing from one temperature to another; in the arts, to calculate the effect of a machine moved by heat or to have a proper knowledge of the expansion of certain bodies; in meteorology, to determine the quantity of water held in solution in the air, a quantity which varies with its temperature and density, according to a law which is not yet known. Finally in constructing tables of refraction for astronomy and in the use of the barometer for measuring heights, it is indispensable to know exactly both the temperature of the air and the law of its expansion.

Although these considerations were sufficient to make it desirable that someone should take up an investigation of such general utility, yet the difficulty of the researches which it involved would have prevented me giving myself to it, if I had not been strongly urged to undertake it by Citizen Berthollet, of whom I have the honor to be a pupil. I owe to him the apparatus needed to carry out this work, in which I have often been guided by his advice and that of Citizen Laplace: such great authorities will increase the confidence with which it may be received.

The researches which I have undertaken on the law of expansion of gases and vapors, and on the progress of the thermometer, are not yet complete. In this memoir my object is only to examine the expansion of gases and vapors for a fixed rise of temperature, and to show that it is the same for all these fluids; but before giving an account of my experiments I think it will be well to give a historical sketch of what has been done in this field; and as I shall introduce at the same time some observations on the different methods which have been employed, I will before entering on the history mention one of the principal causes of uncertainty which can arise in this sort of experiment. Although it is very important and although it seems to have been unrecognized by most of the physicists who have studied the expansion of gases, it will be sufficient for me to mention it to make clear what its influence will be. What I shall say of atmospheric air applies to the other gases also.

This cause of uncertainty is the presence of water in the apparatus. Suppose, in fact, that some drops of water are left in a globe full of air, of which the temperature is raised to that of boiling water; this water will turn into vapor and will occupy a volume about 1800 times greater than its original volume, and thus will drive out a great part of the air contained in the globe. It will then necessarily happen that when these vapors are condensed so that they occupy a volume 1800 times smaller, there will be attributed to the air which remains in the globe much too great an expansion; because it will be assumed that it is this air which, at the temperature of boiling water, occupied the whole volume of the globe. If the temperature is not carried to this degree the same cause of uncertainty will nevertheless exist, and its importance will depend upon the temperature at which the experiment is tried: for in this case the water will not be entirely vaporized, but the air will dissolve more and more of it as its temperature rises, and will receive in consequence a greater and greater increase in volume besides that which it gets from the heat; so that when we go again to the lower temperature, the volume of air which fills the globe will diminish for two reasons, 1. because of the loss of caloric, 2. because of the loss of the water which it holds in solution. We shall then still attribute to the air too great an expansion.

In general whenever we enclose with the gas liquids or even solids, for example muriate of ammonia, which can dissolve or vaporize at the temperature to which they are exposed, there necessarily result errors in the determination of the expansion of these gases.

The Second Section contains the historical sketch of the various methods which had been employed by previous observers to determine the expansion of gases. The Third Section contains the description of Gay-Lussac's apparatus. The Fourth Section contains the experimental results and concludes with the following paragraphs.

The experiments which I have now reported and which have all been made with great care prove incontestably that atmospheric air and the gases oxygen, hydrogen, nitrogen, nitrous oxide, ammonia, muriatic acid, sulphurous acid and carbonic acid all have the same expansion between the same degrees of heat; and thus consequently their greater or less density at the same pressure and temperature, their greater or less solubility in water, and their particular character have no influence on their expansion. On this basis I conclude that all gases in general expand equally between the same degrees of heat provided that they are all brought under the same conditions.

These investigations of the expansion of gases led me naturally to examine the expansion of vapors; but since I expected from the preceding results that they would expand like gases I decided to make my experiments only with one vapor, and I chose by preference the vapor of sulphuric ether as being the easiest to manage.

To determine the expansion of the vapor of ether I used the two tubes of which I have already spoken, taking atmospheric air for comparison. This apparatus was kept for some time in a vessel at the temperature of about 60°. I then introduced ether vapor into one of the tubes and atmospheric air into the other, in such a way that each of them corresponded to the same division. I then raised the temperature of the vessel from 60° to 100°, and I had the satisfaction of seeing that, whether rising or falling, the ether vapor and the atmospheric air always corresponded at the same time to the same divisions. This experiment, which was shown to Citizen Berthollet, was repeated several times, and I never was able to observe any difference in the expansion of the ether compared with that of atmospheric air. I may however remark that when the temperature of the ether is only a little above its boiling point, its condensation is a little more rapid than that of atmospheric air. This fact is related to a phenomenon which is exhibited by a great many bodies when passing from the liquid to the solid state, but which is no longer sensible at temperatures a few degrees above that at which the transition occurs.

This experiment, by showing that ether vapor and the gases expand equally, shows that this property depends in no way on the particular nature of gases and vapors, but only on their elastic state, and leads us in consequence to the conclusion that all gases and all vapors expand equally between the same degrees of heat.

Since all gases are equally expansible by heat and equally compressible, and since these two properties depend on each other, as I shall show in another place, the vapors which are equally expansible with the gases should also be equally compressible; but I may mention that this last conclusion cannot be true except so long as the compressed vapors remain entirely in the elastic state; and this requires that their temperature shall be sufficiently elevated to enable them to resist the pressure which tends to make them assume the liquid state.

I have reported from Saussure and my experiments confirm him, that very dry air and air carrying more or less water in solution are equally expansible; I am therefore authorized to draw from all that I have said the following conclusions.

1. All gases, whatever may be their density and the quantity of water which they hold in solution, and all vapors expand equally between the same degrees of heat.

2. For the permanent gases the increase of volume received by each of them between the temperature of melting ice and that

of boiling water is equal to $^{80}\!/\!_{21333}$ of the original volume for
the thermometer divided into 80 parts, or to $^{100}\!/\!_{26666}$ of the same
volume for the centigrade thermometer.

To complete this work I must determine the law of the expansion
of gases and vapors, so as to obtain the coefficient of expansion
for any determined degree of heat and to establish the true progress
of a thermometer. I shall occupy myself with these new investigations; and when they are finished I shall have the honor of presenting them to the Institute.

Free Expansion of Gases

Starting from the two facts—that all gases are expanded equally
by heat, and that they occupy volumes which are inversely
proportional to the weights which compress them—I thought,
with Mr. Dalton, that by putting them all under the same conditions and then decreasing by the same amount the pressure
common to them all, it would be possible to see from the changes
of temperature produced by the increase in volume whether they
had or had not the same capacities for heat. It was to this end
that I used the following apparatus:

I took two balloon flasks, each having two openings and each
of twelve litres capacity. Into one of the openings of each flask I
fitted a cock, and into the other a very sensitive alcohol thermometer, whose centigrade degrees could easily be read to hundredths.
I at first used an air thermometer, constructed on the principles
of Count Rumford or Mr. Leslie; but although infinitely more
sensitive than the alcohol thermometer, it was, in several respects,
inconvenient. I can remedy these defects now; but they made me
prefer the alcohol thermometer, because it gave me results which
were more comparable among themselves. In order to avoid
effects due to moisture, I introduced dried calcium chloride into
each flask. The arrangement of the apparatus for each experiment was as follows: having exhausted both flasks, and having
assured myself that there was no leakage, I filled one flask with
the gas upon which I wished to experiment. About twelve hours
later I connected the two flasks by a lead tube, and, on opening
the cocks, the gas rushed into the empty flask until equilibrium
of pressure was established. During this time the thermometer
experienced changes which I carefully noted.

Being convinced of this important fact, that the higher the
vacuum in a receiver so much the greater is the amount of heat

set free when the exterior air enters, I sought to determine by exact experiments what relation there was between the heat absorbed in one of the receivers and that set free in the other, and how these changes in temperature depended upon the differences in density of the air. For brevity's sake I shall call "No. 1" the flask in which is enclosed the gas which is made the subject of experiment; and "No. 2" that which is empty. It is in the first that cold is produced; in the second, heat. In each experiment I have noted exactly the thermometer outside and the barometer; but, as one varied only between 19° and 21° C., and the other between $0^m.755$ and $0^m.765$, the corrections which should be made to the results are quite small, and can be neglected. In order to see what connection there was between the densities of the air and the changes of temperature which are due to differences of density, I used in succession air whose density decreased as the numbers 1, $\frac{1}{2}$, $\frac{1}{4}$, etc. In order to do this, after having made air pass from receiver No. 1 into the empty receiver No. 2, I renewed the vacuum in the latter, and waited until there was complete equilibrium of temperature between them both. Since the two receivers had equal volumes, the density of the air was thus reduced one-half. On opening the cocks, the air was again divided between the two flasks, and the density was reduced to one-quarter. I could have carried the reduction in a similar manner to $\frac{1}{8}$, $\frac{1}{16}$, etc.; but I stopped at $\frac{1}{8}$, because below that the changes in temperature, which continued to diminish, could have been observed accurately only with the greatest difficulty. The following table contains the means of the results of six experiments which I made on atmospheric air:

Density of Air Expressed by the Barometer.	Cold Produced in Flask No. 1.	Heat Produced in Flask No. 2.
$0^m.76$	$0°.61$	$0°.58$
$0^m.38$	$0°.34$	$0°.34$
$0^m.19$	$0°.20$	$0°.20$

In this table I give the records of only the means of the results, because the greatest variations above or below this mean have been but 0.05, when the density of the air was that expressed by $0^m.76$; and they were much smaller when the densities were those expressed by $0^m.38$ and $0^m.19$.

On comparing the results, it is seen that the heat absorbed by the air of flask No. 1 in the first experiment is $0°.61$, while that liberated in receiver No. 2 is only 0°.58. The difference between these numbers is of itself sufficiently small to be attributed to some circumstance whose influence one might overlook, or even to errors of observation; but, if we consider the results given in the second and third rows, we see that the temperature changes are exactly equal to each other. I think, therefore, that I am justified in concluding that, when a given volume of air is made to pass from one receiver into another which is empty and of the same volume, the temperature changes in each receiver are the same.

JOULE

The extract which follows is from a paper entitled "On the Changes of Temperature Produced by the Rarefaction and Condensation of Air," published in the *Philosophical Magazine*, Vol. 26, Series 3, p. 369, 1845. In it Joule (p. 203) describes a more accurate method than that of Gay-Lussac for reaching the conclusion upon which Mayer's calculation of the mechanical equivalent was based.

Free Expansion of Gases

I provided another copper receiver (E) (Fig. 38) which had a capacity of 134 cubic inches. Like the former receiver, to

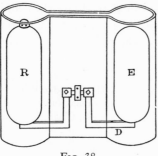

which it could be connected by a coupling nut, it had a piece D attached, in the centre of which there was a bore ⅛ of an inch diameter, which could be closed perfectly by means of a proper stopcock.

Having filled the receiver R with about 22 atmospheres of dry air, and having exhausted the receiver E by means of an air-pump, I screwed

Fig. 38.

them together, and then put them into a tin can containing 16½ lb. of water. The water was first thoroughly stirred, and its temperature taken by the same delicate thermometer which was made use of in the former experiments. The stopcocks were then opened by means of a proper key, and the air allowed to pass from the full into the empty receiver until equilibrium was established between the two. Lastly, the water was again stirred and its temperature carefully noted.

. . . .

The difference between the means of the expansions and alternations being exactly such as was found to be due to the increased effect of the temperature of the room in the latter case, we arrive at the conclusion that *no change of temperature occurs when air is allowed to expand in such a manner as not to develop mechanical power.*

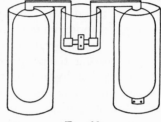

FIG. 39.

In order to analyze the above experiments, I inverted the receivers, as shown in Fig. 39, and immersed them, as well as the connecting piece, into separate cans of water. One of the receivers had 2828 cubic inches of dry air condensed into it, while the other was vacuous. After equilibrium was restored by opening the cocks, I found that 2°.36 of cold per lb. of water had been produced in the receiver from which the air had expanded, while 2°.38 of heat had been produced in the other receiver, and 0°.31 of heat also in the can in which the connecting piece was immersed, the sum of the whole amounting nearly to zero. The slight redundance of heat was owing to the loss of cold during the passage of the air from the charged receiver to the stopcocks, through a part of the pipe which could not be immersed in water.

JOULE AND THOMSON

FREE EXPANSION OF GASES

Joule (p. 203) and Thomson (Lord Kelvin) (p. 236) carried out an elaborate set of experiments to test the accuracy of the general conclusion which had been drawn from Gay-Lussac's and Joule's previous experiments on the free expansion of gases. The abstract of their last paper on the subject gives a general statement of their results. It is taken from the *Proceedings of the Royal Society*, Vol. 12, p. 202, 1862.

A brief notice of some of the experiments contained in this paper has already appeared in the 'Proceedings.' Their object was to ascertain with accuracy the lowering of temperature, in atmospheric air and other gases, which takes place on passing them through a porous plug from a state of high to one of low pressure. Various pressures were employed, with the result that the thermal effect is approximately proportional to the difference of pressure on the two sides of the plug. The experiments were also tried at various temperatures, ranging from 5° to 98° Cent.; and have shown that the thermal effect, if one of cooling, is approximately proportional to the inverse square of the absolute temperature. Thus, for example, the refrigeration at the freezing temperature is about twice that at 100° Cent. In the case of hydrogen, the reverse phenomenon of a rise of temperature on passing through the plug was observed, the rise being doubled in quantity when the temperature of the gas was raised to 100°. This result is conformable with the experiments of Regnault, who found that hydrogen, unlike other gases, has its elasticity increased more rapidly than in the inverse ratio of the volume. The Authors have also made numerous experiments with mixtures of gases, the remarkable result being that the thermal effect (cooling) of the compound gas is less than it would be if the gases after mixture retained in integrity the physical characters they possessed while in a pure state.

FOURIER

Jean-Baptiste-Joseph Fourier was born on March 21, 1768, in Auxerre. He was the son of a poor tailor. By the help of the Bishop of Auxerre he entered the military school of Saint Maur, where from the first he occupied himself with mathematical studies. Since he was not of noble descent he was not admitted to the artillery school, but in 1789 he was made a teacher of mathematics in the military school at Auxerre, and not long after that was transferred to the École Polytechnique. He was in the group of scientific men who were taken on the Egyptian expedition. After his return he occupied various public offices. He was a member of the Academy and served as its permanent secretary. He died on May 16, 1830, in Paris.

Fourier was a distinguished mathematician and the theorem known by his name has been of the greatest importance in the development of theoretical physics. The extract which follows is taken from his great work, *Théorie Analytique de la Chaleur*, 1822, as translated by Alexander Freeman. It is given because in it is presented for the first time the theory of the dimensions of physical quantities. The development of Fourier's theorem, and its applications to various problems in the conduction of heat are too mathematical to admit of insertion.

Theory of Dimensions

157. The investigation of the laws of movement of heat in solids now consists in the integration of the equations which we have constructed; this is the object of the following chapters. We conclude this chapter with general remarks on the nature of the quantities which enter into our analysis.

In order to measure these quantities and express them numerically, they must be compared with different kinds of units, five in number, namely, the unit of length, the unit of time, that of temperature, that of weight, and finally the unit which serves to measure quantities of heat. For the last unit, we might have chosen the quantity of heat which raises a given volume of a certain substance from the temperature 0 to the temperature 1. The choice of this unit would have been preferable in many respects to that of the quantity of heat required to convert a mass of ice of a given weight, into an equal mass of water at 0, without raising its temperature. We have adopted the last unit only because it had been in a manner fixed beforehand in several works on physics; besides, this supposition would introduce no change into the results of analysis.

158. The specific elements which in every body determine the measurable effects of heat are three in number, namely, the conducibility proper to the body, the conducibility relative to the atmospheric air, and the capacity for heat. The numbers which express these quantities are, like the specific gravity, so many natural characters proper to different substances.

We have already remarked, Art. 36, that the conducibility of the surface would be measured in a more exact manner, if we had sufficient observations on the effects of radiant heat in spaces deprived of air.

It may be seen, as has been mentioned in the first section of Chapter 1, Art. 11, that only three specific coefficients, K, h, C, enter into the investigation; they must be determined by observation; and we shall point out in the sequel the experiments adapted to make them known with precision.

These specific coefficients represent conducibility (K), surface conducibility (h), and specific heat (C).

159. The number C which enters into the analysis, is always multiplied by the density D, that is to say, by the number of units

of weight which are equivalent to the weight of unit volume; thus the product CD may be replaced by the coefficient c. In this case we must understand by the specific capacity for heat, the quantity required to raise from temperature 0 to temperature 1 unit of volume of a given substance, and not unit of weight of that substance.

With the view of not departing from the common definition, we have referred the capacity for heat to the weight and not to the volume; but it would be preferable to employ the coefficient c which we have just defined; magnitudes measured by the unit of weight would not then enter into the analytical expressions: we should have to consider only, 1st, the linear dimension x, the temperature v, and the time t; 2nd, the coefficients c, h, and K. The three first quantities are undetermined, and the three others are, for each substance, constant elements which experiment determines. As to the unit of surface and the unit of volume, they are not absolute, but depend on the unit of length.

160. It must now be remarked that every undetermined magnitude or constant has one *dimension* proper to itself, and that the terms of one and the same equation could not be compared, if they had not the same *exponent of dimension*. We have introduced this consideration into the theory of heat, in order to make our definitions more exact, and to serve to verify the analysis; it is derived from primary notions on quantities; for which reason, in geometry and mechanics, it is the equivalent of the fundamental lemmas which the Greeks have left us without proof.

161. In the analytical theory of heat, every equation (E) expresses a necessary relation between the existing magnitudes x, t, v, c, h, K. This relation depends in no respect on the choice of the unit of length, which from its very nature is contingent, that is to say, if we took a different unit to measure the linear dimensions, the equation (E) would still be the same. Suppose then the unit of length to be changed, and its second value to be equal to the first divided by m. Any quantity whatever x which in the equation (E) represents a certain line ab, and which, consequently, denotes a certain number of times the unit of length, becomes mx, corresponding to the same length ab; the value t of the time, and the value v of the temperature will not be changed; the same is not the case with the specific elements h, K, c: the first, h, becomes $\dfrac{h}{m^2}$; for it expresses the quantity of heat which escapes,

during the unit of time, from the unit of surface at the temperature 1.   If we examine attentively the nature of the coefficient K,
as we have defined it in Articles 68 and 135, we perceive that it
becomes K/m; for the flow of heat varies directly as the area
of the surface, and inversely as the distance between two definite
planes (Art. 72).   As to the coefficient c which represents the
product CD, it also depends on the unit of length and becomes
$\frac{c}{m^3}$; hence equation (E) must undergo no change when we write

mx instead of x, and at the same time $\frac{K}{m}, \frac{b}{m^2}, \frac{c}{m^3}$ instead of K, b, c;

the number m disappears after these substitutions: thus the
dimension of x with respect to the unit of length is 1, that of K is
-1, that of b is -2, and that of c is -3.   If we attribute to each
quantity its own *exponent of dimension*, the equation will be
homogeneous, since every term will have the same total exponent.
Numbers such as S, which represent surfaces or solids, are of two
dimensions in the first case, and of three dimensions in the second.
Angles, sines, and other trigonometrical functions, logarithms or
exponents of powers, are, according to the principles of analysis,
absolute numbers which do not change with the unit of length;
their dimensions must therefore be taken equal to 0, which is the
dimension of all abstract numbers.

   If the unit of time, which was at first 1, becomes $1/n$, the number
t will become nt, and the numbers x and v will not change.   The
coefficients K, b, c, will become K/n, b/n, c.   Thus the dimensions
of x, t, v with respect to the unit of time are 0, 1, 0, and those of
K, b, c, are -1, -1, 0.

   If the unit of temperature be changed, so that the temperature
1 becomes that which corresponds to an effect other than the
boiling of water; and if that effect requires a less temperature,
which is to that of boiling water in the ratio of 1 to the number p;
v will become vp, x and t will keep their values, and the coefficients
K, b, c, will become K/p, b/p, c/p.

   The table at the top of page 178 indicates the dimensions of the
three undetermined quantities and the three constants, with
respect to each kind of unit.

   162. If we retain the coefficients C and D, whose product has
been represented by c, we should have to consider the unit of
weight, and we should find that the exponent of dimension, with
respect to the unit of length, is -3 for the density D, and 0 for C.

Quantity of Constant	Length	Duration	Temperature
Exponent of dimension of x.............	1	0	0
Exponent of dimension of t.............	0	1	0
Exponent of dimension of v.............	0	0	1
The specific conducibility, K.............	-1	-1	-1
The surface conducibility, b.............	-2	-1	-1
The capacity for heat, c................	-3	0	-1

On applying the preceding rule to the different equations and their transformations, it will be found that they are homogeneous with respect to each kind of unit, and that the dimension of every angular or exponential quantity is nothing. If this were not the case, some error must have been committed in the analysis, or abridged expressions must have been introduced.

If, for example, we take equation (b) of Art. 105,

$$\frac{dv}{dt} = \frac{K}{CD}\frac{d^2v}{dx^2} - \frac{bl}{CDS}v,$$

we find that, with respect to the unit of length, the dimension of each of the three terms is 0; it is 1 for the unit of temperature, and -1 for the unit of time.

In the equation $v = Ae^{-x\sqrt{\frac{2b}{Kl}}}$ of Art. 76, the linear dimension of each term is 0, and it is evident that the dimension of the exponent $x\sqrt{2b/Kl}$ is always nothing, whatever be the units of length, time, or temperature.

DULONG AND PETIT

Pierre-Louis Dulong was born on February 12, 1785, in Rouen. He became a physician but soon went into academic life as a student and professor of chemistry. He was later a professor of physics at the École Polytechnique. He died on July 19, 1838, in Paris.

Alexis-Thérèse Petit was born on October 2, 1791, in Vesoul. He was a student of the École Polytechnique and became professor of physics at that institution. His principal physical work was done in collaboration with Dulong. He died on June 21, 1820, in Paris.

The extract which follows is taken from a paper entitled, "Recherches sur quelques points importants de la Théorie de la Chaleur," published in the *Annales de Chimie et de Physique*, Vol. 10, p. 395, 1819. The account of the experimental work has been omitted and the extract begins with the statement of the result. The research described was undertaken with a view to discover some connection between the characteristics of the atoms of different substances and their physical properties. The physical property chosen for investigation was the specific heat and the substances examined were elements which occur as solids. The specific heats were determined by the method of cooling, which the authors of the paper had brought to great perfection. The specific heats are referred to that of water and the atomic weights to that of oxygen.

Atomic Heat

. . . .

We shall now present in one table the specific heats of several elementary bodies, limiting ourselves to those determinations about which we no longer feel any doubt.

	Specific heats	Relative weights of the atoms	Products of the weight of each atom by the corresponding capacity
Bismuth	0,0288	13,30	0,3830
Lead	0,0293	12,95	0,3794
Gold	0,0298	12,43	0,3704
Platinum	0,0314	11,16	0,3740
Tin	0,0514	7,35	0,3779
Silver	0,0557	6,75	0,3759
Zinc	0,0927	4,03	0,3736
Tellurium	0,0912	4,03	0,3675
Copper	0,0949	3,957	0,3755
Nickel	0,1035	3,69	0,3819
Iron	0,1100	3,392	0,3731
Cobalt	0,1498	2,46	0,3685
Sulphur	0,1880	2,011	0,3780

To exhibit the law which we propose to announce we have in the preceding table introduced the relative weights of the atoms of the different simple bodies in connection with their specific heats. These weights are deduced, as is known, from the relations which are observed among the weights of the elementary substances which are combined with one another. The care which has been taken for several years in the determination of the proportions of most of the chemical compounds has been such that only slight uncertainty can now exist in the data which we have used. It is true that, as there is no rigorous method for determining the real number of the atoms of each sort which enter into combination, there ought always to be something arbitrary in the assignment of specific weights to the elementary molecules; but the uncertainty which arises extends only to two or three numbers which have among themselves very simple ratios. The reasons which led us to make the choice we did will be sufficiently explained by that which follows. We shall for the moment limit ourselves to the remark that there is no one of the numbers which we have selected which is not in agreement with the best established chemical analogies.

By means of the data contained in the preceding table we may easily calculate the relations which exist among the capacities of atoms of different sorts. For this purpose we remark that to pass from the specific heats given by observation to the specific heats of the particles themselves, it is sufficient to divide the first by the numbers of particles contained in the same weight of the substances which are being compared. Now it is clear that for equal weights of matter these numbers of particles are reciprocally proportional to the densities of the atoms. We therefore reach the desired result by multiplying each of the capacities determined by experiment by the weight of the corresponding atom. We thus obtain the different products which have been collected in the last column of the Table.

A mere glance at these numbers enables us to perceive a relation which is so remarkable in its simplicity as to lead us to recognize at once the existence of a physical law, which can be generalized and extended to all the elementary substances. In fact, these products, which express the capacities of the atoms of different sorts, are so nearly equal to one another that it is impossible that the very slight differences which are noticed arise from anything else than unavoidable errors, either in the measurement of the capacities or in the chemical analysis; especially if we notice that in certain cases the errors arising from these two sources can be in the same sense, and consequently be multiplied in the result. The number and the diversities of the substances with which we have dealt make it impossible to consider the relation which we have just pointed out as a merely accidental one. We are authorized to adopt the following law:

The atoms of all simple bodies have exactly the same capacity for heat.

By recalling that which has been said before about the sort of uncertainty which still exists in the assignment of the specific weights of the atoms it can easily be seen that the law which we have established will be stated differently if a different supposition from that which we have made is adopted about the density of the particles; but in all cases this law will contain the statement of a simple ratio between the weights and the specific heats of the elementary atoms; and we know that, if we have to choose among equally probable hypotheses, we ought to decide in favor of the one which establishes the simplest relation among the elements which are being compared.

Whatever may be the opinion which we adopt about this relation it may serve for a test of the results of chemical analysis and even in certain cases may furnish the most exact method for coming to a knowledge of the proportions of certain combinations. But if, as our work proceeds, no fact is found to weaken the probability of the opinion which we prefer at present, it will have the further advantage of fixing in a definite and uniform way the specific weights of the atoms of all the simple bodies which can be submitted to direct observation.

CAGNIARD DE LA TOUR

Charles Cagniard de la Tour was born in Paris on March 31, 1777, and died in that city on July 5, 1859. He served as an attaché in the Ministry of the Interior and was a member of the French Academy. He occupied himself with research in physics, principally in acoustics.

Cagniard de la Tour's most important discovery is that of the critical temperature, which is described in the following extract from the *Annales de Chimie et de Physique*, Vol. 21, p. 127, 1822. The title of the paper is "Exposé de quelques résultats obtenus par l'action combinée de la chaleur et de la compression sur certains liquides, tel que l'eau, l'alcool, l'éther sulfurique et l'essence de pétrole."

CRITICAL TEMPERATURE

It is well known that by using a Papin's digester we can raise the temperature of liquids far above the temperature at which they ordinarily boil; and we are led to believe that the pressure in the interior, which increases with the temperature, ought to prevent the complete evaporation of the liquid, especially if the space above the liquid is limited.

While reflecting on this, I conceived the idea that the expansion of a volatile liquid has necessarily a limit, above which the liquid ought to pass into vapor, in spite of the compression, so long as the interior volume of the apparatus allows the liquid matter to expand to its maximum dilatation.

To verify this I introduced into a small Papin's vessel, made from the end of a very thick gun-barrel, a certain quantity of alcohol at 36 degrees and a ball or sphere of quartz; the liquid occupied about a third of the capacity of the apparatus. By observing the character of the noise that the ball made when it was rolled about in the barrel, first when it was cold and then as it was warmed little by little on a brazier, I finally came to the point where the ball seemed to bounce about every time it struck as if there were no more liquid in the barrel. This effect, which was even better perceived when the ear was placed at the end of the handle which was used to hold the instrument, ceased to occur when the apparatus was cooled and occurred again as soon as the necessary temperature was restored.

The same experiment tried with water succeeded only imperfectly, because of the high temperature that had to be employed, which prevented the perfect sealing of the apparatus.

It was not so with sulphuric ether or with benzine, which gave the same results as alcohol.

Finally, that I might more easily observe these effects of heat and compression, I introduced the same liquids into small glass tubes, sealed at one end, which were afterwards sealed at the other end by means of a blow-pipe. There was attached to each tube a glass handle by which it could be held in the hand.

One of the tubes, in which alcohol had been placed to about two-fifth of its capacity, was heated with the precautions which were necessary to avoid breaking the tube. As the liquid expanded its mobility became greater and greater. The liquid, after having attained about double its original volume, disappeared completely, and was converted into a vapor so transparent that the tube seemed to be entirely empty; but when it was allowed to cool for a moment, a very thick cloud was formed and then the liquid reappeared in its first condition.

A second tube, in which the liquid occupied about half the volume, gave a similar result; but a third, in which the liquid occupied a little more than a half, was broken.

Similar trials made with benzine giving 42 degrees on the hygrometer scale and with ether gave analogous results except that the ether appeared to need less volume than the benzine to turn into vapor without breaking the tubes, and the benzine needed less than the alcohol. This result seems to indicate that the more a liquid is dilated by its nature the less volume does it need to take to attain its maximum expansion.

All the tubes used in these trials were freed from air before being closed. The experiments tried with tubes in which the air was allowed to remain gave similar results; the progressive expansion of the liquid was even easier to perceive in these last, in which there was no inconvenient boiling such as occurred in the others.

A last trial was made with a glass tube containing water up to about a third of its capacity; this tube lost its transparency and broke a few minutes later. It appears that with a high temperature water becomes capable of decomposing glass by taking up its alkali; which leads one to think that other interesting results for Chemistry might be obtained by using in other cases this process of decomposition.

By observing with care the tubes in which the air was left I noticed that those in which the liquid matter had not all the space necessary for it to attain the dilatation which precedes its transformation into vapor, I noticed, I say, that these tubes did not break always at once after the liquid had seemed completely to fill this space, and that the explosion was delayed longer when the excess of liquid was less.

May we not conclude from this, that liquids which are only slightly compressible at a low temperature, become more compressible at a high temperature? and especially in the case with which we are dealing, in which the liquid is ready to become an elastic fluid under a pressure which, as we may calculate it by theory, seems to be several hundred atmospheres.

In this connection it seems hard to believe that a little tube of glass of three millimeters interior diameter and hardly a millimeter thick can resist so considerable a strain; perhaps we would prefer to suppose that the molecules of an elastic fluid and particularly of a vapor, at a certain degree of compression and heat, are ready to undergo a change of state similar to that of partial melting, and so are able to allow a reduction of volume greater than that due to the pressure itself.

These questions may be cleared up by some further experiments. It seems to me that we can express all that has been reported in the following conclusions:

1. That alcohol at 36 degrees, benzine at 42 degrees, and sulphuric ether, under the action of heat and compression, can

be completely converted into vapor in a volume a little more than double that of each liquid;

2. That the increase of pressure caused by the presence of air in several of the experiments which have been reported offered no obstacle to the evaporation of the liquid in the same space, and that the only effect of the air was to make the dilatation more regular and easier to follow, up to the moment when the liquid seemed to vanish altogether;

3. That water, although it doubtless can be reduced into highly compressed vapor, could not be completely tested from lack of the proper means of completely sealing the vessel, nor could it be tested in the glass tubes, of which it changed the transparency by taking up the alkali which enters into their composition.

I venture to hope that this note may interest particularly those who are dealing with the applications of steam engines, and may also perhaps give some slight indication of the solution of the question relating to the compressibility of liquids, which has recently been proposed as the subject of a prize by the Institute. These considerations determined me to present my work to the Section, my principal ambition being to prove to it that I am trying to render myself more and more worthy of the favorable reception which it has accorded to my preceding work.

Supplement to the Foregoing Memoir

I have tried to determine the pressures which ether and alcohol exert at the moment when they are suddenly converted into vapor. The method which I adopted is as follows:

First Experiment.

I selected a tube *abc* (Fig. 40), with an interior diameter of a millimeter and as uniform as I could find. I sealed this to the tube *def*, of which the interior diameter was about $4\frac{1}{4}$ millimeters. The system then resembled a siphon barometer. While the two ends *a* and *f* were open, I introduced first mercury and then sulphuric ether. The mercury occupied the space *bcde* and the ether the space *ef*; by tilting the apparatus it was easy to change the level of the mercury until it filled the space *ba*; in this way I determined that a variation of a millimeter in the large tube produced one of twenty millimeters in the small tube, a ratio which seemed sufficient for the graduation which was necessary. The space *ba* is that which the mercury can fill; when its level *e* in

the large tube is lowered to the point d, the length ab is 528 millimeters; the space df, which is double of ef, is that
which is occupied by the ether when it is entirely
converted into vapor.

I marked the graduation of the 528 millimeters
on a plank against which the tube could be placed
when necessary. The tube had no graduation
except at its upper part and at certain points of
reference that I found useful.

When the apparatus had been prepared as has
been described and sealed at the ends a, f, it was
warmed with proper precautions above a brazier.
At the instant when the ether was entirely converted into vapor, the level b of the mercury had
risen to the point g, of which the distance from the
point a is 14 millimeters; thus the column of air
whose length was 528 millimeters was reduced
to 14 millimeters; which would indicate a pressure
of from 37 to 38 atmospheres. This trial repeated
three different times gave the same result every time.

Ether may thus be converted into vapor in a
space less than double its original volume, and in
this state of vaporization, it exerts a pressure of
from 37 to 38 atmospheres in the tube which
contains it.

Second Experiment.

I substituted alcohol at 36 degrees in place of
the ether in the apparatus which has been described,
by opening the ends f and a; the alcohol occupied
the space fe', that is to say, the third of the
space df that is supposed necessary for the alcohol
to be entirely converted into vapor. The mercury
occupied the space $b'bcdee'$ and filled the small tube
when by tilting the system the level e' was lowered
to d. The length of the column of air ab' was
476 millimeters. After the ends a and f were
sealed at the lamp, the apparatus was brought
over the fire with the same precautions that were
observed for ether. At the instant at which the alcohol was
converted entirely into vapor, the level b' of the mercury rose

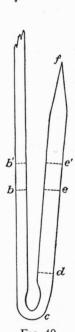

FIG. 40.

to the point g', that is to say, to 4 millimeters from the point a. Thus the column of air 476 millimeters long was reduced to 4 millimeters; which would indicate a pressure of 119 atmospheres.

Alcohol can thus be entirely converted into vapor in a space a little less than three times its original volume and when thus expanded it exerts a pressure of 119 atmospheres in the tube which contains it.

The ends a and f were a little drawn out so as to make it easier to seal them at the lamp; but I was careful to determine the capacity of these parts which were not calibrated by introducing a little mercury, which was then brought into the calibrated part as a means of measuring it.   By taking this precaution I made sure that the length aa', for example, of 10 millimeters should be counted as 2 millimeters, etc.; and this has been done in the results which have been presented.

When the last apparatus was cooled it was noticed that there had been formed above the alcohol a little bubble of gas, but this produced only a difference of level of two millimeters above b' in the mercury of the small tube.

To determine the degree of heat at which ether and alcohol are converted into vapor in these experiments, I heated the tubes containing the liquids in an oil bath in which I introduced a Réaumur thermometer with a mercury column.   I used for holding the oil a cylindrical glass vessel or test tube.   In this way it was easy to recognize the instant at which the liquids in the tubes were converted into vapor; and it was determined that ether needs for this conversion 160 degrees of heat, and alcohol 207.

I also subjected to the heat of the oil bath the apparatus previously described, which was used to determine the pressures exerted by the ether and alcohol; but I previously adjusted to the upper part of the little tube containing the column of air a refrigerant by means of which the temperature of the column was maintained steadily at 18 degrees: the results both in respect to the temperature of the vaporization and to the pressure exerted were found in agreement with those which I have reported.

When I read my memoir to the Academy I announced that water heated in glass tubes changed their transparency so much as to prevent the observation of that which was going on; since then I have found that by adding to the water a small quantity of carbonate of soda the glass was much less damaged.   In this way I have been able to determine, although with difficulty, because of the frequent breaking of the tubes, that, at a temperature not much different from that of melting zinc, water can be completely converted into vapor in a space about four times its original volume.

ANDREWS

Thomas Andrews was born on December 19, 1813, in Belfast and died on November 26, 1885, in the same place. He was first a practising physician but became professor of chemistry at Queen's College, where he taught for many years.

His researches on the liquefaction of gases were carried on for several years, beginning in 1861. They clarified the ideas which had been formerly indistinct about critical temperature. The extract which follows is from his Bakerian Lecture, "On the Continuity of the Gaseous and Liquid States of Matter," published in the *Philosophical Transactions of* 1869, p. 575.

CRITICAL TEMPERATURE

In 1822 M. Cagniard de la Tour observed that certain liquids, such as ether, alcohol, and water, when heated in hermetically sealed glass tubes, became apparently reduced to vapour in a space from twice to four times the original volume of the liquid. He also made a few numerical determinations of the pressures exerted in these experiments. In the following year Faraday succeeded in liquefying, by the aid of pressure alone, chlorine and several other bodies known before only in the gaseous form. A few years later Thilorier obtained solid carbonic acid, and observed that the coefficient of expansion of the liquid for heat is greater than that of any aeriform body. A second memoir by Faraday, published in 1826, greatly extended our knowledge of the effects of cold and pressure on gases. Regnault has examined with care the absolute change of volume in a few gases when exposed to a pressure of twenty atmospheres, and Pouillet has made some observations on the same subject. The experiments of Natterer have carried this inquiry to the enormous pressure of 2790 atmospheres; and although his method is not altogether free from objection, the results he obtained are valuable and deserve more attention than they have hitherto received.

In 1861 a brief notice appeared of some of my early experiments in this direction. Oxygen, hydrogen, nitrogen, carbonic oxide, and nitric oxide were submitted to greater pressures than had previously been attained in glass tubes, and while under these pressures they were exposed to the cold of the carbonic acid and ether-bath. None of the gases exhibited any appearance of liquefaction, although reduced to less than $\frac{1}{500}$ of their ordinary volume by the combined action of cold and pressure. In the third edition of Miller's 'Chemical Physics,' published in 1863, a short account, derived from a private letter addressed by me to Dr. Miller, appeared of some new results I had obtained, under certain fixed conditions of pressure and temperature, with carbonic acid. As these results constitute the foundation of the present investigation and have never been published in a separate form, I may perhaps be permitted to make the following extract from my original communication to Dr. Miller. "On partially liquefying carbonic acid by pressure alone, and gradually raising at the same time the temperature to 88° Fahr., the surface of demarcation between the liquid and gas became fainter, lost its curvature, and at last disappeared. The space was then occupied by a homogeneous fluid, which exhibited, when the pressure was suddenly diminished or the temperature slightly lowered, a peculiar appearance of moving or flickering striae throughout its entire mass. At temperatures above 88° no apparent liquefaction of carbonic acid, or separation into two distinct forms of matter, could be effected, even when a pressure of 300 or 400 atmospheres was applied. Nitrous oxide gave analogous results ".

· · · · · · · · · · · · · · · · · · · ·

I have frequently exposed carbonic acid, without making precise measurements, to much higher pressures than any marked in the Tables, and have made it pass, without break or interruption from what is regarded by every one as the gaseous state, to what is in like manner, universally regarded as the liquid state. Take, for example, a given volume of carbonic acid gas at 50°C., or at a higher temperature, and expose it to increasing pressure till 150 atmospheres have been reached. In this process its volume will steadily diminish as the pressure augments, and no sudden diminution of volume, without the application of external pressure, will occur at any stage of it. When the full pressure has been applied, let the temperature be allowed to fall till the carbonic acid has reached the ordinary temperature of the atmosphere. During the whole of this operation no breach of continuity has occurred. It begins with a gas, and by a series of gradual changes, presenting nowhere any abrupt alteration of volume or sudden evolution of heat, it ends with a liquid. The closest observation fails to discover anywhere indications of a change of condition in the carbonic acid, or evidence, at any period of the process, of part of it being in one physical state and part in another. That the gas has actually changed into a liquid would, indeed, never have been suspected, had it not shown itself to be so changed by entering into ebullition on the removal of the pressure. For convenience this process has been divided into two stages, the compression of the carbonic acid and its subsequent cooling; but these operations might have been performed simultaneously, if care were taken so to arrange the application of the pressure and the rate of cooling, that the pressure should not be less than 76 atmospheres when the carbonic acid had cooled to 31°.

We are now prepared for the consideration of the following important question. What is the condition of carbonic acid when it passes, at temperatures above 31°, from the gaseous state down to the volume of the liquid, without giving evidence at any part of the process of liquefaction having occurred? Does it continue in the gaseous state, or does it liquefy, or have we to deal with a new condition of matter? If the experiment were made at 100°, or at a higher temperature, when all indications of a fall had disappeared, the probable answer which would be given to this question is that the gas preserves its gaseous condition during the compression; and few would hesitate to declare this statement to be true, if the pressure, as in Natterer's experiments, were applied to such gases as hydrogen or nitrogen. On the other hand, when the experiment is made with carbonic acid at temperatures a little above 31°, the great fall which occurs at one period of the process would lead to the conjecture that liquefaction had actually taken place, although optical tests carefully applied failed at any time to discover the presence of a liquid in contact with a gas. But against this view it may be urged with great force, that the fact of additional pressure being always required for a further diminution of volume, is opposed to the known laws which hold in the change of bodies from the gaseous to the liquid state. Besides, the higher the temperature at which the gas is compressed, the less the fall becomes, and at last it disappears.

The answer to the foregoing question, according to what appears to me to be the true interpretation of the experiments already described, is to be found in the close and intimate relations which subsist between the gaseous and liquid states of matter. The ordinary gaseous and ordinary liquid states are, in short, only widely separated forms of the same condition of matter, and may be made to pass into one another by a series of gradations so gentle that the passage shall nowhere present any interruption or breach of continuity. From carbonic acid as a perfect gas to carbonic acid as a perfect liquid, the transition we have seen may be accomplished by a continuous process, and the gas and liquid are only distant stages of a long series of continuous physical changes. Under certain conditions of temperature and pressure, carbonic acid finds itself, it is true, in what may be described as a state of instability, and suddenly passes, with the evolution of heat, and without the application of additional pressure or change of temperature, to the volume, which by the continuous process can only be reached through a long and circuitous route. In the abrupt change which here occurs, a marked difference is exhibited, while the process is going on, in the optical and other physical properties of the carbonic acid which has collapsed into the smaller volume, and of the carbonic acid not yet altered. There is no difficulty here, therefore, in distinguishing between the liquid and the gas. But in other cases the distinction cannot be made; and under many of the conditions I have described it would be vain to attempt to assign carbonic acid to the liquid rather than the gaseous state. Carbonic acid, at the temperature of 35°.5, and under a pressure of 108 atmospheres, is reduced to $\frac{1}{430}$ of the volume it occupied under a pressure of one atmosphere; but if any one ask whether it is now in the gaseous or liquid state, the question does not, I believe, admit of a positive reply. Carbonic acid at 35°.5, and under 108 atmospheres of pressure, stands nearly midway between the gas and the liquid; and we have no valid grounds for assigning it to the one form of matter any more than to the other. The same observation would apply with even greater force to the state in which carbonic acid exists at higher temperatures and under greater pressures than those just mentioned. In the original experiment of Cagniard de la Tour, that distinguished physicist inferred that the liquid had disappeared, and had changed into a gas. A slight modification of the conditions of his experiment would have led him to the opposite conclusion, that what had been before a gas was changed into a liquid. These conditions are, in short, the intermediate states which matter assumes in passing, without sudden change of volume, or abrupt evolution of heat, from the ordinary liquid to the ordinary gaseous state.

In the foregoing observations I have avoided all reference to the molecular forces brought into play in these experiments. The resistance of liquids and gases to external pressure tending to produce a diminution of volume, proves the existence of an internal force of an expansive or resisting character. On the other hand, the sudden diminution of volume, without the application of additional pressure externally, which occurs when a gas is compressed, at any temperature below the critical point, to the volume at which liquefaction begins, can scarcely be explained without assuming that a molecular force of great attractive power comes here into operation, and overcomes the resistance to diminution of volume, which commonly requires the application of external force. When the passage from the gaseous to the liquid state is effected by the continuous process described in the foregoing pages, these molecular forces are so modified as to be unable at any stage of the process to overcome alone the resistance of the fluid to change of volume.

The properties described in this communication, as exhibited by carbonic acid, are not peculiar to it, but are generally true of all bodies which can be obtained as gases and liquids. Nitrous oxide, hydrochloric acid, ammonia, sulphuric ether, and sulphuret of carbon, all exhibited, at fixed pressures and temperatures, critical points, and rapid changes of volume with flickering movements when the temperature or pressure was changed in the neighborhood of those points. The critical points of some of these bodies were above 100°; and in order to make the observations, it was necessary to bend the capillary tube before the commencement of the experiment, and to heat it in a bath of paraffin or oil of vitriol.

The distinction between a gas and vapour has hitherto been founded on principles which are altogether arbitrary. Ether in the state of gas is called a vapour, while sulphurous acid in the same state is called a gas; yet they are both vapours, the one derived from a liquid boiling at 35°, the other from a liquid boiling at −10°. The distinction is thus determined by the trivial condition of the boiling-point of the liquid, under the ordinary pressure of the atmosphere, being higher or lower than the ordinary temperature of the atmosphere. Such a distinction may have some advantages for practical reference, but it has no scientific value. The critical point of temperature affords a criterion for distinguishing a vapour from a gas, if it be considered important to maintain the distinction at all. Many of the properties of vapours depend on the gas and liquid being present in contact with one another; and this, we have seen, can only occur at temperatures below the critical point. We may accordingly define a vapour to be a gas at any temperature under its critical point. According to this definition, a vapour may, by pressure alone, be changed into a liquid, and may therefore exist in presence of its own liquid; while a gas cannot be liquefied by pressure—that is, so changed by pressure as to become a visible liquid distinguished by a surface of demarcation from the gas. If this definition be accepted, carbonic acid will be a vapour below 31°, a gas above that temperature; ether a vapour below 200°, a gas above that temperature.

We have seen that the gaseous and liquid states are only distant stages of the same condition of matter, and are capable of passing into one another by a process of continuous change. A problem of far greater difficulty yet remains to be solved, the possible continuity of the liquid and solid states of matter. The fine discovery made some years ago by James Thomson, of the influence of pressure on the temperature at which liquefaction occurs, and verified experimentally by Sir. W. Thomson, points, as it appears to me, to the direction this inquiry must take; and in the case at least of those bodies which expand in liquefying, and whose melting-points are raised by pressure, the transition may possibly be effected. But this must be a subject for future investigation; and for the present I will not venture to go beyond the conclusion I have already drawn from direct experiment, that the gaseous and liquid forms of matter may be transformed into one another by a series of continuous and unbroken changes.

CAILLETET

Louis Paul Cailletet was born in Châtillon-sur-Seine in September, 1832. He attended the École des Mines. While directing his father's foundry at Châtillon he became interested in the scientific problems presented in the operation of the foundry. In 1877 he carried out the experiments described in the following extract. These experiments preceded by a few weeks only the more elaborate ones of Pictet, and the results of the two investigators were reported to the French Academy at the same session. The paper appeared in *Comptes Rendus*, Vol. 85, p. 1213, 1877, under the title "De la Condensation de l'oxygène et de l'oxyde de carbone." Cailletet was admitted to the French Academy in 1884. He died in Paris on January 5, 1913.

LIQUEFACTION OF OXYGEN, ETC.

If we enclose oxygen or carbon monoxide in a tube such as I have formerly described, placed in the compression apparatus which has been operated before the Academy; if we bring this gas to a temperature of −29 degrees by means of sulphurous acid and to the pressure of about 300 atmospheres, both these gases remain in the gaseous state. But if they are allowed to expand suddenly there will be brought about, according to Poisson's formula, a temperature at least 200 degrees below the original temperature and then we shall see appear immediately a dense mist, produced by the liquefaction and perhaps by the solidification of the oxygen or of the carbon monoxide.

The same phenomenon is observed on the release of carbonic acid, of nitrous oxide and of nitric oxide if they have been forcibly compressed.

This mist is produced in the case of oxygen even when the gas is at ordinary temperatures, provided that we give it time to lose the heat which it acquires from compression alone. This fact has been demonstrated by experiments made on Sunday, the sixteenth of December, at the Chemical Laboratory of the École Normale Supérieure before several learned gentlemen and professors, among whom were some members of the Academy of Sciences.

I had hoped to find in Paris, along with the materials needed to produce extreme cold (nitrogen monoxide or liquid carbonic acid) a pump which could replace the compression apparatus which I had set up at Châtillon-sur-Seine. Unfortunately I have not been able to obtain at Paris a pump properly installed and suited to these sorts of experiments, and I am forced to bring to Châtillon-sur-Seine the refrigerating agents which are needed to collect the condensed matter on the walls of the tube.

To determine if the oxygen and the monoxide of carbon are in the liquid or in the solid state in the mist which was observed, an optical experiment would be sufficient, which is easier to plan than to execute, because of the shape and the thickness of the tubes which contain the substances. Some chemical reactions allow us to feel sure that the oxygen is not transformed into ozone by compression. I reserve for myself the study of all these questions with apparatus which I am now constructing.

In the same conditions of temperature and pressure the release of pure hydrogen, even when it was most rapid, gave no trace of a cloudy matter. There only remains for me to study from this point of view nitrogen, which because of its small solubility may be considered as likely to be very refractory to any change of state.

I am very happy that I have been able to realize in this way the predictions about oxygen expressed by M. Berthelot with a kindness for which I here express my thanks.

PICTET

Raoul Pierre Pictet was born in Geneva on April 4, 1846. He was for some years professor of physics at the University of Geneva. In 1886 he went to Berlin, where he installed an establishment for producing low temperatures. He subsequently took up his residence in Paris. The account of his experiments on the liquefaction of oxygen was presented through M. de Loynes to the French Academy and was published in *Comptes Rendus*, Vol. 85, p. 1214, 1877, in the same number which contained the announcement of the results obtained by Cailletet.

Liquefaction of Oxygen

We have the honor of presenting to the Academy a communication on the subject of an important result which has just been obtained by M. Raoul Pictet at Geneva.

On the twenty-second of December at eight o'clock in the evening we received from him the following despatch:

Oxygen liquefied to-day under 320 atmospheres and 140 of cold by sulphurous acid and carbonic acid in succession.

Raoul Pictet

And since that time we have received in addition a description, which we shall present, of the method used by M. Raoul Pictet to obtain this result, which he had been striving to attain for a long time.

His procedure is as follows:

A and *B* (Fig. 41) are two pumps, each of which withdraw and compress the vapor. They are joined in the way called *compound*, the one acting with the other in such a way as to obtain the greatest difference possible between the pressures at which the vapor is withdrawn and that at which it is recompressed. These pumps act on anhydride sulphurous acid contained in the annular holder *C*.

The pressure in this holder is such that the sulphurous acid evaporates at a temperature of 65 degrees below zero.

The sulphurous acid is recompressed by the pumps and is sent into a condenser *D* cooled by a current of cold water; there it is

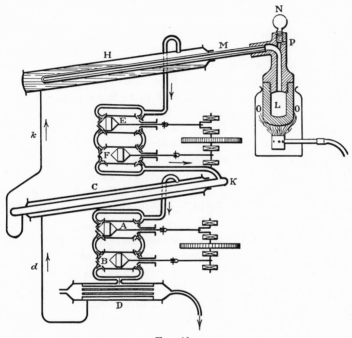

Fig. 41.

liquefied at the temperature of 25 above zero and at the pressure of about 2¾ atmospheres.

The sulphurous acid returns to the holder *C* through a small tube *d* as fast as it liquefies.

E and *F* are two pumps just like the two former ones and joined in the same way. They act on carbonic acid contained in an annular holder *H*.

The pressure in this holder is such that the carbonic acid evaporates at the temperature of 140 degrees below zero.

The carbonic acid is recompressed by the pumps and is sent into the condenser *K* surrounded by the holder *C* for the sulphurous acid, which is at the temperature of 65 degrees below zero: it there liquefies at a pressure of 5 atmospheres.

The carbonic acid passes into the holder *H* through a small tube *k* as fast as it liquefies.

L is a retort of wrought iron, thick enough to sustain a pressure of 500 atmospheres. It contains chlorate of potash and is heated so as to disengage pure oxygen; it is in communication through a narrow tube with an inclined tube *M* of very thick glass, one metre long, which is surrounded by the holder *H* containing the carbonic acid at a temperature of 140 degrees below zero.

A screw stopper *N* set in the small tube of the retort is used to open an orifice *P* which opens to the free air.

After the four pumps have been working for several hours, kept in operation by a steam engine of 15 horse power, and when all the oxygen has been released from the chlorate of potash, the pressure of the oxygen in the glass tube is 320 atmospheres and its temperature is 140 degrees below zero.

When the orifice *P* is suddenly opened, the oxygen escapes with violence and the sudden release produces an absorption of heat sufficient to cause part of it to appear as a liquid in the glass tube and to stream out from the orifice when the apparatus is tilted.

It should be added that the quantity of oxygen liquefied which is contained in the tube one meter long and 0.01 meters in interior diameter occupies a little more than a third of the length and comes out of the orifice *P* in the form of a liquid jet.

We have thought that the importance of the result of these experiments might excite the interest of the Academy of Sciences and for this reason we have undertaken to send this communication to it immediately.

In the discussion which followed at the session of the Academy evidence was presented to show that the results of Cailletet had been obtained at an earlier date than those of Pictet. On the other hand it was recognized that Pictet's results were entirely independent. An impartial critic would probably assign the superiority of method to Pictet, who was able to observe the oxygen in a liquid state.

MAYER

Julius Robert Mayer was born on November 25, 1814, in Heilbron, and died on March 20, 1878. His father was an apothecary and trained his son in that business. He studied medicine and in 1838 was admitted as a physician. To get some experience of the world he served for some months as a ship's surgeon. While on this voyage, in which he had no occasion to practise his art, he conceived the idea of the equivalence between heat and mechanical work. On his return home he practised as a physician in Heilbron. He was still occupied with his theory of the nature of heat, and the paper which he submitted to Liebig was published by him in 1842. Further papers followed at considerable intervals in which the mechanical theory of heat was applied to cosmical relations. He became engaged in disputes about priority with Joule. He suffered during a part of his life from mental disorder, from which he completely recovered. In his later years he was recognized by scientific bodies by election to membership.

The paper which follows, entitled "Remarks on the Forces of Inorganic Nature," appeared in the *Philosophical Magazine*, series 4, Vol. 24, p. 371, 1862. It is a translation by G. C. Foster of the original publication in the *Annalen der Chemie und Pharmacie*, Vol. 42, p. 233, 1842. This is the paper upon which Mayer's claim to priority in the enunciation of the equivalence between heat and mechanical energy is based. Joule's first work on the subject appeared in 1843.

The Conservation of Energy

The following pages are designed as an attempt to answer the questions, What do we understand by "Forces"? and how are different forces related to each other? Whereas the term *matter* implies the possession, by the object to which it is applied, of very definite properties, such as weight and extension; the term *force* conveys for the most part the idea of something unknown, unsearchable, and hypothetical. An attempt to render the notion of force equally exact with that of matter, and so to denote by it only objects of actual investigation, is one which, with the consequences that flow from it, ought not to be unwelcome to those who desire that their views of nature may be clear and unencumbered by hypotheses.

Forces are causes: accordingly, we may in relation to them make full application of the principle—*causa aequat effectum*. If the cause c has the effect e, then $c = e$; if, in its turn, e is the cause of a second effect f, we have $e = f$, and so on: $c = e = f \cdots = c$. In a chain of causes and effects, a term or a part of a term can never, as plainly appears from the nature of an equation, become equal to nothing. This first property of all causes we call their *indestructibility*.

If the given cause c has produced an effect e equal to itself, it has in that very act ceased to be: c has become e; if, after the production of e, c still remained in whole or in part, there must be still further effects corresponding to this remaining cause: the total effect of c would thus be $>e$, which would be contrary to the supposition $c = e$. Accordingly, since c becomes e, and e becomes f, &c., we must regard these various magnitudes as different forms under which one and the same object makes its appearance. This capability of assuming various forms is the second essential property of all causes. Taking both properties together, we may say, causes are (quantitatively) *indestructible* and (qualitatively) *convertible* objects.

Two classes of causes occur in nature, which, so far as experience goes, never pass one into another. The first class consists of such causes as possess the properties of weight and impenetrability; these are kinds of Matter: the other class is made up of causes which are wanting in the properties just mentioned, namely Forces, called also Imponderables, from the negative property that has been indicated. Forces are therefore *indestructible, convertible imponderable objects.*

We will in the first instance take matter, to afford us an example of causes and effects. Explosive gas, H + O, and water HO, are related to each other as cause and effect, therefore H + O = HO. But if H + O becomes HO, heat, *cal.*, makes its appearance as well as water; this heat must likewise have a cause, x, and we have therefore H + O + x = HO + *cal.* It might, however, be asked whether H + O is really =HO, and x = *cal.*, and not perhaps H + O = *cal.*, and x = HO, whence the above equation could equally be deduced; and so in many other cases. The phlogistic chemists recognized the equation between *cal.* and x, or Phlogiston as they called it, and in so doing made a great step in advance; but they involved themselves again in a system of mistakes by putting $-x$ in place of O; thus, for instance, they obtained H = HO + x.

Chemistry, whose problem it is to set forth in equations the causal connexion existing between the different kinds of matter, teaches us that matter, as a cause, has matter for its effect; but we are equally justified in saying that to force as cause, corresponds force as effect. Since $c = e$, and $e = c$, it is unnatural to call one term of an equation a force, and the other an effect of force or phenomenon, and to attach different notions to the expressions Force and Phenomenon. In brief, then, if the cause is matter, the effect is matter; if the cause is a force, the effect is also a force.

A cause which brings about the raising of a weight is a force; its effect (*the raised weight*) is, accordingly, equally *a force*; or, expressing this relation in a more general form, *separation in space of ponderable objects is a force;* since this force causes the fall of bodies, we call it *falling force.* Falling force and fall, or, more generally still, falling force and motion, are forces which are related to each other as cause and effect—forces which are convertible one into the other—two different forms of one and the same object. For example, a weight resting on the ground is not a force: it is neither the cause of motion, nor of the lifting of another weight; it becomes so, however, in proportion as it is raised above the ground: the cause—the distance between a weight and the earth—and the effect—the quantity of motion produced—bear to each other, as we learn from mechanics, a constant relation.

Gravity being regarded as the cause of the falling of bodies, a gravitating force is spoken of, and so the notions of *property* and of *force* are confounded with each other: precisely that which is the essential attribute of every force—the *union* of indestructibility with convertibility—is wanting in every property: between a property and a force, between gravity and motion, it is therefore impossible to establish the equation required for a rightly conceived causal relation. If gravity be called a force, a cause is supposed which produces effects without itself diminishing, and incorrect conceptions of the causal connexion of things are thereby fostered. In order that a body may fall, it is no less necessary that it should be lifted up, than that it should be heavy or possess gravity; the fall of bodies ought not therefore to be ascribed to their gravity alone.

It is the problem of Mechanics to develope the equations which subsist between falling force and motion, motion and falling force, and between different motions: here we will call to mind only one point. The magnitude of the falling force v is directly proportional (the earth's radius being assumed $= \infty$) to the magnitude of the mass m, and the height d to which it is raised; that is, $v = md$. If the height $d = 1$, to which the mass m is raised, is transformed into the final velocity $c = 1$ of this mass, we have also $v = mc$; but from known relations existing between d and c, it results that, for other values of d or of c, the measure of the force v is mc^2; accordingly $v = md = mc^2$: the law of the conservation of *vis viva* is thus found to be based on the general law of the indestructibility of causes.

In numberless cases we see motion cease without having caused another motion or the lifting of a weight; but a force once in existence cannot be annihilated, it can only change its form; and the question therefore arises, What other forms is force, which we have become acquainted with as falling force and motion, capable of assuming? Experience alone can lead us to a conclusion on this point. In order to experiment with advantage, we must select implements which, besides causing a real cessation of motion, are as little as possible altered by the objects to be examined. If, for example, we rub together two metal plates, we see motion disappear, and heat, on the other hand, make its appearance, and we have now only to ask whether *motion* is the cause of heat. In order to come to a decision on this point, we must discuss the question whether, in the numberless cases in which the expenditure of motion is accompanied by the appearance of heat, the motion has not some other effect than the production of heat, and the heat some other cause than the motion.

An attempt to ascertain the effects of ceasing motion has never yet been seriously made; without, therefore, wishing to exclude *a priori* the hypotheses which it may be possible to set up, we observe only that, as a rule, this effect cannot be supposed to be an alteration in the state of aggregation of the moved (that is, rubbing, &c.) bodies. If we assume that a certain quantity of motion v is expended in the conversion of a rubbing substance m into n, we must then have $m + v = n$, and $n = m + v$; and when n is reconverted into m, v must appear again in some form or other. By the friction of two metallic plates continued for a very long time, we can gradually cause the cessation of an immense quantity of movement; but would it ever occur to us to look for even the smallest trace of the force which has disappeared in the metallic dust that we could collect, and to try to regain it thence? We repeat, the motion cannot have been annihilated; and contrary, or positive and negative, motions cannot be regarded as $= 0$, any more than contrary motions can come out of nothing, or a weight can raise itself.

Without the recognition of a causal connexion between motion and heat, it is just as difficult to explain the production of heat as it is to give any account of the motion that disappears. The heat cannot be derived from the diminution of the volume of the rubbing substances. It is well known that two pieces of ice may be melted by rubbing them together *in vacuo;* but let any one try to convert ice into water by pressure, however enormous. Water undergoes, as was found by the author, a rise of temperature when violently shaken. The water so heated (from 12° to 13°C.) has a greater bulk after being shaken than it had before; whence now comes this quantity of heat, which by repeated shaking may be called into existence in the same apparatus as often as we please? The vibratory hypothesis of heat is an approach towards the doctrine of heat being the effect of motion, but it does not favour the admission of this causal relation in its full generality; it rather lays the chief stress on uneasy oscillations (*unbehagliche Schwingungen*).

If it be now considered as established that in many cases (*exceptio confirmat regulam*) no other effect of motion can be traced except heat, and that no other cause than motion can be found for the heat that is produced, we prefer the assumption that heat proceeds from motion, to the assumption of a cause without effect and of an effect without a cause,—just as the chemist, instead of allowing oxygen and hydrogen to disappear without further investigation, and water to be produced in some inexplicable manner, establishes a connexion between oxygen and hydrogen on the one hand and water on the other.

The natural connexion existing between falling force, motion, and heat may be conceived of as follows. We know that heat makes its appearance when the separate particles of a body approach nearer to each other: condensation produces heat. And what applies to the smallest particles of matter, and the smallest intervals between them, must also apply to large masses and to measurable distances. The falling of a weight is a real diminution of the bulk of the earth, and must therefore without doubt be related to the quantity of heat thereby developed; this quantity of heat must be proportional to the greatness of the weight and its distance from the ground. From this point of view we are very easily led to the equations between falling force, motion, and heat, that have already been discussed.

But just as little as the connexion between falling force and motion authorized the conclusion that the essence of falling force is motion, can such a conclusion be adopted in the case of heat. We are, on the contrary, rather inclined to infer that, before it can become heat, motion—whether simple, or vibratory as in the case of light and radiant heat, &c.—must cease to exist as motion.

If falling force and motion are equivalent to heat, heat must also naturally be equivalent to motion and falling force. Just as heat appears as an *effect* of the diminution of bulk and of the cessation of motion, so also does heat disappear as a *cause* when its effects are produced in the shape of motion, expansion, or raising of weight.

In water-mills, the continual diminution in bulk which the earth undergoes, owing to the fall of the water, gives rise to motion, which afterwards disappears again, calling forth unceasingly a great quantity of heat; and inversely, the steam-engine serves to decompose heat again into motion or the raising of weights. A locomotive engine with its train may be compared to a distilling apparatus; the heat applied under the boiler passes off as motion, and this is deposited again as heat at the axles of the wheels.

We will close our disquisition, the propositions of which have resulted as necessary consequences from the principle "*causa aequat effectum*," and which are in accordance with all the phenomena of Nature, with a practical deduction. The solution of the equations subsisting between falling force and motion requires that the space fallen through in a given time, e.g. the first second, should be experimentally determined; in like manner, the solution of the equations subsisting between falling force and motion on the one hand and heat on the other, requires an answer to the question, How great is the quantity of heat which corresponds to a given quantity of motion or falling force? For instance, we must ascertain how high a given weight requires to be raised above the ground in order that its falling force may be equivalent to the raising of the temperature of an equal weight of water from 0° to 1°C. The attempt to show that such an equation is the expression of a physical truth may be regarded as the substance of the foregoing remarks.

By applying the principles that have been set forth to the relations subsisting between the temperature and the volume of gases, we find that the sinking of a mercury column by which a gas is compressed is equivalent to the quantity of heat set free by the compression; and hence it follows, the ratio between the capacity for heat of air under constant pressure and its capacity under constant volume being taken as = 1.421, that the warming of a given weight of water from 0° to 1° C. corresponds to the fall of an equal weight from the height of about 365 metres. If we compare with this result the working of our best steam-engines, we see how small a part only of the heat applied under the boiler is really transformed into motion or the raising of weights; and this may serve as justification for the attempts at the profitable production of motion by some other method than the expenditure of the chemical difference between carbon and oxygen—more particularly by the transformation into motion of electricity obtained by chemical means.

JOULE

James Prescott Joule was born in Manchester on December 24, 1818, and died in Sale on October 11, 1889. His father owned an important brewery and he succeeded to the business. He early began to devote himself to physical investigation. In December of 1840 he presented a paper to the Royal Society on the production of heat by the electric current. His course of thought led him to the consideration of the relation between heat and mechanical work. The results of his investigations were embodied in a series of papers which culminated in his great memoir on the mechanical equivalent of heat, published in 1850. Joule was one of the founders of the principle of the conservation of energy. Some of his work was done in collaboration with Lord Kelvin.

MECHANICAL EQUIVALENT OF HEAT

Joule's first measurement of the mechanical equivalent of heat was published in the *Philosophical Magazine*, Vol. 23, Series 3, pp. 263, 347, and 435, 1843. It was made by comparing the heat generated by the current of a magnetoelectric machine with the excess of work which was used in turning the machine when the circuit was closed above that used when it was open. The experiments are described in great detail and too extensively for quotation. Only some of the results will be given. In the following papers in which the mechanical equivalent was measured in different ways we find the same elaborate description of the experiments and a brief statement of the final results. This is particularly true of the great memoir of 1850 in which Joule's work culminated. Extracts from these memoirs will be given when suitable passages can be found in which the method employed is described or a brief description of the method will be prefixed to the quotation of the results.

Part II.—*On the Mechanical Value of Heat*

Having proved that *heat* is *generated* by the magneto-electrical machine, and that by means of the inductive power of magnetism, we can *diminish* or *increase* at pleasure the *heat* due to chemical changes, it became an object of great interest to inquire

whether a constant ratio existed between it and the mechanical
power gained or lost.   For this purpose it was only necessary to
repeat some of the previous experiments, and to ascertain, at the
same time, the mechanical force necessary in order to turn the
apparatus.

.   .   .   .

1° of heat per lb. of water is therefore equivalent to a mechanical
force capable of raising a weight of 896 lb. to the perpendicular
height of one foot.

Two other experiments, conducted precisely in the same manner,
gave a degree of heat to mechanical forces represented respectively
by 1001 lb. and 1040 lb.

.   .   .   .

The method first employed was modified so as to make it simpler and to
concentrate the heat developed in a smaller body of water. The results of
some of these experiments are given.

In other words, one degree of heat per lb. of water may be
generated by the expenditure of a mechanical power capable of
raising 742 lb. to the height of one foot.

By a similar calculation, I find the result of the last two experiments of the table to be 860 lb.

The foregoing are all the experiments I have hitherto made on
the mechanical value of heat.   I admit that there is a considerable
difference between some of the results, but not, I think, greater
than may be referred with propriety to mere errors of experiment.
I intend to repeat the experiments with a more powerful and
more delicate apparatus.   At present we shall adopt the mean
result of the thirteen experiments given in this paper, and state
generally that,

*The quantity of heat capable of increasing the temperature of a
pound of water by one degree of Fahrenheit's scale is equal to, and
may be converted into, a mechanical force capable of raising 838 lb.
to the perpendicular height of one foot.*

.   .   .   .

P.S.—We shall be obliged to admit that Count Rumford was
right in attributing the heat evolved by boring cannon to friction,
and not (in any considerable degree) to any change in the capacity
of the metal.   I have lately proved experimentally that *heat is*

evolved by the passage of water through narrow tubes. My apparatus consisted of a piston perforated by a number of small holes, working in a cylindrical glass jar containing about 7 lb. of water. I thus obtained one degree of heat per lb. of water from a mechanical force capable of raising about 770 lb. to the height of one foot, a result which will be allowed to be very strongly confirmatory of our previous deductions. I shall lose no time in repeating and extending these experiments, being satisfied that the grand agents of nature are, by the Creator's fiat, *indestructible;* and that wherever mechanical force is expended, an exact equivalent of heat is *always* obtained.

Joule's next paper on the subject was published in the *Philosophical Magazine*, Vol. 26, Series 3, p. 369, 1845. In it he compares the work needed to condense a quantity of air with the heat developed by the condensation. In this connection Joule performed the experiments which have already been quoted to prove that no heat was developed or absorbed by the free expansion of air, so that the heat developed when the air was condensed could be properly considered equivalent to the work done in compressing it.

Hence the equivalent of a degree of heat per lb. of water, as determined by the above series, is 795 lb. raised to the height of one foot.

The mechanical equivalents of heat derived from the foregoing experiments were so near 838 lb., the result of magnetical experiments in which "latent heat" could not be suspected to interfere in any way, as to convince me that the heat evolved was simply the manifestation, in another form, of the mechanical power expended in the act of condensation.

The first experiments in which Joule used the heat developed by friction in water to measure the mechanical equivalent of heat were described in a letter to the editors of the *Philosophical Magazine* which was published in the *Philosophical Magazine*, Vol. 27, Series 3, p. 205, 1845.

The principal part of this letter was brought under the notice of the British Association at its last meeting at Cambridge. I have hitherto hesitated to give it further publication, not because I was in any degree doubtful of the conclusions at which I had arrived, but because I intended to make a slight alteration in the apparatus calculated to give still greater precision to the experiments. Being unable, however, just at present to spare the time necessary to fulfil this design, and being at the same time most anxious to convince the scientific world of the truth of the positions I have maintained, I hope you will do me the favour of publishing this letter in your excellent Magazine.

The apparatus exhibited before the Association consisted of a brass paddle-wheel working *horizontally* in a can of water. Motion could be communicated to this paddle by means of weights, pulleys, &c., exactly in the manner described in a previous paper.

The paddle moved with great resistance in the can of water, so that the weights (each of four pounds) descended at the slow rate of about one foot per second. The height of the pulleys from the ground was twelve yards, and consequently, when the weights had descended throughout that distance, they had to be wound up again in order to renew the motion of the paddle. After this operation had been repeated sixteen times, the increase of the temperature of the water was ascertained by means of a very sensible and accurate thermometer.

A series of nine experiments was performed in the above manner, and nine experiments were made in order to eliminate the cooling or heating effects of the atmosphere. After reducing the result to the capacity for heat of a pound of water, it appeared that for each degree of heat evolved by the friction of water a mechanical power equal to that which can raise a weight of 890 lb. to the height of one foot had been expended.

The equivalents I have already obtained are: 1st, 823 lb., derived from magneto-electrical experiments; 2nd, 795 lb., deduced from the cold produced by the rarefaction of air; and 3rd, 774 lb. from experiments (hitherto unpublished) on the motion of water through narrow tubes. This last class of experiments being similar to that with the paddle-wheel, we may take the mean of 774 and 890, or 832 lb., as the equivalent derived from the friction of water. In such delicate experiments, where one hardly ever collects more than half a degree of heat, greater accordance of the results with one another than that above exhibited could hardly have been expected. I may therefore conclude that the existence of an equivalent relation between heat and the ordinary forms of mechanical power is proved; and assume 817 lb., the mean of the results of three distinct classes of experiments, as the equivalent, until more accurate experiments shall have been made.

Any of your readers who are so fortunate as to reside amid the romantic scenery of Wales or Scotland could, I doubt not, confirm my experiments by trying the temperature of the water at the top and at the bottom of a cascade. If my views be correct, a fall of 817 feet will of course generate one degree of heat, and the temperature of the river Niagara will be raised about one fifth of a degree by its fall of 160 feet.

The more extensive series of a similar sort were described in the *Philosophical Magazine*, Vol. 31, Series 3, p. 173, 1847. The liquids used were water and sperm oil. The results were as follows.

The equivalent of a degree of heat in a pound of water was therefore found to be 781.5 lb., raised to the height of one foot.

. . . .

Hence the equivalent deduced from the friction of sperm-oil was 782.1, a result almost identical with that obtained from the friction of water. The mean of the two results is 781.8, which is the equivalent I shall adopt until further and still more accurate experiments shall have been made.

The memoir in which Joule described the experiments which led him to his final value of the mechanical equivalent of heat was presented to the Royal Society on June 21, 1849, and published in the *Philosophical Transactions*, Vol. 140, p. 61, 1850. In these experiments Joule determined the heat developed by friction in water as before, and also the heat developed by friction in mercury and by the rubbing of iron plates together. The experiments themselves are given in such detail that they do not admit of quotation. There are here presented the Introduction to the memoir, in which Joule reviews his earlier work and states the principle of the equivalence of heat and energy, and then a Table containing the equivalents obtained, followed by his final conclusion.

In accordance with the pledge I gave the Royal Society some years ago, I have now the honour to present it with the results of the experiments I have made in order to determine the mechanical equivalent of heat with exactness. I will commence with a slight sketch of the progress of the mechanical doctrine, endeavouring to confine myself, for the sake of conciseness, to the notice of such researches as are immediately connected with the subject. I shall not therefore be able to review the valuable labours of Mr. Forbes and other illustrious men, whose researches on radiant heat and other subjects do not come exactly within the scope of the present memoir.

For a long time it had been a favorite hypothesis that heat consists of "a force or power belonging to bodies", but it was reserved for Count Rumford to make the first experiments decidedly in favour of that view. That justly celebrated natural philosopher demonstrated by his ingenious experiments that the very great quantity of heat excited by the boring of cannon could not be ascribed to a change taking place in the calorific capacity of the metal; and he therefore concluded that the motion of the borer was communicated to the particles of metal, thus producing the phenomena of heat. "It appears to me," he remarks, "extremely difficult, if not quite impossible, to form any distinct idea of anything capable of being excited and communicated in the manner the heat was excited and communicated in these experiments, except it be motion."

One of the most important parts of Count Rumford's paper, though one to which little attention has hitherto been paid, is that in which he makes an estimate of the quantity of mechanical force required to produce a certain amount of heat. Referring to his third experiment, he remarks that the "total quantity of ice-cold water which with the heat actually generated by friction, and accumulated in 2^h 30^m, might have been heated 180°, or made to boil, = 26.58 lb." In the next page he states that "the machinery used in the experiment could easily be carried round by the force of one horse (though, to render the work lighter, two horses were actually employed in doing it)." Now the power of a horse is estimated by Watt at 33,000 foot-pounds per minute, and therefore if continued for two hours and a half will amount to 4,950,000 foot-pounds, which, according to Count Rumford's experiment, will be equivalent to 26.58 lb. of water raised 180°. Hence the heat required to raise a lb. of water 1° will be equivalent to the force represented by 1034 foot-pounds. This result is not very widely different from that which I have deduced from my own experiments related in this paper, viz. 772 foot-pounds; and it must be observed that the excess of Count Rumford's equivalent is just such as might have been anticipated from the circumstance, which he himself mentions, that "no estimate was made of the heat accumulated in the wooden box, nor of that dispersed during the experiment."

About the end of the last century Sir Humphry Davy communicated a paper to Dr. Beddoes's West Country Contributions, entitled "Researches on Heat, Light, and Respiration," in which he gave ample confirmation to the views of Count Rumford. By rubbing two pieces of ice against one another in the vacuum of an air-pump, part of them was melted, although the temperature of the receiver was kept below the freezing-point. This experiment was the more decisively in favour of the doctrine of the immateriality of heat, inasmuch as the capacity of ice for heat is much less than that of water. It was therefore with good reason that Davy drew the inference that "the immediate cause of the phenomena of heat is motion, and the laws of its communication are precisely the same as the laws of the communication of motion."

The researches of Dulong on the specific heat of elastic fluids were rewarded by the discovery of the remarkable fact that "equal volumes of all the elastic fluids, taken at the same temperature and under the same pressure, being compressed or dilated suddenly to the same fraction of their volume, disengage or absorb the same *absolute quantity of heat.*" This law is of the utmost importance in the development of the theory of heat, inasmuch as it proves that the calorific effect is, under certain conditions, proportional to the force expended.

In 1834 Dr. Faraday demonstrated the "Identity of the Chemical and Electrical Forces." This law, along with others subsequently discovered by that great man, showing the relations which subsist between magnetism, electricity, and light, have enabled him to advance the idea that the so-called imponderable bodies are merely the exponents of different forms of Force. Mr. Grove and M. Mayer have also given their powerful advocacy to similar views.

My own experiments in reference to the subject were commenced in 1840, in which year I communicated to the Royal Society my discovery of the law of the heat evolved by voltaic electricity, a law from which the immediate deductions were drawn,—1st, that the heat evolved by any voltaic pair is proportional, *caeteris paribus,* to its intensity or electromotive force; and 2nd, that the heat evolved by the combustion of a body is proportional to the intensity of its affinity for oxygen. I thus succeeded in establishing relations between heat and chemical affinity. In 1843 I showed that the heat evolved by magneto-electricity is proportional to the force absorbed, and that the force of the electromagnetic engine is derived from the force of chemical affinity in the battery, a force which otherwise would be evolved in the form of heat. From these facts I considered myself justified in announcing "that the quantity of heat capable of increasing the temperature of a lb. of water by one degree of Fahrenheit's scale is equal to, and may be converted into, a mechanical force capable of raising 838 lb. to the perpendicular height of one foot."

In a subsequent paper, read before the Royal Society in 1844, I endeavoured to show that the heat absorbed and evolved by the rarefaction and condensation of air is proportional to the force evolved and absorbed in those operations. The quantitative relation between force and heat deduced from these experiments is almost identical with that derived from the electro-magnetic experiments just referred to, and is confirmed by the experiments of M. Seguin on the dilatation of steam.

From the explanation given by Count Rumford of the heat arising from the friction of solids, one might have anticipated, as a matter of course, that the evolution of heat would also be detected in the friction of liquid and gaseous bodies. Moreover there were many facts, such as, for instance, the warmth of the sea after a few days of stormy weather, which had long been commonly attributed to fluid friction. Nevertheless the scientific world, preoccupied with the hypothesis that heat is a substance, and following the deductions drawn by Pictet from experiments not sufficiently delicate, have almost unanimously denied the possibility of generating heat in that way. The first mention, so far as I am aware, of experiments in which the evolution of heat from fluid friction is asserted was in 1842 by M. Mayer, who states that he has raised the temperature of water from 12° C. to 13° C. by agitating it, without, however, indicating the quantity of force employed, or the precautions taken to secure a correct result. In 1843 I announced the fact that "heat is evolved by the passage of water through narrow tubes," and that each degree of heat per lb. of water required for its evolution in this way a mechanical force represented by 770 foot-pounds. Subsequently, in 1845 and 1847, I employed a paddle-wheel to produce the fluid friction, and obtained the equivalents 781.5, 782.1, and 787.6 respectively from the agitation of water, sperm-oil, and mercury. Results so closely coinciding with one another, and with those previously derived from experiments with elastic fluids and the electro-magnetic machine, left no doubt on my mind as to the existence of an equivalent relation between force and heat; but still it appeared of the highest importance to obtain that relation with still greater accuracy. This I have attempted in the present paper.

. . . .

The following Table contains a summary of the equivalents derived from the experiments above detailed.   In its fourth column I have supplied the results with the correction necessary to reduce them to a vacuum.

TABLE IX.

No. of series.	Material employed.	Equivalent in air.	Equivalent *in vacuo.*	Mean.
1.	Water.................	773.640	772.692	772.692
2.	Mercury...............	773.762	772.814 }	774.083
3.	Mercury...............	776.303	775.352 }	
4.	Cast iron..............	776.997	776.045 }	774.987
5.	Cast iron..............	774.880	773.930 }	

It is highly probable that the equivalent from cast iron was somewhat increased by the abrasion of particles of the metal during friction, which could not occur without the absorption of a certain quantity of force in overcoming the attraction of cohesion.   But since the quantity abraded was not considerable enough to be weighed after the experiments were completed, the error from this source cannot be of much moment.   I consider that 772.692, the equivalent derived from the friction of water, is the most correct, both on account of the number of experiments tried and the great capacity of the apparatus for heat.   And since, even in the friction of fluids, it was impossible entirely to avoid vibration and the production of a slight sound, it is probable that the above number is slightly in excess.   I will therefore conclude by considering it as demonstrated by the experiments contained in this paper,—

1st.   *That the quantity of heat produced by the friction of bodies, whether solid or liquid, is always proportional to the quantity of force expended.*   And,

2nd.   *That the quantity of heat capable of increasing the temperature of a pound of water (weighed in vacuo, and taken at between 55° and 60°) by 1° Fabr. requires for its evolution the expenditure of a mechanical force represented by the fall of 772 lb. through the space of one foot.*

HELMHOLTZ

The following extract contains portions of Helmholtz' (p. 122) great paper, *Ueber die Erhaltung der Kraft*. This paper was presented to the Physical Society of Berlin on July 23, 1847, and appeared in that year as a separate publication. In it the author shows that the conservation of mechanical energy requires that the forces between the least parts of matter must act along the lines joining those parts. The elaborate mathematical discussion of this theorem has necessarily been omitted. The author applies his principle to various mechanical theorems and to studies of what he calls the force-equivalent of heat and the force-equivalent of electric processes. Two of these applications have been selected for presentation, the one on thermoelectric currents and the other a portion of the one on electromagnetism.

The translation is by John Tyndall, and appeared in *Scientific Memoirs, Natural Philosophy*, p. 114, 1853.

THE CONSERVATION OF ENERGY

Introduction

The principal contents of the present memoir show it to be addressed to physicists chiefly, and I have therefore thought it judicious to lay down its fundamental principles purely in the form of a physical premise, and independent of metaphysical considerations,—to develop the consequences of these principles, and to submit them to a comparison with what experience has established in the various branches of physics. The deduction of the propositions contained in the memoir may be based on either of two maxims; either on the maxim that it is not possible by any combination whatever of natural bodies to derive an unlimited amount of mechanical force, or on the assumption that all actions in nature can be ultimately referred to attractive or repulsive forces, the intensity of which depends solely upon the distances between the points by which the forces are exerted. That both these propositions are identical is shown at the commencement of the memoir itself. Meanwhile the important bearing which they have upon the final aim of the physical sciences may with propriety be made the subject of a special introduction.

The problem of the sciences just alluded to is, in the first place, to seek the laws by which the particular processes of nature may be referred to, and deduced from, general rules. These rules,—for example, the law of the reflexion and refraction of light, the law of Mariotte and Gay-Lussac regarding the volumes of gases,—are evidently nothing more than general ideas by which the various phaenomena which belong to them are connected together. The finding out of these is the office of the experimental portion of our science. The theoretic portion seeks, on the contrary, to evolve the unknown causes of the processes from the visible actions which they present; it seeks to comprehend these processes according to the laws of causality. We are justified, and indeed impelled in this proceeding, by the conviction that every change in nature *must* have a sufficient cause. The proximate causes to which we refer phaenomena may, in themselves, be either variable or invariable; in the former case the above conviction impels us to seek for causes to account for the change, and thus we proceed until we at length arrive at final causes which are unchangeable, and which therefore must, in all cases where the exterior conditions are the same, produce the same invariable effects. The final aim of the theoretic natural sciences is therefore to discover the ultimate and unchangeable causes of natural phaenomena. Whether all the processes of nature be actually referrible to such,—whether changes occur which are not subject to the laws of necessary causation, but spring from spontaneity or freedom, this is not the place to decide; it is at all events clear that the science whose object it is to comprehend nature must proceed from the assumption that it is comprehensible, and in accordance with this assumption investigate and conclude until perhaps, she is at length admonished by irrefragable facts that there are limits beyond which she cannot proceed.

Science regards the phaenomena of the exterior world according to two processes of abstraction: in the first place it looks upon them as simple existences, without regard to their action upon our organs of sense or upon each other; in this aspect they are named *matter.* The existence of matter in itself is to us something tranquil and devoid of action: in it we distinguish merely the relations of space and of quantity (mass), which is assumed to be eternally unchangeable. To matter, thus regarded, we must not ascribe qualitative differences, for when we speak of different kinds of matter we refer to differences of action, that is, to differences in the forces of matter. Matter in itself can therefore partake of one change only,—a change which has reference to space, that is, motion. Natural objects are not, however, thus passive; in fact we come to a knowledge of their existence solely from their actions upon our organs of sense, and infer from these actions a something which acts. When, therefore, we wish to make actual application of our idea of matter, we can only do it by means of a second abstraction, and ascribe to it properties which in the first case were excluded from our idea, namely the capability of producing effects, or, in other words, of exerting force. It is evident that in the application of the ideas of matter and force to nature the two former should never be separated: a mass of pure matter would, as far as we and nature are concerned, be a nullity, inasmuch as no action could be wrought by it either upon our organs of sense or upon the remaining portion of nature. A pure force would be something which must have a basis, and yet which has no basis, for the basis we name matter. It would be just as erroneous to define matter as something which has an actual existence, and force as an idea which has no corresponding reality. Both, on the contrary, are abstractions from the actual formed in precisely similar ways. Matter is only discernible by its forces, and not by itself.

We have seen above that the problem before us is to refer back the phaenomena of nature to unchangeable final causes. This requirement may now be expressed by saying that for final causes unchangeable forces must be found. Bodies with unchangeable forces have been named in science (chemistry) elements. Let us suppose the universe decomposed into elements possessing unchangeable qualities, the only alteration possible to such a system is an alteration of position, that is, motion; hence, the forces can be only moving forces dependent in their action upon conditions of space.

To speak more particularly: the phaenomena of nature are to be referred back to motions of material particles possessing unchangeable moving forces, which are dependent upon conditions of space alone.

Motion is the alteration of the conditions of space. Motion, as a matter of experience, can only appear as a change in the relative position of at least two material bodies. Force, which originates motion, can only be conceived of as referring to the relation of at least two material bodies towards each other; it is therefore to be defined as the endeavour of two masses to alter their relative position. But the force which two masses exert upon each other must be resolved into those exerted by all their particles upon each other; hence in mechanics we go back to forces exerted by material points. The relation of one point to another, as regards space, has reference solely to their distance apart: a moving force, therefore, exerted by each upon the other, can only act so as to cause an alteration of their distance, that is, it must be either attractive or repulsive.

Finally, therefore, we discover the problem of physical natural science to be, to refer natural phaenomena back to unchangeable attractive and repulsive forces, whose intensity depends solely upon distance. The solvability of this problem is the condition of the complete comprehensibility of nature. In mechanical calculations this limitation of the idea of moving force has not yet been assumed: a great number, however, of general principles referring to the motion of compound systems of bodies are only valid for the case that these bodies operate upon each other by unchangeable attractive or repulsive forces; for example, the principle of virtual velocities; the conservation of the motion of the centre of gravity; the conservation of the principal plane of rotation; of the moment of rotation of free systems, and the conservation of *vis viva*. In terrestrial matters application is made chiefly of the first and last of these principles, inasmuch as the others refer to systems which are supposed to be completely free; we shall however show that the first is only a special case of the last, which therefore must be regarded as the most general and important consequence of the deduction which we have made.

Theoretical natural science therefore, if she does not rest contented with half views of things, must bring her notions into harmony with the expressed requirements as to the nature of simple forces, and with the consequences which flow from them. Her vocation will be ended as soon as the reduction of natural phaenomena to simple forces is complete, and the proof given that this is the only reduction of which the phaenomena are capable.

I. *The Principle of the Conservation of Vis Viva.*

We will set out with the assumption that it is impossible, by any combination whatever of natural bodies, to produce force continually from nothing. By this proposition Carnot and Clapeyron have deduced theoretically a series of laws, part of which are proved by experiment and part not yet submitted to this test, regarding the latent and specific heats of various natural bodies. The object of the present memoir is to carry the same principle, in the same manner, through all branches of physics; partly for the purpose of showing its applicability in all those cases where the laws of the phaenomena have been sufficiently investigated, partly, supported by the manifold analogies of the known cases, to draw further conclusions regarding laws which are as yet but imperfectly known, and thus to indicate the course which the experimenter must pursue.

The principle mentioned can be represented in the following manner:—Let us imagine a system of natural bodies occupying certain relative positions towards each other, operated upon by forces mutually exerted among themselves, and caused to move until another definite position is attained; we can regard the velocities thus acquired as a certain mechanical work and translate them into such. If now we wish the same forces to act a second time, so as to produce again the same quantity of work, we must, in some way, by means of other forces placed at our disposal, bring the bodies back to their original position, and in effecting this a certain quantity of the latter forces will be consumed. In this case our principle requires that the quantity of work gained by the passage of the system from the first position to the second, and the quantity lost by the passage of the system from the second position back again to the first, are always equal, it matters not in what way or at what velocity the change has been effected. For were the quantity of work greater in one way than another, we might use the former for the production of work and the latter to carry the bodies back to their primitive positions, and in this way procure an indefinite amount of mechanical force. We should thus have built a *perpetuum mobile* which could not only impart motion to itself, but also to exterior bodies.

If we inquire after the mathematical expression of this principle, we shall find it in the known law of the conservation of *vis viva*. The quantity of work which is produced and consumed may, as is known, be expressed by a weight m, which is raised to a certain height h; it is then mgh, where g represents the force of gravity. To rise perpendicularly to the height h, the body m requires the velocity $v = \sqrt{2gh}$, and attains the same by falling through the same height. Hence we have $\frac{1}{2}mv^2 = mgh$; and hence we can set the half of the product mv^2, which is known in mechanics under the name of the *vis viva* of the body m, in the place of the quantity of work. For the sake of better agreement with the customary manner of measuring the intensity of force, I propose calling the quantity $\frac{1}{2}mv^2$ the quantity of *vis viva*, by which it is rendered

identical with the quantity of work. For the applications of the
doctrine of *vis viva* which have been hitherto made this alteration
is of no importance, but we shall derive much advantage from it
in the following. The principle of the conservation of *vis viva*, as is
known, declares that when any number whatever of material
points are set in motion, solely by such forces as they exert upon
each other, or as are directed against fixed centres, the total sum
of the *vires vivae*, at all times when the points occupy the same
relative position, is the same, whatever may have been their paths
or their velocities during the intervening times. Let us suppose
the *vires vivae* applied to raise the parts of the system or their
equivalent masses to a certain height, it follows from what has
just been shown, that the quantities of work, which are represented
in a similar manner, must also be equal under the conditions
mentioned. This principle however is not applicable to all
possible kinds of forces; in mechanics it is generally derived from
the principle of virtual velocities, and the latter can only be proved
in the case of material points endowed with attractive or repulsive
forces. We will now show that the principle of the conservation of
vis viva is alone valid where the forces in action may be resolved
into those of material points which act in the direction of the
lines which unite them, and the intensity of which depends only
upon the distance. In mechanics such forces are generally named
central forces. Hence, conversely, it follows that in all actions
of natural bodies upon each other, where the above principle is
capable of general application, even to the ultimate particles of
these bodies, such central forces must be regarded as the simplest
fundamental ones.

There follows a mathematical argument, based upon the principle of conservation, by which the conclusions are reached which are given in the following statements.

1. Whenever natural bodies act upon each other by attractive
or repulsive forces, which are independent of time and velocity,
the sum of their *vires vivae* and tensions must be constant; the
maximum quantity of work which can be obtained is therefore a
limited quantity.

2. If, on the contrary, natural bodies are possessed of forces
which depend upon time and velocity, or which act in other
directions than the lines which unite each two separate material

points, for example, rotatory forces, then combinations of such
bodies would be possible in which force might be either lost or
gained *ad infinitum.*

3. In the case of the equilibrium of a system of bodies under
the operation of central forces, the exterior and the interior forces
must, each system for itself, be in equilibrium, if we suppose that
the bodies of the system cannot be displaced, the whole system
only being moveable in regard to bodies which lie without it.
A rigid system of such bodies can therefore never be set in motion
by the action of its interior forces, but only by the operation of
exterior forces.   If, however, other than central forces had an
existence, rigid combinations of natural bodies might be formed
which could move of themselves without needing any relation
whatever to other bodies.

In the following sections are given the application of the principle to Mechanical Theorems, and studies of the Force-equivalent of Heat, and of the
Force-equivalent of Electric Processes.   Of the various Electric Processes,
there are here presented the treatment of Thermo-electric currents and a
portion of that of Electromagnetism.

Thermo-electric Currents.—In these currents we must seek the
origin of force in the actions discovered by Peltier at the place of
contact, by which a current opposed to the given one is developed.
Let us suppose the case of a constant hydro-electric current into
the conducting wire of which a piece of another metal is soldered,
the temperatures of the places of union being $t_{,}$ and $t_{,,}$, the electric
current will then, during the element of time dt, generate in the
entire conduction the heat I^2Rdt; besides this, at one of the points
where the metals are soldered together, the quantity $q_{,}dt$ will be
developed, and at the other the quantity $q_{,,}dt$ absorbed.   Let the
electromotive force of the entire circuit be A, hence $AIdt$ the heat
to be generated chemically, it then follows from the law of the
conservation of force.

$$AI = I^2R + q_{,} - q_{,,} \dots\dots\dots\dots\dots\dots(1)$$

Let the electromotive force of the thermo-circuit be B_t, when
one of the soldered junctions possesses the temperature t, and the
other any constant temperature whatever, for example 0°; then,
for the entire circuit, we have

$$I = \frac{A - B_{t,} + B_{t,,}}{R} \dots\dots\dots\dots\dots\dots(2)$$

When $t_{,} = t_{,,}$, we have

$$I = \frac{A}{R},$$

This set in equation (1) gives

$$q_{,} = q_{,,},$$

that is, when the temperatures of the places of soldering are both the same and the intensity of the current constant, the heat developed and that absorbed must be equal, independently of the cross section.   If we assumed that the process is the same in every point of the cross section, it would follow that the heat developed in equal spaces of different cross sections is proportional to the density of the current, and from this again, that the quantities generated by different currents in the whole of the transverse sections are directly proportional to the intensity of the current.

When the solderings are of different temperatures, it follows from equations (1) and (2), that

$$(B_{t,} - B_{t,,}) I = q_{,} - q_{,,},$$

that is to say, with the same intensity of current both the force which generates and which absorbs the heat increases with the temperature, in the same proportion as the electromotive force.

I am thus far unacquainted with any quantitative experiments with which either of the inferences might be compared.

Portion of the Discussion of the Force-equivalent of Electro-magnetism.

When a magnet moves under the influence of a current, the *vis viva* gained thereby must be furnished by the tensions consumed in the current.   During the portion of time dt, according to the notation before made use of, these are, $AIdt$ in units of heat, or $aAIdt$ in mechanical units, where a is the mechanical equivalent of the unit of heat.   The *vis viva* generated in the path of the current is aI^2Rdt, that gained by the magnet $I\frac{dV}{dt}dt$, where V represents its potential towards the conductor through which the unit of current passes.   Hence

$$aAIdt = aI^2Rdt + I\frac{dV}{dt}dt$$

consequently

$$I = \frac{A - \frac{1}{a} \cdot \frac{dV}{dt}}{R}.$$

We can distinguish the quantity $\frac{1}{a}\frac{dV}{dt}$ as a new electromotive force, that of the induced current. It always acts against that which moves the magnet in the direction which it follows, or which would increase its velocity. As this force is independent of the intensity of the current, it must remain the same, when before the motion of the magnet no current existed.

If the intensity be changeable, the whole induced current during a certain time is

$$\int I dt = -\frac{1}{aR}\int \frac{dV}{dt} dt = \frac{I}{a}\frac{V_{\prime} - V_{\prime\prime}}{R},$$

where $V_{\prime}$ denotes the potential at the beginning, and $V_{\prime\prime}$ at the end of the motion. If the magnet comes from a very great distance, we have

$$\int I dt = -\frac{\frac{1}{a}V_{\prime\prime}}{R}$$

independent of the route or the velocity of the magnet.

We can express the law thus:—The entire electromotive force of the induced current, generated by a change of position of a magnet relative to a closed conductor, is equal to the change which thereby takes place in the potential of the magnet towards the conductor, when the latter is traversed by the current $-1/a$. The unit of the electromotive force is here regarded as that by which the arbitrary unit of current is generated in the unit of resistance, the latter being that in which the above unit of current develops the unit of heat in the unit of time.

CARNOT

Nicolas Leonard Sadi Carnot was born on June 1, 1796, in Paris. He was the son of Lazare Carnot, who was prominent in the French Revolution and who was himself a mathematician of great ability. Sadi Carnot studied in the École Polytechnique. He joined the army in 1813 and was captain of engineers when he died of cholera on August 24, 1832.

In 1824 Carnot published a paper entitled *Reflexions sur la Puissance Motrice du Feu,* which is fundamental in the history of thermodynamics. In it he introduced the conception of a cycle of operations and proved that the most efficient thermodynamic engine is one in which all the operations of the cycle are reversible. His proof is erroneous because of his assumption of the materiality of heat. The principle is true, as was proved later by Clausius and Lord Kelvin, and when Carnot was able to use it without introducing the materiality of heat the conclusions which he drew from it were correct. In Carnot's notebooks from which extracts were published in 1878 there are passages which show that he had begun to doubt the truth of the doctrine of the materiality of heat and that he was planning experiments similar to those of Joule with a view of testing it. The translation by W. F. Magie appears in Harper's *Scientific Memoirs.*

THE MOTIVE POWER OF HEAT

The production of motion in the steam-engine is always accompanied by a circumstance which we should particularly notice. This circumstance is the re-establishment of equilibrium in the caloric—that is, its passage from one body where the temperature is more or less elevated to another where it is lower. What happens, in fact, in a steam-engine at work? The caloric developed in the fire-box as an effect of combustion passes through the wall of the boiler and produces steam, incorporating itself with the steam in some way. This steam, carrying the caloric with it, transports it first into the cylinder, where it fulfils some function, and thence into the condenser, where the steam is precipitated by coming in contact with cold water. As a last result the cold water in the condenser receives the caloric developed by combustion. It is warmed by means of the steam, as if it had been placed directly on the fire-box. The steam is here only a means of transporting caloric; it thus fulfils the same office as in the heating of baths by steam, with the exception that in the case in hand its motion is rendered useful.

We can easily perceive, in the operation which we have just described, the re-establishment of equilibrium in the caloric and its passage from a hotter to a colder body. The first of these bodies is the heated air of the fire-box; the second, the water of condensation. The re-establishment of equilibrium of the caloric is accomplished between them—if not completely, at least in part; for, on the one hand, the heated air after having done its work escapes through the smoke-stack at a much lower temperature than that which it had acquired by the combustion; and, on the other hand, the water of the condenser, after having precipitated the steam, leaves the engine with a higher temperature than that which it had when it entered.

The production of motive power in the steam-engine is therefore not due to a real consumption of the caloric, but to its transfer from a hotter to a colder body—that is to say, to the reestablishment of its equilibrium, which is assumed to have been destroyed by a chemical action such as combustion, or by some other cause. We shall soon see that this principle is applicable to all engines operated by heat.

. . . .

At this point we naturally raise an interesting and important question: Is the motive power of heat invariable in quantity, or does it vary with the agent which one uses to obtain it—that is, with the intermediate body chosen as the subject of the action of heat?

It is clear that the question thus raised supposes given a certain quantity of caloric and a certain difference of temperature. For example, we suppose that we have at our disposal a body, *A*, maintained at the temperature 100 degrees, and another body, *B*, at 0 degrees, and inquire what quantity of motive power will be produced by the transfer of a given quantity of caloric—for example, of so much as is necessary to melt a kilogram of ice—from the first of these bodies to the second; we inquire if this quantity of motive power is necessarily limited; if it varies with the substance used to obtain it; if water vapor offers in this respect more or less advantage than vapor of alcohol or of mercury, than a permanent gas or than any other substance. We shall try to answer these questions in the light of the considerations already advanced.

We have previously called attention to the fact, which is self-evident, or at least becomes so if we take into consideration the changes of volume occasioned by heat, that wherever there is a difference of temperature the production of motive power is possible. Conversely, wherever this power can be employed, it is possible to produce a difference of temperature or to destroy the equilibrium of the caloric. Percussion and friction of bodies are means of raising their temperature spontaneously to a higher degree than that of surrounding bodies, and consequently of destroying that equilibrium in the caloric which had previously existed.   It is an experimental fact that the temperature of gaseous fluids is raised by compression and lowered by expansion.   This is a sure method of changing the temperature of bodies, and thus of destroying the equilibrium of the caloric in the same substance, as often as we please.   Steam, when used in a reverse way from that in which it is used in the steam-engine, can thus be considered as a means of destroying the equilibrium of the caloric.   To be convinced of this, it is only necessary to notice attentively the way in which motive power is developed by the action of heat on water vapor.   Let us consider two bodies, *A* and *B*, each maintained at a constant temperature, that of *A* being higher than that of *B*; these two bodies, which can either give up or receive heat without a change of temperature, perform the functions of two indefinitely great reservoirs of caloric.   We will call the first body the source and the second the refrigerator.

If we desire to produce motive power by the transfer of a certain quantity of heat from the body *A* to the body *B* we may proceed in the following way:

1. We take from the body *A* a quantity of caloric to make steam—that is, we cause *A* to serve as the fire-pot, or rather as the metal of the boiler in an ordinary engine; we assume the steam produced to be at the same temperature as the body *A*.

2. The steam is received into an envelope capable of enlargement, such as a cylinder furnished with a piston.   We then increase the volume of this envelope, and consequently also the volume of the steam.   The temperature of the steam falls when it is thus rarefied, as is the case with all elastic fluids; let us assume that the rarefaction is carried to the point where the temperature becomes precisely that of the body *B*.

3. We condense the steam by bringing it in contact with *B* and exerting on it at the same time a constant pressure until it becomes entirely condensed.   The body *B* here performs the function of the injected water in an ordinary engine, with the difference that it condenses the steam without mixing with it and without changing its own temperature.   The operations which we have just described could have been performed in a reverse sense and order. There is nothing to prevent the formation of vapor by means of the caloric of the body *B*, and its compression from the temperature of *B*, in such a way that it acquires the temperature of the body *A*, and then its condensation in contact with *A*, under a pressure which is maintained constant until it is completely liquefied.

In the first series of operations there is at the same time a production of motive power and a transfer of caloric from the body *A* to the body *B*; in the reverse series there is at the same time an expenditure of motive power and a return of the caloric from *B* to *A*. But if in each case the same quantity of vapor has been used, if there is no loss of motive power or of caloric, the quantity of motive power produced in the first case will equal the quantity expended in the second, and the quantity of caloric which in the first case passed from *A* to *B* will equal the quantity which in the second case returns from *B* to *A*, so that an indefinite number of such alternating operations can be effected without the production of motive power or the transfer of caloric from one body to the other. Now if there were any method of using heat preferable to that which we have employed, that is to say, if it were possible that the caloric should produce, by any process whatever, a larger quantity of motive power than that produced in our first series of operations, it would be possible, by diverting a portion of this power, to effect a return of caloric, by the method just indicated, from the body *B* to the body *A*—that is, from the refrigerator to the source—and thus to re-establish things in their original state, and to put them in position to recommence an operation exactly similar to the first one, and so on: there would thus result not only the perpetual motion, but an indefinite creation of motive power without consumption of caloric or of any other agent whatsoever. Such a creation is entirely contrary to the ideas now accepted, to the laws of mechanics and of sound physics; it is inadmissible. We may hence conclude that the maximum motive power resulting from the use of steam is also the maximum motive power which can be obtained by any other means. We shall soon give a second and more rigorous demonstration of the law. What has been given should only be regarded as a sketch.

It may properly be asked, in connection with the proposition just stated, what is the meaning of the word maximum? How can we know that this maximum is reached and that the steam is used in the most advantageous way possible to produce motive power?

Since any re-establishment of equilibrium in the caloric can be used to produce motive power, any re-establishment of equilibrium which is effected without producing motive power should be considered as a veritable loss: now, with little reflection, we can see that any change of temperature which is not due to a change of volume of the body can be only a useless re-establishment of equilibrium in the caloric. The necessary condition of the maximum is, then, that in bodies used to obtain the motive power of heat, no change of temperature occurs which is not due to a change of volume. Conversely, every time that this condition is fulfilled, the maximum is attained.

. . . .

We shall give here a second demonstration of the fundamental proposition stated on page 224 and present this proposition in a more general form than we have before.

When a gaseous fluid is rapidly compressed its temperature rises, and when it is rapidly expanded its temperature falls. This is one of the best established facts of experience; we shall take it as the basis of our demonstration. When the temperature of a gas is raised and we wish to bring it back to its original temperature without again changing its volume, it is necessary to remove caloric from it. This caloric may also be removed as the compression is effected, so that the temperature of the gas remains constant. In the same way, if the gas is rarefied, we can prevent its temperature from falling, by furnishing it with a certain quantity of caloric. We shall call the caloric used in such cases,

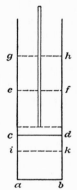

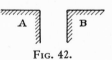

FIG. 42.

when it occasions no change of temperature, caloric due to a change of volume. This name does not indicate that the caloric belongs to the volume; it does not belong to it any more than it does to the pressure, and it might equally well be called caloric due to a change of pressure. We are ignorant of what laws it obeys in respect to changes of volume: it is possible that its quantity changes with the nature of the gas, or with its density or with its temperature. Experiment has taught us nothing on this subject; it has taught us only that this caloric is developed in greater or less quantity by the compression of elastic fluids.

This preliminary idea having been stated, let us imagine an elastic fluid—atmospheric air, for example—enclosed in a cylindrical vessel *abcd* (Fig. 42) furnished with a movable diaphragm or piston *cd*; let us assume also the two bodies *A*, *B*, both at constant temperatures, that of *A* being higher than that of *B*, and let us consider the series of operations which follows:

1. Contact of the body *A* with the air contained in the vessel *abcd* or with the wall of this vessel, which wall is supposed to be a good conductor of caloric. By means of this contact the air attains the same temperature as the body *A*; *cd* is the position of the piston.

2. The piston rises gradually until it takes the position *ef*. Contact is always maintained between the air and the body *A*, and the temperature thus remains constant during the rarefaction. The body *A* furnishes the caloric necessary to maintain a constant temperature.

3. The body *A* is removed and the air is no longer in contact with any body capable of supplying it with caloric; the piston however, continues to move and passes from the position *ef* to the position *gh*. The air is rarefied without receiving caloric and its temperature falls. Let us suppose that it falls until it becomes equal to that of the body *B*; at this instant the piston ceases to move and occupies the position *gh*.

4. The air is brought in contact with the body *B*; it is compressed by the piston as it returns from the position *gh* to the position *cd*. The air, however, remains at a constant temperature on account of its contact with the body *B*, to which it gives up its caloric.

5. The body *B* is removed and the compression of the air continued. The temperature of the air, which is now isolated, rises. The compression is continued until the air acquires the temperature of the body *A*. The piston during this time passes from the position *cd* to the position *ik*.

6. The air is again brought in contact with the body *A*; the piston returns from the position *ik* to the position *ef*, and the temperature remains constant.

7. The operation described in No. 3 is repeated, and then the operations 4, 5, 6, 3, 4, 5, 6, 3, 4, 5, and so on, successively.

In these various operations a pressure is exerted upon the piston by the air contained in the cylinder; the elastic force of this air varies with the changes of volume as well as with the changes of temperature; but we should notice that at equal volumes—that is, for similar positions of the piston—the temperature is higher during the expansions than during the compressions. During the former, therefore, the elastic force of the air is greater, and consequently the quantity of motive power produced by the expansions is greater than that which is consumed in effecting the compressions. Thus there remains an excess of motive power, which we can dispose of for any purpose whatsoever. The air has therefore served as a heat-engine; and it has been used in the most advantageous way possible, for there has been no useless re-establishment of equilibrium in the caloric.

All the operations described above can be carried out in a direct and in a reverse order. Let us suppose that after the sixth step, when the piston is at *ef*, it is brought back to the position *ik*, and that, at the same time, the air is kept in contact with the body *A*; the caloric furnished by this body during the sixth operation returns to its source—that is, to the body *A*, and the condition of things is the same as at the end of the fifth operation. If now we remove the body *A* and move the piston from *ef* to *cd*, the temperature of the air will fall as many degrees as it rose during the fifth operation and will equal that of the body *B*. A series of reverse operations to those above described could evidently be carried out; it is only necessary to bring the system into the same initial state and in each operation to carry out an expansion instead of a compression, and conversely.

The result of the first operation was the production of a certain quantity of motive power and the transfer of the caloric from the body *A* to the body *B*; the result of the reverse operation would be the consumption of the motive power produced and the return of the caloric from the body *B* to the body *A*; so that the two series of operations in a sense annul or neutralize each other.

The impossibility of making the caloric produce a larger quantity of motive power than that which we obtained in our first series of operations is now easy to prove. It may be demonstrated by an argument similar to that used on page 224. The argument will have even a greater degree of rigor; the air which serves to develop the motive power is brought back, at the end of each cycle of operations, to its original condition which was, as we noticed, not quite the case with the steam.

We have chosen atmospheric air as the agency employed to develop the motive power of heat; but it is evident that the same reasoning would hold for any other gaseous substance, and even for all other bodies susceptible of changes of temperature by successive contractions and expansions—that is, for all bodies in Nature, at least, all those which are capable of developing the motive power of heat.   Thus we are led to establish this general proposition:

The motive power of heat is independent of the agents employed to develop it; its quantity is determined solely by the temperatures of the bodies between which, in the final result, the transfer of the caloric occurs.

CLAUSIUS

Rudolph Julius Emmanuel Clausius was born on January 2, 1822, at Koslin in Pomerania.   He studied for four years in the University of Berlin and was graduated at Halle in 1848.   In 1850 he became professor of physics at the Royal Artillery and Engineering School in Berlin.   He then passed successively as professor of physics to Zurich, Würzburg, and Bonn.   He was professor of physics at Bonn from 1869 until his death, which occurred on August 24, 1888.

Clausius was one of the founders of the kinetic theory of gases and of the science of thermodynamics.   He and Lord Kelvin at about the same time and independently announced the Second Law of thermodynamics.   Clausius particularly developed the theory of thermodynamics by applying it to the study of gases and vapors.

The extract which follows is taken from a paper entitled "Ueber die bewegende Kraft der Wärme," published in the *Annalen der Physik und Chemie*, Vol. 79, pp. 368 and 500, 1850.   The translation by W. F. Magie appears in Harper's *Scientific Memoirs*.

THE SECOND LAW OF THERMODYNAMICS

Since heat was first used as a motive power in the steam-engine, thereby suggesting from practice that a certain quantity of work may be treated as equivalent to the heat needed to produce it, it was natural to assume also in theory a definite relation between a quantity of heat and the work which in any possible way can be produced by it, and to use this relation in drawing conclusions about the nature and the laws of heat itself.   In fact, several fruitful investigations of this sort have already been made; yet I think that the subject is not yet exhausted, but on the other hand deserves the earnest attention of physicists, partly because serious objections can be raised to the conclusions that have already been reached, partly because other conclusions, which may readily be drawn and which will essentially contribute to the establishment and completion of the theory of heat, still remain entirely unnoticed or have not yet been stated with sufficient definiteness.

The most important of the researches here referred to was that of S. Carnot, and the ideas of this author were afterwards given analytical form in a very skilful way by Clapeyron. Carnot showed that whenever work is done by heat and no permanent change occurs in the condition of the working body, a certain quantity of heat passes from a hotter to a colder body. In the steam-engine, for example, by means of the steam which is developed in the boiler and precipitated in the condenser, heat is transferred from the grate to the condenser. This transfer he considered as the heat change, corresponding to the work done. He says expressly that no heat is lost in the process, but that the quantity of heat remains unchanged, and adds: "This fact is not doubted; it was assumed at first without investigation, and then established in many cases by calorimetric measurements. To deny it would overthrow the whole theory of heat, of which it is the foundation." I am not aware, however, that it has been sufficiently proved by experiment that no loss of heat occurs when work is done; it may, perhaps, on the contrary, be asserted with more correctness that even if such a loss has not been proved directly, it has yet been shown by other facts to be not only admissible, but even highly probable. If it be assumed that heat, like a substance, can not diminish in quantity, it must also be assumed that it can not increase. It is, however, almost impossible to explain the heat produced by friction except as an increase in the quantity of heat. The careful investigations of Joule, in which heat is produced in several different ways by the application of mechanical work, have almost certainly proved not only the possibility of increasing the quantity of heat in any circumstances but also the law that the quantity of heat developed is proportional to the work expended in the operation. To this it must be added that other facts have lately become known which support the view, that heat is not a substance, but consists in a motion of the least parts of bodies. If this view is correct, it is admissible to apply to heat the general mechanical principle that a motion may be transformed into work, and in such a manner that the loss of *vis viva* is proportional to the work accomplished.

These facts, with which Carnot also was well acquainted and the importance of which he has expressly recognized, almost compel us to accept the equivalence between heat and work, on the modified hypothesis that the accomplishment of work requires not merely a change in the distribution of heat, but also an actual consumption of heat, and that, conversely, heat can be developed again by the expenditure of work.

.   .   .   .

If any body changes its volume, mechanical work will in general be either produced or expended. It is, however, in most cases impossible to determine this exactly, since besides the external work there is generally an unknown amount of internal work done. To avoid this difficulty, Carnot employed the ingenious method already referred to of allowing the body to undergo its various changes in succession, which are so arranged that it returns

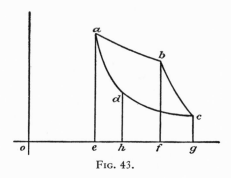

Fig. 43.

at last exactly to its original condition. In this case, if internal work is done in some of the changes, it is exactly compensated for in the others, and we may be sure that the external work, which remains over after the changes are completed, is all the work that has been done.   Clapeyron has represented this process graphically in a very clear way, and we shall follow his presentation now for the permanent gases, with a slight alteration rendered necessary by our principle.

In the figure (Fig. 43), let the abscissa oe represent the volume and the ordinate ea the pressure on a unit weight of gas, in a condition in which its temperature $= t$. We assume that the gas is contained in an extensible envelope, which, however, cannot exchange heat with it. If, now, it is allowed to expand in this envelope, its temperature would fall if no heat were imparted to it. To avoid this, let it be put in contact, during its expansion, with a body *A*, which is kept at the constant temperature *t*, and which imparts just so much heat to the gas that its temperature also remains equal to *t*. During this expansion at constant temperature, its pressure diminishes according to Mariotte's law, and may be represented by the ordinate of a curve, *ab*, which is a portion of an equilateral hyperbola. When the volume of the gas has increased in this way from *oe* to *of*, the body *A* is removed, and the expansion is allowed to continue without the introduction of more heat. The temperature will then fall, and the pressure diminish more rapidly than before. The law which is followed in this part of the process may be represented by the curve *bc*. When the volume of the gas has increased in this way from *of* to *og*, and its temperature has fallen from *t* to *τ*, we begin to compress it, in order to restore it again to its original volume *oe*. If it were left to itself its temperature would again rise. This, however, we do not permit, but bring it in contact with a body *B*, at the constant temperature *τ*, to which it at once gives up the heat that is produced, so that it keeps the temperature *τ*; and while it is in contact with this body we compress it so far (by the amount *gh*) that the remaining compression *he* is exactly sufficient to raise its temperature from *τ* to *t*, if during this last compression it gives up no heat. During the former compression the pressure increases according to Mariotte's law, and is represented by the portion *cd* of an equilateral hyperbola. During the latter, on the other hand, the increase is more rapid and is represented by the curve *da*. This curve must end exactly at *a*, for since at the end of the operation the volume and temperature have again their original values, the same must be true of the pressure also, which is a function of them both. The gas is therefore in the same condition again as it was at the beginning.

Now, to determine the work produced by these changes, for the reasons already given, we need to direct our attention only to the external work. During the expansion the gas does work, which is determined by the integral of the product of the differential of the volume into the corresponding pressure, and is therefore represented geometrically by the quadrilaterals *eabf* and *fbcg*. During the compression, on the other hand, work is expended, which is represented similarly by the quadrilaterals *gcdh* and *hdae*. The excess of the former quantity of work over the latter is to be looked on as the whole work produced during the changes, and this is represented by the quadrilateral *abcd*.

If the process above described is carried out in the reverse order, the same magnitude, *abcd*, is obtained as the excess of the work expended over the work done.

.   .   .   .

Carnot assumed, as has already been mentioned, that *the equivalent of the work done by heat is found in the mere transfer of heat from a hotter to a colder body, while the quantity of heat remains undiminished.*

The latter part of this assumption—namely, that the quantity of heat remains undiminished—contradicts our former principle, and must therefore be rejected if we are to retain that principle. On the other hand, the first part may still obtain in all its essentials. For though we do not need a special equivalent for the work done, since we have assumed as such an actual consumption of heat, it still may well be possible that such a transfer of heat occurs at the same time as the consumption of heat, and also stands in a definite relation to the work done.   It becomes important, therefore, to consider whether this assumption, besides the mere possibility, has also a sufficient probability in its favor.

A transfer of heat from a hotter to a colder body always occurs in those cases in which work is done by heat, and in which also the condition is fulfilled that the working substance is in the same state at the end as at the beginning of the operation. . . . Yet, in order to establish a relation between the heat transferred and the work done, a certain restriction is necessary.   For since a transfer of heat can take place without mechanical effect if a hotter and a colder body are immediately in contact and heat passes from one to the other by conduction, the way in which the transfer of a certain quantity of heat between two bodies at the temperatures t and τ can be made to do the maximum of work is so to carry out the process, as was done in the above cases, that two bodies of different temperatures never come in contact.

It is this maximum of work which must be compared with the heat transferred.   When this is done it appears that there is in fact ground for asserting, with Carnot, that it depends only on the quantity of the heat transferred and on the temperatures t and τ of the two bodies A and B, but not on the nature of the substance by means of which the work is done.   This maximum has, namely, the property that by expending it as great a quantity of heat can be transferred from the cold body B to the hot body A

as passes from *A* to *B* when it is produced. This may easily be seen, if we think of the whole process formerly described as carried out in the reverse order, so that, for example, in the first case the gas first expands by itself, until its temperature falls from *t* to *τ*, is then expanded in connection with *B*, is then compressed by itself until its temperature is again *t*, and finally is compressed in connection with *A*. In this case more work will be employed during the compression than is produced during the expansion, so that on the whole there is a loss of work, which is exactly as great as the gain of work in the former process. Further, there will be just as much heat taken from the body *B* as was before given to it, and just as much given to the body *A* as was before taken from it, whence it follows not only that the same amount of heat is produced as was formerly consumed, but also that the heat which in the former process was transferred from *A* to *B* now passes from *B* to *A*.

If we now suppose that there are two substances of which the one can produce more work than the other by the transfer of a given amount of heat, or, what comes to the same thing, needs to transfer less heat from *A* to *B* to produce a given quantity of work, we may use these two substances alternately by producing work with one of them in the above process, and by expending work upon the other in the reverse process. At the end of the operations both bodies are in their original condition; further, the work produced will have exactly counterbalanced the work done, and therefore, by our former principle, the quantity of heat can have neither increased nor diminished. The only change will occur in the distribution of the heat, since more heat will be transferred from *B* to *A* than from *A* to *B*, and so on the whole heat will be transferred from *B* to *A*. By repeating these two processes alternately it would be possible, without any expenditure of force or any other change, to transfer as much heat as we please from a cold to a hot body, and this is not in accord with the other relations of heat, since it always shows a tendency to equalize temperature differences and therefore to pass from hotter to colder bodies.

It seems, therefore, to be theoretically admissible to retain the first and the really essential part of Carnot's assumptions, and to apply it as a second principle in conjunction with the first; and the correctness of this method is, as we shall soon see, established already in many cases by its consequences.

Entropy

The function which was introduced into thermodynamics by Clausius under the name of entropy had already been used and some of its properties studied by Rankine and Lord Kelvin, but owing to his giving it its name and to his development of its properties it is commonly ascribed to him. It is not necessary to present the thermodynamic and mathematical arguments by which Clausius arrived at the equation with which the extract opens. The extract as given deals with the application of entropy to the universe. The paper from which it is taken appeared in the *Annalen der Physik und Chemie*, Vol. 125, p. 353, 1865, under the title "Ueber verschiedene für die Anwendung bequeme Formen der Hauptgleichungen der mechanischen Wärmetheorie." The symbols in the first equation represent quantity of heat (Q), absolute temperature (T), and entropy (S).

We obtain the equation

$$\int \frac{dQ}{T} = S - S_0$$

which, while somewhat differently arranged, is the same as that which was formerly used to determine S.

If we wish to designate S by a proper name we can say of it that it is the *transformation content* of the body, in the same way that we say of the quantity U that it is the *heat and work content* of the body. However, since I think it is better to take the names of such quantities as these, which are important for science, from the ancient languages, so that they can be introduced without change into all the modern languages, I propose to name the magnitude S the *entropy* of the body, from the Greek word ἡ τροπή, a transformation. I have intentionally formed the word *entropy* so as to be as similar as possible to the word *energy*, since both these quantities, which are to be known by these names, are so nearly related to each other in their physical significance that a certain similarity in their names seemed to me advantageous.

. . . .

Finally I may allow myself to touch on a matter whose complete treatment would not be in place here, because the statements necessary for that purpose would take up too much room, but of which I believe that even the following short indication will not be without interest, in that it will contribute to the recognition of the importance of the quantities which I have introduced into the formulation of the second law of the mechanical theory of heat.

The second law, in the form which I have given it, states the fact that all transformations which occur in nature occur in a certain sense which I have taken as positive, of themselves, that is, without compensation, but that they can only occur in the opposite or negative sense in such a way that they are compensated by positive transformations which occur at the same time. The application of this law to the universe leads to a conclusion to which W. Thomson first called attention and about which I have already spoken in a recently published paper. This conclusion is that if among all the changes of state which occur in the universe the transformations in one sense exceed in magnitude those in the opposite sense, then the general condition of the universe will change more and more in the former sense, and the universe will thus persistently approach a final state.

The question now arises whether this final state can be characterised in a simple and also a definite way. This can be done by treating the transformations, as I have done, as mathematical quantities, whose equivalent values can be calculated and united in a sum by algebraic addition.

In my papers so far published I have carried out such calculations with respect to the heat present in bodies and to the arrangement of the constituents of the bodies. For each body there are found two quantities, the transformation value of its heat content and its disgregration, the sum of which is its entropy. This however does not complete the business. The discussion must also be extended to the radiant heat, or otherwise expressed, to the heat transmitted through the universe in the form of advancing vibrations of the ether, and also to such motions as cannot be comprehended under the name heat.

The treatment of these latter motions, at least as far as they are the motions of ponderable masses, can be briefly settled, since we come by a simple argument to the following conclusion: If a mass, which is so great that in comparison with it an atom may be considered as vanishingly small, moves as a whole, the transformation value of this motion is to be looked on as vanishingly small in the same way in comparison with its kinetic energy; from which it follows that if such a motion is transformed into heat by a passive resistance, then the equivalent value of the uncompensated transformation which then occurs is simply represented by the transformation value of the heat produced. The radiant heat, however, cannot be treated so briefly, since there is need still of a certain special treatment in order to find out how its transformation value is to be determined.   Although, in the paper which was recently published and to which I have previously referred, I have already discussed radiant heat in its connection with the mechanical theory of heat, yet I have not as yet treated the question which has here come up, since it was then only my purpose to prove that there was no contradiction between the laws of radiant heat and a fundamental law which I assumed in the mechanical theory of heat.   I reserve for future consideration the more particular application of the mechanical theory of heat and especially of the law of equivalents of transformation to radiant heat.

For the present I will confine myself to announcing as a result of my argument that if we think of that quantity which with reference to a single body I have called its entropy, as formed in a consistent way, with consideration of all the circumstances, for the whole universe, and if we use in connection with it the other simpler concept of energy, we can express the fundamental laws of the universe which correspond to the two fundamental laws of the mechanical theory of heat in the following simple form.

1. The energy of the universe is constant.

2. The entropy of the universe tends toward a maximum.

KELVIN

William Thomson, Lord Kelvin, was born on June 26, 1824, in Belfast, where his father was professor of mathematics at the University.   When he was eight years old he went to Glasgow, to which city his father had been called as professor in the University.   He remained a resident of that city through the rest of his long life.   He studied at Cambridge, where he was graduated Second Wrangler and First Smith's Prizeman in 1845.   After a year's experimental work with Regnault in Paris he was called in 1846 to the chair of Natural Philosophy in the University of Glasgow, where he served as professor for fifty-three years.   After his retirement in 1899 he lived on his estate near Largs, where he died on December 17, 1907.

Besides being one of the founders of the science of thermodynamics, Thomson was the consulting engineer for the company that was attempting to lay the first Atlantic telegraph cables, and it was largely through his inventions that the successful construction and operation of the cables were made possible. He also invented and used in various researches a number of important instruments for the making of accurate electrical measurements.   In 1866 he was knighted and in 1892 he was raised to the peerage as Lord Kelvin of Largs.

The first of the following papers by Lord Kelvin is the one in which he proposes the adoption of an absolute thermometric scale founded on Carnot's theory of the motive power of heat. It appeared in the *Cambridge Philosophical Society Proceedings* for June 5, 1848, and in the *Philosophical Magazine,* October, 1848. It antedates the discovery of the second law of thermodynamics and the scale proposed in it is unsatisfactory, but the principle upon which the scale is based is correct. The scale as it is now used is described in Kelvin's paper on the "Dynamical Theory of Heat," published in the *Transactions of the Royal Society of Edinburgh,* March, 1851, and in the *Philosophical Magazine,* Vol. 4, 1852. An extract from this paper describing the scale is also given. It follows the second extract which is from that part of the paper on the "Dynamical Theory of Heat" in which Kelvin enunciates the second law of thermodynamics.

AN ABSOLUTE SCALE OF TEMPERATURE

The determination of temperature has long been recognized as a problem of the greatest importance in physical science. It has accordingly been made a subject of most careful attention, and, especially in late years, of very elaborate and refined experimental researches; and we are thus at present in possession of as complete a practical solution of the problem as can be desired, even for the most accurate investigations. The theory of thermometry is however as yet far from being in so satisfactory a state. The principle to be followed in constructing a thermometric scale might at first sight seem to be obvious, as it might appear that a perfect thermometer would indicate equal additions of heat, as corresponding to equal elevations of temperature, estimated by the numbered divisions of its scale. It is however now recognized (from the variations in the specific heats of bodies) as an experimentally demonstrated fact that thermometry under this condition is impossible, and we are left without any principle on which to found an absolute thermometric scale.

Next in importance to the primary establishment of an absolute scale, independently of the properties of any particular kind of matter, is the fixing upon an arbitrary system of thermometry, according to which results of observations made by different experimenters, in various positions and circumstances, may be exactly compared. This object is very fully attained by means of thermometers constructed and graduated according to the clearly defined methods adopted by the best instrument-makers of the present day, when the rigorous experimental processes which have been indicated, especially by Regnault, for interpreting their indications in a comparable way, are followed. The particular kind of thermometer which is least liable to uncertain variations of any kind is that founded on the expansion of air, and this is therefore generally adopted as the standard for the comparison of thermometers of all constructions. Hence the scale which is at present employed for estimating temperature is that of the air-thermometer; and in accurate researches care is always taken to reduce to this scale the indications of the instrument actually used, whatever may be its specific construction and graduation.

The principle according to which the scale of the air-thermometer is graduated, is simply that equal absolute expansions of the mass of air or gas in the instrument, under a constant pressure, shall indicate equal differences of the numbers on the scale; the length of a "degree" being determined by allowing a given number for the interval between the freezing- and the boiling-points. Now it is found by Regnault that various thermometers, constructed with air under different pressures, or with different gases, give indications which coincide so closely, that, unless when certain gases, such as sulphurous acid, which approach the physical condition of vapours at saturation, are made use of, the variations are inappreciable. This remarkable circumstance enhances very much the practical value of the air-thermometer; but still a rigorous standard can only be defined by fixing upon a certain gas at a determinate pressure, as the thermometric substance. Although we have thus a strict principle for constructing a *definite* system for the estimation of temperature, yet as reference is essentially made to a specific body as the standard thermometric substance, we cannot consider that we have arrived at an *absolute* scale, and we can only regard, in strictness, the scale actually adopted as *an arbitrary series of numbered points of reference sufficiently close for the requirements of practical thermometry.*

In the present state of physical science, therefore, a question of extreme interest arises: *Is there any principle on which an absolute thermometric scale can be founded?* It appears to me that Carnot's theory of the motive power of heat enables us to give an affirmative answer.

The relation between motive power and heat, as established by Carnot, is such that *quantities of heat,* and *intervals of temperature,* are involved as the sole elements in the expression for the amount of mechanical effect to be obtained through the agency of heat; and since we have, independently, a definite system for the measurement of quantities of heat, we are thus furnished with a measure for intervals according to which absolute differences of temperature may be estimated. To make this intelligible, a few words in explanation of Carnot's theory must be given; but for a full account of this most valuable contribution to physical science, the reader is referred to either of the works mentioned above.

The original treatise by Carnot, and Clapeyron's paper on the same subject.

In the present state of science no operation is known by which heat can be absorbed, without either elevating the temperature of matter, or becoming latent and producing some alteration in the physical condition of the body into which it is absorbed; and the conversion of heat (or *caloric*) into mechanical effect is probably impossible, certainly undiscovered. In actual engines for obtaining mechanical effect through the agency of heat, we must consequently look for the source of power, not in any absorption and conversion, but merely in a transmission of heat. Now Carnot, starting from universally acknowledged physical principles, demonstrates that it is by the *letting down* of heat from a hot body to a cold body, through the medium of an engine (a steam-engine, or an air-engine for instance), that mechanical effect is to be obtained; and conversely, he proves that the same amount of heat may, by the expenditure of an equal amount of labouring force, be *raised* from the cold to the hot body (the engine being in this case *worked backwards*); just as mechanical effect may be obtained by the descent of water let down by a water-wheel, and by spending labouring force in turning the wheel backwards, or in working a pump, water may be elevated to a higher level. The amount of mechanical effect to be obtained by the transmission of a given quantity of heat, through the medium of any kind of engine in which the economy is perfect, will depend, as Carnot demonstrates, not on the specific nature of the substance employed as the medium of transmission of heat in the engine, but solely on the interval between the temperature of the two bodies between which the heat is transferred.

Carnot examines in detail the ideal construction of an air-engine and of a steam-engine, in which, besides the condition of perfect economy being satisfied, the machine is so arranged, that at the close of a complete operation the substance (air in one case and water in the other) employed is restored to precisely the same physical condition as at the commencement. He thus shews on what elements, capable of experimental determination, either with reference to air, or with reference to a liquid and its vapour, the absolute amount of mechanical effect due to the transmission of a unit of heat from a hot body to a cold body, through any given interval of the thermometric scale, may be ascertained. In M. Clapeyron's paper various experimental data, confessedly very imperfect, are brought forward, and the amounts of mechanical effect due to a unit of heat descending a degree of the air-thermometer, in various parts of the scale, are calculated from them, according to Carnot's expressions. The results so obtained indicate very decidedly, that what we may with much propriety call *the value of a degree* (estimated by the mechanical effect to be obtained from the descent of a unit of heat through it) of the air-thermometer depends on the part of the scale in which it is taken, being less for high than for low temperatures.

The characteristic property of the scale which I now propose is, that all degrees have the same value; that is, that a unit of heat descending from a body A at the temperature $T°$ of this scale, to a body B at the temperature $(T - 1)°$, would give out the same mechanical effect, whatever be the number T. This may justly be termed an absolute scale, since its characteristic is quite independent of the physical properties of any specific substance.

To compare this scale with that of the air-thermometer, the *values* (according to the principle of estimation stated above) of degrees of the air-thermometer must be known. Now an expression, obtained by Carnot from the consideration of his ideal steam-engine, enables us to calculate these values, when the latent heat of a given volume and the pressure of saturated vapour at any temperature are experimentally determined. The determination of these elements is the principal object of Regnault's great work, already referred to, but at present his researches are not complete. In the first part, which alone has been as yet published, the latent heats of a given *weight*, and the pressures of saturated vapour at all temperatures between 0° and 230° (Cent. of the air-thermometer), have been ascertained; but it would be necessary in addition to know the densities of saturated vapour at different temperatures, to enable us to determine the latent heat of a given volume at any temperature. M. Regnault announces his intention of instituting researches for this object; but till the results are made known, we have no way of completing the data necessary for the present problem, except by estimating the density of saturated vapour at any temperature (the corresponding pressure being known by Regnault's researches already published) according to the approximate laws of compressibility and expansion (the laws of Mariotte and Gay-Lussac, or Boyle and Dalton). Within the limits of natural temperature in ordinary climates, the density of saturated vapour is actually found by Regnault (*Études Hygrométriques* in the *Annales de Chimie*) to verify very closely these laws; and we have reason to believe from experiments which have been made by Gay-Lussac and others, that as high as the temperature 100° there can be no considerable deviation; but our estimate of the density of saturated vapour, founded on these laws, may be very erroneous at such high temperatures as 230°. Hence a completely satisfactory calculation of the proposed scale cannot be made till after the additional experimental data shall have been obtained; but with the data which we actually possess, we may make an approximate comparison of the new scale with that of the air-thermometer, which at least between 0° and 100° will be tolerably satisfactory.

The labour of performing the necessary calculations for effecting a comparison of the proposed scale with that of the air-thermometer, between the limits 0° and 230° of the latter, has been kindly undertaken by Mr. William Steele, lately of Glasgow College. now of St. Peter's College, Cambridge. His results in tabulated forms were laid before the Society, with a diagram, in which the comparison between the two scales is represented graphically. In the first table, the amounts of mechanical effect due to the descent of a unit of heat through the successive degrees of the air-thermometer are exhibited. The unit of heat adopted is the quantity necessary to elevate the temperature of a kilogramme of water from 0° to 1° of the air-thermometer; and the unit of mechanical effect is a metre-kilogramme; that is, a kilogramme raised a metre high.

In the second table, the temperatures according to the proposed scale, which correspond to the different degrees of the air-thermometer from 0° to 230°, are exhibited. [The arbitrary points which coincide on the two scales are 0° and 100°.]

Note.—If we add together the first hundred numbers given in the first table, we find 135.7 for the amount of work due to a unit of heat descending from a body *A* at 100° to *B* at 0°. Now 79 such units of heat would, according to Dr. Black (his result being very slightly corrected by Regnault), melt a kilogramme of ice. Hence if the heat necessary to melt a pound of ice be now taken as unity, and if a *metre-pound* be taken as the unit of mechanical effect, the amount of work to be obtained by the descent of a unit of heat from 100° to 0° is 79 × 135.7, or 10,700 nearly. This is the same as 35,100 foot pounds, which is a little more than the work of a one-horse-power engine (33,000 foot pounds) in a minute; and consequently, if we had a steam-engine working with perfect economy at one-horse-power, the boiler being at the temperature 100°, and the condenser kept at 0° by a constant supply of ice, rather less than a pound of ice would be melted in a minute.

The Second Law of Thermodynamics

According to an obvious principle, first introduced, however, into the theory of the motive power of heat by Carnot, mechanical effect produced in any process cannot be said to have been derived from a purely thermal source, unless at the end of the process all the materials used are in precisely the same physical and mechanical circumstances as they were at the beginning. In some conceivable "thermo-dynamic engines," as, for instance, Faraday's floating magnet, or Barlow's "wheel and axle," made to rotate and perform work uniformly by means of a current continuously excited by heat communicated to two metals in contact, or the thermo-electric rotatory apparatus devised by Marsh, which has been actually constructed, this condition is fulfilled at every instant. On the other hand, in all thermo-dynamic engines, founded on electrical agency, in which discontinuous galvanic currents, or pieces of soft iron in a variable state of magnetization, are used, and in all engines founded on the alternate expansions and contractions of media, there are really alterations in the condition of materials; but, in accordance with the principle stated above, these alterations must be strictly periodical. In any such engine the series of motions performed during a period, at the end of which the materials are restored to precisely the same condition as that in which they existed at the beginning, constitutes what will be called a complete cycle of its operations. Whenever in what follows, the work done or the mechanical effect produced by a thermo-dynamic engine is mentioned without qualification, it must be understood that the mechanical effect produced, either in a non-varying engine, or in a complete cycle, or any number of complete cycles of a periodical engine, is meant.

The source of heat will always be supposed to be a hot body at a given constant temperature put in contact with some part of the engine; and when any part of the engine is to be kept from rising in temperature (which can only be done by drawing off whatever heat is deposited in it), this will be supposed to be done by putting a cold body, which will be called the refrigerator, at a given constant temperature in contact with it.

The whole theory of the motive power of heat is founded on the two following propositions, due respectively to Joule, and to Carnot and Clausius.

Prop. I. (Joule).—When equal quantities of mechanical effect are produced by any means whatever from purely thermal sources, or lost in purely thermal effects, equal quantities of heat are put out of existence or are generated.

Prop. II. (Carnot and Clausius).—If an engine be such that, when it is worked backwards, the physical and mechanical agencies in every part of its motions are all reversed, it produces as much mechanical effect as can be produced by any thermo-dynamic engine, with the same temperatures of source and refrigerator, from a given quantity of heat.

The former proposition is shown to be included in the general "principle of mechanical effect," and is so established beyond all doubt by the following demonstration.

By whatever direct effect the heat gained or lost by a body in any conceivable circumstances is tested, the measurement of its quantity may always be founded on a determination of the quantity of some standard substance, which it or any equal quantity of heat could raise from one standard temperature to another; the test of equality between two quantities of heat being their capability of raising equal quantities of any substance from any temperature to the same higher temperatures. Now, according to the dynamical theory of heat, the temperature of a substance can only be raised by working upon it in some way so as to produce increased thermal motions within it, besides effecting any modifications in the mutual distances or arrangements of its particles which may accompany a change of temperature. The work necessary to produce this total mechanical effect is of course proportional to the quantity of the substance raised from one standard temperature to another; and therefore when a body, or a group of bodies, or a machine, parts with or receives heat, there is in reality mechanical effect produced from it, or taken into it, to an extent precisely proportional to the quantity of heat which it emits or absorbs.   But the work which any external forces do upon it, the work done by its own molecular forces, and the amount by which the half *vis viva* of the thermal motions of all its parts is diminished, must together be equal to the mechanical effect produced from it; and, consequently, to the mechanical equivalent of the heat which it emits (which will be positive or negative, according as the sum of those terms is positive or negative).   Now let there be either no molecular change or alteration of temperature in any part of the body, or, by a cycle of operations, let the temperature and physical condition be restored exactly to what they were at the beginning; the second and third of the three parts of the work which it has to produce vanish; and we conclude that the heat which it emits or absorbs will be the thermal equivalent of the work done upon it by external forces, or done by it against external forces; which is the proposition to be proved.

The demonstration of the second proposition is founded on the following axiom:

It is impossible, by means of inanimate material agency, to derive mechanical effect from any portion of matter by cooling it below the temperature of the coldest of the surrounding objects.

To demonstrate the second proposition, let A and B be two thermo-dynamic engines, of which B satisfies the conditions expressed in the enunciation; and let, if possible A derive more work from a given quantity of heat than B, when their sources and refrigerators are at the same temperatures, respectively. Then on account of the condition of complete reversibility in all its operations which it fulfills, B may be worked backwards, and made to restore any quantity of heat to its source, by the expenditure of the amount of work which, by its forward action, it would derive from the same quantity of heat.   If, therefore, B be worked backwards, and made to restore to the source of A (which we may suppose to be adjustable to the engine B) as much heat as has been drawn from it during a certain period of the working of A, a smaller amount of work will be spent thus than was gained by the working of A.   Hence, if such a series of operations of A forwards and of B backwards be continued, either alternately or simultaneously, there will result a continued production of work without any continued abstraction of heat from the source; and, by Prop. I., it follows that there must be more heat abstracted from the refrigerator by the working of B backwards than is deposited in it by *A*.   Now it is obvious that *A* might be made to spend part of its work in working *B* backwards, and the whole might be made self-acting.   Also, there being no heat either taken from or given to the source on the whole, all the surrounding bodies and space except the refrigerator might, without interfering with any of the conditions which have been assumed, be made of the same temperature as the source, whatever that may be.   We should thus have a self acting machine, capable of drawing heat constantly from a body surrounded by others of a higher temperature, and converting it into mechanical effect.   But this is contrary to the axiom, and therefore we conclude that the hypothesis that *A* derives more mechanical effect from the same quantity of heat drawn from the source than *B* is false.   Hence no engine whatever, with source and refrigerator at the same temperatures, can get more work from a given quantity of heat introduced than any engine which satisfies the condition of reversibility, which was to be proved.

This proposition was first enunciated by Carnot, being the expression of his criterion of a perfect thermo-dynamic engine. He proved it by demonstrating that a negation of it would require the admission that there might be a self-acting machine constructed which would produce mechanical effect indefinitely, without any source either in heat or the consumption of materials, or any other physical agency; but this demonstration involves, fundamentally, the assumption that, in "a complete cycle of operations," the medium parts with exactly the same quantity of heat as it receives.   A very strong expression of doubt regarding the truth of this assumption, as a universal principle, is given by Carnot himself; and that it is false, where mechanical work is, on the whole, either gained or spent in the operations, may (as I have tried to show above) be considered to be perfectly certain. It must then be admitted that Carnot's original demonstration utterly fails, but we cannot infer that the proposition concluded is false.   The truth of the conclusion appeared to me, indeed so probable that I took it in connection with Joule's principle, on account of which Carnot's demonstration of it fails, as the foundation of an investigation of the motive power of heat in air-engines or steam-engines through finite ranges of temperature, and obtained about a year ago results, of which the substance is given in the second part of the paper at present communicated to the Royal Society.   It was not until the commencement of the present year that I found the demonstration given above, by which the truth of the proposition is established upon an axiom, which I think will be generally admitted.   It is with no wish to claim priority that I make these statements, as the merit of first establishing the proposition upon correct principles is entirely due to Clausius, who published his demonstration of it in the month of May last year, in the second part of his paper on the motive power of heat.   I may be allowed to add that I have given the demonstration exactly as it occurred to me before I knew that Clausius had either enunciated or demonstrated the proposition.   The following is the axiom on which Clausius's demonstration is founded:

It is impossible for a self-acting machine, unaided by any external agency, to convey heat from one body to another at a higher temperature.

It is easily shown that, although this and the axiom I have used are different in form, either is a consequence of the other.   The reasoning in each demonstration is strictly analogous to that which Carnot originally gave.

An Absolute Scale of Temperature

Definition of temperature and *general thermometric assumption.* If two bodies be put in contact, and neither gives heat to the other, their temperatures are said to be the same; but if one gives heat to the other, its temperature is said to be higher.

The temperatures of two bodies are proportional to the quantities of heat respectively taken in and given out in localities at one temperature and at the other, respectively, by a material system subjected to a complete cycle of perfectly reversible thermodynamic operations, and not allowed to part with or take in heat at any other temperature: or, the absolute values of two temperatures are to one another in the proportion of the heat taken in to the heat rejected in a perfect thermo-dynamic engine working with a source and refrigerator at the higher and lower of the temperatures respectively.

Convention for thermometric unit, and determination of absolute temperatures of fixed points in terms of it.

Two fixed points of temperature being chosen according to Sir Isaac Newton's suggestions, by particular effects on a particular substance or substances, the difference of these temperatures is to be called unity, or any number of units or degrees as may be found convenient.   The particular convention is, that the difference of temperatures between the freezing- and boiling-points of water under standard atmospheric pressure shall be called 100 degrees. The determination of the absolute temperatures of the fixed points is then to be effected by means of observations indicating the economy of a perfect thermo-dynamic engine, with the higher and the lower respectively as the temperatures of its source and refrigerator. The kind of observation best adapted for this object was originated by Mr. Joule, whose work in 1844 laid the foundation of the theory, and opened the experimental investigation; and it has been carried out by him, in conjunction with myself, within the last two years, in accordance with the plan proposed in Part IV of the present series. The best result, as regards this determination, which we have yet been able to obtain is, that the temperature of freezing water is 273.7 on the absolute scale; that of the boiling-point being consequently 373.7. Further details regarding the new thermometric system will be found in a joint communication to be made by Mr. Joule and myself to the Royal Society of London before the close of the present session.

D. BERNOULLI

Daniel Bernoulli was born on February 9, 1700 in Groningen, and died in Basel on March 17, 1782. He was of a family distinguished for its mathematical attainments. His father was John Bernoulli. He studied medicine, but his gift for mathematical and physical pursuits led to his being called to St. Petersburg to occupy a place in the Academy. There he remained for several years, but finally returned to his home in Basel, where he lived until his death. Besides many memoirs he wrote the first treatise on hydrodynamics. This work was published in 1738. In it he applies the principle of the conservation of *vis viva* to the solution of hydrodynamical problems.

The extract which follows is taken from the *Hydrodynamica*, tenth section. It contains the first successful application of the kinetic theory of gases.

KINETIC THEORY OF GASES

1. In the consideration of elastic fluids we may assign to them such a constitution as will be consistent with all their known properties, that so we may approach the study of their other properties, which have not yet been sufficiently investigated. The particular properties of elastic fluids are as follows: 1. They are heavy; 2. they expand in all directions unless they are restrained; and 3. they are continually more and more compressed when the force of compression increases. Air is a body of this sort, to which especially the present investigation pertains.

2. Consider a cylindrical vessel *ACDB* (Fig. 44) set vertically, and a movable piston *EF* in it, on which is placed a weight *P*: let the cavity *ECDF* contain very minute corpuscles, which are driven hither and thither with a very rapid motion; so that these corpuscles, when they strike against the piston *EF* and sustain it by their repeated impacts, form an elastic fluid which will expand of itself if the weight *P* is removed or diminished, which will be condensed if the weight is increased, and which gravitates toward the horizontal bottom *CD* just as if it were endowed with no elastic powers: for whether the corpuscles are at rest or are agitated they do not lose their weight, so that the bottom sustains not only the weight but the elasticity of the fluid. Such therefore is the fluid which we shall substitute for air. Its properties agree with those which we have already assumed for elastic fluids, and by them we shall explain other properties which have been found for air and shall point out others which have not yet been sufficiently considered.

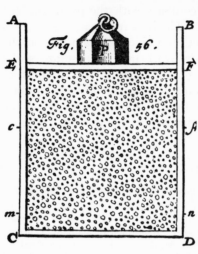

FIG. 44.

3. We consider the corpuscles which are contained in the cylindrical cavity as practically infinite in number, and when they occupy the space *ECDF* we assume that they constitute ordinary air, to which as a standard all our measurements are to be referred: and so the weight *P* holding the piston in the position *EF* does not differ from the pressure of the superincumbent atmosphere, which therefore we shall designate by *P* in what follows.

It should be noticed that this pressure is not exactly equal to the absolute weight of a vertical cylinder of air resting on the piston *EF*, as hitherto most authors have asserted without sufficient consideration; rather it is equal to the fourth proportional to the surface of the earth, to the size of the piston *EF*, and to the weight of all the atmosphere on the surface of the earth.

4. We shall now investigate the weight π, which is sufficient to condense the air *ECDF* into the space *eCDf*, on the assumption

that the velocity of the particles is the same in both conditions of the air, the natural condition as well as the condensed. Let $EC = 1$ and $eC = s$. When the piston EF is moved to ef, it appears that a greater effort is made by the fluid for two reasons: first, because the number of particles is now greater in the ratio of the space in which they are contained, and secondly, because each particle repeats its impacts more often. That we may properly calculate the increment which depends on the first cause we may consider the particles as if they were at rest. We shall set the number of them which are contiguous to the piston in the position $EF = n$; then the like number when the piston is in the position ef will be $= n : \left(\dfrac{eC}{EC} \right)^{\frac{2}{3}}$ or $= n : s^{\frac{2}{3}}$.

It should be noticed that the fluid is no more condensed in the lower part than in the upper part, because the weight P is infinitely greater than the weight of the fluid itself: hence it is plain that for this reason the force of the fluid is in the ratio of the numbers n and $n : s^{\frac{2}{3}}$ that is, as $s^{\frac{2}{3}}$ is to 1. Now in reference to the other increment arising from the second cause, this is found by considering the motion of the particles, and it appears that their impacts are made more often by as much as the particles are closer together: therefore the numbers of the impacts will be reciprocally as the mean distances between the surfaces of the particles, and these mean distances will be thus determined.

We assume that the particles are spheres. We represent by D the mean distance between the centers of the spheres when the piston is in the position EF, and by d the diameter of a sphere. Then the mean distance between the surfaces of the spheres will be $D - d$. But it is evident that when the piston is in the position ef, the mean distance between the centers of the spheres $= D \sqrt[3]{s}$ and therefore the mean distance between the surfaces of the spheres $= D \sqrt[3]{s} - d$. Therefore, with respect to the second cause, the force of the natural air in $ECDF$ will be to the force of the compressed air in $eCDf$ as $\dfrac{1}{D - d}$ to $\dfrac{1}{D \sqrt[3]{s} - d}$, or as $D \sqrt[3]{s} - d$ to $D - d$. When both causes are joined the predicted forces will be as $s^{\frac{2}{3}} \times (D \sqrt[3]{s} - d)$ to $D - d$.

For the ratio of D to d we may substitute one which is easier to understand: for if we think of the piston EF as depressed by an infinite weight, so that it descends to the position mn, in which all the particles are in contact, and if we represent the line mC

by m, we shall have D is to d as 1 is to $\sqrt[3]{m}$. If we substitute this in the ratio above, we shall find that the force of the natural air in $ECDF$ is to the force of the compressed air in $eCDf$ as $s^{2/3} \times (\sqrt[3]{s} - \sqrt[3]{m})$ is to $1 - \sqrt[3]{m}$, or as $s - \sqrt[3]{mss}$ is to $1 - \sqrt[3]{m}$. Therefore $\pi = \dfrac{1 - \sqrt[3]{m}}{s - \sqrt[3]{mss}} \times P$.

5. From all the facts known we may conclude that natural air can be very much condensed and compressed into a practically infinitely small space; so that we may set $m = 0$, and hence $\pi = P/s$; so that the compressing weights are almost in the inverse ratio of the spaces which air occupies when compressed by different amounts. This law has been proved by many experiments. It certainly may be safely adopted for air that is less dense than natural air; whether it holds for considerably denser air I have not sufficiently investigated: nor have there yet been experiments instituted with the accuracy which is necessary in this case. There is special need of an experiment to find the value of m, but this experiment must be most accurately carried out and with air under very high pressure; and the temperature of the air while it is being compressed must be carefully kept constant.

6. The elasticity of air is not only increased by condensation but by heat supplied to it, and since it is admitted that heat may be considered as an increasing internal motion of the particles, it follows that if the elasticity of air of which the volume does not change is increased, this indicates a more intense motion in the particles of air; which fits in well with our hypothesis; for it is plain that so much the greater weight P is needed to keep the air in the condition $ECDF$, as the aerial particles are agitated by the greater velocity. It is not difficult to see that the weight P should be in the duplicate ratio of this velocity because, when the velocity increases, not only the number of impacts but also the intensity of each of them increases equally, and each of them is proportional to the weight P.

Therefore, if the velocity of the particles is called v, the weight which is able to sustain the piston in the position $EF = vvP$ and in the position $ef = \dfrac{1 - \sqrt[3]{m}}{s - \sqrt[3]{mss}} \times vvP$, or approximately $= \dfrac{vvP}{s}$, because as we have seen the number m is very small in comparison with unity or with the number s.

7. This theorem, as I have presented it in the preceding paragraph, in which it is shown that in air of any density but at a fixed

temperature, the elasticities are proportional to the densities, and further that the increments of elasticity which are produced by equal changes of temperature are proportional to the densities, this theorem, I say, D. Amontons discovered by experiment and presented it in the Memoirs of the Royal Academy of Sciences of Paris in 1702.

BROWN

Robert Brown was born on December 21, 1773, in Montrose, and died in London on June 10, 1858. He studied at Marischal College, Aberdeen and at the University of Edinburgh. Between 1801 and 1805 he served as naturalist on an expedition which surveyed the coasts of Australia. On his return he became librarian of the Linnean Society, and in 1810 librarian to Sir Joseph Banks. Sir Joseph so appreciated his abilities and services that by his will he left him his books and collections for life. These were transferred to the British Museum in 1827.

Brown was a distinguished botanist and was interested particularly in the sexual relations of plants. He contributed to physics by his discovery of the Brownian movements, as they are called, which are described in the following extract. The paper containing this discovery was published as a separate pamphlet with the title "A Brief Account of Microscopical Observations made in the Months of June, July, and August, 1827, on the Particles Contained in the Pollen of Plants; and on the General Existence of Active Molecules in Organic and Inorganic Bodies." It appears also in the *Miscellaneous Botanical Works* of Robert Brown, Esq., Vol. 1, p. 465, 1866. A kinetic theory of these motions has been given by Smoluchowski and by Einstein and this theory has been verified by the experimental work of Perrin.

THE BROWNIAN MOVEMENTS

The observations, of which it is my intention to give a summary in the following pages, have all been made with a simple microscope, and indeed with one and the same lens, the focal length of which is about ⅟₃₂nd of an inch.

The author gives the reasons which led him to the examination of pollen and particularly of the particles contained in the grains of pollen. He then proceeds as follows:

My inquiry on this point was commenced in June 1827, and the first plant examined proved in some respects remarkably well adapted to the object in view.

This plant was *Clarckia pulchella,* of which the grains of pollen, taken from antherae full grown, but before bursting, were filled with particles or granules of unusually large size, varying from nearly $\frac{1}{4000}$th to about $\frac{1}{5000}$th of an inch in length, and of a figure between cylindrical and oblong, perhaps slightly flattened, and having rounded and equal extremities. While examining the form of these particles immersed in water, I observed many of them very evidently in motion; their motion consisting not only of a change of place in the fluid, manifested by alterations in their relative positions, but also not unfrequently of a change of form in the particle itself; a contraction or curvature taking place repeatedly about the middle of one side, accompanied by a corresponding swelling or convexity on the opposite side of the particle. In a few instances the particle was seen to turn on its longer axis. These motions were such as to satisfy me, after frequently repeated observation, that they arose neither from currents in the fluid, nor from its gradual evaporation, but belonged to the particle itself.

Grains of pollen of the same plant taken from antherae immediately after bursting, contained similar subcylindrical particles, in reduced numbers, however, and mixed with other particles, at least as numerous, of much smaller size, apparently spherical, and in rapid oscillatory motion.

These smaller particles, or Molecules as I shall term them, when first seen, I considered to be some of the cylindrical particles swimming vertically in the fluid. But frequent and careful examination lessened my confidence in this supposition; and on continuing to observe them until the water had entirely evaporated, both the cylindrical particles and spherical molecules were found on the stage of the miscroscope.

The author describes the discovery of similar movements in the particles obtained from other living plants. He then extended his observations to other substances.

Having found motion in the particles of the pollen of all the living plants which I had examined, I was led next to inquire whether this property continued after the death of the plant, and for what length of time it was retained.

In plants, either dried or immersed in spirit for a few days only, the particles of pollen of both kinds were found in motion equally evident with that observed in the living plant; specimens of several plants, some of which had been dried and preserved in an herbarium for upwards of twenty years, and others not less than a century, still exhibited the molecules or smaller spherical particles in considerable numbers, and in evident motion, along with a few of the larger particles, whose motions were much less manifest, and in some cases not observable.

In this stage of the investigation having found, as I believed, a peculiar character in the motions of the particles of pollen in water, it occurred to me to appeal to this peculiarity as a test in certain families of Cryptogamous plants, namely, Mosses, and the genus Equisetum, in which the existence of sexual organs had not been universally admitted.

In the supposed stamina of both these families, namely, in the cylindrical antherae or pollen of Mosses, and on the surface of the four spathulate bodies surrounding the naked ovulum, as it may be considered, of Equisetum, I found minute spherical particles, apparently of the same size with the molecule described in Onagrariae, and having equally vivid motion on immersion in water; and this motion was still observable in specimens both of Mosses and of Equiseta, which had been dried upwards of one hundred years.

The very unexpected fact of seeming vitality retained by these minute particles so long after the death of the plant would not perhaps have materially lessened my confidence in the supposed peculiarity. But I at the same time observed, that on bruising the ovula or seeds of Equisetum, which at first happened accidentally, I so greatly increased the number of moving particles, that the source of the added quantity could not be doubted. I found also that on bruising first the floral leaves of Mosses, and then all other parts of those plants, that I readily obtained similar particles, not in equal quantity indeed, but equally in motion. My supposed test of the male organ was therefore necessarily abandoned.

Reflecting on all the facts with which I had now become acquainted, I was disposed to believe that the minute spherical particles or Molecules of apparently uniform size, first seen in the advanced state of the pollen of Onagrariae, and most other Phaenogamous plants,—then in the antherae of Mosses and on the surface of the bodies regarded as the stamina of Equisetum,— and lastly in bruised portions of other parts of the same plants, were in reality the supposed constituent or elementary Molecules of organic bodies, first so considered by Buffon and Needham, then by Wrisberg with greater precision, soon after and still more particularly by Müller, and, very recently, by Dr. Milne Edwards, who has revived the doctrine and supported it with much interesting detail. I now therefore expected to find these molecules in all organic bodies: and accordingly on examining the various animal and vegetable tissues, whether living or dead, they were always found to exist; and merely by bruising these substances in water, I never failed to disengage the molecules in sufficient numbers to ascertain their apparent identity in size, form, and motion, with the smaller particles of the grains of pollen.

I examined also various products of organic bodies, particularly the gum resins, and substances of vegetable origin, extending my inquiry even to pit-coal; and in all these bodies Molecules were found in abundance. I remark here also, partly as a caution to those who may hereafter engage in the same inquiry, that the dust or soot deposited on all bodies in such quantity, especially in London, is entirely composed of these molecules.

One of the substances examined, was a specimen of fossil wood, found in Wiltshire oolite, in a state to burn with flame; and as I found these molecules abundantly, and in motion in this specimen, I supposed that their existence, though in smaller quantity, might be ascertained in mineralized vegetable remains. With this view a minute portion of silicified wood, which exhibited the structure of Coniferae, was bruised, and spherical particles, or molecules in all respects like those so frequently mentioned, were readily obtained from it; in such quantity, however, that the whole substance of the petrifaction seemed to be formed of them. But hence I inferred that these molecules were not limited to organic bodies, nor even to their products.

To establish the correctness of the inference, and to ascertain to what extent the molecules existed in mineral bodies, became the next object of inquiry. The first substance examined was a minute fragment of window-glass, from which, when merely bruised on the stage of the microscope, I readily and copiously obtained molecules agreeing in size, form, and motion with those which I had already seen.

I then proceeded to examine, and with similar results, such minerals as I either had at hand or could readily obtain, including several of the simple earths and metals, with many of their combinations.

Rocks of all ages, including those in which organic remains have never been found, yielded the molecules in abundance. Their existence was ascertained in each of the constituent minerals of granite, a fragment of the Sphinx being one of the specimens examined.

To mention all the mineral substances in which I have found these molecules, would be tedious; and I shall confine myself in this summary to an enumeration of a few of the most remarkable. These were both of aqueous and igneous origin, as travertine, stalactites, lava, obsidian, pumice, volcanic ashes, and meteorites from various localities. Of metals I may mention manganese, nickel, plumbago, bismuth, antimony, and arsenic. In a word, in every mineral which I could reduce to a powder, sufficiently fine to be temporarily suspended in water, I found these molecules more or less copiously; and in some cases, more particularly in siliceous crystals, the whole body submitted to examination appeared to be composed of them.

. . . .

JOULE

The paper which follows, under the title "Some Remarks on Heat and the Constitution of Elastic Fluids," was published by Joule (p. 203) in the *Philosophical Magazine*, Series 4, Vol. 14, p. 211, 1857. It was originally published in the *Memoirs of the Manchester Literary and Philosophical Society*, November, 1851. In this paper Joule gives the first published calculation of the velocity of gaseous molecules.

THE VELOCITY OF GASEOUS MOLECULES

I have myself endeavoured to prove that a rotary motion, such as that described by Sir H. Davy, can account for the law of Boyle and Mariotte, and other phenomena presented by elastic fluids; nevertheless, since the hypothesis of Herapath—in which it is assumed that the particles of a gas are constantly flying about in every direction with great velocity, the pressure of the gas being owing to the impact of the particles against any surface presented to them—is somewhat simpler, I shall employ it in the following remarks on the constitution of elastic fluids, premising, however, that the hypothesis of a rotary motion accords equally well with the phenomena.

Let us suppose an envelope of the size and shape of a cubic foot to be filled with hydrogen gas, which, at 60° temperature and 30 inches barometrical pressure, will weigh 36.927 grs.   Further, let us suppose the above quantity to be divided into three equal and indefinitely small elastic particles, each weighing 12.309 grs.; and, further, that each of these particles vibrates between opposite sides of the cube, and maintains a uniform velocity except at the instant of impact; it is required to find the velocity at which each particle must move so as to produce the atmospherical pressure of 14,831,712 grs. on each of the sides of the cube.   In the first place, it is known that if a body moving with the velocity of 32⅙ feet per second be opposed, during one second, by a pressure equal to its weight its motion will be stopped, and that, if the pressure be continued one second longer, the particle will acquire the velocity of 32⅙ feet per second in the contrary direction.   At this velocity there will be 32⅙ collisions of a particle of 12.309 grs. against each side of the cubical vessel in every two seconds of time; and the pressure occasioned thereby will be 12.309 × 32⅙ = 395.938 grs.   Therefore, since it is manifest that the pressure will be proportional to the square of the velocity of the particles, we shall have for the velocity of the particles requisite to produce the pressure of 14,831,712 grs. on each side of the cubical vessel,

$$v = \sqrt{\left(\frac{14,831,712}{395.938}\right)} 32\tfrac{1}{6} = 6225 \text{ feet per second}$$

The above velocity will be found equal to produce the atmospheric pressure, whether the particles strike each other before they arrive at the sides of the cubical vessel, whether they strike the sides obliquely, and, thirdly, into whatever number of particles the 36.927 grs. of hydrogen are divided.

If only one half the weight of hydrogen, or 18.4635 grs., be enclosed in the cubical vessel, and the velocity of the particles be, as before, 6225 feet per second, the pressure will manifestly be only one half of what it was previously; which shows that the law of Boyle and Mariotte flows naturally from the hypothesis.

The velocity above named is that of hydrogen at the temperature of 60°; but we know that the pressure of an elastic fluid at 60° is to that at 32° as 519 is to 491.   Therefore the velocity of the particles at 60° will be to that at 32° as $\sqrt{519} : \sqrt{491}$; which shows that the velocity at the freezing temperature of water is 6055 feet per second.

In the above calculations it is supposed that the particles of hydrogen have no sensible magnitude, otherwise the velocity corresponding to the same pressure would be lessened.

Since the pressure of a gas increases with its temperature in arithmetical progression, and since the pressure is proportional to the squares of the velocity of the particles, in other words to their *vis viva*, it follows that the absolute temperature, pressure, and *vis viva* are proportional to one another, and that the zero of temperature is 491° below the freezing-point of water. Further, the absolute heat of the gas, or, in other words, its capacity, will be represented by the whole amount of *vis viva* at a given temperature. The specific heat may therefore be determined in the following simple manner:—

The velocity of the particles of hydrogen, at the temperature of 60°, has been stated to be 6225 feet per second, a velocity equivalent to a fall from the perpendicular height of 602,342 feet. The velocity at 61° will be $6225\sqrt{520/519} = 6230.93$ feet per second, which is equivalent to a fall of 603,502 feet. The difference between the above falls is 1160 feet, which is therefore the space through which 1 lb. of pressure must operate upon each lb. of hydrogen, in order to elevate its temperature one degree. But our mechanical equivalent of heat shows that 770 feet is the altitude representing the force required to raise the temperature of water one degree; consequently the specific heat of hydrogen will be $1160/778 = 1.506$, calling that of water unity.

The specific heats of other gases will be easily deduced from that of hydrogen; for the whole *vis viva* and capacity of equal bulks of the various gases will be equal to one another; and the velocity of the particles will be inversely as the square root of the specific gravity. Hence the specific heat will be inversely proportional to the specific gravity, a law which has been arrived at experimentally by De la Rive and Marcet.

MAXWELL

James Clerk Maxwell was born in Edinburgh on November 13, 1831. His father was the possessor of an estate and a man of great intellectual power. Maxwell attended the Edinburgh Academy and the University of Edinburgh. While there he already engaged in mathematical and physical investigations, some of which were published. In 1850 he went to Cambridge, entering first at Peterhouse, but soon transferring to Trinity. He was graduated in 1854 as Second Wrangler. He shared the Smith's prize with Routh. In 1856 he became professor of natural philosophy at Marischal College, Aberdeen, where he remained four years. In the eight years following he was professor of physics and astronomy at Kings College, London. He then retired to his estate at Glenlair in Kirkcudbrightshire, where he remained until he was called in 1871 to take the chair of experimental physics at Cambridge and to install the Cavendish Laboratory. He died in Cambridge on November 5, 1879.

In 1859 Maxwell was given the Adams prize for an essay on Saturn's rings, in which he proved that on mechanical principles these rings could only consist of small separate bodies revolving like satellites around the planet. Not long afterward he began the study of the Kinetic Theory of Gases, in which he applied the theory of probabilities to the discovery of the law of distribution of molecular velocities and to the explanation of many of the properties of gases. He also for many years studied color vision by skillfully devised experiments. His greatest work, however, was in the field of electricity. He early adopted Faraday's view that electric and magnetic actions are transmitted through a medium and he set himself the task of devising the mode of action of this medium which would account for all electric and magnetic phenomena. The paper in which his theory was first extensively presented was published in 1867, and this was followed in 1873 by his *Treatise on Electricity and Magnetism*. In this great work he did for the ether what Newton did for matter in the *Principia*, and though the modern theories of relativity and the quantum may compel us to modify the views which we have obtained from Newton and from Maxwell yet the influence of their work will always abide.

The extract which follows is from Maxwell's paper entitled "Illustrations of the Dynamical Theory of Gases." It was published in the *Philosophical Magazine*, Series 4, Vol. 19, p. 19, 1860. It contains Maxwell's first proof of the distribution law. This proof was felt to be imperfect, and more elaborate discussions, leading however to the same general results, have been given by Boltzmann, by Maxwell himself and by Jeans.

THE DISTRIBUTION OF MOLECULAR VELOCITIES

So many of the properties of matter, especially when in the gaseous form, can be deduced from the hypothesis that their minute parts are in rapid motion, the velocity increasing with the temperature, that the precise nature of this motion becomes a subject of rational curiosity. Daniel Bernoulli, Herapath, Joule, Krönig, Clausius, &c. have shown that the relations between pressure, temperature, and density in a perfect gas can be explained by supposing the particles to move with uniform velocity in straight lines, striking against the sides of the containing vessel and thus producing pressure. It is not necessary to suppose each particle to travel to any great distance in the same straight line;

for the effect in producing pressure will be the same if the particles
strike against each other; so that the straight line described may
be very short. M. Clausius has determined the mean length
of path in terms of the average distance of the particles, and the
distance between the centres of two particles when collision takes
place. We have at present no means of ascertaining either of
these distances; but certain phaenomena, such as the internal
friction of gases, the conduction of heat through a gas, and the
diffusion of one gas through another, seem to indicate the possibility of determining accurately the mean length of path which a
particle describes between two successive collisions. In order to
lay the foundation of such investigations on strict mechanical
principles, I shall demonstrate the laws of motion of an indefinite
number of small, hard, and perfectly elastic spheres acting on one
another only during impact.

If the properties of such a system of bodies are found to correspond to those of gases, an important physical analogy will be
established, which may lead to more accurate knowledge of the
properties of matter. If experiments on gases are inconsistent
with the hypothesis of these propositions, then our theory, though
consistent with itself, is proved to be incapable of explaining the
phaenomena of gases. In either case it is necessary to follow
out the consequences of the hypothesis.

Instead of saying that the particles are hard, spherical, and
elastic, we may if we please say that the particles are centres of
force, of which the action is insensible except at a certain small
distance, when it suddenly appears as a repulsive force of very
great intensity. It is evident that either assumption will lead
to the same results. For the sake of avoiding the repetition of a
long phrase about these repulsive forces, I shall proceed upon the
assumption of perfectly elastic spherical bodies. If we suppose
those aggregate molecules which move together to have a bounding
surface which is not spherical, then the rotatory motion of the
system will store up a certain proportion of the whole *vis viva*, as
has been shown by Clausius, and in this way we may account
for the value of the specific heat being greater than on the more
simple hypothesis.

. . . .

Prop. IV. To find the average number of particles whose
velocities lie between given limits, after a great number of collisions
among a great number of equal particles.

Let N be the whole number of particles. Let x, y, z, be the components of the velocity of each particle in three rectangular directions, and let the number of particles for which x lies between x and $x + dx$ be $Nf(x)dx$, where $f(x)$ is a function of x to be determined.

The number of particles for which y lies between y and $y + dy$ will be $Nf(y)dy$; and the number for which z lies between z and $z + dz$ will be $Nf(z)dz$, where f always stands for the same function.

Now the existence of the velocity x does not in any way affect that of the velocities y or z, since these are all at right angles to each other and independent, so that the number of particles whose velocity lies between x and $x + dx$, and also between y and $y + dy$, and also between z and $z + dz$, is

$$Nf(x)f(y)f(z)dx \ dy \ dz.$$

If we suppose the N particles to start from the origin at the same instant, then this will be the number in the element of volume $(dx \ dy \ dz)$ after unit of time, and the number referred to unit of volume will be

$$Nf(x)f(y)f(z).$$

But the directions of the coordinates are perfectly arbitrary, and therefore this number must depend on the distance from the origin alone, that is

$$f(x)f(y)f(z) \ = \ \phi(x^2 + y^2 + z^2).$$

Solving this functional equation, we find

$$f(x) \ = \ Ce^{Ax^2}, \ \phi(r^2) \ = \ C^3 e^{Ar^2}$$

If we make A positive, the number of particles will increase with the velocity, and we should find the whole number of particles infinite. We therefore make A negative and equal to $-1/\alpha^2$, so that the number between x and $x + dx$ is

$$NCe^{-\frac{x^2}{\alpha^2}} \ dx.$$

Integrating from $x = -\infty$ to $x = +\infty$, we find the whole number of particles,

$$NC\sqrt{\pi}\alpha \ = \ N, \qquad \therefore C = \frac{1}{\alpha\sqrt{\pi}},$$

$f(x)$ is therefore

$$\frac{1}{\alpha\sqrt{\pi}}e^{-\frac{x^2}{\alpha^2}}.$$

Whence we may draw the following conclusions:—

1st. The number of particles whose velocity, resolved in a certain direction, lies between x and $x + dx$ is

$$N\frac{1}{\alpha\sqrt{\pi}}e^{-\frac{x^2}{\alpha^2}}dx \quad . \quad . \quad . \quad . \quad . \quad . \quad . \quad (1)$$

2nd. The number whose actual velocity lies between v and $v + dv$ is

$$N\frac{4}{\alpha^3\sqrt{\pi}}v^2e^{-\frac{v^2}{\alpha^2}}dv \quad . \quad . \quad . \quad . \quad . \quad . \quad (2)$$

3rd. To find the mean value of v, add the velocities of all the particles together and divide by the number of particles; the result is

$$\text{Mean velocity} = \frac{2\alpha}{\sqrt{\pi}} \quad . \quad . \quad . \quad . \quad . \quad . \quad (3)$$

4th. To find the mean value of v^2, add all the values together and divide by N,

$$\text{mean value of } v^2 = \tfrac{3}{2}\alpha^2 \quad . \quad . \quad . \quad . \quad . \quad . \quad (4)$$

This is greater than the square of the mean velocity, as it ought to be.

It appears from this proposition that the velocities are distributed among the particles according to the same law as the errors are distributed among the observations in the theory of the "method of least squares." The velocities range from 0 to ∞, but the number of those having great velocities is comparatively small. In addition to these velocities, which are in all directions equally, there may be a general motion of translation of the entire system of particles which must be compounded with the motion of the particles relatively to one another. We may call the one the motion of translation, and the other the motion of agitation.

BOLTZMANN

Ludwig Boltzmann was born on February 20, 1844, in Vienna. He studied in that city and was graduated from the university in 1866. From 1876 to 1890 he was professor of experimental and theoretical physics in Graz. He then passed successively to the universities of Munich, of Vienna, of Leipzig, and again of Vienna. He died on September 5, 1906.

Boltzmann was distinguished particularly for his researches in the kinetic theory of gases, which he developed contemporaneously with Maxwell. The short extract which follows, taken from the *Sitzungsberichte der Akademie der Wissenschaften in Wien: Mathematisch-naturwissenschaftliche Classe*, Vol. 76 II, p. 373, 1877, is a portion of his paper, "Ueber die Beziehung zwischen dem zweiten Hauptsatze der mechanischen Wärmetheorie und der Wahrscheinlichkeitsrechnung, respective den Sätzen über das Wärmegleichgewicht." It contains Boltzmann's statement of the general relation between the entropy and probability. His subsequent development of the idea, which led him to a formula of fundamental importance connecting entropy and probability, is too extensive to admit of quotation.

ENTROPY AND PROBABILITY

A relation between the second law of thermodynamics and the theory of probabilities was first shown when I proved that an analytical proof of that law can be erected only on a foundation which is taken from the theory of probabilities. This relation is further confirmed by the proof that an exact proof of the laws of equilibrium of heat is most easily obtained by showing that a certain quantity, which I will again designate by E, can only diminish by the exchange of kinetic energy between the molecules of the gas, and therefore will have a minimum value in the condition of equilibrium of heat. The connection between the second law and the laws of equilibrium of heat becomes clearer still by the developments in section II of my "Remarks on Some Problems of the Mechanical Theory of Heat." In that place I also first suggested the possibility of a special way of calculating the equilibrium of heat, in the following words: "It is clear that any individual uniform state which occurs after the lapse of a definite time from a definite initial state is just as improbable as any particular nonuniform state, just as in the game of lotto each individual quintern is just as improbable as the quintern 12345. It is therefore only because there are many more uniform states than nonuniform states that the probability is greater that the state becomes uniform in the progress of time"; further: "We might even calculate from the relation of the number of the different states their probability, which perhaps would lead to an interesting method of calculating the equilibrium of heat." The belief is therefore expressed that we can calculate the state of equilibrium by investigating the probability of the different possible states of the system. The initial state will in most cases be a very improbable one and from it the system will progress toward more probable states, until it at last reaches the most probable state, that is, that of equilibrium of heat. If we apply this to the second law we can identify that quantity which we commonly designate as entropy with the probability of the actual state. We think of a system of bodies which is isolated and makes no exchanges with other bodies, for example, a body of higher temperature and one of lower temperature and a so called intermediate body, which permits a transfer of heat between the two; or, to choose another example, a vessel with absolutely smooth and rigid walls, one half of which is filled with air at a lower temperature or pressure while the other half is filled with air at a higher temperature or pressure. The system of bodies which we have thought of may have at the beginning of the time any state; by exchange between the bodies this state changes; in accordance with the second law this change must always occur in such a way that the total entropy of all the bodies increases; according to our present interpretation this means nothing else than that the probability of the totality of the states of all these bodies becomes greater and greater; the system of bodies goes from a more improbable to a more probable state. What is meant by this will appear more clearly later.

·   ·   ·   ·   ·   ·   ,   ·   ·   ·   ·   ·   ·   ·   ·   ·   ·   ·

LIGHT

DESCARTES

The following extract is from Descartes' (p. 50) work, *La Dioptrique*, which was published in 1637 as a supplement to the *Discours de la Methode*, to illustrate the method of philosophizing proposed in that book. It is interesting as an illustration of Descartes' method of thinking and because it brings out the fact that in the corpuscular theory of light the velocity of the corpuscles must necessarily be greater in the more highly refracting medium.

Refraction of Light

Since we shall need hereafter to know exactly the quantity of this refraction, and that it may be conveniently understood by the comparison which I am going to use, I believe that it is proper

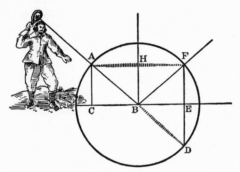

Fig. 45.

that I attempt here a complete explanation and that I speak first therefore of reflection, so as to make it easier to understand the explanation of refraction. Let us suppose therefore that a ball driven from A to B (Fig. 45) encounters at the point B the surface of the earth C B E; which prevents its going on and is the reason that it changes its direction. Let us see toward what side. But first, in order that we shall not embarrass ourselves with new difficulties, let us suppose that the earth is perfectly flat and hard, and that the ball proceeds always with constant velocity, both as it descends and as it rises again, not inquiring in any way about the

265

force which maintains its motion after it is no longer in contact with the racquet, and not considering any effect of its weight or its size or its shape.   For there is no question here of looking so closely into the matter, and none of these things come into the action of the light, to which this motion is to be compared.   It only needs to be noticed that the force, whatever it may be, which keeps up the motion of the ball, is different from that which makes it move in one direction rather than in another, as it is very easy to see from this, that it is the force by which it has been driven by the racquet on which its motion depends, and that this same force would have been able to make it move in any other direction as easily as toward *B*, while in fact it is the position of the racquet which makes it move toward *B*, and which would have been able to make it move in the same way, even if another force had moved it.   This shows already that it is not impossible that the ball may be turned in its path by its encounter with the earth, and that the tendency which it had to go to *B* may be changed without anything being changed in the force of its motion, since they are two different things; and consequently that we should not imagine that it is necessary that it should stop for a moment at the point *B*, before turning toward *F*, as several of our philosophers would have it do; for if its motion were once interrupted by this check, there would be no cause which would thereafter make it start off again.   Furthermore, it must be remarked that the tendency to move itself in any direction, just as well as the motion itself, and generally as any other sort of quantity, may be divided into all the parts of which we may imagine that it is compounded, and that we may easily imagine that this motion of the ball which moves it from *A* to *B* is compounded of two others, one of which would make it descend from the line *AF* toward the line *CE*, and the other would, at the same time, make it go from the left-hand line *AC* toward the right-hand line *FE*, so that these two motions together carry it to *B* along the line *AB*.   And further it is easy to understand that the encounter with the earth can only prevent one of these two motions, and cannot affect the other in any way.   For it certainly ought to prevent that motion which would make the ball descend from *AF* toward *CE*, because it occupies all the space which is below *CE*; but why should it prevent the other motion, which would make it advance toward the right, seeing that it is not opposed in any way in that sense?   Therefore, to find out correctly in what direction the ball ought to rebound, we describe a circle with the center *B*, which passes through the point *A*, and we say that, in the same time that it would take to move from *A* to *B*, it infallibly should return from *B* to some point of the circumference of this circle, in as much as all the points which are as distant from the center *B* as *A* is, are found in this circumference, and we suppose the motion of the ball to be always equally swift. Then finally, to find out precisely to which one of all the points of this circumference it ought to return, we draw three straight lines *AC*, *HB*, and *FE*, perpendicular to *CE*, and in such a way that there is neither more nor less distance between *AC* and *HB* than between *HB* and *FE*: and we say that in the same time that the ball has advanced toward the right from *A*, one of the points of the line *AC*, to *B*, one of the points of the line *HB*, it should also move from the line *HB* as far as some point of the line *FE*; for all the points of this line *FE* are as far away from the line *HB* in this sense on the one side as those of the line *AC* are on the other, and it is also as ready to move in this direction as it was before. Now it cannot in the same time reach some point in the line *FE*, and also some point in the circumference of the circle *AFD*, except at the point *D*, or at the point *F*, since there are only these two points where these lines cut each other; so that since the earth prevents its passing toward *D*, we must conclude that it must go infallibly toward *F*, and so you see easily how reflection occurs, that is, with an angle equal to that which we call the angle of incidence. Thus if a ray coming from the point *A* falls at the point *B* on the surface of a plane mirror *CBE*, it is reflected toward *F* in such a way that the angle of reflection *FBE* is neither greater nor less than the angle of incidence *ABC*.

We now come to refraction. And, first, we suppose that a ball, driven from *A* to *B*, (Fig. 46) encounters at the point *B*, not now the surface of the earth, but a cloth *CBE*, which is so weak and so thin that the ball can break it and pass entirely through it, losing only a part of its velocity, for example, half of it. Now, this being supposed, in order to determine what path it should follow, we notice first that its motion differs entirely from its tendency to move in one direction rather than in another; from which it follows that the quantities of these motions should be considered separately. And we notice also that of the two parts of which we may imagine that this tendency is compounded, only that part which would make the ball move from above downward can be changed in any way by encountering the cloth; and that the tendency which made it move toward the right should always remain the same as it has been, because the cloth is in no way opposed to it in that sense. Then, having described from the center B the circle AFD and drawn at right angles to CBE the three straight lines AC, HB, FE, in such a way that there is twice as much distance between FE and HB as between HB and AC, we see that the ball ought to move toward the point I. For since it loses half of its velocity when it passes through the cloth CDE, it ought to take twice as long to move downward from B to some point of the circumference of the circle AFD as it has taken above it to pass from A to B. And, since it loses none of the tendency which it had previously to move toward the right, in twice the

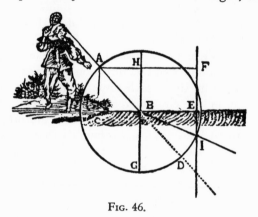

Fig. 46.

time that it has taken to pass from the line AC to HB it ought to travel twice as far in this same direction, and consequently should reach some point of the straight line FE at the same instant that it reaches also some point of the circumference of the circle AFD. This would be impossible if it did not go to I, since that is the only point below the cloth CBE where the circle AFD and the straight line FE cut each other.

Let us now think of the ball which moves from A toward D, as encountering at the point B, no longer a cloth, but water, of which the surface CBE deprives it of half of its velocity just as the cloth did. And supposing everything else to be the same as before, I say that the ball ought to pass from B in a straight line, not toward D but toward I. For, first, it is certain that the surface of the water ought to turn it toward that point in the same way that the cloth did, seeing that it deprives it of just as much of its force and is opposed to it in the same sense. Considering the body of water which fills all the space between *B* and *I*, while it may resist its motion more or less than the air did which we supposed before, we cannot say nevertheless that it ought to change its path; for it may open up to give it passage as easily in one direction as in another. At least that is so, if we suppose always, as we have done, that neither the weight nor the lightness of the ball, nor its size, nor its shape, nor any other such cause changes its course. And we may here remark that it is so much more changed in direction by the surface of the water or by the cloth as it encounters it more obliquely; so that if it encounters the surface at right-angles, as when it is driven from *H* toward *B*, (Fig. 47) it ought to go on in a straight line toward *G* without turning out

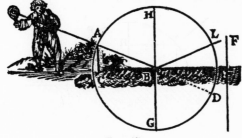

Fig. 47.

of it all. But if it is driven along a line such as *AB*, which is so much inclined to the surface of the water or to the cloth *CBE*, that when the line *FE* is drawn as before it does not cut the circle *AD*, then the ball does not penetrate the surface at all but rebounds from the surface *B* toward the air *L*, just as if it had encountered the earth. This effect has sometimes produced the regrettable result that when cannon have been shot for fun toward the surface of a river, men have been wounded who were on the bank on the other side.

But now let us make here another supposition, and assume that the ball which has been first driven from *A* to *B* is driven just when it is at the point *B* by the racquet *CBE* which increases the force of its motion, for example, by a third, so that it can afterwards move over as great a distance in two moments as it did in three before. This will have the same effect as if the ball had encountered at the point *B* a body of such a nature that it passes through the surface *CBE* a third more easily than through air. It follows manifestly from that which has already been demonstrated, that if we describe the circle AD as before (Fig. 48) and the lines AC, HB, FE, in such a way that there is a third less distance between FE and HB than between HB and AC, the point I, in which the straight line FE and the circular line AD cut each other, will determine the place toward which the ball which is at the point B should turn.

Now we may also take the reverse of this conclusion, and say that, since the ball which comes from A moves in a straight line as far as B, and at the point B turns and proceeds to the point I, this means that the force of facility with which it enters the body $CBEI$ is to that with which it leaves the body $ACBE$, as the distance between AC and HB is to that which is between HB and FI, that is to say, as the line CB is to BE.

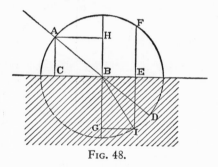

Fig. 48.

To conclude, inasmuch as the action of light follows in this respect the same laws as the motion of the ball, we must say that when its rays pass obliquely from one transparent body into another, which receives them more or less easily than the first body, they turn in such a way that they are always less inclined to the surface separating these bodies on the side where that body is which receives them more easily than on the side where the other body is, and this just in proportion to that which receives them more easily than the other does. Only we must take notice that this inclination should be measured by the magnitudes of the straight lines, like CB or AH, and EB or IG, and others like them, compared one to the other; not by the magnitudes of the angles, such as ABH or GBI; and much less by the magnitudes of the angles, such as DBI, which are called angles of refraction. For the ratio or proportion between these angles changes for all the different inclinations of the rays, while that which holds between the lines *AH* and *IG*, or the like, remains the same for all the refractions which are caused by the same bodies. So, for example, (Fig. 49) if a ray passes in air from *A* to *B* and encounters at the point *B* the surface of glass *CBR*, so that it turns toward *I* in the glass; and if another one comes from *K* to *B* which turns toward *L*; and another from *P* toward *R* which turns toward *S*, there ought to be the same proportion between the lines *KM* and *LN*, or *PQ* and *ST*, as between *AH* and *IG*, but not the same proportion between the angles *KBM* and *LBN*, or *PRQ* and *SRT*, as between *ABH* and *IBG*.

Now that you see in what way refraction should be measured and further that to determine the quantity of refraction, in so

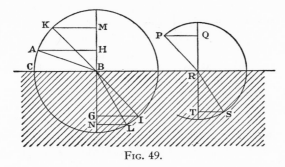

F<small>IG</small>. 49.

far as it depends on the particular nature of the bodies in which it occurs, there is need of proceeding by experiment, there is found to be no difficulty in doing this with sufficient certainty and facility, since all refractions are thus reduced to the same measure; for it is only necessary to determine them for a single ray to determine all those which occur at the same surface, and we can avoid all error if we examine in addition some others. Thus if we wish to know the measure of the refractions which occur in the surface *CBR*, which separates air *AKP* from glass *LIS*, we have only to test the refraction of the ray *ABI*, by finding the ratio between the lines *AH* and *IG*. Then if we fear we have made some mistake in this experiment, we can test our result by using other rays, such as *KBL* or *PRS*, and when we find the same ratio of *KL* to *LM* and of *PQ* to *ST* as that of *AH* to *IG*, we shall have no further reason to question the accuracy of our experiment.

But perhaps you will be astonished when you make these experiments to find that the rays of light are more inclined in air than in water to the surface where the refractions occur; and still more in water than in glass, exactly the opposite from the course of a ball, which is more inclined to the surface in water than in air and can not enter glass at all. For example, if a ball, which is driven in air from *A* to *B* (Fig. 50), encounters the surface of water *CBE* at the point *B*, it will be deflected from *B* toward *V*; and if it is a ray of light it will go on the contrary from *B* toward *I*. You will cease, however, to find this a strange effect, if you recall the nature that I have attributed to light, when I said that it is nothing other than a certain motion or an action conceived in a

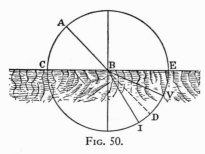

FIG. 50.

very subtle matter, which fills the pores of all other bodies; and when you consider that as a ball loses more of its motion when it strikes against a soft body than against a hard one, and that it rolls less easily on a table-cloth than on a bare table; so the action of this subtle matter may be much more restrained by the parts of the air which, being as they are soft and loosely joined together, do not offer much resistance to it than by the parts of the water, which offer more resistance, and still more by the parts of the water than by those of glass or crystal. Thus it happens that so much as the small parts of a transparent body are harder and firmer so much the more do they allow the light to pass more easily; for the light should not drive any of them out of their places, as a ball ought to drive out the parts of the water to find passage among them.

Further, as we now know the cause of the refractions which occur in water and in glass, and generally in all other transparent bodies which exist about us, we may remark that they should be in all respects similar when the rays come out from the bodies and when they enter them. Thus, if the ray which passes from *A* toward *B* is bent at *B* toward *I* in passing from air into glass, the ray which will come back from *I* toward *B* should also bend at *B* toward *A*. It may possibly be that other bodies may be found, principally in the skies, where refractions proceed from other causes and are not thus reciprocal. And there may also be other certain cases in which the rays ought to bend even though they pass only through a single transparent body; just as the motion of a ball is often a curved motion because it is turned in one direction by its weight, and in another by the action which has set it going; or for divers other reasons. In fact I dare to say that the three comparisons which I have just employed are so suitable that all the particularities which can be noticed are comparable with others which are found just like them in light; but I have not tried to explain those which are not of the most importance to my subject. And I shall not ask you to consider anything further except this, that the surfaces of transparent bodies which are curved bend the rays which pass at each of their points in the same way that plane surfaces would that we may imagine touching these bodies at the same points. Thus for example, the refraction of the rays *AB, AC, AD,* (Fig. 51) which come from the flame *A*

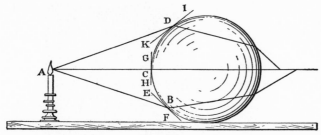

Fig. 51.

and fall on the curved surface of the crystal ball *BCD* ought to be treated as if *AB* fell on the plane surface *EBF* and *AC* on *GCH* and *AD* on *IDK*, and so for the others. Thus you may see that these rays may be brought together or may separate in different ways, according as they fall on surfaces which are differently curved. It is time that I begin the description of the structure of the eye, that I may make you understand how the rays which enter within it conduct themselves so as to cause the sensation of sight.

The Rainbow

The following extract is from Descartes' treatise, *Les Météores*, which was also published in 1637 as a supplement to the *Discours de la Metbode*. In it Descartes explains the position of the primary and secondary rainbows. He was not able to account for the colors of the bows in any satisfactory way. This was first done by Newton.

The rainbow is such a remarkable natural wonder and its cause has been so zealously sought by able men and is so little understood, that I thought that there was nothing I could choose which is better suited to show how, by the method which I employ, we can arrive at knowledge which those whose writings we possess have not had. In the first place, considering that this bow appears not only in the sky, but also in the air near us, wherever there are drops of water illuminated by the sun, as we can see in certain fountains, I readily decided that it arose only from the way in which the rays of light act on these drops and pass from them to our eyes. Further, knowing that the drops are round, as has been formerly proved, and seeing that whether they are larger or smaller, the appearance of the bow is not changed in any way, I had the idea of making a very large one, so that I could examine it better. For this purpose I filled with water a large glass phial, perfectly spherical in shape and very transparent, and then found that if the sunlight came, for example, from the part of the sky which is

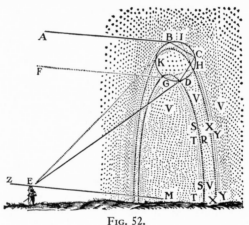

FIG. 52.

marked *AFZ* (Fig. 52), and my eye was at the point *E*, when I put the globe in the position *BCD*, its part *D* appeared all red, and much more brilliant than the rest of it; and that whether I approached it, or receded from it, or put it on my right or my left, or even turned it round about my head, provided that the line *DE* always made an angle of about forty-two degrees with the line *EM*, which we are to think of as drawn from the center of the sun to the eye, the part *D* appeared always similarly red; but that as soon as I made this angle *DEM* even a little larger, the red color disappeared; and if I made the angle a little smaller, the color did not disappear so all at once, but divided itself first as if into two parts, less brilliant, and in which I could see yellow, blue, and other colors. Next, when I looked at that part of the globe which is marked *K*, I saw that, if I made the angle *KEM* about fifty-two degrees, the part K appeared also of a red color, but not so bright as that at D; and if I made the angle a little larger, there appeared other less brilliant colors, but if I made it even a little smaller, or much larger, no colors at all appeared. From which I clearly perceived that if all the air which is near M is filled with such globes, or instead of them with drops of water, there ought to appear a bright red point in every one of the drops so placed that the lines drawn from them to the eye at E make an angle of about forty-two degrees with the line EM, as I suppose those do which are marked R; and that if we look at all these points together, without any consideration of their exact position except of the angle at which they are viewed, they should appear as a continuous circle of a red color; and that something similar ought to appear at the points marked S and T, the lines drawn from which to E make more acute angles with EM, where there will be circles of less brilliant colors. This constitutes the first and principal rainbow. And further if the angle MEK is fifty-two degrees, there should appear a red circle in the drops marked X; and other circles of less brilliant colors in the drops marked Y. This constitutes the second and less important rainbow. And finally in all the other drops marked V no colors at all should appear. When I examined more particularly, in the globe BCD, what it was which made the part D appear red, I found that it was the rays of the sun which, coming from A to B, bend on entering the water at the point B, and pass to C, where they are reflected to D, and bending there again as they pass out of the water, proceed to the point E; for when I put an opaque body or screen in any part of the lines AB, BC, CD, or DE, the red color disappeared. And although I covered all the globe except the two points B and D, and set up screens everywhere else, provided that I did not interfere with the rays $ABCDE$, the red never failed to appear. Then when I sought also for the cause of the red that appeared at K, I found that it was the rays which come from F to G, where they bend towards H, and at H are reflected to I, where they are again reflected to K, and finally bend at K and proceed to E. So that the first bow is caused by the rays which come to the eye after two refractions and one reflection, and the second by other rays which reach the eye only after two refractions and two reflections, so that it does not appear so often as the first one.

But the principal difficulty still remained, which was to determine why, since there are many other rays which can reach the eye after two refractions and one or two reflections when the globe is in some other position, it is only those of which I have spoken which exhibit the colors. And to answer that question I asked myself if there were not some other method of making the colors appear, so that by a comparison of the two I might better determine the reason for them. Then remembering that a prism or triangle of glass shows similar colors, I considered such a prism as that represented at *MNP* (Fig. 53), whose two surfaces *MN* and *NP* are plane and inclined to each other, at an angle of 30 or 40 degrees, so that if the rays of the sun *ABC* traverse *MN* per-

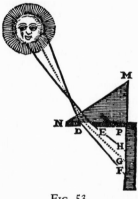

pendicularly or almost perpendicularly, so that they experience no appreciable refraction, they will experience a considerable refraction as they pass out through *NP*. And when I covered one of these surfaces with a screen, in which there was a small opening *DE*, I observed that the rays which pass through this opening and are received on a white cloth or sheet of paper, show all the colors of the rainbow; and that the red always appears at *F* and the blue or violet at *H*. From which I learned, in the first place, that the curvature of

FIG. 53.

the surface of the drops of water is not necessary for the production of the colors; for the surfaces of the crystal are plane; and that the size of the angle at which the colors appear is not important, since that can be changed without changing the colors; and even if we make the rays which go to *F* bend sometimes more and sometimes less than those which go to *H*, they still never fail to give red, and those which go to *H* to give blue; nor is reflection necessary, for there is none; nor finally the plurality of refractions, for in this case there is only one. I decided however that at least one refraction is necessary, and one the effect of which is not destroyed by another contrary one; for experiment shows that if the surfaces *MN* and *NP* are parallel, the rays which are bent at the one surface return to their original direction at the other, and produce no colors. I had no doubt that light was necessary, for without it we should see nothing. And further I observed that a shadow or a limitation of the light was necessary; for if we remove the screen on *NP* the colors *FGH* no longer appear; and if we make the opening *DE* large enough, the red, the orange, and the yellow which go to *F* do not move farther out, nor do the green, the blue, and the violet, which go to *H*, but all the rest of the space between them at *G* remains white.

An explanation which Descartes gives of the sending of the different colors in different directions, based upon his speculations on the nature and structure of the medium which transmits light through bodies, is omitted.

However I was in doubt whether the colors of the rainbow are produced in the same way as they are in the crystal *MNP*; for I saw no shadow there to limit the light, and did not understand why the colors appeared only at certain angles; until I took my pen and made an accurate calculation of the paths of the rays which fall on the different points of a globe of water, to determine at what angles, after two refractions and one or two reflections they will come to the eye, and then I found that after one reflection and two refractions there are many more rays which can be seen at an angle of from forty-one to forty-two degrees than at any smaller angle; and that there are none which can be seen at a larger angle. I found also that, after two reflections and two refractions, there are many more rays which come to the eye at an angle of from fifty-one to fifty-two degrees than at any larger angle, and none which come at a smaller angle. Thus there is a shadow on one side and the other, which limits the light which, after having passed through an infinity of drops of rain illuminated by the sun, comes to the eye, at the angle of forty-two degrees or a little less, and thus causes the first and principal rainbow; and there is also a shadow which limits the light which comes at the angle of fifty-one degrees or a little greater, and causes the exterior bow; for to receive no rays of light in the eye, or to receive notably less light from an object than from another one which is near it, is to see a shadow. This shows clearly that the colors of these rainbows are produced by the same cause as that which produces them when we use the crystal *MNP*, and that the semi-diameter of the interior bow should not be greater than forty-two degrees or that of the exterior bow less than fifty-one degrees; and finally that the former should be more sharply limited at its outer edge than at its inner edge; and exactly the contrary with the latter, as is verified by observation.

Descartes then presents his geometrical method of calculating the deviations of the rays of light which fall on different selected points on the surface of a spherical drop, and gives tables of these deviations which illustrate the preceding statements.

FERMAT

Pierre Fermat was born near Toulouse in 1608. He was for most of his life a magistrate in Toulouse, where he died on January 12, 1665. He was a mathematician of extraordinary ability. By some he has been ranked as the first discoverer of the differential calculus. He certainly was one of the founders of the theory of probabilities, which owed much to the correspondence between him and Pascal on that subject. His contributions to the theory of numbers are also important.

The following paper on the analytical treatment of refraction was published in his collected works in 1679. In it he shows that if his principle of least time applies to the transmission of light the velocity of light would necessarily be less in the more highly refracting medium.

REFRACTION OF LIGHT

Let $ACBI$ (Fig. 54) represent a circle whose diameter $AFDB$ separates two media of different nature, the rarer medium being on the side ACB, the denser on the side AIB.

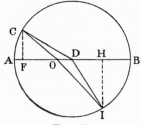

Fig. 54.

Let D be the center of the circle and CD an incident ray falling on this center from a given point C; it is required to find the refracted ray DI, or otherwise the point I through which the ray passes after refraction.

We drop the perpendiculars CF, IH on the diameter. Since the point C is given as well as the diameter AB and the center D, the point F and the line FB are also given. Let us suppose that the ratio of the resistance of the denser medium to that of the rarer medium is equal to the ratio of the given straight line DF to another line m drawn outside the figure. We ought to have $m < DF$, because the resistance of the rarer medium ought to be less than that of the denser medium, by an axiom which it is natural to adopt.

We now have to measure by means of the lines m and DF the motions along the lines CD and DI; we can thus represent proportionally the whole of the motion along these two lines by the sum of the two products: $CD \cdot m + DI \cdot DF$.

The question is thus reduced to that of dividing the diameter
AB at a point H in such a way that if at this point we erect a
perpendicular HI and then join DI, the area $CD \cdot m + DI \cdot DF$
shall be a minimum.

To accomplish this we shall employ our method which is already
distributed among mathematicians and was presented about
twenty years ago by Hérigone in his *Cursus Mathematicus*. We
use n to represent the radius CD or its equal DI, b to represent
the line DF, and we set $DH = a$. The quantity $nm + nb$ should
be a minimum.

For the unknown e we take an arbitrary line DO; we join CO,
OI. In analytical notation we have $CO^2 = n^2 + e^2 - 2be$, and
$OI^2 = n^2 + e^2 + 2ae$; then

$$CO \cdot m = \sqrt{m^2n^2 + m^2e^2 - 2m^2be},$$

and

$$IO \cdot b = \sqrt{b^2n^2 + b^2e^2 + 2b^2ae}.$$

The sum of these two radicals ought to be set equal, according
to the rules of the art, to the sum $mn + bn$.

To get rid of the radicals we square the sides of the equation,
we cancel the common terms, and transpose in such a way that
on one side of the equation only the remaining radical is left;
then we square the sides again; after another cancellation of
common terms on both sides, by dividing all the terms by e and
throwing out all those in which e still remains, according to the
rules of our method, which has been generally known for some
time, we shall finally, by cancelling common factors, reach the
simplest possible equation between a and m, that is to say, when
we have got rid of the difficulties presented by the radicals, we
find that the line DH of the figure is equal to the line m.

It follows that to find the point of refraction we must, when
we have drawn the lines CD and CF, take the lines DF and DH
in the ratio of the resistance of the denser medium to that of the
rarer medium, or in the ratio of b to m. We then erect at H
the line HI perpendicular to the diameter; it will cut the circle at
I, the point through which the refracted ray will pass; and thus
further the ray on passing from a rarer to a denser medium will
be bent toward the perpendicular. This result agrees exactly
and without exception with the theorem discovered by Descartes;

the previous analysis based on our principle thus gives to this theorem a demonstration which is rigorously exact.

BARTHOLINUS

Erasmus Bartholinus was born in Roeskilde, Denmark, on August 16, 1625. His father was a doctor of medicine and a professor at Copenhagen. After obtaining his medical degree he journeyed extensively in foreign countries, not returning to his native land until 1656. He was then made professor of mathematics and a year later also professor of medicine at the University of Copenhagen. He died on November 4, 1692.

Bartholinus discovered the double refraction of Iceland spar. He described his studies of that crystal in a short memoir entitled *Experimentis Crystalli Islandici Disdiaclastici, quibus Mira et Insolita Refractio Detegitur*, 1669, from which the following extract is taken.

DOUBLE REFRACTION

Greatly prized by all men is the diamond, and many are the joys which similar treasures bring, such as precious stones and pearls, though they serve only for decoration and adornment of the finger and the neck; but he, who, on the other hand, prefers the knowledge of unusual phenomena to these delights, he will,

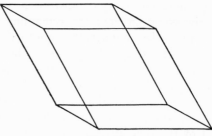

Fig. 55.

I hope, have no less joy in a new sort of body, namely, a transparent crystal, recently brought to us from Iceland, which perhaps is one of the greatest wonders that nature has produced. I have occupied myself for a long time with this remarkable body and carried out a number of investigations with it, which I gladly publish, since I believe that they can serve lovers of nature and other interested persons for instruction, or at least for pleasure.

First, the external form of this body is not less unusual and peculiar than that of snow, of salt or of other minerals and crystals.

It is composed of quadrilaterals with plane surfaces and sides, which are equi-distant from each other, measured from the middle point of the quadrilateral, but they enclose unequal angles; the plane surface which thus occurs is called in geometry a Rhombus, or a Rhomboid, the whole body is called a Rhomboid. This shape appears not only in the unbroken body, but it is maintained in all the separate parts into which it may be broken up. Figure 55 and the following figures represent it approximately, with the exception of some cases in which the natural form is a three-sided pyramid.

The author presents the results of his investigation of the electrical properties of the crystal, of its hardness, and of some of its chemical relations. He then presents his study of the angles contained by the edges and by the sides of the crystal, and proceeds to his seventh investigation as follows:

As my investigation of this crystal proceeded there showed itself a wonderful and extraordinary phenomenon: objects which are looked at through the crystal do not show, as in the case of other transparent bodies, a single refracted image, but they appear double. This discovery and its explanation occupied me for a long time, so that I neglected other things for it; I recognized that I had come upon a fundamental question

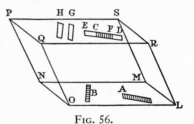

Fig. 56.

in refraction. In a superficial examination it is easy to miss seeing this phenomenon yet it can be easily exhibited in the following way: We place on a table or on clean paper any object, for example, a point or something similar of the size of *B* or *A* (Fig. 56) and place the lower surface *LMNO* of the Rhomboid upon it. Then we look through the upper surface *RSPQ* at the object *B* or *A* by directing the eye through the whole mass of the prism *RSPQOLMN* first on *B* and then on *A*. If we look through other transparent bodies, like glass, water, etc., the image of an object will appear only once, while in this case we see each of them in a double image on the surface *RSPQ*, that is *B* in *G* and *H*, *A* in *CD* and *FE*, as shown in the Figure. It is to be noticed that the distance between the images *H* and *G* which are given from the object *B*, is greater or smaller according to the size of the prism used; with thin pieces

it can hardly be noticed and it increases in proportion to the size
of the crystal.

Several investigations of less interest are omitted.

Investigation 13. A special property of our crystal is that it
gives a double image of an object; another property must be
mentioned which is peculiar to it and makes this crystal of special
interest among all the minerals. If we look at objects through
transparent media, the image remains fixed and immovable in

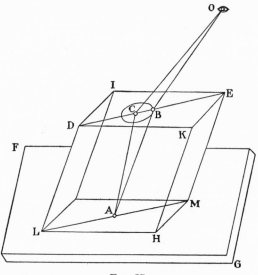

Fig. 57.

the same position however we move the medium to and fro, and
only if the object itself is moved does the image also change its
position on the surface of the transparent medium; in this case,
on the other hand, we can observe that one of the two images
is movable, and this can be established in the following way:
In Fig. 57 let the eye and the object *A* remain at rest; we then turn
the prism which is placed on *A* in such a manner that its lower
surface *MHL* always rests on the table, but so that the edge *EM*
which originally leaned toward *G*, finally is directed toward *F*.
We can then perceive that one of the images follows the motion of
the prism, that is, the image *B*; for while originally it was seen
in the direction of *G*, it appears, after the prism is rotated, in

the direction of *F*. The image *C*, however, remains fixed and unmoved. In the future, therefore, the former image will be called the movable one, the latter the immovable or fixed one.

The other investigations are not so important and are omitted. A paragraph from Investigation 16 is inserted for the sake of the names which are suggested in it.

We now know that an image of an object through two transparent bodies of different nature can only be produced by refraction and that an image requires a refraction; by assuming that the refraction is the cause of the phenomenon in question, it is admissible to draw the conclusion that for the doubled image there should be double refraction. Further, we noticed that the two images produced by our Iceland crystal are not exactly alike, but are distinguished from each other by one of them remaining at rest while the other is movable. We therefore conclude that we can distinguish the two kinds of refraction, and we designate that one which gives us the fixed image as ordinary refraction and the other, which gives the movable image, as extraordinary refraction. The crystal itself we call doubly refracting, on account of the extraordinary and peculiar properties of double refraction.

HUYGENS

The following extracts from Huygens' (p. 27) *Traité de la Lumière*, 1690 present what is known as Huygens' Principle and the use which he makes of it in explaining reflection and refraction, and also something from his study of double refraction and his discovery of the polarization of light.

It is impossible to present all that is novel and interesting in this great work of Huygens. All that can be done is to give some specimens of his theory and of the way in which he applied it.

Huygens starts with the idea that light is transmitted through an ether, which fills all space and which passes quite freely through material bodies, by a sort of elasticity called into action by shocks or impulses, as a shock is transmitted through a row of elastic balls. After discussing in general the way in which a shock would be transmitted, he proceeds as follows.

HUYGENS' PRINCIPLE

We have still to consider, in studying the spreading out of these waves, that each particle of the matter in which a wave proceeds not only communicates its motion to the next particle to it, which is on the straight line drawn from the luminous point, but that it also necessarily gives a motion to all the others which touch it and which oppose its motion. The result is that around each particle there arises a wave of which this particle is the center. So if *DCF* is a wave coming from the luminous point *A* (Fig. 58), which is its center, the particle *B*, one of those which belong to the sphere *DCF*, will set up its particular wave *KCL* which will touch the wave *DCF* in *C* at the same instant that the principal wave coming from the point *A* has reached *DCF*; and it is plain that the only point of the wave *KCL* which will touch the wave *DCF*, is the point *C* which is on the straight line drawn to *AB*. In the same way the other particles contained in the sphere *DCF*, such as *bb*, *dd*, and so forth, will each have made its own wave.

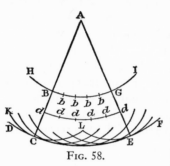

Fig. 58.

But each of these waves can be only infinitely feeble compared to the wave *DCF*, to the composition of which all the others contribute by that part of their surface which is the most distant from the center *A*.

We see further that the wave *DCF* is determined by the extreme limit of the motion which has gone out from the point *A* in a certain period of time, there being no motion beyond this wave, although there is motion in the space that it encloses, that is in the parts of the particular waves which do not touch the sphere *DCF*. This assumption should not be thought labored or over-subtle, since we shall see in the future that all the properties of light and those which pertain to reflection and refraction are explained fundamentally in this manner. This has not been known by those who have hitherto considered waves of light, among whom are Mr. Hooke in his *Micrographia*, and Pére Pardies, who, in a treatise of which he showed me a part and which he could not finish, for he died shortly afterwards, undertook to prove by waves the effects of reflection and refraction. But the fundamental principle which is stated in the remarks that I have just made is wanting in his demonstration, and he had furthermore opinions very different from mine, as will be seen perhaps some day if his writings have been preserved.

To come to the properties of light, we may first remark that every part of a wave ought to proceed in such a way that the ends of it are always contained between the same straight lines drawn from the luminous point. Thus the part of the wave *BG* of which the luminous point *A* is the center spreads out into the arc *CE* limited by the straight lines *ABC, AGE*. For while the particular waves produced by the particles which are in the space *CAE* spread also out of this space, they never conspire at the same instant to make up a wave which is the limit of the motion except precisely in the circumference *CE*, which is their common tangent.

And thus we can see the reason why light, at least when its rays are not reflected or refracted, proceeds only in straight lines in such a way that it does not illuminate any object except when the path from the source to the object is open along such a line. For if, for example, there were an opening *BG* bounded by opaque bodies *BH, GI*, the wave of light which starts from the point *A* will be always bounded by the straight lines *AC, AE*, as has been demonstrated; the parts of the particular waves which spread out of the space *ACE* are too feeble to produce light.

. . . .

Reflection of Light

Having explained the effects of waves of light which proceed in homogeneous matter, we shall next consider what happens to them when they meet other bodies. We shall first show how reflection of light is explained by the help of these waves and why it makes the angles equal. Suppose a plane and polished surface of some metal, glass, or any other body *AB* (Fig. 59), which at first I shall consider perfectly smooth (reserving for the end of this discussion the consideration of irregularities, of which the body cannot be free), and suppose that

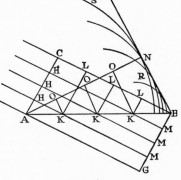

FIG. 59.

a line *AC*, inclined to *AB*, represents a part of a wave of light, of which the center is so far away that the part *AC* can be considered a straight line. I consider everything in a plane by supposing that the plane of the figure cuts the sphere of the wave in its center, and the plane *AB* at right angles, which is here stated once for all.

The part C of the wave AC in a certain time will advance to the point B in the plane AB along the straight line CB, which we may suppose coming from the luminous center, and which consequently is perpendicular to AC. Now in the same time the part A in the same wave, which has been prevented from communicating its motion to the body below the plane AB, at least in part, should have continued its motion in the matter which is above this plane and that through a distance equal to CB, making its particular spherical wave according to what has been already said. This wave is here represented by the circumference SNR, of which the center is A and the radius AN equal to CB.

Now if we go on to consider the other parts H of the wave AC, it appears that they will not only have reached the surface AB by the straight lines HK, parallel to CB, but that further they will have set up from the centers K particular spherical waves in the medium, represented here by circumferences whose radii are equal to KM, that is to say, to the prolongations of HK to the straight line BG parallel to AC.

But all these circumferences have for a common tangent the straight line BN, that is, the line which is tangent from B to the first of these circles, of which A is the center and AN the radius equal to BC, as it is easy to see.

It is therefore the line BN (limited by the point B and the point N on which falls the perpendicular from the point A) which is, as it were, formed by all these circumferences, and which terminates the motion which is set up by the reflection of the wave AC; and it is also the place where this motion is much greater than anywhere else. This is why, according to that which has been explained, BN is the place reached by the wave AC at the instant that its part C has reached B. For there is no other line except BN which is a common tangent to all these circles, unless it is BG below the plane AB; which BG would be the place reached by the wave if the motion could go on in a matter homogeneous with that which is above the plane. If we wish to see how the wave AC has come successively to BN, we have only to draw in the same figure the lines KO parallel to BN and the lines KL parallel to AC. Then we shall see that the wave AC, originally a straight line, is broken in all the lines OKL successively and becomes a straight line again in NB.

It appears, further, that the angle of reflection is equal to the angle of incidence. For the triangles ACB, BNA being right-angled triangles and having the side AB common and the side CB

equal to *NA*, it follows that the angles opposite these sides will be equal and therefore also the angles *CBA*, *NAB*. But since *CB*, perpendicular to *CA*, represents the direction of the incident ray, so *AN*, perpendicular to the wave *BN*, represents the direction of the reflected ray; these rays are therefore at equal angles with the plane *AB*.

The consideration of a beam of light illuminating a small area on the reflecting surface is omitted, as also speculations on the way in which reflection can be explained consistently with the real roughness of the reflecting surface.

Refraction of Light

Speculations on the way in which the ether is associated with matter, and on the way in which the impulses of light pass through matter are omitted. A statement of the law of refraction is also omitted.

To give a reason for these phenomena on our principles let us consider the straight line *AB* (Fig. 60), which represents a plane surface separating the transparent bodies on the side *C* and the side *N*. When I say plane I do not mean that the surface is perfectly smooth, but is such as has been understood in our treatment of reflection, and for the same reason. Let the line *AC* represent a part of a wave of light, of which the center is so far

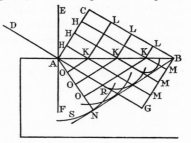

Fig. 60.

away that this part can be considered a straight line. The part *C* then of the wave *AC* in a certain time will progress to the plane *AB* along the line *CB*, which we should imagine coming from the luminous center and which consequently will cut *AC* at a right angle. Now in the same time the part *A* will come to *G* along the line *AG*, equal and parallel to *CB*, and all the part *AC* of the wave will be in *GB*, if the matter of the transparent body transmits the motion of the wave as quickly as that of the ether does. But let us suppose that it transmits this motion less quickly, for example, by a third. Motion will then be spread out from the point *A* in the matter of the transparent body through an extent equal to two-thirds of *CB*, making its particular spherical wave according to what has been said before; this wave is then repre-

sented by the circumference *SNR*, of which the center is *A* and
the radius equal to two-thirds of *CB*. Now if we consider further
the other parts *H* of the wave *AC*, it appears that, in the time
taken by the part *C* to come to *B*, they will not only have reached
the surface *AB* by the lines *HK* parallel to *CB*, but that further
they will have set up at the centers *K* particular waves in the
transparent body represented here by circumferences whose
radii are equal to two-thirds of the lines *KM*, that is to say to
two-thirds of the prolongation of *HK* to the line *BG*; for these radii
would be equal to the lines *KM* if the two bodies were equally
penetrable.

Now all these circumferences have for a common tangent
the straight line *BN*, that is, the line which from the point *B* is
tangent to the circumference *SNR*, which is the first that we
considered. For it is easy to see that all the other circumferences
will touch the same line *BN* from *B* to the point of contact *N*,
which is the point on which *AN* falls perpendicularly to *BN*.

It is then *BN* which is, as it were, formed by the minute arcs
of these circumferences, which terminates the motion that the
wave *AC* has communicated to the transparent body, and in which
this motion occurs in much greater quantity than anywhere else.
And for that reason this line, according to what has been said
more than once, is the place reached by the wave *AC* at the instant
that its part *C* has reached *B*. For there is no other line except *BN*
below the plane *AB* which is a common tangent to all the particular
waves. Now if we wish to know how the wave *AC* has come
successively to *BN*, we need only draw in the same figure the
lines *KO* parallel to *BN* and the lines *KL* parallel to *AC*. Then
we shall see that the wave *CA*, which was a straight line, has
become broken into all the lines *LKO* successively and that it
becomes a straight line again in *BN*. This being evident by what
has been already demonstrated, there is no need of explaining it
further.

Now in the same figure, if we take *EAF*, which cuts the plane
AB at right angles at the point *A*, and draw *AD* perpendicular to
the wave *AC*, then *DA* will represent the ray of incident light and
AN, which is perpendicular to *BN*, the refracted ray; since the
rays are nothing other than the straight lines along which the parts
of the waves proceed.

From this it is easy to deduce the principal property of refraction,
namely, that the sine of the angle *DAE* is always in the same ratio

to the sine of the angle *NAF*, whatever may be the inclination of the ray *DA*, and that this ratio is the same as that of the velocity of the waves in the body which is on the side of *AE* to the velocity of those in the body on the side of *AF*. For if we consider *AB* as the radius of a circle, the sine of the angle *BAC* is *BC* and the sine of the angle *ABN* is *AN*. But the angle *BAC* is equal to *DAE*, since each of them added to *CAE* makes a right angle. And the angle *ABN* is equal to *NAF*, since each of them with *BAN* makes a right angle. Therefore the sine of the angle *DAE* is to the sine of the angle *NAF* as *BC* to *AN*. But the ratio of *BC* to *AN* is the same as that of the velocities of light in the matter on the side of *AE* and in that on the side of *AF*; therefore also, the sine of the angle *DAE* will be to the sine of the angle *NAF* as these velocities of the light.

There follows a discussion of the case of light emerging from a medium in which its velocity is less into one in which its velocity is greater. This leads to a discussion of total internal reflection, which is explained on the principles already employed. There is given also a demonstration of Fermat's theorem of least time.

DOUBLE REFRACTION AND POLARIZATION

Huygens begins his study of double refraction by a description of crystals of Iceland spar very much like that which had already been given by Bartholinus, although Huygens made more accurate measurements of the angles of the crystal. He then proceeds as follows

It was after having explained the refraction of ordinary transparent bodies by means of spherical emanations of light that I took up again the examination of the nature of this crystal, about which I had formerly been able to discover nothing.

As there are two different refractions, I conceived also that there are two different emanations of the waves of light, and that one of them could occur in the ethereal matter distributed in the body of the crystal. This matter, being in much greater quantity than is that of the particles which compose the body, is alone capable of causing the transparence of the body, according to that which has been explained before. I attributed to this emanation of waves the regular refraction which is observed in this crystal, by supposing that these waves are as usual of spherical form, and that they proceed more slowly within the crystal than they do outside it; from which supposition I have shown that refraction will follow.

As to the other emanation, which produces the irregular refraction, I determined to try what elliptical waves would do or better spheroidal waves; which I supposed spread themselves out indifferently not only in the ethereal matter distributed in the crystal, but also in the particles of which it is composed, in accordance with the last way in which I have explained transparence. It seemed to me that the position, or the regular arrangement of these particles, might contribute to form spheroidal waves (the only requisite for that being that the motion of the light proceeds a little more rapidly in one direction than in another), and I did not doubt that there was in this crystal such an arrangement of equal and similar particles, because of its shape and because of the certain and invariable size of its angles. As to these particles and their form and arrangement, I shall at the end of this treatise present my conjectures and some experiments which confirm them.

The two-fold emanation of waves of light, which I have supposed, seemed to me more probable after I had observed a certain phenomenon in ordinary crystal, which develops a hexagonal shape and which because of this regularity seems also to be composed of particles of a certain shape and arranged in order. The fact is, that this crystal also exhibits double refraction, as well as the Iceland spar, although less evidently. When I had cut some well-polished prisms by different sections, I noticed in all of them, when I looked through them at the flame of a candle or at the leads between the window panes, that the objects seen all appeared double, although the images were only a little distance apart. From this I understood the reason why this body, though it is so transparent, is useless for telescopes when they have any considerable length.

Now this double refraction, according to my previously established theory, seems to require a double emanation of waves of light, both of them spherical (for the two refractions are regular) and one of them only a little slower than the other. For in this way this phenomenon is very naturally explained, by assuming material media which serve to carry these waves, just as I did in the case of the Iceland spar. After this discovery, it was less difficult to admit two emanations of waves in the same body. And as it might be objected that if these crystals are made up of equal particles of a certain shape and arranged regularly, the interstices that these particles leave between them and which contain the ethereal matter, could scarcely be sufficient to transmit the waves of light as I have supposed, I avoid this difficulty by supposing that these particles are of a highly rarified material or possibly are composed of other much smaller particles, among which the ethereal matter passes very freely. This conception otherwise follows necessarily from that which has been demonstrated before concerning the small amount of matter of which bodies are composed.

Supposing then these spheroidal waves, in addition to spherical waves, I began to examine the question whether they could serve to explain the phenomena of irregular refraction and how, by these phenomena themselves, I could determine the figure and the position of the spheroids; in which investigation I at last obtained the desired success, by proceeding as follows:

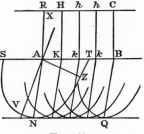

FIG. 61.

I first considered waves of this form in the case of rays which fall perpendicularly on the plane surface of a transparent body, in which they proceed in this way: (Fig. 61)

I take *AB* to represent the surface. And since a ray perpendicular to a plane and coming from a very distant source is equivalent, by our former theory, to the incidence of a portion of a wave parallel to this plane, I suppose the line *RC*, parallel and equal to *AB*, to be a portion of a wave of light, of which the infinitely many points like *RHbC* meet the surface *AB* at points like *AKkB*. Then in place of the particular hemispherical waves, which, in an ordinary refracting body, ought to proceed from each of these last points, as we have explained before in treating of refraction, the waves ought to be here hemispheroids, of which I assume that their axes or their greatest diameters are oblique to the plane *AB*, as represented by the line *AV*, the semi-axis or the half of the greatest diameter of the spheroid *SVT*, which represents the particular wave which comes from the point *A*, after that the wave *RC* has reached *AB*. I say either the axis or the greatest diameter, because the same ellipse *SVT* may be considered as the section of the spheroid of which the axis is *AZ*, perpendicular to *AV*. But for the present, without deciding as yet for one or the other, we shall consider these spheroids only in their sections, which make ellipses in the plane of the figure. Now in a certain length of time, during which the wave

SVT has gone out from the point *A*, it will happen that all the other points *KkB* send out waves similar and similarly placed to *SVT*. And the common tangent *NQ* of all these ellipses represents the progress of the wave *RC* in the transparent body, as proposed in our former theory, because this line is that which bounds, at the same instant, the motion which has been caused by the wave *RC* in falling on *AB*, and in that line this motion is in much greater quantity than anywhere else, since it is made up of the infinitely many arcs of the ellipses whose centers are along the line *AB*.

Now it appears that this common tangent *NQ* is parallel to *AB*, and of the same length, but that it is not directly opposite to it, since it is contained between the lines *AN*, *BQ*, which are conjugate diameters of the ellipses whose centers are *A* and *B* to the diameters which lie in the line *AB*. In this way I understood that which had appeared so difficult, that a ray perpendicular to a surface can undergo refraction when it enters a transparent body, seeing that the wave *RC* which came from the opening *AB* continues to advance between the parallels *AN*, *BQ*, while still itself always remaining parallel to *AB*, so that in this case the light does not proceed by lines perpendicular to its waves as in ordinary refraction, but these lines cut the waves obliquely.

The remaining investigations of Huygens, by which he determines the form and position of the spheroidal wave in the crystal, have been omitted. The following sections describe the discovery of the polarization of light.

Before closing my treatment of this crystal, I will add an account of a wonderful phenomenon which I discovered after I had written all that has gone before. For, although I have not been able to find the reason for it, I will not refrain from pointing it out, so as to give an opportunity to others to investigate it. It seems necessary to make other assumptions in addition to those that I have made, which, nevertheless, retain all their probability, having been confirmed by so many proofs.

The phenomenon is this: if we take two blocks of the crystal (Fig. 62) and place them one over the other, or hold them separated by some distance, so that the sides of one of them are parallel to those of the other, then a ray of light, such as *AB*, which is divided into two in the first block, as represented by *BD* and *BC*, according to the two refractions, regular and irregular, on passing from one block to the other, so proceeds that each ray passes in it without again dividing into two. The one which has been made by regular refraction, as *DG*, will make only a regular refraction in *GH*, and the other *CE* an irregular refraction in *EF*. The same thing happens not only with this arrangement, but also in all cases in which the principal sections of the two blocks are in the same plane. It is not necessary that the two surfaces which face each other should be parallel. Now it is wonderful that the rays *CE* and *DG*, coming from air to the lower crystal, do not divide in the same way that the first ray *AB* does. We might say that the ray *DG*, by passing through the upper block has

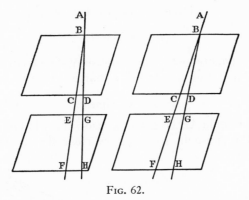

Fig. 62.

lost some property by which it can move the matter which takes part in the irregular refraction, and that *CE* has similarly lost that property by which it moves the matter which serves for the regular refraction, but there is one thing which overthrows this hypothesis. When we place the two crystals in such a way that the planes of the principal sections cut each other at right angles, whether the surfaces opposite to each other are parallel or not, then the ray which comes from the regular refraction, such as *DG*, makes only an irregular refraction in the lower block, and on the contrary the ray which comes from the irregular refraction, such as *CE*, makes now only a regular refraction.

But in all the other possible positions except those which I have indicated the rays *DG*, *CE* divide each into two by refraction in the lower crystal; so that from the single ray *AB* there are made four rays, sometimes of equal brightness, sometimes so that some are brighter than others, according to the different

positions of the crystals, but they do not appear to have more
light altogether than the single ray *AB*.

GRIMALDI

Francesco Maria Grimaldi was born in Bologna in 1618. He belonged to
the order of Jesuits and was professor of mathematics at Bologna. He was
an ardent experimenter in natural science and is particularly known as the
discoverer of the diffraction of light. He died in 1663 in Bologna.

The extract which follows is taken from his book entitled, *Physico-mathesis
de Lumine, Coloribus et Iride,* which was published after his death in 1665.

Diffraction of Light

Proposition I.

Light is propagated or diffused not only directly, by refraction,
and by reflection, but also in still a fourth way,—by diffraction.

First Experiment.

A very small hole *AB* (Fig. 63) is made in a window shutter and
through it the sun-light from a very clear sky is admitted into

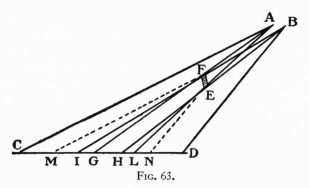

Fig. 63.

a room which is otherwise closed so as to be dark. The diffusion
of this light will be in a cone or what is nearly a cone, *ACDB*,
and becomes visible if the air is filled with dust or if some smoke
is set up in it. In this cone there is placed an opaque body *EF*,
at a great distance from the hole *AB*, and so that at least one end
of the opaque body is illuminated. Then the before mentioned
cone is received on a white board or on a sheet of white paper

laid on the floor, and there is seen its illuminated base *CD* with the shadow *GH* cast by the opaque body *EF* which has been placed in the cone and illuminated at either end *E* or *F*. This shadow, according to the laws of optics, will not be exactly defined and terminated at a point *G* on the one side and at another point *H* on the other side; but because of the width of the hole *AB* and also because of the lateral extension of the sun and for another reason, the edge of the shadow will be to some degree uncertain, on account of the so called penumbra, and with a perceptible decrease, or, as they say, a shading off of light through the distance *IG* between the certain shadow and the bright light on one side of the base and through the distance *HL* on the other side.

But what should be especially noticed is that the shadow *IL* appears considerably larger in fact than it ought to be, if the whole thing is supposed to act by straight lines from the ends *AB* produced through either end *EF* as shown in the figure, and greater than would be deduced by calculation from the given distances *EF* and *FI* and from the sizes of *AB* and *EF*, and from all the angles necessary for the solution of the triangles shown in the figure, as we ourselves have often proved by trial. To deal with this subject briefly, from the three sides of the triangle *AFE* we get by trigonometry the angle *A*, with which in the triangle *AGL*, with the side *AG*, or *AL*, and with the observed angle *G* we obtain *GL*. Then in the triangle *AFB*, which can be taken as isosceles, from the three given sides we know the angle *F* and its vertical and equal angle *IFG*; with which in the triangle *IGF*, together with the distance *FI* and the observed angle *I*, we obtain the line *IG*, which added to the line *GL* already found gives us the desired base *IL* of such a size as it ought to be if the whole diffusion went on by straight lines in a luminous cone interrupted by the opaque body *EF*. These triangles are very acute so that tables extended to a great radius are needed, and yet their solution is not impossible. And so in the figure we may place the shadow by calculation, and when deduced from the straight lines drawn as above it will be *IL*; the true shadow however, which appears by observation, is *MN*.

In addition, by observing in the part *CM* or *ND* of the field which is brightly and strongly illuminated, there will be seen certain tracts or bands of colored light, of such a sort that in any one of these, in the middle of it, the light is very pure and clear, but at its sides there is some color, always bluish at the side which is nearer the shadow MN and reddish on the further side. These bright bands apparently depend on the size of the hole AB, because they are not seen if it is very large, yet they are not determined by it. Nor are they determined by the magnitude of the sun's diameter, as will appear from what is to be said.

It may further be observed of the beforementioned tracts or bands of colored light, as they extend from M toward C (and we may say the same of the others between N and D) that the first is wider than the second, and the second is wider than the third (it has never yet happened that more than three have been seen). These decrease in the intensity of the light and of the colors in the same order in which they recede from the shadow. Yet the individual bands are wider and further apart, the further the white board on which they are received is from the opaque body

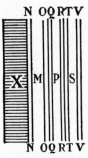 casting the shadows, and further also by how much the more this is set obliquely to the solar radiation. This reason itself demands, because they are made by the rays which are part of the luminous cone and which become more and more separated from one another the further they proceed.

There may be someone perhaps who, because of his failure to consider this observation, is unwilling to recognize that the beforementioned bands are bands of light, as we have said they are, but will contend that they should rather be called bands of shadow, because he has not paid enough attention to those obscure colors which we have said appear at the sides of these light bands. That we may more clearly explain this, in the foregoing figure, (Fig. 64) alongside of the shadow X cast by the opaque body as before, and falling on a board or a sheet of white paper, there are represented three of these luminous bands, each one consisting of three strips. The first and broadest band is NMO, in the middle of which is M, the broadest and brightest region of all, which shows no color, but is bounded by two lesser strips which are colored, one of which, N, nearer the shadow is bluish, the other O is reddish. The second band QPR is narrower than the first, and in the middle of it is P, a bright uncolored strip which is bounded by two strips, colored but not bright; the one Q, nearer the shadow, is bluish, and the other, R, reddish. The third band TSV is the narrowest of all. It has in its middle a bright strip S,

FIG. 64.

and on the sides two less conspicuous colored strips, of which *T* is bluish and *V* reddish.

Second Experiment

An opening is made, perhaps a finger breadth wide, in the window shutter of a well darkened room, and in this opening is placed a thin opaque plate *AB* (Fig. 65), and through a very narrow opening *CD* in this plate sunlight is admitted and forms a cone of light. At a great distance from the plate *AB* another plate *EF* is placed so as to cut this cone at right angles. In it

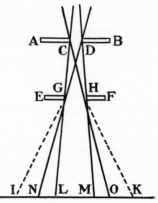

FIG. 65.

there is also a small hole *GH*, through which some of the before mentioned cone of light, which is interrupted by the plate *EF*, is admitted. This plate is so placed that the base of the cone considerably exceeds the size of the hole *GH*, so that the hole is all lighted or filled with light. Then the light which enters through the hole *GH* will again form a cone or almost a cone. When this is cut orthogonally or is terminated by a smooth white surface it will show on this surface an illuminated base *IK*, notably greater than the rays would make which are transmitted in straight lines through the two holes, not only those which pass the edges of the holes on the same side, like *CGL* and *DHM*, but even those which pass on the opposite sides, like *DGN* and *CHO*.

This is proved, as often as the experiment is tried, by observing how great the base *IK* is in fact, and deducing by calculation how great the base *NO* ought to be, which is formed by the direct rays.

. . . .

For the experiment to succeed there should be bright sunlight because, as has been said, the holes should be very small, especially the first one *CD*, and also the white surface on which the base *IK* is received should be at a great distance from the hole *GH*, otherwise the base observed is either not at all or only a little larger than the base *NO* deduced by calculation.

The details of the observations are omitted.

When the observation was made in the middle of the day in summer and with a clear sky the observed base *IK* was so greatly in excess of the calculated base *NO* that it would be out of the question to ascribe it to errors in the observation.

Further it should not be omitted that the illuminated base *IK* appears in the middle suffused with pure light, and at either extremity its light is colored, partly reddish, partly also strongly bluish.

NEWTON

The following extract presents a part of Newton's (p. 30) paper entitled "A New Theory About Light and Colours," published in the *Philosophical Transactions, Abridged,* Vol. I, p. 128, under date of February, 1672. The paper describes his demonstration of the composite nature of white light. Newton subsequently presented the same general discussion in his *Treatise on Optics,* published in 1704.

The next extract is from the Fourth Edition of Newton's *Optics.* When Newton found that he could not complete to his own satisfaction his study of diffraction he closed his exposition and added a set of Queries, which were increased by successive additions as one edition after another of his work was published. Some of these Queries are presented here, in which Newton suggests the corpuscular theory of light which, largely owing to his eminence, remained dominant in physics for over a hundred years.

DISPERSION OF LIGHT

In the year 1666, (at which time I applied myself to the grinding of optick glasses of other figures than spherical) I procured me a triangular glass prism, to try therewith the celebrated phaenomena of colours. And in order thereto, having darkened my chamber, and made a small hole in my window-shuts, to let in a convenient quantity of the sun's light, I placed my prism at its entrance, that it might be thereby refracted to the opposite wall. It was at first a very pleasing divertisement, to view the vivid and intense colours produced thereby; but after a while applying my self to consider them more circumspectly, I became surprised, to see them in an oblong form; which, according to the received laws of refraction, I expected should have been circular. They were terminated at the sides with straight lines, but at the ends, the decay of light was so gradual that it was difficult to determine justly, what was their figure; yet they seemed semicircular.

Comparing the length of this colour'd Spectrum with its breadth, I found it about five times greater, a disproportion so extravagant, that it excited me to a more than ordinary curiosity to examining from whence it might proceed. I could scarce think, that the various thicknesses of the glass, or the termination with shadow or darkness, could have any influence on light to produce such an effect; yet I thought it not amiss, first to examine those circumstances, and so try'd what would happen by transmitting light through parts of the glass of divers thicknesses, or through holes in the window of diverse bignesses, or by setting the prism without, so that the light might pass through it, and be refracted, before it was terminated by the hole: But I found none of those circumstances material. The fashion of the colours was in all these cases the same.

Then I suspected, whether by any unevenness in the glass or other contingent irregularity, these colours might be thus dilated. And to try this, I took another prism like the former, and so placed it, that the light passing through them both might be refracted contrary ways, and so by the latter returned into that course from which the former had diverted it. For by this means I thought the regular effects of the first prism would be destroyed by the second prism, but the irregular ones more augmented, by the multiplicity of refractions. The event was, that the light, which by the first prism was diffused into an oblong form was by the second reduced into an orbicular one, with as much regularity as when it did not at all pass through them.

. . . .

Then I began to suspect, whether the rays, after their trajection through the prism, did not move in curve lines, and according to their more or less curvity tend to divers parts of the wall. And it increased my suspicion, when I remembered that I had often seen a tennis ball struck with an oblique racket, describe such a curve line. For, a circular as well as a progressive motion being communicated to it by that stroke, its parts on that side where the motions conspire, must press and beat the contiguous air more violently than on the other, and there excite a reluctancy and reaction of the air proportionably greater. And for the same reason, if the rays of light should possibly be globular bodies, and by their oblique passage out of one medium into another, acquire a circulating motion, they ought to feel the greater resistance from the ambient ether, on that side, where the motions conspire, and thence be continually bowed to the other. But notwithstanding this plausible ground of suspicion, when I came to examine it, I could observe no such curvity in them. And besides (which was enough for my purpose) I observed, that the difference 'twixt the length of the image, and the diameter of the hole, through which the light was transmitted, was proportionable to their distance.

The gradual removal of these suspicions at length led me to the *Experimentum Crucis*, which was this: I took two boards, and placed one of them close behind the prism at the window, so that the light might pass through a small hole, made in it for the purpose, and fall on the other board, which I placed at about 12 feet distance, having first made a small hole in it also, for some of the incident light to pass through. Then I placed another prism behind this second board, so that the light trajected through both the boards might pass through that also, and be again refracted before it arrived at the wall. This done, I took the first prism in my hand, and turned it to and fro slowly about its axis, so much as to make the several parts of the image cast, on the second board, successively pass through the hole in it, that I might observe to what places on the wall the second prism would refract them. And I saw by the variation of those places, that the light, tending to that end of the image, towards which the refraction of the first prism was made, did in the second prism suffer a refraction considerably greater than the light tending to the other end. And so the true cause of the length of that image was detected to be no other, than that light is not similar or homogenial, but consists of *Difform Rays, some of which are more Refrangible than others;* so that without any difference in their incidence on the same medium, some shall be more Refracted than others; and therefore that, according to their *particular Degrees of Refrangibility*, they were transmitted through the prism to divers parts of the opposite wall.

Now I shall proceed to acquaint you with another more notable *Difformity* in its rays, wherein the origin of colours is unfolded: concerning which I shall lay down the doctrine first, and then for its examination give you an instance or two of the experiments, as a specimen of the rest.

The doctrine you will find comprehended and illustrated in the following propositions:

1. As the rays of light differ in degrees of refrangibility so they also differ in their disposition to exhibit, this or that particular colour. Colours are not qualifications of light, derived from refractions, or reflections of natural bodies (as 'tis generally believed) but original and connate properties, which in divers rays are divers. Some rays are disposed to exhibit a red colour and no other; some a yellow and no other, some a green and no other, and so of the rest. Nor are there only rays proper and particular to the more eminent colours, but even to all their intermediate gradations.

2. To the same degree of refrangibility ever belongs the same colour, and to the same colour ever belongs the same degree of refrangibility. The least refrangible rays are all disposed to exhibit a red colour, and contrarily those rays which are disposed to exhibit a red colour, are all the least refrangible: so the most refrangible rays are all disposed to exhibit a deep violet colour, and contrarily those which are apt to exhibit such a violet colour are all the most refrangible. And so to all the intermediate colours in a continued series belong intermediate degrees of refrangibility. And this Analogy 'twixt colours and refrangibility is very precise and strict; the rays always either exactly agreeing in both, or proportionally disagreeing in both.

3. The species of colour, and degree of refrangibility proper to any particular sort of rays, is not mutable by refraction, nor by reflection from natural bodies, nor by any other cause that I could yet observe. When any one sort of rays hath been well parted from those of other kinds, it hath afterwards obstinately retained its colour, notwithstanding my utmost endeavors to change it. I have refracted it with prisms, and reflected it with bodies, which in daylight were of other colours; I have intercepted it with the coloured film of air, interceeded two compressed plates of glass; transmitted it through coloured mediums, and through mediums irradiated with other sorts of rays, and diversely terminated it; and yet could never produce any new colour out of it. It would by contracting or dilating become more brisk, or faint, and by the loss of many rays, in some cases very obscure and dark; but I could never see it changed in specie.

4. Yet seeming transmutations of colours may be made, where there is any mixture of divers sorts of rays. For in such mixtures, the component colours appear not, but, by their mutual allaying each other, constitute a midling colour. And therefore, if by

refraction, or any other of the aforesaid causes, the difform rays, latent in such a mixture, be separated, there shall emerge colours different from the colour of the composition. Which colours are not new generated, but only made apparent by being parted; for if they be again entirely mixt and blended together, they will again compose that colour, which they did before separation. And for the same reason, transmutations made by the convening of divers colours are not real; for when the difform rays are again severed, they will exhibit the very same colours which they did before they entered the composition; as you see blue and yellow powders, when finely mixed, appear to the naked eye, green, and yet the colours of the component corpuscles are not thereby really transmuted, but only blended. For when viewed with a good microscope they still appear blue and yellow interspersedly.

5. There are therefore two sorts of colours. The one original and simple, and the other compounded of these. The original or primary colours are red, yellow, green, blue, and a violet-purple, together with orange, indico, and an indefinite variety of intermediate gradations.

6. The same colours in specie with these primary ones, may be also produced by composition. For a mixture of yellow and blue makes green; of red and yellow makes orange; of orange and yellowish green makes yellow. And in general, if any two colours be mixed, which in the series of those generated by the prism are not too far distant one from another, they by their mutual alloy compound that colour, which in the said series appeareth in the midway between them. But those which are situated at too great a distance, do not so. Orange and indico produce not the intermediate green, nor scarlet and green the intermediate yellow.

7. But the most surprising, and wonderful composition was that of whiteness. There is no one sort of rays which alone can exhibit this. 'Tis ever compounded, and to its composition, are requisite all the aforesaid primary colours, mixed in a due proportion. I have often with admiration beheld that all the colours of the prism being made to converge, and thereby to be again mixed, as they were in the light before it was incident upon the prism, reproduced light, entirely and perfectly white, and not at all sensibly differing from a direct light of the sun, unless when the glasses, I used, were not sufficiently clear; for then they would a little incline it to their colour.

8. Hence therefore it comes to pass, that whiteness is the usual colour of light; for light is a confused aggregate of rays indued with all sorts of colours, as they were promiscuously darted from the various parts of luminous bodies. And of such a confused aggregate, as I said, is generated whiteness, if there be a due proportion of the ingredients; but if any one predominate, the light must incline to that colour; as it happens in the blue flame of brimstone; the yellow flame of a candle; and the various colours of the fixed stars.

9. These things considered, the manner how colours are produced by the prism is evident. For, of the rays, constituting the incident light, since those which differ in colour proportionally differ in refrangibility, they by their unequal refractions must be severed and dispersed into an oblong form in an orderly succession, from the least refracted scarlet, to the most refracted violet. And for the same reason it is, that objects when looked upon through a prism, appear coloured. For the difform rays, by their unequal refractions, are made to diverge towards several parts of the Retina, and these express the images of things coloured, as in the former case they did the sun's image upon a wall. And by this inequality of refractions, they become not only coloured, but also very confused and indistinct.

10. Why the colours of the rainbow appear in falling drops of rain, is also from hence evident. For those drops which refract the rays, disposed to appear purple, in greatest quantity to the spectator's eye, refract the rays of other sorts so much less, as to make them pass beside it; and such are the drops on the inside of the primary bow, and on the outside of the secondary or exterior one. So those drops, which refract in greatest plenty the rays, apt to appear red, toward the spectator's eye, refract those of other sorts so much more, as to make them pass beside it; and such are the drops on the exterior part of the primary, and interior part of the secondary bow.

11. The odd phaenomena of an infusion of *Lignum Nephriticum,* leaf-gold, fragments of coloured glass, and some other transparently coloured bodies, appearing in one position of one colour, and of another in another, are on these grounds no longer riddles. For those are substances apt to reflect one sort of light, and transmit another; as may be seen in a dark room, by illuminating them with familiar or uncompounded light. For then they appear of that colour only, with which they are illuminated, but yet in one

position more vivid and luminous than in another, accordingly as they are disposed more or less to reflect or transmit the incident colour.

12. From hence also is manifest the reason of an unexpected experiment which Mr. *Hook*, somewhere in his *Micrography* relates to have made with two wedge-like transparent vessels, filled the one with a red, the other with a blue liquor: namely, that though they were severally transparent enough, yet both together became opake; for if one transmitted only red, and the other only blue, no rays could pass through both.

13. I might add more instances of this nature, but I shall conclude with this general one. That the colours of all natural bodies have no other origin than is, that they are variously qualified, to reflect one sort of light in greater plenty than another. And this I have experimented in a dark room, by illuminating those bodies with uncompounded light of divers colours. For by that means any body may be made to appear of any colour. They have there no appropriate colour, but ever appear of the colour of the light cast upon them, but yet with this difference, that they are most brisk and vivid in the light of their own daylight colour. Minium appeareth there of any colour indifferently, with which it is illustrated, but yet most luminous in red, and so bise appeareth indifferently of any colour, but yet most luminous in blue. And therefore minium reflecteth rays of any colour, but most copiously those endowed with red, and consequently when illustrated with daylight; that is, with all sorts of rays promiscuously blended, those qualified with red shall abound most in the reflected light, and by their prevalence cause it to appear of that colour. And for the same reason bise, reflecting blue most copiously, shall appear blue by the excess of those rays in its reflected light; and the like of other bodies. And that this is the entire and adequate cause of their colours, is manifest, because they have no power to change or alter the colours of any sort of rays incident apart, but put on all colours indifferently, with which they are enlightened.

These things being so, it can be no longer disputed, whether there be colours in the dark, or whether they be the qualities of the objects we see, no nor perhaps, whether light be a body. For, since colours are the qualities of light, having its rays for their entire and immediate subject, how can we think those rays qualities also, unless one quality may be the subject of, and sustain

another; which in effect is to call it substance. We should not
know bodies for substances; were it not for their sensible qualities,
and the principal of those being now found due to something else,
we have as good reason to believe that to be a substance also.

Besides, who ever thought any quality to be a heterogeneous
aggregate, such as light is discovered to be? But to determine
more absolutely what light is, after what manner refracted, and
by what modes or actions it produceth in our minds the phantasms
of colours, is not so easie; and I shall not mingle conjectures with
certainties.

The rest of the paper contains suggestions of various experiments by which
the views presented can be tested.

THE NATURE OF LIGHT

Qu. 28. Are not all Hypotheses erroneous, in which Light is
supposed to consist in Pression or Motion, propagated through a
fluid Medium? For in all these Hypotheses the Phaenomena
of Light have been hitherto explained by supposing that they arise
from new Modifications of the Rays; which is an erroneous
Supposition.

If Light consisted only in Pression propagated without actual
Motion, it would not be able to agitate and heat the Bodies which
refract and reflect it. If it consisted in Motion propagated to
all distances in an instant, it would require an infinite force every
moment, in every shining Particle, to generate that Motion.
And if it consisted in Pression or Motion, propagated either in
an instant or in time, it would bend into the Shadow. For Pression
or Motion cannot be propagated in a Fluid in right Lines, beyond
an Obstacle which stops part of the Motion, but will bend and
spread every way into the quiescent Medium which lies beyond
the Obstacle. Gravity tends downward, but the Pressure of
Water arising from Gravity tends every way with equal Force,
and is propagated as readily, and with as much force sideways
as downwards, and through crooked passages as through strait
ones. The Waves on the surface of stagnating Water, passing
by the sides of a broad Obstacle which stops part of them, bend
afterwards and dilate themselves gradually into the quiet Water
behind the Obstacle. The Waves, Pulses or Vibrations of the Air,
wherein Sounds consist, bend manifestly, though not so much as
the Waves of Water. For a Bell or a Cannon may be heard
beyond a Hill which intercepts the sight of the sounding Body,

and Sounds are propagated as readily through crooked Pipes as through streight ones. But Light is never known to follow crooked Passages nor to bend into the Shadow. For the fix'd Stars by the Interposition of any of the Planets cease to be seen. And so do the Parts of the Sun by the Interposition of the Moon, *Mercury* or *Venus.* The Rays which pass very near to the edges of any Body, are bent a little by the action of the Body, as we shew'd above; but this bending is not towards but from the Shadow, and is perform'd only in the passage of the Ray by the Body, and at a very small distance from it. So soon as the Ray is past the Body, it goes right on.

The rest of this Query is omitted.

Qu. 29. Are not the rays of Light very small Bodies emitted from shining Substances? For such Bodies will pass through uniform Mediums in right Lines without bending into the Shadow, which is the Nature of the Rays of Light. They will also be capable of several Properties, and be able to conserve their Properties unchanged in passing through several Mediums, which is another Condition of the Rays of Light. Pellucid Substances act upon the Rays of Light at a distance in refracting, reflecting, and inflecting them, and the Rays mutually agitate the Parts of those Substances at a distance for heating them; and this Action and Re-action at a distance very much resembles an attractive Force between Bodies. If Refraction be perform'd by Attraction of the Rays, the Sines of Incidence must be to the Sines of Refraction in a given Proportion, as we show'd in our Principles of Philosophy: And this Rule is true by Experience. The Rays of Light in going out of Glass into a *Vacuum,* are bent towards the Glass; and if they fall too obliquely on the *Vacuum,* they are bent backwards into the Glass, and totally reflected; and this Reflexion cannot be ascribed to the Resistance of an absolute *Vacuum,* but must be caused by the Power of the Glass attracting the Rays at their going out of it into the Vacuum, and bringing them back. For if the farther Surface of the Glass be moisten'd with Water or clear Oil, or liquid and clear Honey, the Rays which would otherwise be reflected will go into the Water, Oil, or Honey; and therefore are not reflected before they arrive at the farther Surface of the Glass, and begin to go out of it. If they go out of it into the Water, Oil, or Honey, they go on, because the Attraction of the Glass is almost balanced and rendered ineffectual by the contrary Attraction of the Liquor.   But if they go out of it into a *Vacuum* which has not Attraction to balance that of the Glass, the Attraction of the Glass either bends and refracts them, or brings them back and reflects them.   And this is still more evident by laying together two Prisms of Glass, or two Object-glasses of very long Telescopes, the one plane, the other a little convex, and so compressing them that they do not fully touch, nor are too far asunder.   For the Light which falls upon the farther Surface of the first Glass where the Interval between the Glasses is not above the ten hundred thousandth Part of an Inch, will go through that Surface, and through the Air or *Vacuum* between the Glasses, and enter into the second Glass, as was explain'd in the first, fourth, and eighth Observations of the first Part of the second book.   But, if the second Glass be taken away, the Light which goes out of the second Surface of the first Glass into the Air or *Vacuum*, will not go on forwards, but turns back into the first Glass, and is reflected; and therefore it is drawn back by the Power of the first Glass, there being nothing else to turn it back.   Nothing more is requisite for producing all the variety of Colours, and degrees of Refrangibility, than that the Rays of Light be Bodies of different Sizes, the least of which may make violet, the weakest and darkest of the Colours, and be more easily diverted by refracting Surfaces from the right Course; and the rest as they are bigger and bigger, may make the stronger and more lucid Colours, blue, green, yellow, and red, and be more and more difficultly diverted. Nothing more is requisite for putting the Rays of Light into Fits of easy Reflexion and easy Transmission, than that they be small Bodies which by their attractive Powers, or some other Force, stir up Vibrations in what they act upon, which Vibrations being swifter than the Rays, overtake them successively, and agitate them so as by turns to increase and decrease their Velocities, and thereby put them into those Fits.   And lastly, the unusual Refraction of Island-Crystal looks very much as if it were perform'd by some kind of attractive virtue lodged in certain Sides both of the Rays, and of the Particles of the Crystal.   For were it not for some kind of Disposition or Virtue lodged in some Sides of the Particles of the Crystal, and not in their other Sides, and which inclines and bends the Rays towards the Coast of unusual Refraction, the Rays which fall perpendicularly on the Crystal, would not be refracted towards that Coast rather than towards any other Coast, both at their Incidence and at their Emergence, so as to emerge perpendicularly by a contrary Situation of the Coast of unusual Refraction at the second Surface; the Crystal acting upon the Rays after they have pass'd through it, and are emerging into the Air; or, if you please, into a *Vacuum*. And since the Crystal by this Disposition or Virtue does not act upon the Rays, unless when one of their Sides of unusual Refraction looks towards that Coast, this argues a Virtue or Disposition in those Sides of the Rays, which answers to, and sympathizes with that Virtue or Disposition of the Crystal, as the Poles of two Magnets answer to one another. And as Magnetism may be intended and remitted, and is found only in the Magnet and in Iron: So this Virtue of refracting the perpendicular Rays is greater in Island-Crystal, less in Crystal of the Rock, and is not yet found in other Bodies. I do not say that this Virtue is magnetical: It seems to be of another kind. I only say, that whatever it be, it's difficult to conceive how the Rays of Light, unless they be Bodies, can have a permanent Virtue in two of their Sides which is not in their other Sides, and this without any regard to their Position to the Space or Medium through which they pass.

What I mean in this Question by a *Vacuum*, and by the Attractions of the Rays of Light towards Glass or Crystal, may be understood by what was said in the 18th, 19th, and 20th Questions.

YOUNG

The following extracts present Thomas Young's (p. 59) discovery of the interference of light. The first is taken from the *Philosophical Transactions*, from a paper entitled "Experiments and Calculations Relative to Physical Optics," dated November 24, 1803.

The second extract is from Lecture XXXIX of Young's *Course of Lectures on Natural Philosophy and the Mechanical Arts*, published in 1807. It contains Young's exposition of the interference of light.

INTERFERENCE OF LIGHT

1. Experimental Demonstration of the General Law of the Interference of Light.

In making some experiments on the fringes of colours accompanying shadows, I have found so simple and so demonstrative a proof of the general law of the interference of two portions of light, which I have already endeavoured to establish, that I think it right to lay before the Royal Society a short statement of the facts, which appear to me to be thus decisive. The proposition, on which I mean to insist at present, is simply this, that fringes of colours are produced by the interference of two portions of light; and I think it will not be denied by the most prejudiced, that the assertion is proved by the experiments I am about to relate, which may be repeated with great ease, whenever the sun shines, and without any other apparatus than is at hand to every one.

Exper. 1.　I made a small hole in a window shutter, and covered it with a piece of thick paper, which I perforated with a fine needle. For greater convenience of observation, I placed a small looking glass without the window shutter, in such a position as to reflect the sun's light, in a direction nearly horizontal, upon the opposite wall, and to cause the cone of diverging light to pass over a table, on which were several little screens of card paper.　I brought into the sunbeam a slip of card, about one thirtieth of an inch in breadth, and observed its shadow, either on the wall, or on other cards held at different distances.　Besides the fringes of colours on each side of the shadow, the shadow itself was divided by similar parallel fringes, of smaller dimensions, differing in number, according to the distance at which the shadow was observed, but leaving the middle of the shadow always white.　Now these fringes were the joint effects of the portions of light passing on each side of the slip of card, and inflected, or rather diffracted, into the shadow.　For, a little screen being placed either before the card, or a few inches behind it, so as either to throw the edge of its shadow on the margin of the card, or to receive on its own margin the extremity of the shadow of the card, all the fringes which had before been observed in the shadow on the wall immediately disappeared, although the light inflected on the other side was allowed to retain its course, and although this light must have undergone any modification that the proximity of the other edge of the slip of card might have been capable of occasioning.　When the interposed screen was at a greater distance behind the narrow card, it was necessary to plunge it more deeply into the shadow, in order to extinguish the parallel lines; for here the light, diffracted from the edge of the object, had entered further into the shadow, in its way towards the fringes.　Nor was it for want of a sufficient intensity of light, that one of the two portions was incapable of producing the fringes alone; for, when they were both uninterrupted, the lines appeared, even if the intensity was reduced to one tenth or one twentieth.

INTERFERENCE OF LIGHT

Supposing the light of any given colour to consist of undulations, of a given breadth, or of a given frequency, it follows that these undulations must be liable to those effects which we have already examined in the case of the waves of water, and the pulses of sound. It has been shown that two equal series of waves, proceeding from centres near each other, may be seen to destroy each other's effects at certain points, and at other points to redouble them; and the

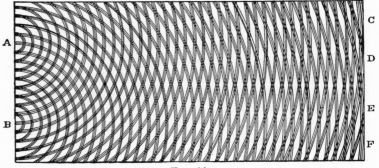

FIG. 66.

beating of two sounds has been explained from a similar interference. We are now to apply the same principles to the alternate union and extinction of colours. (Fig. 66.)

In order that the effects of two portions of light may be thus combined, it is necessary that they be derived from the same origin, and that they arrive at the same point by different paths, in directions not much deviating from each other. This deviation may be produced in one or both of the portions by diffraction, by reflection, by refraction, or by any of these effects combined: but the simplest case appears to be, when a beam of homogeneous light falls on a screen in which there are two very small holes or slits, which may be considered as centres of divergence, from whence the light is diffracted in every direction. In this case, when the two newly formed beams are received on a surface placed so as to intercept them, their light is divided by dark stripes into portions nearly equal, but becoming wider as the surface is more remote from the apertures, so as to subtend very nearly equal

angles from the apertures at all distances, and wider also in the
same proportion as the apertures are closer to each other. The
middle of the two portions is always light, and the bright stripes
on each side are at such distances, that the light, coming to them
from one of the apertures, must have passed through a longer
space than that which comes from the other, by an interval which
is equal to the breadth of one, two, three, or more of the supposed
undulations, while the intervening dark spaces correspond to a
difference of half a supposed undulation, of one and a half, of two
and a half, or more.

From a comparison of various experiments, it appears that the
breadth of the undulations constituting the extreme red light must
be supposed to be, in air, about one 36 thousandth of an inch,
and those of the extreme violet about one 60 thousandth; the mean
of the whole spectrum, with respect to the intensity of light, being
about one 45 thousandth. From these dimensions it follows,
calculating upon the known velocity of light, that almost 500 millions of millions of the slowest of such undulations must enter the
eye in a single second. The combination of two portions of white
or mixed light, when viewed at a great distance, exhibits a few
white and black stripes, corresponding to this interval; although,
upon closer inspection, the distinct effects of an infinite number
of stripes of different breadths appear to be compounded together,
so as to produce a beautiful diversity of tints, passing by degrees
into each other. The central whiteness is first changed to a
yellowish, and then to a tawny colour, succeeded by crimson,
and by violet and blue, which together appear, when seen at a
distance, as a dark stripe; after this a green light appears, and the
dark space beyond it has a crimson hue; the subsequent lights are
all more or less green, the dark spaces purple and reddish; and
the red light appears so far to predominate in all these effects,
that the red or purple stripes occupy nearly the same place in the
mixed fringes as if their light were received separately.

The comparison of the results of this theory with experiments
fully establishes their general coincidence; it indicates, however, a
slight correction in some of the measures, on account of some
unknown cause, perhaps connected with the intimate nature of
diffraction, which uniformly occasions the portions of light,
proceeding in a direction very nearly rectilinear, to be divided
into stripes or fringes a little wider than the external stripes,
formed by the light which is more bent.

When the parallel slits are enlarged, and leave only the intervening substance to cast its shadow, the divergence from its opposite margins still continues to produce the same fringes as before, but they are not easily visible, except within the extent of its shadow, being overpowered in other parts by a stronger light; but if the light thus diffracted be allowed to fall on the eye, either within the shadow, or in its neighbourhood, the stripes will still appear; and in this manner the colours of small fibres are probably formed. Hence if a collection of equal fibres, for example a lock of wool, be held before the eye when we look at a luminous object, the series of stripes belonging to each fibre combine their effects, in such a manner, as to be converted into circular fringes or coronae. This is probably the origin of the coloured circles or coronae sometimes seen round the sun and moon, two or three of them appearing together, nearly at equal distances from each other and from the luminary, the internal ones being, however, like the stripes, a little dilated. It is only necessary that the air should be loaded with globules of moisture, nearly of equal size among themselves, not much exceeding one two thousandth of an inch in diameter, in order that a series of such coronae, at the distance of two or three degrees from each other, may be exhibited.

If, on the other hand, we remove the portion of the screen which separates the parallel slits from each other, their external margins will still continue to exhibit the effects of diffracted light in the shadow on each side; and the experiment will assume the form of those which were made by Newton on the light passing between the edges of two knives, brought very nearly into contact; although some of these experiments appear to show the influence of a portion of light reflected by a remoter part of the polished edge of the knives, which indeed must unavoidably constitute a part of the light concerned in the appearance of fringes, wherever their whole breadth exceeds that of the aperture, or of the shadow of the fibre.

The edges of two knives, placed very near each other, may represent the opposite margins of a minute furrow, cut in the surface of a polished substance of any kind, which, when viewed with different degrees of obliquity, present a series of colours nearly resembling those which are exhibited within the shadows of the knives: in this case, however, the paths of the two portions of light before their incidence are also to be considered, and the whole difference of these paths will be found to determine the appearance of colour in the usual manner; thus when the surface is so situated, that the image of the luminous point would be seen in it by regular reflection, the difference will vanish, and the light will remain perfectly white, but in other cases various colours will appear, according to the degree of obliquity. These colours may easily be seen, in an irregular form, by looking at any metal, coarsely polished, in the sunshine; but they become more distinct and conspicuous, when a number of fine lines of equal strength are drawn parallel to each other, so as to conspire in their effects.

It sometimes happens that an object, of which a shadow is formed in a beam of light, admitted through a small aperture, is not terminated by parallel sides; thus the two portions of light, which are diffracted from two sides of an object, at right angles with each other, frequently form a short series of curved fringes within the shadow, situated on each side of the diagonal, which were first observed by Grimaldi, and which are completely explicable from the general principle, of the interference of the two portions encroaching perpendicularly on the shadow.

But the most obvious of all the appearances of this kind is that of the fringes, which are usually seen beyond the termination of any shadow, formed in a beam of light, admitted through a small aperture: in white light three of these fringes are usually visible, and sometimes four; but in light of one colour only, their number is greater; and they are always much narrower as they are remoter from the shadow. Their origin is easily deduced from the interference of the direct light with a portion of light reflected from the margin of the object which produces them, the obliquity of its incidence causing a reflection so copious as to exhibit a visible effect, however narrow that margin may be; the fringes are, however, rendered more obvious as the quantity of this reflected light is greater. Upon this theory it follows that the distance of the first dark fringe from the shadow should be half as great as that of the fourth, the difference of the lengths of the different paths of the light being as the squares of those distances; and the experiment precisely confirms this calculation, with the same slight correction only as is required in all other cases; the distances of the first fringes being always a little increased. It may also be observed, that the extent of the shadow itself is always augmented, and nearly in an equal degree with that of the fringes: the reason of this circumstance appears to be the gradual loss of light at the edges of every separate beam, which is so strongly analogous to the phenomena visible in waves of water. The same cause may also perhaps have some effect in producing the general modification or correction of the place of the first fringes, although it appears to be scarcely sufficient for explaining the whole of it.

A still more common and convenient method, of exhibiting the effects of the mutual interference of light, is afforded us by the colours of the thin plates of transparent substances. The lights are here derived from the successive partial reflections produced by the upper and under surface of the plate, or when the plate is viewed by transmitted light, from the direct beam which is simply refracted, and that portion of it which is twice reflected within the plate. The appearance in the latter case is much less striking than in the former, because the light thus affected is only a small portion of the whole beam, with which it is mixed; while in the former the two reflected portions are nearly of equal intensity, and may be separated from all other light tending to overpower them. In both cases, when the plate is gradually reduced in thickness to an extremely thin edge, the order of colours may be precisely the same as in the stripes and coronae already described; their distance only varying when the surfaces of the plate, instead of being plane, are concave, as it frequently happens in such experiments. The scale of an oxid, which is often formed by the effect of heat on the surface of a metal, in particular of iron, affords us an example of such a series formed in reflected light: this scale is at first inconceivably thin, and destroys none of the light reflected, it soon, however, begins to be of a dull yellow, which changes to red, and then to crimson and blue, after which the effect is destroyed by the opacity which the oxid acquires. Usually, however, the series of colours produced in reflected light follows an order somewhat different: the scale of oxid is denser than the air, and the iron below than the oxid; but where the mediums above and below the plate are either both rarer or both denser than itself, the different natures of the reflections at its different surfaces appear to produce a modification in the state of the undulations, and the infinitely thin edge of the plate becomes black instead of white, one of the portions of light at once destroying the other, instead of cooperating with it. Thus when a film of soapy water is stretched over a wine glass, and placed in a vertical position, its upper edge becomes extremely thin, and appears nearly black, while the parts below are divided by horizontal lines into a series of coloured bands; and when two glasses, one of which is slightly convex, are pressed together with some force, the plate of air between them exhibits the appearance of coloured rings, beginning from a black spot at the centre, and becoming narrower and narrower, as the curved figure of the glass causes the thickness of the plate of air to increase more and more rapidly. The black is succeeded by a violet, so faint as to be scarcely perceptible; next to this is an orange yellow, and then crimson and blue. When water, or any other fluid, is substituted for the air between the glasses, the rings appear where the thickness is as much less than that of the plate of air, as the refractive density of the fluid is greater; a circumstance which necessarily follows from the proportion of the velocities with which light must, upon the Huygenian hypothesis, be supposed to move in different mediums. It is also a consequence equally necessary in this theory, and equally inconsistent with all others, that when the direction of the light is oblique, the effect of a thicker plate must be the same as that of a thinner plate, when the light falls perpendicularly upon it; the difference of the paths described by the different portions of light precisely corresponding with the observed phenomena.

MALUS

Etienne-Louis Malus was born on June 23, 1775, in Paris. His father was a public official. After a careful early education he entered the school of military engineering. He served for a while as a common soldier during the Revolution, but when the École Polytechnique was founded he was sent there as one of its students. He reentered the army as an officer of engineers in 1796, served in several campaigns and in Egypt. After his return, while he still continued in the service, he was connected as an examiner with the École Polytechnique. It was while in Paris that he made the discovery of polarization by reflection. Biot relates that his first observation of this phenomenon was made when he observed through a crystal of Iceland spar the sunlight reflected from the windows of the Luxembourg. He died in Paris on February 23, 1812.

The extract in which polarization by reflection is described was published in the *Mémoires de la Société d'Arcueil*, Vol. 2, p. 143. The paper begins with a general account of the modifications impressed on rays of light when they are sent through crystals of Iceland spar, as exhibited when the two rays formed by refraction are received on another similar crystal. He then proceeds as follows:

POLARIZATION BY REFLECTION

Thus the characteristic which distinguishes direct light from that which has been subjected to the action of a crystal, is that the one may always be divided into two beams, while in the other this division depends on the angle contained between the plane of incidence and that of the principal section.

This power of changing the character of light and of giving to it a new property, which it carries with it, is not peculiar to Iceland spar; I have found it in all known substances which give double images. It is a remarkable thing about this phenomenon that it is not necessary, when we produce it, to use two crystals of the same sort. Thus the second crystal, for example, may be carbonate of lead, or sulphate of baryta: the first crystal may be a crystal of sulphur and the second rock crystal. All these substances behave in the same way as two rhomboids of calcspar. In general this disposition of light to be refracted in two beams or in one only depends only on the particular position of the axis of the integrating molecules of the crystals which we use, whatever may be otherwise their chemical origins and the natural or artificial faces at which refraction takes place. This result proves that the modification that light receives from these different bodies is perfectly identical.

In order to exhibit more plainly the phenomena which I have described, we may look at the flame of a candle through two prisms of different materials which give double refraction, set one over the other. We shall find in general four images of the flame; but if we slowly turn one of the prisms about the visual ray as axis, the four images are reduced to two, whenever the principal sections of the contiguous faces are parallel or perpendicular. The two images which disappear do not blend with the other two. We see them gradually extinguished while the others increase in intensity. When the two principal sections are parallel, one of the images is formed by the rays undergoing ordinary refraction in the two prisms, and the other image by the rays undergoing extraordinary refraction. When the two principal sections are at right angles, one of the images is formed by the rays refracted ordinarily by the first crystal and extraordinarily by the second, and the other image is formed by the rays refracted extraordinarily by the first crystal and ordinarily by the second.

Not only can all the crystals which give double images give to light this property of being refracted in two beams or in one according to the position of the refracting crystal, but all transparent bodies, solid or liquid, and even opaque bodies can impress on the molecules of light this singular disposition, which seems to be one of the effects of double refraction.

When a beam of light traverses a transparent substance, a part of the rays is reflected by the refracting surface and another part by the surface of emergence. The cause of this partial reflection, which hitherto has escaped the investigations of physicists, seems to have, in some respects, some analogy with the forces which produce double refraction.

For example, light reflected by the surface of water at an angle of 52° 45′ with the vertical has all the characteristics of one of the beams produced by the double refraction of a crystal of calcspar, of which the principal section is parallel or perpendicular to the plane which passes through the incident ray and the reflected ray and which we will call the plane of reflection.

If we receive this reflected ray on any crystal which has the property of doubling the images, and whose principal section is parallel to the plane of reflection, it will not be divided into two beams as a ray of direct light would have been, but it will be entirely refracted according to the ordinary law, as if the crystal had lost the power of doubling the images. If, on the other hand, the principal section of the crystal is perpendicular to the plane of reflection, the reflected ray will be entirely refracted according to the extraordinary law. In intermediate positions it will be divided into two beams following the same law and in the same proportion as if it had acquired its new character by double refraction. The ray reflected from the surface of the liquid has then in these conditions all the characteristics of an ordinary ray formed by a crystal whose principal section is perpendicular to the plane of reflection.

To analyze this phenomenon completely, I arranged a crystal with its principal section vertical and after I had divided a luminous ray by the aid of double refraction I received the two beams on the surface of water at the angle of 52° 45′. The ordinary ray, when it is refracted, gives up to partial reflection a part of its molecules, as a beam of direct light would have done, but the extraordinary ray enters the liquid completely; none of its molecules escape refraction. On the other hand, when the principal section of the crystal is perpendicular to the plane of incidence, the extraordinary ray only gives rise to partial reflection, and the ordinary ray is completely refracted.

The angle at which light experiences this modification when it is reflected at the surfaces of transparent bodies is different for each of them. In general it is greater for bodies which refract light more.   Above or below this angle a part of the ray is more or less modified in a way analogous to that which occurs when light passes through two crystals whose principal sections are neither parallel nor perpendicular.

If we simply wish to examine this phenomenon without measuring it exactly, we place in front of a candle either the transparent body or the vessel containing the liquid upon which we are going to experiment.   We look through a block of crystal at the image of the flame reflected at the surface of the body or of the liquid. We see in general two images; but by turning the crystal about the visual ray as an axis, we perceive that one of the images diminishes as the other increases in brightness.   Beyond a certain limit the image which was enfeebled begins to increase in brightness at the expense of the other.   We must determine approximately the point at which the intensity of the light is a minimum, and then move the candle nearer or further away from the reflecting body until the angle of incidence is such that one of the two images completely disappears; when this distance is determined, if we continue to turn the crystal slowly, we shall perceive that one of the two images is extinguished alternately at each quarter of a revolution.

FRESNEL

Augustin Jean Fresnel was born at Broglie in Normandy on May 10, 1788. His father was an architect.   Owing to feeble health his education proceeded slowly.   He showed little interest in linguistics but his mathematical abilities were noticed by his teachers when he entered the École Polytechnique.   He soon transferred to the École des Ponts et Chaussées, from which he was graduated as an engineer.   For several years he was fully occupied with the duties of his office.   Apparently in about 1814 he turned his attention to the study of light, and in the course of the next twelve years he published the memoirs which established the undulatory theory of light.   Some of Fresnel's work was done in collaboration with Arago.   He died of consumption in the country near Paris on July 14, 1827.

Fresnel's work is too extensive to make a satisfactory presentation of it possible.   His work largely consisted in elaborate experiments, the results of which he compared with the predictions of the formulas developed by him from the undulatory theory.   The extracts which follow present this theory as Fresnel describes it by the aid of diagrams.   The first is taken from the "*Supplément au deuxième mémoire sur la diffraction de la lumière,*" §38 presented to the French Academy of Sciences on July 15, 1816.   The second is from the *Mémoire Courronné* by the French Academy of Sciences and published in Vol. 5 of the *Mémoires* in 1826.

Diffraction of Light

Fresnel had proved by observation that the diffraction bands could not be explained by supposing them to arise from the interference of the direct light with rays reflected from the edge of the solid obstacle. His observations showed that the bands must arise from points in space which are at considerable distances from the edge of the obstacle. He argues further that the phenomena which he had investigated could not be explained on the Newtonian hypothesis of emission. He then proceeds to present a sketch of the way in which he applies the wave theory to explain the phenomena.

On the wave theory, on the contrary, it seems to me that we can explain how it is that the rays bent into the shadow have their source in the direct light out to a sensible distance from the opaque body. When nothing interferes with the regularity of the undulatory motion produced by a luminous point, it is plain that all the waves ought to be exactly spherical and should have for their center the luminous point. In fact, at every point in space where the ether is condensed, it presses and tends to expand in all directions; but this expansion cannot take place except in a direction perpendicular to the spherical surface to which the point

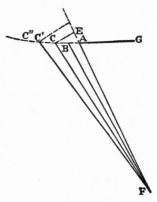

Fig. 67.

belongs, because a similar pressure arises at the same instant throughout the whole extent of the wave. It is no longer so when the vibratory motion is intercepted in a part of the space; and we may imagine that the ends of the waves can give rise to new undulations; but these do not become sensible except in the directions in which they are mutually added and cannot be transmitted in those directions in which the motions oppose each other.

Let *A* be the end of a body *AG*, *F* a point situated inside its shadow, and *ACC'C''* the wave of light of which the body *AG* has intercepted a part. The question is, what portion of the end of this wave can send light to the point *F*.

From the point *F* as center (Fig. 67), and with a radius equal to *AF* plus a half wave length, I describe the arc *EC*, which cuts the wave front at the point *C*; the rays *CF* and *AF* differ by a half wave length. I suppose the point *C'* belonging to the direct wave so situated that $C'F$ is equal to CF plus a half wave length.   Then all the vibrations which leave the arc CC' in this oblique direction will be completely in disaccord with the vibrations coming from the corresponding points of AC.   But all those which arise from CC' are already very much enfeebled by the action of those of the next arc $C'C''$, so that probably they cannot produce a diminution of more than a half in the undulatory motions which originate in AC: with the exception of this extreme arc, each part of the direct wave is contained between two others which destroy the oblique rays which it tends to produce.   It is therefore the middle point B of the arc AC which should be considered as the principal center of the waves which are perceived at the point F.   I suppose here that the obliquity of the rays is so great that the line BF fills sensibly the same conditions in almost all its extent, so that the wave has had time to build itself up in this direction by successive additions.   It results also from this pronounced obliquity that the arc AC is very small and that thus the ray BF which comes from the middle of this arc is almost exactly the mean between the two extreme rays CF and AF.   We see thus that the effective ray BF and consequently the path traversed by the inflected light will be longer by a quarter of a wave length than the path reckoned from the actual edge of the body AG.   We may prove by a similar argument that when the rays are bent outside the shadow the effective ray is shorter by a quarter of a wave length than that which would come from the body.   I consider here considerable inflections as I have already said, and it is natural to suppose that the intermediate rays in the vicinity of the tangent will pass gradually from the increase to the diminution of a quarter of a wave length; but I have not as yet been able to determine by what law. The explanation which I have just given of these variations considered only in the limit, still doubtless leaves much to be desired and is not perhaps free from objection.   However that may be, it seems to me clear that the path traversed by the effective rays when their obliquity becomes sensible differs by a quarter of a wave length from the path reckoned from the edge of the opaque body, being sometimes greater, sometimes less, according to the sense of the inflection; at least the phenomena appear as if that were so.

In fact we have seen that, in the fringes produced by a sufficiently narrow slit, the interval between the two dark bands of the first order is double that of the others, and that thus the position of the dark and bright bands is absolutely the inverse of that which would result from the theory, if we reckon the paths traversed as if they were measured from the edges of the slit. Now this is a consequence of the principle which I have just laid down. Let *A* and *B* (Fig. 68) be the two edges of an opening so small that at the distance at which we observe the fringes the dark band of the first order is situated well outside the nearer tangent, in such a way that the rays which produce it are very sensibly inflected by the edges in contrary senses, one of them within and the other without. I take *F* for the point which the dark band of the first order would occupy if the points *A* and *B* were the centers of the waves; that is to say, if *AF* and *BF* differ by a half wave length. The effective rays of the edges *A* and *B* blend, in this case, into a single one which leaves the middle of *AB* and there is no complete discordance, except between the two extreme rays. The point *F* then should not be dark. I now suppose that *F* is a point where there is complete discordance of any order for the rays *AF* and *BF*: it will be a point of accord for the effective rays *CF* and *DF*; for *CF* is longer than *AF* by a quarter of a wave length while *DF* is shorter than *BF* by the same amount, from which there arises a total difference of a half wave length.

Fig. 68.

I pass now to the fringes which arise from the meeting of the rays which are inflected by the two sides of an opaque body. So long as they are within the shadow and sufficiently distant from the tangent or from the edge of the geometrical shadow, the two effective rays which concur in producing them, being both inflected within the shadow, are longer by a quarter of a wave length than the rays which leave the edges of the body; and since this difference is equal and in the same sense, the dark and light bands ought to be placed in the same way as they would be if the undulations had their centers at the edges of the body; I therefore found in my first observations results which conformed to this hypothesis. However, as the band which we consider approaches one of the two tangents *AE* (Fig. 69) the difference of length diminishes between the effective ray inflected on the side *A* of the opaque body *AB*, and the ray which leaves *A*, while the other effective ray continues to have its quarter of a wave length more than the ray coming from *B*. Thus the difference of the paths passed over increases more rapidly between the two effective rays than between those which leave *A* and *B*; and consequently the width of the fringes ought to diminish. Finally, when the point of intersection *F* of the inflected rays has come out of the shadow and is sufficiently far from *AE* to make the angle *FAE* fairly large, the effective ray *CF* becomes shorter by a quarter of a wave length than *AF*, while *FD* is

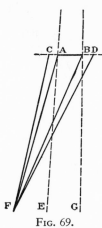

FIG. 69.

always longer than *BF* by the same amount; from which there results a total difference of a half wave length and consequently of a half interval in the position of the dark and bright bands which are sufficiently removed from the edge of the shadow. This is indicated by the observations.

This theory of effective rays which I have described, incomplete as it is, can already give a very simple explanation of the rapid falling off of light which bends around into the interior of shadows. As the inclination of the ray *BF* increases the arc *CA* diminishes (Fig. 67), since *AE* should always be equal to a quarter of a wave length: now it is from the arc *AC* alone

that the vibrations come which are perceived at the point *F*. The intensity of the light will therefore diminish as rapidly as the length of this arc. Suppose first that the inclination of *AF*, or the angle *ACE*, is five minutes, for example, and to simplify the calculations, that the wave *ACC′* is sensibly a straight line; since *AE* should be equal to a quarter of a wave length or to 0.000000144m the length of the arc *AC* which gives the light will be 0.000099m, that is to say nearly one-tenth of a millimeter. Now if I suppose that the obliquity of *BF* is equal to 1 degree, the arc giving the light will be no larger than 0.000008m, that is to say will be less than one-hundredth of a millimeter. We see by these two examples that the source of the undulatory movement of the inflected rays becomes very small as soon as the inflection is at all considerable.

THE DIFFRACTION OF LIGHT

Fresnel considers generally the theory of emission and also the wave theory on the supposition that the rays bent by passing near an obstacle are bent at the edge and shows that both these theories lead to results that are inconsistent with experiment. He then proceeds to develop his own theory.

If we assume that light consists in vibrations in the ether similar to those of sound waves, it is easy to explain the inflection of luminous rays at sensible distances from a screen. In fact, when a small portion of an elastic fluid undergoes a condensation, for example, it tends to dilate in all directions; and if in a complete wave the molecules only move parallel to the normal, this comes from the fact that all parts of the wave which are situated on the same spherical surface simultaneously experience the same condensation or expansion, so that the transverse pressures are in equilibrium. But if a portion of the luminous wave is intercepted, or retarded in its progress by the interposition of a screen either opaque or transparent, it would appear that this transverse equilibrium should be destroyed, and that different parts of the wave ought to acquire the faculty of sending rays in new directions.

It would doubtless be very difficult to follow by mechanical analysis all the modifications which the luminous wave experiences successively from the moment when the encounter with the screen intercepts a part of it: it is not in this way, however, that we shall proceed to attempt to determine the laws of diffraction. We shall not endeavor to discover what is happening in the neighborhood of the opaque body, where these laws are, without doubt, very complicated, and where the form of the edges of the screen should have a notable influence on the position and intensity of the fringes. We propose to calculate the relative intensity of the different points of the luminous wave only after it has passed the screen by a large number of wave lengths, so that the positions of the wave which we shall consider will always be thought of as removed from the screen by a distance which is very large in proportion to the length of a wave.

We shall not look at the problem of the vibrations of an elastic fluid from the same point of view as that ordinarily taken by mathematicians, that is to say, by considering only a single displacement. In nature vibrations are never isolated; they repeat themselves always a great many times, as we can notice in the oscillations of a pendulum, or the vibrations of sounding bodies. We shall suppose that the vibrations of the luminiferous particles are executed in the same way, succeeding each other regularly in numerous series; a hypothesis to which we are led by analogy and

which furthermore seems to be a consequence of the forces which
hold the molecules of a body in equilibrium. To conceive of a
numerous succession of almost equal oscillations from the same
luminous particle, it is sufficient to suppose that its density is
much greater than that of the fluid in which it oscillates. This
we already have a right to conclude from the regularity with which
the planets move through this same fluid, which fills the celestial
spaces. It is very probable also that the optic nerve is not disturbed in such a way as to produce the sensation of sight, except
by a certain number of successive shocks:

However extended we suppose all the systems of luminous
waves, it is clear that they have limits, and that when we consider their interferences we cannot say about their extremities,
what is true of the space in which they are superposed. Thus,
for example, two systems of waves of equal length and of the
same intensity which differ in their progress by a half wave
length, destroy each other mutually only at the points in the
ether where they meet, and the two extreme half waves experience
no interference.

We shall suppose nevertheless that the systems of waves undergo
the same modification throughout their whole extent, the difference
between this hypothesis and reality being inappreciable by our
senses; or, what comes to the same thing, we shall consider these
series of luminous undulations as without limit, and as general
vibrations of the ether, in the calculation of their interferences.

Fresnel introduces next a discussion of the combination of two waves of the
same wave length which meet after passing over nearly equal distances. He
obtains an expression for the amplitude of the resulting vibration and other
properties of the resulting disturbance. He then proceeds to apply Huygens's
principle to the problems of diffraction.

ARAGO AND FRESNEL

Dominique-François Jean Arago was born on February 26, 1786, in Estagel.
In 1803 he entered the École Polytechnique and in the next year became connected with the observatory at Paris. In 1806 he went to Spain to assist
in the determination of the length of a degree on the meridian. His operations
were interrupted by the war which broke out between France and Spain.
In June, 1808, he was imprisoned but soon afterwards escaped to Algiers.
After various adventures he finally succeeded in reaching Marseilles in June,
1809, bringing back with him the records of his observations. He was at once

elected to the French Academy and soon after was appointed to the chair of analytical geometry in the École Polytechnique and was also named as one of the astronomers of the Royal Observatory. He resided there until his death on October 2, 1853. Arago was an ardent liberal and took part in the revolutions of 1830 and of 1848. After the second revolution he was for a time Minister of War and Marine. As permanent secretary of the Academy he published a number of biographies of famous scientific men.

Arago formed a close friendship with Fresnel, whose views on the nature of light he ardently supported. He assisted Fresnel in some of his most important work and made original discoveries in the same field.

The following "Mémoire sur l'action que les rayons de lumière polarisée exercent les uns sur les autres" was published in the *Annales de Chimie et de Physique*, Vol. 10, Series 2, p. 288, 1819. In it are described the experiments by Arago and Fresnel (p. 318) from which it can be concluded that the vibrations of light are transverse to the line of progress.

Interference of Polarized Light

Before reporting the experiments which are the subject of this memoir, it will perhaps be worth while to recall some of the beautiful results that Dr. Thomas Young has already obtained by studying, with the rare sagacity which characterizes him, the influence which, in certain circumstances, the rays of light exert on one another.

1. Two rays of homogeneous light coming from the same source, which reach a certain point in space by two different routes which are slightly unequal, enhance each other or destroy each other, forming on the screen which receives them a bright or dark point according as the difference of their routes has one value or another:

2. Two rays enhance each other always when they have passed over equal distances. If we find that they enhance each other again when the difference of the two distances is equal to a certain quantity d, they enhance each other again for all differences which are contained in the series $2d$, $3d$, $4d$, etc. The intermediate values $0 + \frac{1}{2}d$, $d + \frac{1}{2}d$, $2d + \frac{1}{2}d$, etc. determine the cases in which the rays mutually annul each other.

3. The quantity d has not the same value for all the homogeneous rays: in the air it is equal to $\frac{67}{100000}$ millimeters for the extreme red rays of the spectrum and to $\frac{42}{100000}$ only for the violet rays. The values corresponding to the other colors are between these limits.

The periodic colors of colored rings or of halos, etc. appear to depend on the influence that the rays which are first separated exert on one another when they again coincide: in any case, in order that the laws which we have stated should conform to these various phenomena we must assume that the difference in path is not the only cause of the action of the two rays at the point where they cross, except when they are both of them always in the same medium; and that if there is any difference between the refractive powers or the thicknesses of the transparent bodies through which the rays severally pass, that will produce an effect equivalent to a difference in path. There has been elsewhere reported a direct experiment tried by M. Arago which gives the same results, and from which there can also be drawn this conclusion, that a transparent body diminishes the velocity of the light which traverses it in the ratio of the sine of the angle of incidence to the sine of the angle of refraction: so that in all the phenomena of interference two different media produce equal effects when their thicknesses are in the inverse ratio of their indices of refraction. These considerations lead us, as we can see, to a new method for measuring small differences of refrangibility.

During the trials that we made together to test the degree of precision of which this method is capable, one of us (M. Arago) conceived that it would be interesting to see if the actions which the ordinary rays habitually exert on one another would not be modified when the two luminous beams were made to interfere only after they had previously been polarized.

It is known that if we illuminate a narrow strip of a body by light which comes from a radiant point, its shadow is bordered on the outside by a set of fringes formed by the interference of the direct light and of the rays which have been bent in the neighborhood of the opaque body; and that a part of the same light, which enters the geometrical shadow past the two edges of the body, gives rise to fringes of the same sort. We found, to start with, very easily that these two systems of fringes are absolutely similar whether the incident light has not been modified in any way or has reached the body only after having been previously polarized. Thus rays polarized in the same sense influence each other when they meet in the same way as natural rays.

It still remained to try whether two rays originally polarized at right angles, or to use an accepted expression, in opposite senses, would produce phenomena of the same sort when they meet in the interior of the geometrical shadow of an opaque body.

To try this we placed before the radiating source sometimes a rhomboid of calcspar, sometimes an achromatized prism of rock crystal and we obtained thus two luminous points. From each of these proceeded a divergent beam: these two beams, as we see, were polarized in opposite senses. A metallic cylinder was then placed between the two radiant points so as to correspond precisely with the middle of the interval which separated them. With this arrangement a part of the polarized rays of the first beam penetrated from the right into the space behind the cylinder; and a part of the rays of the second beam, polarized in the opposite sense, entered it from the left. Some rays of these two groups met each other near the line which joined the center of the cylinder and the middle of the straight line which passed through the two radiant points. These rays had passed over paths which were equal or only slightly different: it seemed then that they ought to have formed fringes; but we did not see the least trace of them even with a lens, in a word, the rays had crossed without influencing each other. The only system of fringes that we perceived in this experiment, came from the interference of the rays which penetrated into the shadow by the two edges of the cylinder, having started from each radiant point considered separately. Those that we tried to produce by crossing rays polarized in opposite senses would have evidently occupied a position between these.

The crystal which we used separated the images only slightly, and the two rays, the ordinary and extraordinary, traversed it through almost equal thicknesses. However, we had already too often noticed, when we made similar experiments, how much the smallest difference in the velocity of the rays, in the thickness, or in the refracting power of the media that they traversed, modified sensibly the phenomena of interference, not to be convinced of the necessity of repeating our test while avoiding all the doubtful circumstances to which we have called attention. Each of us sought to find the way to do this.

M. Fresnel proposed first for this two different methods. The principle of interference shows that rays coming from two different foci, originating from the same source, will form when they cross dark and bright bands, without its being necessary to introduce in their path any opaque body.

To settle the question it is therefore sufficient to try if the two images formed by placing a rhomboid of calcspar in the path of the rays from a luminous point would give a similar result;

but since, according to the theory of double refraction, the extra-ordinary ray moves in the calcspar more quickly than the ordinary ray, we must artificially compensate for this excess of velocity before allowing the rays to cross. To do this, utilizing an experiment of M. Arago, which has been cited, M. Fresnel placed in the path of the extraordinary beam a glass plate whose thickness had been determined by calculation in such a way that by traversing it with perpendicular incidence this beam lost almost all the advance which it had made in the crystal on the ordinary beam; further, by slightly tilting the plate we could obtain in this respect an exact compensation. In spite of this, the crossing of the two beams, polarized in opposite senses, did not give the bands.

In another experiment, to compensate the difference of velocity of the two rays, M. Fresnel let both of them fall on an unsilvered sheet of plate glass, whose thickness had been calculated in such a way that the extraordinary ray when reflected perpendicularly at the second face, lost by passing twice through the glass more than it gained in traversing the crystal; a gradual change of inclination led finally to a perfect compensation: nevertheless, under any incidence, the ordinary rays reflected at the upper surface of the glass gave no perceptible bands when they met the rays reflected from the second surface.

M. Fresnel avoided the defect of the preceding experiment, of depending on theoretical considerations, and furthermore pre-served all the intensity of the incident light, by the following procedure. He cut a rhomboid of calcspar through the middle and placed the two parts one before the other in such a way that the principal sections were perpendicular: in this situation the ordinary beam of the first crystal experienced extraordinary refraction in the second; and reciprocally, the beam which first followed the extraordinary route experienced next ordinary refraction. In looking through this apparatus there was seen only a double image of the luminous point; each beam had experienced the two sorts of refraction; the sums of the paths traversed by each of them in the two crystals at once ought then to be equal, since by hypothesis these crystals had both the same thickness; all was then compensated in respect of the velocities and the routes traversed; and nevertheless the two systems of rays polarized in opposite senses did not produce any perceptible fringes by interference. We may add, that for fear that the two pieces of the rhomboid had not exactly the same thickness he took pains in each test to make slight and gradual changes of the angle at which the incident rays encountered the second crystal.

The method that M. Arago proposed, on his part, for trying the same experiment was independent of double refraction. It has been known for a long time that if in a thin plate we cut two fine slits near each other and if we illuminate them by light from a single luminous point, there are formed behind the plate very bright fringes resulting from the action that the rays from the slit on the right exert on the rays of the other slit in the points where they meet. To polarize in opposite senses the rays coming from these two openings M. Arago had first proposed to use a thin piece of agate, to cut it through the middle and to place each half before one of the slits in such a way that the portions of the agate which at first were contiguous, would be at right angles to each other. This arrangement should evidently produce the expected effect: but since he did not have at the moment a suitable agate, M. Arago proposed to substitute for it two piles of plates, and in order to preserve the thinness that was necessary for the success of the experiment, to make them of sheets of mica.

To do this we chose fifteen of these sheets, as transparent as possible, and superposed them. Then with a sharp instrument this single pile was divided through the middle. It is clear that the two partial piles which were made by this process should have closely the same thickness, at least in the parts which were at first contiguous, even if the sheets composing the pile were sensibly prismatic. These piles polarized almost completely the light which traversed them when the incidence measured from the surface was 30°. It is precisely at this angle that each of them was placed behind one of the slits of the copper plate.

When the two planes of incidence were parallel, that is, when the two piles were inclined in the same sense, from above downward for example, we clearly saw the bands formed by the interference of the two polarized rays, just as when two ordinary rays of light act on each other; but if, by turning one of the piles about the incident ray, the two planes of incidence became perpendicular; if while the first pile always remained inclined from above downward, the second was inclined, for example, from left to right, the two emergent beams, then polarized at right angles or in opposite senses, no longer formed any perceptible bands when they encountered each other.

The precautions that we had taken to give the same thickness to the two piles show clearly enough that when we placed them behind the slits we took pains to have them traversed by the light in those parts which were contiguous before the pile was cut. It has otherwise been seen, and this circumstance removes all the difficulties which could have been suggested in this regard, that the fringes showed themselves as in ordinary light when the rays were polarized in the same sense; we may nevertheless add that a slow and gradual change in the inclination of one of the piles never made the bands appear, when the planes of incidence were at right angles.

On the same day on which we tried the system of the two piles we made an experiment suggested by M. Fresnel, really less direct than the preceding one but also easier to carry out, which also demonstrates the impossibility of producing fringes by the crossing of luminous rays at right angles.

We placed behind the plate of copper, pierced with its two slits, a thin plate of selenite, for example: since this crystal is doubly refracting, there come from each slit two rays polarized in opposite senses; now if the rays of one sort can act on the rays of the opposite sort we ought to see with this apparatus three distinct systems of fringes. The ordinary rays from the right combined with the ordinary rays from the left should give a first system corresponding exactly to the middle of the interval between the two slits; the bands formed by the interference of the two extraordinary beams should occupy the same place as the others, and should increase their intensity but should not be distinguishable from them. As to those which result from the action of the ordinary rays on the right and the extraordinary rays on the left, and reciprocally, they should be placed on the right and on the left of the central fringes and so much further apart as the plate employed is thicker: for we have seen that a difference of velocity changes the position of the fringes as well as a difference of route. Now since the fringes in the middle are the only ones visible, even when the interposed plate is so thin that the two other systems should not be far removed, we must conclude that rays of different names or polarized in opposite senses do not affect each other.

To further confirm this conclusion, let us suppose that we cut in two our plate of selenite; that one of the halves is set up at the first slit; that the other is placed behind the other slit and that the axes, instead of being parallel, as they were when the plate was uncut, are now perpendicular. With this arrangement the ordinary ray coming from the slit on the right will be polarized in the same sense as the extraordinary ray coming from the slit on the left, and reciprocally. These rays will then form fringes; but since their velocities in the crystal are not equal these fringes do not correspond to the center of the interval between the two openings; only the ordinary or the extraordinary rays from one of the slits, by meeting the rays of the same name coming from the other slit, can give central images; but since from the particular arrangement that has been made of the two pieces of crystal these rays are polarized in opposite senses they ought not to affect each other, so we see only the first two systems of fringes separated by an interval which is uniformly white.

If, without changing any other arrangements of the preceding experiment, we set the two plates of selenite in such a way that their axes, instead of being perpendicular, make an angle of 45°, we then perceive all at once three systems of fringes; for each pencil from the right acts on the two pencils on the left and reciprocally, since their planes of polarization are no longer perpendicular. We should here remark that the system in the middle is the most intense and results from the perfect superposition of the bands formed by the interference of the beams of the same name.

Let us take up again the arrangement with the piles and suppose that the planes of incidence are perpendicular and that the beams transmitted through the two slits are polarized in opposite senses. Let us further place between the sheet of copper and the eye a doubly refracting crystal, whose principal section makes an angle of 45° with the planes of incidence. From the known laws of double refraction the rays transmitted by the piles will each be divided in the crystal into two rays of the same intensity and polarized in two perpendicular directions, one of which is precisely that of the principal section. One might then expect to observe in this experiment a series of fringes produced by the action of the ordinary beam on the right and the ordinary beam on the left, and a second similar series coming from the interference of the two extraordinary beams; nevertheless, we do not perceive the slightest trace of it and the four beams when they cross give only continuous light.

This experiment, which originated with M. Arago, proves that two rays which have been originally polarized in opposite senses can then be brought to the same plane of polarization without thereby acquiring the power of affecting each other.

In order that two rays polarized in opposite senses and then brought to the same state of polarization can mutually affect each other, it is necessary that they should have first started from the same plane of polarization.   This results from the experiment devised by M. Fresnel, which we proceed to describe.

We place in a pencil of polarized light coming from a radiant point a plate of selenite cut parallel to the axis, and covered with a thin sheet of copper pierced with two openings: the incidence is perpendicular and the axis of the plate makes an angle of 45° with the original plane of polarization.   As in all the analogous experiments we observe the shadow of the sheet with a lens; but this time we place in addition in front of its focus a rhomboid of calcspar which gives a sensible double refraction and whose principal section makes in its turn with that of the plate an angle of 45°.   Then we see in each image three systems of fringes; the one of them corresponds exactly to the middle of the shadow; the other systems are to the left and right of the first.

Let us now examine how these three systems of fringes arise in one of the two images, in the ordinary image, for example.

The beams polarized in the same sense which pass through the two slits are each divided when they traverse the plate of selenite into two beams polarized in opposite senses.   Since the double refraction of the plate is insensible, the ordinary and extraordinary parts of each beam follow the same route but with different velocities.

One of these double beams, that of the slit on the right, for example, is divided when it traverses the rhomboid into four beams, two ordinary and two extraordinary; but in fact we only see two, since the component parts of the beams of the same name coincide. It is further evident, from the known laws of double refraction and the positions which we have given to the plate of selenite and to the rhomboid, that when the ordinary beam leaves this last crystal, it is composed of half of the ray which was ordinary in the plate and of half of the extraordinary ray; and that the two other halves of these same rays pass to the extraordinary image, which we have agreed to neglect for the present.   The beam coming from the slit on the left behaves in the same way.   We see, in a word, that after having traversed the two crystals in this new apparatus, the ordinary beams coming from the slit on the right or from that on the left are both of them composed of a portion of light which has always followed the ordinary route in the two crystals and of a second portion which at first was extraordinary.

Those rays coming from the two slits which, when they traverse the plate of selenite and the rhomboid, always follow the ordinary route traverse equal paths with the same velocity and therefore, when they reunite, should give rise to central fringes. The same is true of the rays which, while extraordinary in the plate of selenite, have become at the same time ordinary by the action of the rhomboid; the fringes of the middle of the shadow therefore result from the superposition of two different systems.

As to the portion of light on the right, which is extraordinary, for example, in the plate of selenite and becomes ordinary by traversing the rhomboid, it will have traversed a path equal to that of the portion of the beam on the left, which is always refracted ordinarily; but as these rays in the plate have slightly unequal velocities, the points where they form sensible fringes when they meet, instead of corresponding to the middle of the interval between the two slits, will be on the right, that is to say, on the side opposite to the ray which is for a moment extraordinary and so moves more slowly. There comes finally for the last combination the interference of the part of the beam from the right, which is ordinary in the two crystals, with the portion of the beam from the left which is extraordinary in the plate and ordinary in the rhomboid and which therefore gives rise to bands situated on the left of the center.

We have now explained the passage of the rays which take part in the formation of the three systems of fringes in the apparatus in question; and we may notice that the systems on the right and left result from the interference of rays first polarized in opposite senses in the plate of selenite and brought back finally to a similar polarization by the action of the rhomboid. Two rays polarized in opposite senses and brought back to the same plane of polarization can give fringes when they meet; but that this may happen it is indispensable that they have been first polarized in the same sense.

We have left out of consideration hitherto the mutual action of the two pencils which undergo extraordinary refraction in the rhomboid. These pencils furnish also three systems of fringes, but they are separated from the former ones. If, while everything remains in the same condition, we substitute for the rhomboid a plate of selenite or of rock crystal, which does not give two distinct images, the six systems, instead of producing three by superposition, are reduced to that of the middle one. This remarkable result demonstrates: 1. That fringes resulting from the interference of ordinary rays are complements to the fringes produced by the extraordinary rays; and, 2. That these two systems are so placed that a bright fringe of the first system corresponds to a dark fringe of the second and reciprocally; without these two conditions we would perceive nothing but a uniform and continuous light on the two sides of the central fringes. We encounter here the difference of a half wave, as in the phenomena of colored rings.

The experiments that we have reported lead us then definitely to the following conclusions:

1. In the same condition in which two rays of ordinary light seem to destroy each other mutually, two rays polarized at right angles or in opposite senses exert on each other no appreciable action;

2. The rays of light polarized in one sense act on one another like natural rays: so that in these two sorts of light the phenomena of interference are absolutely the same;

3. Two rays originally polarized in opposite senses can be brought to the same plane of polarization without acquiring thereby the power of affecting each other;

4. Two rays polarized in opposite senses and brought to similar states of polarization affect each other like natural rays, if they come from a pencil originally polarized in one sense;

5. In the phenomena of interference produced by rays which have experienced double refraction, the place of the fringes is not determined uniquely by the difference of the paths and of the velocities; and in certain circumstances which we have pointed out we must take account in addition of a difference of half a wave.

All these laws are deduced, as we have seen, from direct experiment. We might reach them more simply still by starting from the phenomena which are presented by crystalline plates; but then we should have had to admit that the tints with which these plates are colored when they are illuminated by a beam of polarized light arise from the interference of several systems of waves. The demonstrations which we have reported have the advantage of establishing the same laws independently of all hypothesis.

ROEMER

Ole Roemer was born on September 25, 1644, at Aarhuus, Denmark. He was a pupil and an assistant of Bartholinus (p. 280). He removed to Paris, where for ten years he was a tutor of the Dauphin. He was chosen a member of the Academy. On the revocation of the Edict of Nantes he returned home and became professor of mathematics at the University of Copenhagen. He also occupied several official positions. He died at Copenhagen on September 19, 1710.

His observations of the eclipses of Jupiter's satellites made while he was in Paris led him to the discovery of the velocity of light. The following account of his discovery is based upon a memoir presented to the French Academy in 1666. The abstract of this memoir here given is found in the *Mémoires de l'Académie Royale des Sciences*, 1730. This paper is also given in the *Source Book in Astronomy*. It is inserted here because it serves to make complete the account of the measurements of the velocity of light.

THE VELOCITY OF LIGHT

For a long time philosophers have been endeavoring to decide by some experiment whether the action of light is transmitted in an instant to any distance whatever, or re-quires time. M. Roemer of the Royal Academy of Sciences has thought out a way of doing this, based upon observations of the first satellite of Jupiter, by which he has shown that to traverse a distance of about 3000 leagues, which is almost equal to the diameter of the earth, light does not need a second of time.

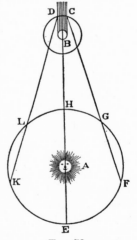

FIG. 70.

Let *A* (Fig. 70) represent the sun, *B*, Jupiter, *C*, the first Satellite as it enters Jupiter's shadow, to come out again at *D*, and let *EFGHLK* represent the Earth at different distances from Jupiter.

Now suppose that when the Earth is in *L* in the second quadrature of Jupiter, the first satellite has been viewed just at its emergence from the shadow at *D*; and that about 42 and one half hours later, that is after one revolution of this satellite, the Earth then being at *K*, it is seen again as it returns to *D*. It is then clear that if light requires time to traverse the distance *LK*, the satellite will seem to return to *D* later than it would have if the

Earth had remained at K, so that the revolution of this satellite, determined thus by its emergences, will be made longer by as much time as light would take to pass from L to K, and that on the contrary in the other quadrature FG, in which the Earth is approaching or going toward the light, the revolutions determined by the immersions will appear diminished by as much as those determined from the emergences appeared increased; and because in the 42 hours and one half that the satellite takes for each revolution the distance between the Earth and Jupiter, in either of the quadratures, changes at least by 210 diameters of the Earth, it will follow that if a second of time was needed by light to traverse the diameter of the Earth, light would take 3 and one-half minutes to traverse each of the intervals FG, KL, and this would result in a difference of about half a quarter of an hour in the times of revolution of the first satellite, one of which was observed in FG and the other in KL, whereas no sensible difference can be detected.

It does not, however, follow that light needs no time; for after having examined the matter more closely he has found that this difference, which was not sensible in two revolutions, became very considerable when several of them were taken together, and that, for example, 40 revolutions observed from the side F were sensibly shorter than 40 others observed from the other side, in whatever position Jupiter happened to be, and this in the ratio of 22 (minutes of time) for the whole distance HE, which is twice the distance from the Earth to the Sun.

The necessity for this new equation of the retardation of light is established by all the observations which have been made at the Royal Academy and at the Observatory for the past eight years, and recently it has been confirmed by an emergence of the first Satellite observed at Paris on November 9th last at 5 hours 34 minutes 45 seconds in the evening, 10 minutes later than one would have expected it by deducing it from those observed in the month of August, when the Earth was much nearer Jupiter, as M. Roemer had predicted to the Academy early in September; and to remove all doubt that this inequality was caused by the retardation of light, he demonstrates that it cannot come from any eccentricity or any other of the causes which are ordinarily suggested to explain the irregularities of the moon and of the other planets. Nevertheless, he did notice that the first satellite of Jupiter was eccentric and that further its revolutions were advanced or retarded according as Jupiter approached the sun oı receded from it, and even that the revolutions of this first satellite were unequal, and yet all these last three causes of inequality do not prevent the first cause plainly showing itself.

BRADLEY

James Bradley was born at Shireborn in Gloucestershire in 1692.   In 1791 he became professor of astronomy at Oxford.   Later he was the astronomer of the Greenwich observatory.   He died in Chalford on July 13, 1762.

Bradley was the discoverer of the aberration of the fixed stars.   He explained aberration by the assumption that light has a finite velocity.   The extract which follows is taken from his paper entitled "An Account of a New Discovered Motion of the Fixed Stars," published in the *Philosophical Transactions*, Vol. 35, p. 637, 1729.

Bradley was engaged in a series of observations designed to detect a parallax of the fixed stars.   That a parallax might be detected in the case of stars that were not too remote from the earth followed as a consequence of the Copernican theory and the detection of a parallax was sought for as an evidence of the truth of that theory.   Bradley discovered an apparent motion of the stars which was so systematic in character and was exhibited in such a way by all of them that he found it necessary to assign a common cause for these apparent motions.   After testing various hypotheses without success, he resolved to accumulate a considerable number of observations before proceeding further to speculate on the cause of the motions observed.   The hypothesis which he finally adopted and a demonstration of the satisfactory way in which it accounted for his observations are given in the following paragraphs.

The Velocity of Light

When the year was completed, I began to examine and compare my observations, and having pretty well satisfied myself as to the general laws of the Phenomena, I then endeavored to find out the cause of them.   I was already convinced, that the apparent motion of the stars was not owing to a nutation of the earth's axis.   The next thing that offered itself, was an alteration in the direction of the plumb-line, with which the instrument was constantly rectified; but this upon trial proved insufficient. Then I considered what refraction might do, but here also nothing satisfactory occurred.   At last I conjectured, that all the Phenomena hitherto mentioned, proceeded from the progressive motion of light and the earth's annual motion in its orbit.   For I perceived, that, if light was propagated in time, the apparent place of a fixt object would not be the same when the eye is at rest, as when it is moving in any other direction than that of the line passing through the eye and object; and that, when the eye is moving in different directions, the apparent place of the object would be different.

I considered this matter in the following manner; I imagined *CA* (Fig. 71) to be a ray of light, falling perpendicularly upon the line *BD*; then if the eye is at rest at *A*, the object must appear

FIG. 71.

in the direction *AC*, whether light be propagated in time or in an instant. But if the eye is moving from *B* towards *A*, and light is propagated in time, with a velocity that is to the velocity of the eye, as *CA* to *BA*; then light moving from *C* to *A*, whilst the eye moves from *B* to *A*, that particle of it, by which the object will be discerned, when the eye in its motion comes to *A*, is at *C* when the eye is at *B*. Joining the points *B*, *C*, I supposed the line *CB*, to be a tube (inclined to the line *BD* in the angle *DBC*) of such a diameter, as to admit of but one particle of light; then it was easy to conceive, that the particle of light at *C* (by which the object must be seen when the eye, as it moves along, arrives at *A*) would pass through the tube *BC*, if it is inclined to *BD* in the angle *DBC*, and accompanies the eye in its motion from *B* to *A*; and that it could not come to the eye, placed behind such a tube, if it had any other inclination to the line *BD*. If instead of supposing *CB* so small a tube, we imagine it to be the axis of a larger; then for the same reason, the particle of light at *C*, could not pass through that axis, unless it is inclined to *BD*, in the angle *CBD*. In like manner, if the eye moved the contrary way, from *D* towards *A*, with the same velocity; then the tube must be inclined in the angle *BDC*. Although therefore the true or real place of an object is perpendicular to the line in which the eye is moving, yet the visible place will not be so, since that, no doubt, must be in the direction of the tube; but the difference between the true and apparent place will be (*caeteris paribus*) greater or less, according to the different proportion between the velocity of light and that of the eye. So that if we could suppose that light was propagated in an instant, then there would be no difference between the real and visible place of an object, although the eye were in motion, for in that case, *AC* being infinite with respect to *AB*, the angle ACB (the difference between the true and visible place) vanishes. But if light be propagated in time (which I presume will readily be allowed by most of the philosophers of this age) then it is evident from the foregoing considerations, that there will be always a difference between the real and visible place of an object, unless the eye is moving either directly towards or from the object. And in all cases, the sine of the difference between the real and visible place of the object, will be to the sine of the visible inclination of the object to the line in which the eye is moving, as the Velocity of the eye to the Velocity of light.

Bradley illustrates the consequences of this hypothesis by a numerical example and shows that it can account for the aberration of the stars discovered by him. He then proceeds as follows:

This being premised, I shall now proceed to determine from the observations, what the real proportion is between the velocity of light and the velocity of the earth's annual motion in its orbit; upon supposition that the Phaenomena before mentioned do depend upon the causes I have here assigned. But I must first let you know, that in all the observations hereafter mentioned, I have made an allowance for the change of the star's declination on account of the precession of the equinox, upon supposition that the alteration from this cause is proportional to the time, and regular through all the parts of the year. I had deduced the real annual alteration of declination of each star from the observations themselves; and I the rather choose to depend upon them in this article, because all which I have yet made, concur to prove, that the stars near the Equinoctial Colure, change their declination at this time $1''\frac{1}{2}$ or $2''$ in a year more than they would do if the precession was only $50''$, as is now generally supposed. I have likewise met with some small varieties in the declination of other stars in different years, which do not seem to proceed from the same cause, particularly in those that are near the solstitial Colure, which on the contrary have altered their declination less than they ought, if the precession was $50''$. But whether these small alterations proceed from a regular cause, or are occasioned by any change in the materials, etc., of my instrument, I am not yet able fully to determine. However, I thought it might not be amiss just to mention to you how I have endeavored to allow for them, though the result would have been nearly the same, if I had not considered them at all. What that is, I will show, first from the observations of Draconis, which was found to be 39″ more southerly in the beginning of March, than in September.

From what hath been premised, it will appear that the greatest alteration of the apparent declination of Draconis, on account of the successive propagation of light, would be to the diameter of the little circle which a star (as was before remarked) would seem to describe about the pole of the ecliptick, as 39″ to 40″,4. The half of this is the angle ACB (as represented in the Fig.). This therefore being 20″,2, AC will be to AB, that is, the velocity of light to the velocity of the eye (which in this case may be supposed the same as the velocity of the earth's annual motion in its orbit) as 10210 to one, from whence it would follow that light moves, or is propagated as far as from the Sun to the Earth in 8′ 12″.

It is well known, that Mr. Roemer, who first attempted to account for an apparent inequality in the times of the eclipses of Jupiter's satellites, by the hypothesis of the progressive motion of light, supposed that it spent about 11 minutes of time in its passage from the sun to us: but it hath since been concluded by others from the like eclipses, that it is propagated as far in about 7 minutes. The velocity of light therefore deduced from the foregoing hypothesis, is as it were a mean betwixt what had at different times been determined from the eclipses of Jupiter's satellites.

FIZEAU

Armand Hyppolyte Louis Fizeau was born in Paris on September 23, 1819. His early scientific work was connected with the improvement of photographic processes. He studied interference of light and invented one form of the interferometer. He also studied the expansion of bodies by heat. Much of his earlier work was done in collaboration with Foucault. He died at Venteuil on September 18, 1896.

The extract which follows, under the title, "Sur un expérience relative à la vitesse de propagation de la lumière," appeared in the *Comptes Rendus*, Vol. 29, p. 90, 1849. It contains an account of the first successful attempt to measure the velocity of light by observations that do not involve astronomical constants. Fizeau's method was employed later by Cornu and by Forbes and Young, though it is generally held that more trustworthy results are to be obtained by the method introduced by Foucault.

The Velocity of Light

I have succeeded in making perceptible the velocity of transmission of light by a method which seems to me to furnish a new way of studying with precision this important phenomenon. This method is founded on the following principles:

When a disc turns in its own plane about its center of figure with great rapidity, we may determine the time taken by a point on its circumference in passing through a very small angular distance, $\frac{1}{1000}$ of the circumference, for example.

When the velocity of rotation is sufficiently great this time is, in general, very short; for ten and for one hundred turns per second it is only $\frac{1}{10000}$ and $\frac{1}{100000}$ of a second. If the disc is divided at its circumference, in the same way as cogwheels are, into equal intervals alternately empty and full, we shall have for the time of passage of each interval past the same point in space the same very small fraction.

During these short times light passes through distances fairly limited, 31 kilometers for the first fraction, 3 kilometers for the second.

When we consider the effects produced when a ray of light traverses the openings of such a disc in motion, we arrive at this result, that if the ray, after it has passed through, is reflected by a mirror and sent back to the disc in such a way as to meet it again at the same point in space, the velocity of light will enter in such a way that the ray will pass through or will be intercepted, depending upon the velocity of the disc and the distance at which the reflection occurs.

Further, a system of two telescopes, directed the one toward the other in such a way that the image of the objective of each of them is formed at the focus of the other one, possesses properties which make it possible to realize these conditions in a simple way. All that is needed is to place a mirror at the focus of one of them and to modify the ocular system of the other by introducing between the focus and the eye-piece a transparent sheet of glass inclined to the axis by 45 degrees, so arranged as to receive from the side the light of a lamp, or of the sun, which it reflects toward the focus. With this arrangement the light which traverses the focus in the area, supposed very small, of the image which represents the objective of the second telescope, is sent out toward this, is reflected at its focus, and comes back again, after traversing the same distance, to pass anew through the focus of the first telescope, where it can be observed through the glass by means of the eye-piece.

This arrangement succeeds very well even when the telescopes are separated by a considerable distance; with telescopes of six centimeters aperture the distance may be eight kilometers without the light becoming too feeble. We see then a luminous point like a star formed by the light which has gone from this point and after traversing a distance of sixteen kilometers has come back again to pass exactly through the same point before reaching the eye.

It is at this very point that we must allow the teeth of a rotating disc to pass in order to produce the effects described. The experiment succeeds very well, and we observe that, according as the velocity of rotation is greater or less, the luminous point shines out brilliantly or is totally eclipsed. In the circumstances in which the experiment has been tried, the first eclipse occurs with 12.6 turns per second. With twice the velocity the point shines out again; with three times the velocity a second eclipse occurs; with four times the velocity the point shines out again, and so on.

The first telescope was placed in the belvedere of a house at Suresnes, the second was on the hill of Montmartre at an approximate distance away of 8633 meters.

The disc, with 720 teeth, was mounted on a system of wheels moved by weights and constructed by M. Froment; a counter made it possible to measure the rate of rotation. The light was taken from a lamp so arranged so as to afford a brilliant source of light.

These first trials furnished a value of the velocity of light only a little different from that which is accepted by astronomers. The mean value determined from twenty-eight observations which have been made up to this time gives for this value 70948 leagues of 25 to the degree. I shall have the honor of submitting to the judgment of the Academy a detailed memoir when all the features of the experiment have been studied more completely.

FOUCAULT

Jean Bernard Léon Foucault was born in Paris on September 18, 1819. His father was a publisher. His first studies were in medicine, but he soon abandoned that field for the study of physics. In 1851 he made the famous experiment of the Foucault pendulum to demonstrate the rotation of the earth, and in the next year constructed the gyroscope. In 1855 he became assistant in physics at the Imperial Observatory at Paris. He died in Paris on February 11, 1868.

The paper which follows is entitled "Determination expérimentale de la vitesse de la lumière: parallaxe du Soleil." It appeared in the *Comptes Rendus*, Vol. 55, p. 501, 1862. It gives an account of the determination of the velocity of light by the revolving mirror, a method which has since been developed by Michelson and others.

The Velocity of Light

In the session of May 6, 1850, I presented the result of a differential experiment on the velocity of light in two media of unequal densities; and at the same time I suggested that the same method, based on the use of a rotating mirror, might be used to measure the absolute velocity of light in space.

After a careful consideration of this project, the Director of the Observatory determined to push it forward and put at my disposal the necessary resources. At the beginning of the summer the apparatus was in condition to be used, but the bad weather has made it impossible for me to give myself as promptly as I should have desired to these observations, which require light from the sun. However, the sky has at last become clear, and by taking advantage of the recent fine weather, I have obtained results which seem to me to come very close to the expression of the truth.

The apparatus used does not differ essentially from that which has been previously described, except by the addition of a system of wheels arranged to move a circular indented screen so as to get an exact measure of the velocity of the mirror, and by the enlargement of the distance used in the experiment, which, by means of several reflections, has been carried from 4 to 20 meters. By thus increasing the distance traversed by the light and by introducing greater exactness in the measurement of time, I have obtained determinations of which the extreme differences do not pass $\frac{1}{100}$ and which, combined by taking the mean, very soon give series which are in agreement to about $\frac{1}{500}$.

As a final result, the velocity of light seems to be notably less than it was supposed to be. According to the ordinarily received values this velocity is 308 million meters per second, while the new experiment of the revolving mirror gives in round numbers 298 million.

We may, it seems to me, count on the correctness of this number in this sense, that the corrections which should be applied to it, ought not to be greater than 500000 meters.

If we accept this new number and combine it with the constant of aberration 20″.45 to deduce from it the parallax of the sun, which evidently is a function of these two numbers, we find instead of 8″.57 the noticeably larger value 8″.86; thus the mean distance of the earth from the sun is diminished by about $\frac{1}{30}$.

To give an idea of the degree of confidence which can be had in the system of observations which has been followed, I will transcribe here a series of original determinations chosen among those whose mean agrees the best with the general mean.

1024	1026
1025	1026
1029	1026
1028	1025
1027	1026
1026	1028
1027	1028
1025	1027
1026	1026,5
1027	1027

Mean....1026,47

This number 1026.47 is referred to an arbitrary length which comes in the apparatus and which is changed at each determination in such a way as to produce a constant displacement of the image displaced by the revolving mirror.

In another communication I will give a sufficient description of my apparatus to afford a basis for discussion, and to recognize properly the talents and the services of the distinguished artisans who have assisted me.

STOKES

Sir George Gabriel Stokes was born on August 13, 1819, at Skreen in Ireland. After attending schools at Dublin and Bristol he was entered at Pembroke College, Cambridge, where he was graduated in 1841 as Senior Wrangler and first Smith's Prizeman. He was at once elected to a fellowship at Pembroke, which he held until his marriage in 1857. Twelve years later under new statutes he was reëlected fellow and continued in that connection until he was elected master of Pembroke in 1902. In 1849 he was made Lucasian professor of mathematics in the university. He held this position for over fifty years. He was made a baronet in 1849. He died in Cambridge on February 1, 1903.

Stokes was distinguished for his researches in many branches of physics. He combined in a peculiarly fortunate way mathematical ability and knowledge with experimental skill. The paper which follows is an abstract prepared by himself of an elaborate experimental paper "On the Change of Refrangibility of Light," taken from the *Proceedings of the Royal Society*, Vol. 6, p. 195, 1852.

FLUORESCENCE

The author was led into the researches detailed in this paper by considering a very singular phenomenon which Sir John Herschel had discovered in the case of a weak solution of sulphate of quinine, and various other salts of the same alkaloid. This fluid appears colourless and transparent, like water, when viewed by transmitted light, but exhibits in certain aspects a peculiar blue colour. Sir John Herschel found that when the fluid was illuminated by a beam of ordinary daylight, the blue light was produced only throughout a very thin stratum of fluid adjacent to the surface by which the light entered. It was unpolarized. It passed freely through many inches of the fluid. The incident beam, after having passed through the stratum from which the blue light came, was not sensibly enfeebled nor coloured, but yet it had lost the power of producing the usual blue colour when admitted into a solution of sulphate of quinine. A beam of light modified in this mysterious manner was called by Sir John Herschel *epipolized*.

Several years before Sir David Brewster had discovered in the case of an alcoholic solution of the green colouring matter of leaves a very remarkable phenomenon, which he has designated as *internal dispersion*. On admitting into this fluid a beam of sunlight condensed by a lens, he was surprised by finding the path of the rays within the fluid marked by a bright light of a bloodred colour, strangely contrasting with the beautiful green of the fluid itself when seen in moderate thickness. Sir David afterwards observed the same phenomenon in various vegetable solutions and essential oils, and in some solids. He conceived it to be due to coloured particles held in suspension. But there was one circumstance attending the phenomenon which seemed very difficult of explanation on such a supposition, namely, that the whole or a great part of the dispersed beam was unpolarized, whereas a beam reflected from suspended particles might be expected to be polarized by reflexion. And such was, in fact, the case with those beams which were plainly due to nothing but particles held in suspension. From the general identity of the circumstances attending the two phenomena, Sir David Brewster was led to conclude that epipolic was merely a particular case of internal dispersion, peculiar only in this respect, that the rays capable of dispersion were dispersed with unusual rapidity. But what rays they were which were capable of affecting a solution of sulphate of quinine, why the active rays were so quickly used up, while the dispersed rays which they produced passed freely through the fluid, why the transmitted light when subjected to prismatic analysis showed no deficiencies in those regions to which, with respect to refrangibility, the dispersed rays chiefly belonged, were questions to which the answers appeared to be involved in as much mystery as ever.

After having repeated some of the experiments of Sir David Brewster and Sir John Herschel, the author could not fail to take a most lively interest in the phenomenon. The firm conviction which he felt that two portions of light were not distinguishable as to their nature otherwise than by refrangibility and state of polarization, left him but few hypotheses to choose between, respecting the explanation of the phenomenon. In fact, having regarded it at first as an axiom that dispersed light of any particular refrangibility could only have arisen from light of the same refrangibility contained in the incident beam, he was led by necessity to adopt hypotheses of so artificial a character as to render them wholly improbable. He was thus compelled to adopt the other alternative, namely, to suppose that in the process of internal dispersion the refrangibility of light had been changed. Startling as such a supposition might appear at first sight, the ease with which it accounted for the whole phenomenon was such as already to produce a strong probability of its truth. Accordingly the author determined to put this hypothesis to the test of experiment.

The experiments soon placed the fact of a change of refrangibility beyond all doubt. It would exceed the limits of an abstract like the present to describe the various experiments. It will be sufficient to mention some of the more remarkable results.

A pure spectrum from sunlight having been formed in air in the usual manner, a glass vessel containing a weak solution of sulphate of quinine was placed in it. The rays belonging to the greater part of the visible spectrum passed freely through the fluid, just as if it had been water, being merely reflected here and there from motes. But from a point about half-way between the fixed lines *G* and *H* to far beyond the extreme violet the incident rays gave rise to light of a sky-blue colour, which emanated in all directions from the portion of the fluid which was under the influence of the incident rays. The anterior surface of the blue space coincided of course with the inner surface of the vessel in which the fluid was contained. The posterior surface marked the distance to which the incident rays were able to penetrate before they were absorbed. This distance was at first considerable, greater than the diameter of the vessel, but it decreased with great rapidity as the refrangibility of the incident rays increased, so that from a little beyond the extreme violet to the end the blue space was reduced to an excessively thin stratum adjacent to the surface by which the incident rays entered. It appears therefore that this fluid, which is so transparent with respect to nearly the whole of the visible rays, is of an inky blackness with respect to the invisible rays more refrangible than the extreme violet. The fixed lines belonging to the violet and the invisible region beyond were beautifully represented by dark planes interrupting the blue space. When the eye was properly placed, these planes were of course projected into lines. The author has made a sketch of these fixed lines, which accompanies the paper. They may be readily identified with the fixed lines represented in M. Becquerel's map of the fixed lines of the chemical spectrum. The last line seen in a solution of sulphate of quinine appears to be the line next beyond the last represented in M. Becquerel's map. Under very favourable circumstances two dusky bands were seen still further on. Several circumstances led the author to conclude that in all probability fixed lines might be readily seen corresponding to still more refrangible rays, were it not for the opacity of glass with respect to those rays of very high refrangibility.

It is very easy to prove experimentally that the blue dispersed light corresponding to any particular part of the incident spectrum is not homogeneous light, having a refrangibility equal to that of the incident rays, and rendered visible in consequence of its complete isolation; but that it is in fact heterogeneous light, consisting of rays extending over a wide range of refrangibility, and not passing beyond the limits of refrangibility of the spectrum visible under ordinary circumstances. To show this it is sufficient to isolate a part of the incident spectrum, and view the narrow beam of dispersed light which it produces through a prism held to the eye.

In Sir David Brewster's mode of observation, the beam of light which was of the same nature as the blue light exhibited by a solution of sulphate of quinine was necessarily mixed with the beam due merely to reflexion from suspended particles; and in the case of vegetable solutions, a beam of the latter kind almost always exists, to a greater or less degree.   But in the method of observation employed by the author, to which he was led by the discovery of the change of refrangibility, the two beams are exhibited quite distinct from one another.   The author proposes to call the two kinds of internal dispersion just mentioned *true internal dispersion* and *false internal dispersion*, the latter being nothing more than the scattering of light which is produced by suspended particles, and having, as is now perfectly plain, nothing to do with the remarkable phenomenon of true internal dispersion.

Now that the nature of the latter phenomenon is better known, it is of course possible to employ methods of observation by which it may be detected even when only feebly exhibited.   It proves to be almost universal in vegetable solutions, that is, in solutions made directly from various parts of vegetables.   When vegetable products are obtained in a state of isolation, their solutions sometimes exhibit the phenomenon and sometimes do not, or at least exhibit it so feebly that it is impossible to say whether what they do show may not be due to some impurity.   Among fluids which exhibit the phenomenon in a high degree, or according to the author's expression are highly *sensitive*, may be mentioned a weak decoction of the bark of the horse-chestnut, an alcoholic extract from the seeds of the *Datura stramonium*, weak tincture of turmeric, and a decoction of madder in a solution of alum.   In these cases the general character of the dispersion resembles that exhibited by a solution of sulphate of quinine, but the tint of the dispersed light, and the part of the spectrum at which the dispersion begins, are different in different cases.   In the last fluid, for example, the dispersion commences somewhere about the fixed line *D*, and continues from thence onwards far beyond the extreme violet. The dispersed light is yellow, or yellowish orange.

In the case of other fluids, however, some of them sensitive in a very high degree, the mode in which light is dispersed internally presents some very remarkable peculiarities.   One of the most singular examples occurs in the case of an alcoholic solution of the green colouring matter of leaves. This fluid disperses a rich red light. The dispersion commences abruptly about the fixed line B, and continues from thence onwards throughout the visible spectrum and a little beyond. The dispersion is subject to fluctuations intimately connected with the singular absorption bands exhibited by this medium.

In order that a medium should be capable of changing the refrangibility of light incident upon it, it is not necessary that the medium should be a fluid, or a clear solid. Washed papers and other opaque substances produce the same effect, but of course the mode of observation must be changed. The author has observed the change of refrangibility in various ways. It will be sufficient to mention here that which was found most generally useful, which he calls the method of observing by a *linear spectrum*. The method is as follows.

A series of prisms and a lens are arranged in the usual manner for forming a pure spectrum, but the slit by which the light enters, instead of being parallel, is placed in a direction perpendicular to the edges of the prisms. A linear spectrum is thus formed at the focus of the lens, consisting of an infinite succession of images of the slit arranged one after the other in the order of refrangibility, and of course overlapping each other to a certain extent. The substance to be examined is placed in the linear spectrum, and the line of light seen upon it is viewed through a prism held to the eye. In this way it is found that almost all common organic substances, such as wood, cork, paper, calico, bone, ivory, horn, wool, quills, feathers, leather, the skin of the hand, the nails, are sensitive in a greater or less degree. Organic substances which are dark-coloured are frequently found to be insensible, but, on the other hand, scarlet cloth and various other dyed articles are highly sensitive. By means of a linear spectrum the peculiar dispersion of a red light produced by chlorophyll, or some of its modifications, may be observed not only in a solution, but in a green leaf, or on a washed paper, or in a sea-weed.

The highly sensitive papers obtained by washing paper with tincture of turmeric, or a solution of sulphate of quinine, or some other highly sensitive medium, display their sensibility in a remarkable manner when they are examined in a linear spectrum. In these cases, however, the paper produces a very striking effect when merely held so as to receive a pure spectrum formed in the usual manner, that is, with a slit parallel to the edges of the prisms. Such a paper may be used as a screen for showing the fixed lines belonging to the invisible rays, though they are not thus shown quite so well as by using a solution. The extraordinary prolongation of the spectrum seen when it is received on turmeric paper has been already observed by Sir John Herschel, by whom it was attributed to a peculiarity in the reflecting power of the substance. Of course it now appears that the true explanation is very different.

A high degree of sensibility appears to be rather rare among inorganic compounds. Certain specimens of fluor spar, as is already known, give a copious internal dispersion of a deep blue light; but this is plainly due to some foreign ingredient, the nature of which is at present unknown. But there is one class of inorganic compounds which are very remarkable for their sensibility, namely, certain compounds of peroxide of uranium, including the ornamental glass called canary glass, and the natural mineral yellow uranite. In these compounds the dispersed light is found on analysis to consist of bright bands arranged at regular intervals. A very remarkable system of absorption bands is also found among these compounds, which is plainly connected with the system of bright bands seen in the spectrum of the dispersed light. The connection between the absorption and internal dispersion exhibited by these compounds is very singular, and is of a totally different nature from the connection which has been already mentioned as occurring in solutions of the green colouring matter of leaves.

There is one law relating to the change of refrangibility which appears to be quite universal, namely, that the refrangibility of light is *always lowered* by internal dispersion. The incident rays being homogeneous, the dispersed light is found to be more or less composite. Its colour depends simply on its refrangibility, having no relation to the colour of the incident light, or to the circumstance that the incident rays were visible or invisible. The dispersed light appears to emanate in all directions, as if the solid or fluid were self-luminous while under the influence of the incident rays.

The phenomenon of the change of refrangibility of light admits of several important applications. In the first place it enables us to determine instantaneously the transparency or opacity of a solid or fluid with respect to the invisible rays more refrangible than the violet, and that, not only for these rays as a whole, but for the rays of each refrangibility in particular. For this purpose it is sufficient to form a pure spectrum with sunlight as usual, employing instead of a screen a vessel containing a decoction of the bark of the horse-chestnut, or a slab of canary glass, or some other highly sensitive medium, and then to interpose the medium to be examined, which, if fluid, would have to be contained in a vessel with parallel sides of glass. Glass itself ceases to be transparent about the region corresponding to the end of the author's map, and to carry on these experiments with respect to invisible rays of still higher refrangibility would require the substitution of quartz for glass. The reflecting power of a surface with respect to the invisible rays may be examined in a similar manner.

The effect produced on sensitive media leads to interesting information respecting the nature of various flames. Thus, for example, it appears that the feeble flame of alcohol is extremely brilliant with regard to invisible rays of very high refrangibility. The flame of hydrogen appears to abound in invisible rays of still higher refrangibility.

By means of the phenomena relating to the change of refrangibility, the independent existence of one or more sensitive substances may frequently be observed in a mixture of various compounds. In this way the phenomenon seems likely to prove of value in the separation of organic compounds. The phenomena sometimes also afford curious evidence of chemical combinations; but this subject cannot here be further dwelt upon.

The appearance which the rays from an electric spark produce in a solution of sulphate of quinine, shows that the spark is very rich in invisible rays of excessively high refrangibility, such as would plainly put them far beyond the limits of the maps which have hitherto been made of the fixed lines in the chemical part of the solar spectrum. These rays are stopped by glass, but transmitted through quartz. These circumstances render it probable that the phosphorogenic rays of an electric spark are nothing more than rays of same nature as those of light, but which are invisible, and not only so, but of excessively high refrangibility. If so, they ought to be stopped by a very small quantity of a substance known to absorb those rays with great energy. Accordingly the author found that while the rays from an electric spark which excite the phosphorescence of Canton's

phosphorus pass freely through water and quartz, they are stopped
on adding to the water an excessively small quantity of sulphate
of quinine.

At the end of the paper the author explains what he conceives
to be the cause of the change of refrangibility, and enters into
some speculations to account for the law according to which the
refrangibility of light is always lowered in the process of internal
dispersion.

FARADAY

After a short introduction, in which he refers to the convertibility of the
various natural forces and to the failure of all previous attempts to discover
any direct relation between light and electricity, Faraday (p. 472) proceeds
to describe the experiment by which he showed that a magnetic field will
rotate the plane of polarization of light. The extract which contains this
description is taken from the *Philosophical Transactions* of 1846, p. 1, or from
Faraday's *Experimental Researches in Electricity*, Vol. 3, p. 1.

MAGNETIC ROTATION OF THE PLANE OF POLARIZATION

But before I proceed to them, I will define the meaning I connect
with certain terms which I shall have occasion to use:—thus by
line of magnetic force, or magnetic line of force, or magnetic curve,
I mean that exercise of magnetic force which is exerted in the
lines usually called magnetic curves, and which equally exist
as passing from or to magnetic poles, or forming concentric circles
round an electric current. By line of electric force, I mean the
force exerted in the lines joining two bodies, acting on each other
according to the principles of static electric induction, which
may also be either in curved or straight lines. By diamagnetic,
I mean a body through which lines of magnetic force are passing,
and which does not by their action assume the usual magnetic
state of iron or loadstone.

A ray of light issuing from an Argand lamp, was polarized in a
horizontal plane by reflexion from a surface of glass, and the
polarized ray passed through a Nichol's eye-piece revolving on a
horizontal axis, so as to be easily examined by the latter. Between
the polarizing mirror and the eye-piece, two powerful electromagnetic poles were arranged, being either the poles of a horse-shoe
magnet, or the contrary poles of two cylinder magnets; they were
separated from each other about 2 inches in the direction of the

line of the ray, and so placed, that, if on the same side of the polarized ray, it might pass near them; or if on contrary sides, it might go between them, its direction being always parallel, or nearly so, to the magnetic lines of force. After that, any transparent substance placed between the two poles, would have passing through it, both the polarized ray and the magnetic lines of force at the same time and in the same direction.

Sixteen years ago I published certain experiments made upon optical glass, and described the formation and general characters of one variety of heavy glass, which, from its materials, was called silicated borate of lead. It was this glass which first gave me the discovery of the relation between light and magnetism, and it has power to illustrate it in a degree beyond that of any other body; for the sake of perspicuity I will first describe the phaenomena as presented by this substance.

A piece of this glass, about 2 inches square and 0.5 of an inch thick, having flat and polished edges, was placed as a diamagnetic between the poles (not as yet magnetized by the electric current), so that the polarized ray should pass through its length; the glass acted as air, water, or any other indifferent substance would do; and if the eye-piece were previously turned into such a position that the polarized ray was extinguished, or rather the image produced by it rendered invisible, then the introduction of this glass made no alteration in that respect. In this state of circumstances the force of the electro-magnet was developed, by sending an electric current through its coils, and immediately the image of the lamp-flame became visible, and continued so as long as the arrangement continued magnetic. On stopping the electric current, and so causing the magnetic force to cease, the light instantly disappeared; these phaenomena could be renewed at pleasure, at any instant of time, and upon any occasion, showing a perfect dependence of cause and effect.

The voltaic current which I used upon this occasion, was that of five pair of Grove's construction, and the electro-magnets were of such power that the poles would singly sustain a weight of from twenty-eight to fifty-six, or more, pounds. A person looking for the phaenomenon for the first time would not be able to see it with a weak magnet.

The character of the force thus impressed upon the diamagnetic is that of rotation; for when the image of the lamp-flame has thus been rendered visible, revolution of the eye-piece to the right or left, more or less, will cause its extinction; and the further motion of the eye-piece to the one side or other of this position will produce the reappearance of the light, and that with complementary tints, according as this further motion is to the right- or left-hand.

When the pole nearest to the observer was a marked pole, i.e. the same as the north end of a magnetic needle, and the further pole was unmarked, the rotation of the ray was right-handed; for the eye-piece had to be turned to the right-hand, or clock fashion, to overtake the ray and restore the image to its first condition. When the poles were reversed, which was instantly done by changing the direction of the electric current, the rotation was changed also and became left-handed, the alteration being to an equal degree in extent as before. The direction was always the same for the same line of magnetic force.

. . . .

KIRCHHOFF

Gustav Robert Kirchhoff was born in Königsberg on March 12, 1824. He studied at the university of his native town and was graduated in 1847. He was professor of physics for a time at Breslau but in 1854 was transferred to Heidelberg, where he formed the association with Bunsen which led to their discoveries in spectrum analysis. In 1875 he was called to Berlin. He died there on October 17, 1887.

The extracts from Kirchhoff's works which follow are from the *Monatsbericht der Akademie der Wissenchaften zu Berlin*. In the first of these, dated October, 1859, is described the discovery of the absorption of light of definite wave lengths by glowing vapors which led to an explanation of the Fraunhofer lines. In the second, dated December, 1859, is given a simple theory of the connection between the emission and absorption of light and heat.

THE FRAUNHOFER LINES

While engaged in a research carried out by Bunsen and myself in common on the spectra of colored flames, by which it became possible to recognize the qualitative composition of complicated mixtures from the appearance of their spectra in the flame of the blow pipe, I made some observations which give an unexpected explanation of the origin of the Fraunhofer Lines and allow us to draw conclusions from them about the composition of the sun's atmosphere and perhaps also of that of the brighter fixed stars.

Fraunhofer noticed that in the spectrum of a candle flame two bright lines occur which coincide with the two dark lines D of the solar spectrum. We obtain the same bright lines in greater intensity from a flame in which common salt is introduced. I arranged a solar spectrum and allowed the sun's rays, before they fell on the slit, to pass through a flame heavily charged with salt. When the sunlight was sufficiently weakened there appeared, in place of the two dark D lines, two bright lines; if its intensity, however, exceeded a certain limit the two dark D lines showed much more plainly than when the flame charged with salt was not present.

The spectrum of the Drummond light generally contains both the bright sodium lines, if the glowing part of the lime cylinder has not been long exposed to the heat; if the cylinder remains unbroken these lines become weaker and finally disappear. If they have disappeared or are very weak, and if an alcohol flame in which salt is introduced is placed between the lime cylinder and the slit, then in place of the bright lines two dark lines appear, remarkably sharp and fine, which in every respect correspond with the D lines of the solar spectrum. Thus the D lines of the solar spectrum have been artificially produced in a spectrum in which they do not naturally occur.

If we introduce lithium chloride into the flame of a Bunsen burner, its spectrum shows a very bright, sharply defined line which lies between the Fraunhofer lines B and C. If we allow the sun's rays of moderate intensity to pass through the flame and fall on the slit, we shall see in the place indicated the lines bright on a darker ground; when the sunlight is stronger there appears at that place a dark line which has exactly the same character as the Fraunhofer lines. If we remove the flame the line disappears completely, so far as I can see.

I conclude from these observations that a colored flame in whose spectrum bright sharp lines occur so weakens rays of the color of these lines, if they pass through it, that dark lines appear in place of the bright ones, whenever a source of light of sufficient intensity, in whose spectrum these lines are otherwise absent, is brought behind the flame. I conclude further that the dark lines of the solar spectrum, which are not produced by the earth's atmosphere, occur because of the presence of those elements in the glowing atmosphere of the sun which would produce in the spectrum of a flame bright lines in the same position. We may assume that the bright lines corresponding with the D lines in the spectrum of a flame always arise from the presence of sodium;

the dark D lines in the solar spectrum permit us to conclude that sodium is present in the sun's atmosphere. Brewster has found in the spectrum of a flame charged with saltpeter bright lines in the position of the Fraunhofer lines A, a, B; these lines indicate that potassium is present in the sun's atmosphere. From my observations, according to which there is no dark line in the solar spectrum coinciding with the red line of lithium, it seems probable that lithium either is not present in the sun's atmosphere or is there in relatively small quantity.

The investigation of the spectra of colored flames has thus acquired a new and great importance; together with Bunsen I will carry it on as far as our means permit. In this connection we will investigate the weakening of the light rays in flames which has been established by my observations. In the investigations which have already been instituted by us on this subject an effect has appeared which seems to us to be of great importance. In order to bring out the dark D lines when the Drummond light is used, we must employ a flame charged with salt of a low temperature. The flame from alcohol containing water is fitted for this purpose, but the flame of the Bunsen burner is not. When we use the Bunsen burner the smallest quantity of salt, as soon as its effect can be noticed, causes the bright sodium lines to show themselves. We propose to develop the consequences which follow from this effect.

EMISSION AND ABSORPTION

A few weeks ago I had the honor to communicate to the Academy some observations which seem to me to be of interest, because they permit conclusions to be drawn about the chemical constitution of the sun's atmosphere. Starting out from these observations I have now come, by a very simple theoretical treatment, to a general law, which seems to me to be of importance in several respects and which I therefore take leave to present to the Academy. It expresses a property of all bodies, which is connected with the emission and absorption of heat and light.

If we introduce sodium chloride or lithium chloride into the non-luminous flame of the Bunsen burner we obtain a glowing body which sends out light only of certain wave lengths and absorbs light only of the same wave lengths. In this way the result of the observations which have been referred to may be expressed. No one knows how the dark rays of heat behave with respect to emission and absorption; but it seems to be admissible to imagine a possible body which, from among all the rays of heat, the luminous ones as well as the dark ones, emits only rays of one wave-length and absorbs only rays of the same wave-length. If this be admitted, and if we suppose further that it is possible to obtain a mirror which completely reflects all the rays, we can very easily prove, from the general principles of the mechanical theory of heat, that *for rays of the same wave length at the same temperature the ratio of the emissive power to the absorptive power is the same for all bodies.*

We consider a body C in the form of an unbounded plate, which sends out only rays of the wave-length Λ and absorbs only such rays; over against this is set a body c in the form of a similar plate which emits and absorbs rays of all possible wave-lengths; the outer surfaces of these plates are backed by the perfect mirrors R and r. If uniformity of temperature has been once set up in this system each of the two bodies must retain the same temperature, and therefore take up by absorption as much heat as it loses by radiation. Now from the rays which c emits we first consider those of a wave-length λ which is different from Λ. On these rays the body C has no influence; they are reflected from the mirror R as if C was not present; a certain part of them will then be absorbed by c, the remainder will pass a second time to the mirror R, will be again reflected from it, again partly absorbed by c, etc. All the rays of wave-length λ which the body c emits will be gradually taken up by it again. Since this holds for all values of λ which are different from Λ, the invariability of the temperature of the body c requires that this body absorb as much of the rays of wave-length Λ as it emits. For this wave-length let e represent the emissive power, a the absorptive power of the body c, E and A the corresponding magnitudes for the body C. From the quantity E of the rays which C emits, c absorbs the quantity aE and sends back $(1-a)E$; from this C absorbs the quantity $A(1-a)E$ and sends back $(1-A)(1-a)E$ to c, which absorbs from it $a(1-A)(1-a)E$. If we carry out this process further we see that c takes up from E a quantity of radiation which if we set for brevity

$$(1-A)(1-a) = k$$
$$= aE(1 + k + k^2 + k^3 + \cdots),$$

that is

$$= \frac{aE}{1-k}.$$

From the quantity e of radiation which c emits, c itself will absorb the quantity

$$\frac{a(1-A)e}{1-k}$$

as a similar treatment will show. Therefore the condition that the temperature of c does not change is the equation

$$e = \frac{aE}{1-k} + \frac{a(1-A)e}{1-k}$$

that is, the equation:

$$\frac{e}{a} = \frac{E}{A}.$$

We arrive at the same equation if we develop the condition that the temperature of C remains constant. If we suppose that the body C is replaced by another one at the same temperature, we find by repeating the discussion the same value for the ratio of the emissive power to the absorptive power of this body for the rays of the same wave-length Λ. Now the wave-length Λ and the temperature are arbitrary. The law therefore follows that for waves of the same wave-length at the same temperature the ratio of the emissive power to the absorptive power is the same for all bodies.

The concepts of emissive power and absorptive power are here referred to the case that the body forms an unbounded plate which is shielded on one side by a perfect mirror. But the quantity of radiation which a plate standing free sends out toward one side is just as great as the quantity of radiation which a plate of half the thickness, furnished with such a mirror, sends out, and both these plates absorb equally the rays which fall upon them. We may therefore, in the law which has been stated, define the emissive power of a body as the quantity of radiation which an unbounded plate standing free and made of the substance sends out toward one side and the absorptive power as the quantity of radiation which the same plate absorbs from the unit of radiation which falls upon it.

The ratio common to all bodies of the emissive power to the absorptive power e/a is a function of the wave-length and the temperature. At low temperature this function $= 0$ for the wave-lengths of visible light and is different from 0 for the greater values of the wave-length; at higher temperatures the function has finite values for the wave-lengths of visible light also. At the temperature at which the function ceases to be $= 0$ for the wave-length of a certain visible ray, all bodies begin to send out light of the color of that ray, except those which have a vanishingly small absorptive power for that color and that temperature; the greater the absorptive power the more is the light which radiates from the body. The experimental fact that opaque bodies all glow at the same temperature, while transparent gases require for this a much higher temperature and even at that temperature always radiate more weakly than other bodies is thus explained. It also follows that if a glowing gas gives a discontinuous spectrum, and if we pass through it rays of sufficient intensity, which by themselves would furnish a spectrum without dark or bright lines, dark lines must appear at those places in the spectrum in which the bright lines appear in the spectrum of the glowing gas. The method which I pointed out in my former communication as one which leads to the chemical analysis of the solar atmosphere has therefore received a theoretical foundation.

I take this opportunity to refer to a result which I believe I have attained in this way since my former communication. From the investigations of Wheatstone, Masson, Ångström and others we know that in the spectrum of an electric spark bright lines show themselves, which depend on the nature of the metals between which the spark passes, and we may assume that these lines correspond with those which would appear in the spectrum of a flame at very high temperature, if we could bring the same metal in a suitable form into this flame. I have investigated the green part of the spectrum of the electric spark between iron electrodes, and found in it a great many bright lines which seem to coincide with dark lines of the solar spectrum. In the case of single lines the coincidence can scarcely be established with certainty; but I believe that I have seen it in the case of many groups in such a way that the brighter lines in the spark spectrum corresponded with the darker lines in the solar spectrum; from which I believe I may conclude that these coincidences are not merely apparent. When the spark is formed between other metals as, for example, between copper electrodes, these bright lines do not appear. I consider myself justified in drawing the conclusion that iron is present as one of the constituents of the glowing atmosphere of the sun, a conclusion which is otherwise very plausible, if we remember the frequent occurrence of iron in the earth and in meteors. Of the dark lines of the solar spectrum which seem to coincide with the bright lines of the iron spectrum, I can only refer to a few by the help of Fraunhofer's drawing of the solar spectrum; among them is the line E, a few fainter lines near E toward the violet end of the spectrum, and a line which stands between the two nearest of the three very strongly marked lines which Fraunhofer designates by b.

BALMER

Johann Jakob Balmer was born on May 1, 1825, at Lausen, Switzerland. He received his doctor's degree from the University of Basel in 1869. He was a school teacher and privatdocent at the University of Basel. He died in 1898.

The paper which follows, entitled "Notiz über die Spectrallinien des Wasserstoffs," is taken from the *Annalen der Physik und Chemie*, Vol. 25, p. 80, 1885. In it is described his discovery of the formula by which one series of the spectral lines of hydrogen can be represented. All the subsequent analysis of spectra in series had its origin in this discovery.

The Hydrogen Spectral Series

On the basis of the measurements by H. W. Vogel and by Huggins of the ultra-violet lines of the hydrogen spectrum I have undertaken to determine a formula which will represent the wave lengths of the different lines in a satisfactory manner. I was urged to take up this work by Professor E. Hagenbach. Ångström's very exact measurements of the four hydrogen lines made it possible to determine for their wave lengths a common factor which was in as simple a numerical relation as possible to these wave lengths. I gradually arrived at a formula which, at least for these four lines, serves as the expression of a law by which their wave lengths can be represented with striking precision. The common factor in this formula, as it has been deduced from Ångström's measurements, is

$$\left(b = 3645.6\frac{mm}{10^7} \right)$$

This number may be called the fundamental number of hydrogen; and if it should result that corresponding fundamental numbers can be found for the spectral lines of other elements, the hypothesis might be permitted that relations which can be expressed by some function exist between these fundamental numbers and the corresponding atomic weights.

The wave lengths of the first four hydrogen lines are obtained by multiplying the fundamental number $b = 3645.6$ in succession by the coefficients $\frac{9}{5}$; $\frac{4}{3}$; $\frac{25}{21}$ and $\frac{9}{8}$. At first sight these four coefficients do not form a regular series; but if we multiply the numbers in the second and the fourth by 4 a consistent regularity appears and the coefficients have for numerators the numbers 3^2, 4^2, 5^2, 6^2 and for denominators a number that is less by 4.

For several reasons it seems to me probable that the four coefficients which have just been given belong to two series, so that the second series includes again the terms of the first series; and so finally I am able to present the formula for the coefficients in the more general form: $(m^2/m^2 - n^2)$, in which m and n are whole numbers.

For $n = 1$ we obtain the series $\frac{4}{3}$, $\frac{9}{8}$, $\frac{16}{15}$, $\frac{25}{24}$ etc. for $n = 2$ the series $\frac{9}{5}$, $\frac{16}{12}$, $\frac{25}{21}$, $\frac{36}{32}$, $\frac{49}{45}$, $\frac{64}{60}$, $\frac{81}{77}$, $\frac{100}{96}$ etc. In this second series the second term is already in the first series but in a reduced form.

If, with these coefficients and the fundamental number 3645.6, we carry out the calculation of the wave lengths, we obtain the following numbers in 10^{-7} mm. for them.

According to formula			Ångström gives	Difference
$H\alpha$ (C-line) $= \frac{9}{5}b$	$= 6562.08$		6562.10	$+0.02$
$H\beta$ (F-line) $= \frac{4}{3}b$	$= 4860.8$		4860.74	-0.06
$H\gamma$ (near G) $= \frac{25}{21}b$	$= 4340$		4340.1	$+0.1$
$H\delta$ (b-line) $= \frac{9}{8}b$	$= 4101.3$		4101.2	-0.1

The variations of the formula from Ångström's observations amount in the most unfavorable case to not more than $\frac{1}{40000}$ of a wave length, a variation which very likely is within the limits of the possible errors of observation and is really a striking evidence for the great scientific skill and care with which Ångström must have gone to work.

From the formula we obtained for a fifth hydrogen line

$$\frac{49}{45} \cdot 3645.6 = 3969.65.10^{-7} \text{ mm.}$$

I knew nothing of such a fifth line, which must lie within the

362 A SOURCE BOOK IN PHYSICS

visible part of the spectrum just before H_I (which according to Ångström has a wave length 3968.1); and I was compelled to assume that the temperature relations were not favorable to the development of this line or that the formula was not generally applicable.

On reference to Professor Hagenbach he informed me that many more hydrogen lines are known, which have been measured by Vogel and by Huggins in the violet and the ultra-violet parts of the hydrogen spectrum and of the spectrum of the white stars; he was kind enough himself to make a comparison of the wave lengths thus determined with my formula and to communicate to me the result.

While the formula in general gives somewhat larger numbers than those contained in the published lists of Vogel and of Huggins, yet the difference between the calculated and the observed wavelengths is so small that the correspondence is in the highest degree striking. Comparisons of wave lengths measured by different investigators show in general no exact agreement; and yet the observations of one man may be brought to correspond with those of another by a slight reduction in an entirely satisfactory way.

A table in which the wave lengths of the hydrogen lines, as given by several observers, are compared with the numbers calculated by the formula is omitted.

From these comparisons it appears that the formula holds also for the fifth hydrogen line, which lies just before the first Fraunhofer H-line (which belongs to calcium). It further appears that Vogel's hydrogen lines and the corresponding Huggins lines of the white stars can be presented by the formula very satisfactorily. We may almost certainly assume that the other lines of the white stars which Huggins found further on in the ultra-violet part of the spectrum will be expressed by the formula. I lack a knowledge of the wave lengths. According to the formula and with the fundamental number 3645.6 we obtain for the ninth and the following hydrogen lines up to the fifteenth:

$$\tfrac{121}{117}b = 3770.24 \qquad \tfrac{225}{221}b = 3711.58$$
$$\tfrac{36}{35}b = 3749.76 \qquad \tfrac{64}{63}b = 3703.46$$
$$\tfrac{169}{165}b = 3733.98 \qquad \tfrac{289}{285}b = 3696.76$$
$$\tfrac{49}{48}b = 3721.55$$

Whether the hydrogen lines of the white stars confirm the formula out to this point or whether other numerical relations gradually take its place can only be determined by the facts.

I add to what I have said a few questions and consequences.

Does the above formula hold only for the single chemical element hydrogen, and will there not be found other fundamental numbers in the spectral lines of other elements which are peculiar to those elements? If not, we might perhaps assume that the formula that holds for hydrogen is a special case of a more general formula which under certain conditions goes over into the formula for the hydrogen lines.

None of the hydrogen lines which correspond to the formula when $n = 3, 4$ etc., and which may be called lines of the third or fourth order, are found in any spectrum as yet known; they must develop themselves under entirely new relations of temperature and pressure if they are to become perceptible.

If the formula holds for all the principal lines of the hydrogen spectrum with $n = 2$ it follows that these spectral lines on the ultra-violet side approach the wave length 3645.6 in a more closely packed series, but they can never pass over this boundary, while the C-line also is the extreme line on the red side. Only if lines of higher orders are present would lines be found on the infra-red side.

The formula has no relation, so far as can be shown, with the very numerous lines of the second hydrogen spectrum which Hasselberg has published in the Mémoires de l'Académie des Sciences de St. Pétersbourg, 1882. In certain conditions of pressure and temperature hydrogen may easily change in such a way that the law of formation of its spectral lines becomes something entirely different.

There are great difficulties in the way of determining the fundamental number for other chemical elements, such as oxygen or carbon, by means of which their principal spectral lines can be determined from the formula. Only extremely exact determinations of wave lengths of the most prominent lines of an element can furnish a common measure of these wave lengths, and without such a measure it seems as if all trials and guesses will be in vain. Perhaps by using a different graphical construction of the map of the spectrum a way will be found to make progress in such investigations.

Postscript. From a notice by Huggins of September 14, 1884 which I have received through a friendly communication from Professor Hagenbach it appears that other hydrogen lines also occur in the spectrum of the white stars on which Huggins reports. All of these lines correspond to the formula $m^2/(m^2 - 4) \cdot b$, although a small difference, which increases with the coefficient m, becomes evident between the observed wave lengths and those calculated by the formula. Professor Hagenbach first called my attention to this circumstance. He finds by comparing the wave lengths calculated with $b = 3645$ and those observed by Huggins the following differences.

		Calculated from Formula	Observed by Huggins	Difference (Obs-Huggins)
H_γ $m = 5$		4339.3	4340.1	−0.8
b	6	4100.6	4101.2	−0.6
H_I	7	3969.0	3968.1	+0.9
α	8	3888.0	3887.5	+0.5
β	9	3834.3	3834.0	+0.4
γ	10	3796.9	3795.0	+1.9 ⎫
δ	11	3769.6	3767.5	+2.1 ⎬
ϵ	12	3749.1	3745.5	+3.6 ⎫
ζ	13	3733.3	3730.0	+3.3 ⎪
η	14	3720.9	3717.5	+3.4 ⎬
ϑ	15	3711.0	3707.5	+3.5 ⎪
ι	16	3702.9	3699.0	+3.9 ⎭

If the lines $H\alpha$ and $H\beta$ are added to these we have on the whole 14 lines which are represented by the proposed formula, if we give to the m the values of all the whole numbers from 3 to 16. Whether the above differences show that the formula gives the law only approximately, or whether they can be explained from errors of observation, it is hard to decide. Huggins' measurements have been carried out with extreme care, yet the difficulty of obtaining exactly similar conditions during the photographing of the comparison spectra and the stellar spectra, of which Huggins speaks, may perhaps explain the discrepancies. The circumstance that the wave lengths measured by Huggins are referred to air is of no importance, for if by the use of the refractive indices calculated from Lorenz's observations by Cauchy's formula we reduce the wave lengths to vacuum we find only an insignificant increase, for example, in the case of line ι from 3699.0 to 3700.1.

Huggins quotes in a note to his paper a calculation communicated to him by one of his friends of the so-called harmonic relations of the vibration numbers of the hydrogen lines. In this calculation, for the same lines it is found necessary to assume three separate groups of such harmonic terms, and further the common factor of each group is quite small, and the whole-number coefficients form no regular series. Both these considerations make me doubt whether this investigation, so interesting as it is, can really give the inner relations of the phenomena.

ROWLAND

Henry Augustus Rowland was born at Homesdale, Pennsylvania, on November 27, 1848. He was graduated in 1870 from the Rensselaer Polytechnic Institute, Troy, New York. He taught natural sciences as an instructor at Wooster College and was assistant professor of physics in Troy. In 1876 he was called to the newly founded Johns Hopkins University as the professor of physics. He died on April 16, 1901.

The extract which follows, taken from the paper published in the *Philosophical Magazine*, Series 4, Vol. 13, p. 469, 1882, is part of Rowland's account of his construction of gratings and of his invention of the curved grating.

PLANE AND CONCAVE GRATINGS

It is not many years since physicists considered that a spectroscope constructed of a large number of prisms was the best and only instrument for viewing the spectrum, where great power was required. These instruments were large and expensive, so that few physicists could possess them. Professor Young was the first to discover that some of the gratings of Mr. Rutherfurd showed more than any prism spectroscope which had then been constructed. But all the gratings which had been made up to that time were quite small, say one inch square, whereas the power of a grating in resolving the lines of the spectrum increases with the size. Mr. Rutherfurd then attempted to make as large gratings as his machine would allow, and produced some which were nearly two inches square, though he was rarely successful above an inch and three-quarters, having about thirty thousand lines. These gratings were on speculum metal and showed more of the spectrum than had ever before been seen, and have, in the hands of Young, Rutherfurd, Lockyer and others, done much good work for science. Many mechanics in this country and in France and Germany, have sought to equal Mr. Rutherfurd's gratings, but without success.

Under these circumstances, I have taken up the subject with the resources at command in the physical laboratory of the Johns Hopkins University.

One of the problems to be solved in making a machine is to make a perfect screw, and this, mechanics of all countries have sought to do for over a hundred years and have failed. On thinking over the matter, I devised a plan whose details I shall soon publish, by which I hope to make a practically perfect screw, and so important did the problem seem that I immediately set Mr. Schneider, the instrument maker of the university, at work at one. The operation seemed so successful that I immediately designed the remainder of the machine, and have now had the pleasure since Christmas of trying it. The screw is practically perfect, not by accident, but because of the new process for making it, and I have not yet been able to detect an error so great as one one-hundred-thousandth part of an inch at any part. Neither has it any appreciable periodic error. By means of this machine I have been able to make gratings with 43,000 lines to the inch, and have made a ruled surface with 160,000 lines on it, having about 29,000 lines to the inch. The capacity of the machine is to rule a surface $6\frac{1}{4} \times 4\frac{1}{4}$ inches with any required number of lines to the inch, the number only being limited by the wear of the diamond. The machine can be set to almost any number of lines to the inch, but I have not hitherto attempted more than 43,000 lines to the inch. It ruled so perfectly at this figure that I see no reason to doubt that at least two or three times that number might be ruled in one inch, though it would be useless for making gratings.

All gratings hitherto made have been ruled on flat surfaces. Such gratings require a pair of telescopes for viewing the spectrum; these telescopes interfere with many experiments, absorbing the extremities of the spectrum strongly; besides, two telescopes of sufficient size to use with six inch gratings would be very expensive and clumsy affairs. In thinking over what would happen were the grating ruled on a surface not flat, I thought of a new method of attacking the problem, and soon found that if the lines were ruled on a spherical surface the spectrum would be brought to a focus without any telescope. This discovery of concave gratings is important for many physical investigations, such as the photographing of the spectrum both in the ultra-violet and the ultra-red, the determination of the heating effect of the different rays, and the determination of the relative wave lengths of the lines of the spectrum. Furthermore it reduces the spectro-scope to its simplest proportions, so that spectroscopes of the highest power may be made at a cost which can place them in the hands of all observers. With one of my new concave gratings I have been able to detect double lines in the spectrum which were never before seen.

The laws of the concave grating are very beautiful on account of their simplicity, especially in the case where it will be used most. Draw the radius of curvature of the mirror to the centre of the mirror, and from its central point with a radius equal to half the radius of curvature draw a circle; this circle thus passes through the centre of curvature of the mirror and touches the mirror at its centre. Now if the source of light is anywhere in this circle, the image of this source and the different orders of the spectra are all brought to focus on this circle. The word focus is hardly applicable to the case, however, for if the source of light is a point the light is not brought to a single point on the circle but is drawn out into a straight line with its length parallel to the axis of the circle. As the object is to see lines in the spectrum only, this fact is of little consequence provided the slit which is the source of light is parallel to the axis of the circle. Indeed it adds to the beauty of the spectra, as the horizontal lines due to dust in the slit are never present, as the dust has a different focal length from the lines of the spectrum. This action of the concave grating, however, somewhat impairs the light, especially of the higher orders, but the introduction of a cylindrical lens greatly obviates this inconvenience.

The beautiful simplicity of the fact that the line of foci of the different orders of the spectra are on the circle described above leads immediately to a mechanical contrivance by which we can move from one spectrum to the next and yet have the apparatus always in focus; for we only have to attach the slit, the eye-piece and the grating to three arms of equal length, which are pivoted together at their other ends and the conditions are satisfied. However we move the three arms the spectra are always in focus. The most interesting case of this contrivance is when the bars carrying the eye-piece and grating are attached end to end, thus forming a diameter of the circle with the eye-piece at the centre of curvature of the mirror, and the rod carrying the slit alone movable. In this case the spectrum as viewed by the eye-piece is normal, and when a micrometer is used the value of a division of its head in wave-lengths does not depend on the position of the slit, but is simply proportional to the order of the spectrum, so that it need be determined once only. Furthermore, if the eye-piece is replaced by a photographic camera the photographic spectrum is a normal one. The mechanical means of keeping the focus is especially important when investigating the ultra-violet and ultra-red portions of the solar spectrum.

Another important property of the concave grating is that all the superimposed spectra are in exactly the same focus. When viewing such superimposed spectra it is a most beautiful sight to see the lines appear colored on a nearly white ground. By micrometric measurement of such superimposed spectra we have a most beautiful method of determining the relative wave lengths of the different portions of the spectrum, which far exceeds in accuracy any other method yet devised. In working in the ultra-violet or ultra-red portions of the spectrum we can also focus on the superimposed spectrum and so get the focus for the portion experimented on.

The fact that the light has to pass through no glass in the concave grating makes it important in the examination of the extremities of the spectrum where the glass might absorb very much.

There is one important research in which the concave grating in its present form does not seem to be of much use, and that is in the examination of the solar protuberances; an instrument can only be used for this purpose in which the dust in the slit and the lines of the spectrum are in focus at once. It might be possible to introduce a cylindrical lens in such a way as to obviate this difficulty. But for other work on the sun the concave grating will be found very useful. But its principal use will be to get the relative wave lengths of the lines of the spectrum, and so to map the spectrum; to divide lines of the spectrum which are very near together, and so to see as much as possible of the spectrum; to photograph the spectrum so that it shall be normal; to investigate the portions of the spectrum beyond the range of vision; and lastly to put in the hands of any physicist at a moderate cost such a powerful instrument as could only hitherto be purchased by wealthy individuals or institutions.

MICHELSON AND MORLEY

Albert Abraham Michelson was born in Strelno, Germany, on December 19, 1852. He was graduated from the United States Naval Academy in 1873 and also studied for a while at Berlin and Heidelberg. From 1875 to 1879 he taught physics and chemistry at the Naval Academy. In 1883 he became professor of physics at the Case School of Applied Science in Cleveland, Ohio. It was while there that he and Professor Morley made the experiment which is known by their names and which has had so important a part in the development of the theory of relativity. For three years he was professor of physics at Clarke University and then in 1892 he became professor of physics in the University of Chicago. He was connected with that institution until his death, which occurred at Pasadena, California, on May 9, 1931.

Edward Williams Morley was born on January 29, 1838, in Newark, New Jersey. He was graduated from Williams College in 1860. He was a chemist, and from 1869 was professor of chemistry in Western Reserve University in Cleveland, Ohio, and also for some years professor of chemistry in the Cleveland Medical College. He died on February 24, 1923.

The extract which follows contains a description of the Michelson-Morley experiment, by which it was hoped to discover a relative motion of the earth and the ether. The negative result of this experiment led to the introduction of the theory of relativity. The paper appeared in the *Philosophical Magazine*, Series 5, December, 1887.

THE MICHELSON-MORLEY EXPERIMENT

The discovery of the aberration of light was soon followed by an explanation according to the emission theory. The effect was attributed to a simple composition of the velocity of light with the velocity of the earth in its orbit. The difficulties in this apparently sufficient explanation were overlooked until after an explanation on the undulatory theory of light was proposed. This new explanation was at first almost as simple as the former. But it failed to account for the fact proved by experiment that the aberration was unchanged when observations were made with a telescope filled with water. For if the tangent of the angle of aberration is the ratio of the velocity of the earth to the velocity of light, then, since the latter velocity in water is three-fourths its velocity in a vacuum, the aberration observed with a water telescope should be four-thirds of its true value.

On the undulatory theory, according to Fresnel, first, the aether is supposed to be at rest, except in the interior of transparent media, in which, secondly, it is supposed to move with a velocity less than the velocity of the medium in the ratio $\dfrac{n^2 - 1}{n^2}$, where n is the index of refraction. These two hypotheses give a complete and satisfactory explanation of aberration. The second hypothesis, notwithstanding its seeming improbability, must be considered as fully proved, first, by the celebrated experiment of Fizeau, and secondly, by the ample confirmation of our own work. The experimental trial of the first hypothesis forms the subject of the present paper.

If the earth were a transparent body, it might perhaps be conceded, in view of the experiments just cited, that the inter-molecular aether was at rest in space, notwithstanding the motion of the earth in its orbit; but we have no right to extend the con-clusion from these experiments to opaque bodies. But there can hardly be any question that the aether can and does pass through metals. Lorentz cites the illustration of a metallic barometer tube. When the tube is inclined, the aether in the space above the mercury is certainly forced out, for it is incom-pressible. But again we have no right to assume that it makes its escape with perfect freedom, and if there be any resistance, however slight, we certainly could not assume an opaque body such as the whole earth to offer free passage through its entire mass. But as Lorentz aptly remarks: "Quoi qu'il en soit, on fera bien, à mon avis, de ne pas se laisser guider, dans une question aussi importante, par des considérations sur le degré de prob-abilité ou de simplicité de l'une ou de l'autre hypothèse, mais de s'addresser a l'expérience pour apprendre à connaitre l'état, de repos ou de mouvement, dans lequel se trouve l'éther à la surface terrestre."

In April, 1881, a method was proposed and carried out for testing the question experimentally.

In deducing the formula for the quantity to be measured, the effect of the motion of the earth through the aether on the path of the ray at right angles to this motion was overlooked. The discussion of this oversight and of the entire experiment forms the subject of a very searching analysis by H. A. Lorentz, who finds that this effect can by no means be disregarded. In con-sequence, the quantity to be measured had in fact but half the value supposed, and as it was already barely beyond the limits of errors of experiment, the conclusion drawn from the result of the experiment might well be questioned; since, however, the main portion of the theory remains unquestioned, it was decided to repeat the experiment with such modifications as would insure a theoretical result much too large to be masked by experimental errors. The theory of the method may be briefly stated as follows:

Let *sa*, (Fig. 72), be a ray of light which is partly reflected in *ab*, and partly transmitted in *ac*, being returned by the mirrors *b* and *c* along *ba* and *ca*. *ba* is partly transmitted along *ad*, and *ca* is partly reflected along *ad*. If then the paths *ab* and *ac* are equal, the two rays interfere along *ad*. Suppose now, the aether being at rest, that the whole apparatus moves in the direction *sc*, with the velocity of the earth in its orbit, the directions and distances traversed by the rays will be altered thus:—The ray *sa* is reflected

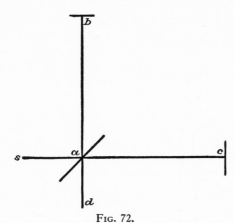

FIG. 72.

along *ab*, Fig. 73; the angle *bab₁*, being equal to the aberration = α is returned along *ba₁*, (*aba₁* = 2α), and goes to the focus of the telescope, whose direction is unaltered. The transmitted ray goes along *ac*, is returned along *ca₁*, and is reflected at *a₁*, making *ca₁e*, equal 90 − α, and therefore still coinciding with the first ray. It may be remarked that the rays *ba₁* and *ca₁* do not now meet exactly in the same point *a₁*, though the difference is of the second order; this does not affect the validity of the reasoning. Let it now be required to find the difference in the two paths *aba₁*, and *aca₁*.

Let V = velocity of light.

v = velocity of the earth in its orbit.

D = distance *ab* or *ac*, Fig. 72.

T = time light occupies to pass from *a* to *c*.

T_1 = time light occupies to return from *c* to *a₁* (Fig. 73).

Then

$$T = \frac{D}{V - v} \qquad T_1 = \frac{D}{V + v}$$

The whole time of going and coming is

$$T + T_1 = 2D\frac{V}{V^2 - v^2},$$

and the distance travelled in this time is

$$2D\frac{V^2}{V^2 - v^2} = 2D\left(1 + \frac{v^2}{V^2}\right),$$

neglecting terms of the fourth order. The length of the other

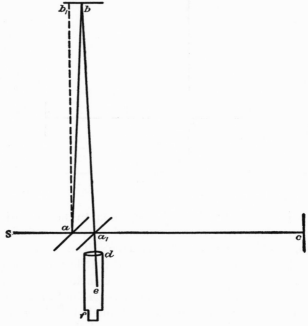

Fig. 73.

path is evidently $2D\sqrt{1 + \frac{v^2}{V^2}}$, or to the same degree of accuracy,

$2D\left(1 + \frac{v^2}{2V^2}\right)$. The difference is therefore $D\frac{v^2}{V^2}$. If now the whole apparatus be turned through 90°, the difference will be in the opposite direction, hence the displacement of the interference fringes should be $2D\dfrac{v^2}{V^2}$. Considering only the velocity of the earth in its orbit, this would be $2D \times 10^{-8}$. If, as was the case in the first experiment, $D = 2 \times 10^6$ waves of yellow light, the displacement to be expected would be 0.04 of the distance between the interference-fringes.

In the first experiment, one of the principal difficulties encountered was that of revolving the apparatus without producing distortion; and another was its extreme sensitiveness to vibration. This was so great that it was impossible to see the interference-fringes except at brief intervals when working in the city, even

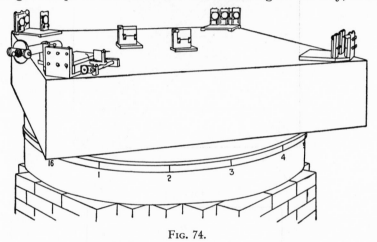

FIG. 74.

at two o'clock in the morning. Finally, as before remarked, the quantity to be observed, namely, a displacement of something less than a twentieth of the distance between the interference-fringes, may have been too small to be detected when masked by experimental errors.

The first-named difficulties were entirely overcome by mounting the apparatus on a massive stone floating on mercury; and the second by increasing, by repeated reflexion, the path of the light to about ten times its former value.

The apparatus is represented in perspective in Fig. 74, in plan in Fig. 75, and in vertical section in Fig. 76. The stone a (Fig. 76) is about 1.5 metre square and 0.3 metre thick. It rests on an annular wooden float bb, 1.5 metre outside diameter, 0.7 metre inside diameter, and 0.25 metre thick. The float rests on mercury contained in the cast-iron trough cc, 1.5 centimetre thick, and of such dimensions as to leave a clearance of about one centimetre around the float. A pin *d*, guided by arms *gggg*, fits into a socket *e* attached to the float. The pin may be pushed into the socket or be withdrawn, by a lever pivoted at *f*. This pin keeps the float concentric with the trough, but does not bear any part of the

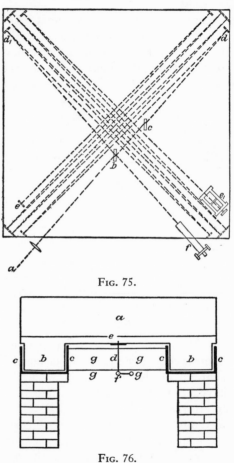

FIG. 75.

FIG. 76.

weight of the stone. The annular iron trough rests on a bed of cement on a low brick pier built in the form of a hollow octagon.

At each corner of the stone were placed four mirrors *dd ee*, Fig. 75. Near the centre of the stone was a plane parallel glass *b*. These were so disposed that light from an argand burner *a*, passing through a lens, fell on *b* so as to be in part reflected to *d*; the two pencils followed the paths indicated in the figure, *bdedbf* and *bd₁e₁d₁bf* respectively, and were observed by the telescope *f*. Both *f* and *a* revolved with the stone. The mirrors were of speculum metal carefully worked to optically plane surfaces five centimetres in diameter, and the glasses *b* and *c* were plane parallel of the same thickness, 1.25 centimetre; their surfaces measured 5.0 by 7.5 centimetres. The second of these was placed in the path of one of the pencils to compensate for the passage of the other through the same thickness of glass. The whole of the optical portion of the apparatus was kept covered with a wooden cover to prevent air-currents and rapid changes of temperature.

The adjustment was effected as follows:—The mirrors having been adjusted by screws in the castings which held the mirrors, against which they were pressed by springs, till light from both pencils could be seen in the telescope, the lengths of the two paths were measured by a light wooden rod reaching diagonally from mirror to mirror, the distance being read from a small steel scale to tenths of millimetres. The difference in the lengths of the two paths was then annulled by moving the mirror e_1. This mirror had three adjustments: it had an adjustment in altitude and one in azimuth, like all the other mirrors, but finer; it also had an adjustment in the direction of the incident ray, sliding forward or backward, but keeping very accurately parallel to its former plane. The three adjustments of this mirror could be made with the wooden cover in position.

The paths being now approximately equal, the two images of the source of light or of some well-defined object placed in front of the condensing lens, were made to coincide, the telescope was now adjusted for distinct vision of the expected interference-bands, and sodium light was substituted for white light, when the interference-bands appeared. These were now made as clear as possible by adjusting the mirror e_1; then white light was restored, the screw altering the length of path was very slowly moved (one turn of a screw of one hundred threads to the inch altering the path nearly 1000 wave-lengths) till the coloured interference-fringes reappeared in white light. These were now given a convenient width and position, and the apparatus was ready for observation.

The observations were conducted as follows:—Around the cast-iron trough were sixteen equidistant marks. The apparatus was revolved very slowly (one turn in six minutes) and after a

few minutes the cross wire of the micrometer was set on the clearest
of the interference-fringes at the instant of passing one of the
marks. The motion was so slow that this could be done readily
and accurately. The reading of the screw-head on the micrometer
was noted, and a very slight and gradual impulse was given to
keep up the motion of the stone; on passing the second mark,
the same process was repeated, and this was continued till the
apparatus had completed six revolutions. It was found that by
keeping the apparatus in slow uniform motion, the results were
much more uniform and consistent than when the stone was
brought to rest for every observation; for the effects of strains
could be noted for at least half a minute after the stone came to
rest, and during this time effects of change of temperature came
into action.

The tables containing the details of the observations are omitted. They
are embodied in the curves which follow.

The results of the observations are expressed graphically in
Fig. 77. The upper is the curve for the observations at noon,
and the lower that for the evening observations. The dotted

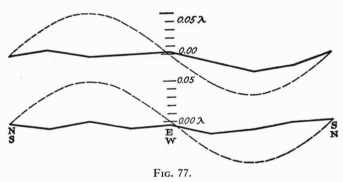

Fig. 77.

curves represent *one eighth* of the theoretical displacements. It
seems fair to conclude from the figure that if there is any displacement due to the relative motion of the earth and the luminiferous
aether, this cannot be much greater than 0.01 of the distance
between the fringes.

Considering the motion of the earth in its orbit only, this
displacement should be

$$2D\frac{v^2}{V^2} = 2D \times 10^{-8}$$

The distance D was about eleven metres, or 2×10^7 wave-lengths of yellow light; hence the displacement to be expected was 0.4 fringe. The actual displacement was certainly less than the twentieth part of this, and probably less than the fortieth part. But since the displacement is proportional to the square of the velocity, the relative velocity of the earth and the aether is probably less than one sixth the earth's orbital velocity, and certainly less than one fourth.

In what precedes, only the orbital motion of the earth is considered. If this is combined with the motion of the solar system, concerning which but little is known with certainty, the result would have to be modified; and it is just possible that the resultant velocity at the time of the observations was small, though the chances are much against it. The experiment will therefore be repeated at intervals of three months, and thus all uncertainty will be avoided.

It appears from all that precedes reasonably certain that if there be any relative motion between the earth and the luminiferous aether, it must be small; quite small enough entirely to refute Fresnel's explanation of aberration. Stokes has given a theory of aberration which assumes the aether at the earth's surface to be at rest with regard to the latter, and only requires in addition that the relative velocity have a potential; but Lorentz shows that these conditions are incompatible. Lorentz then proposes a modification which combines some ideas of Stokes and Fresnel, and assumes the existence of a potential, together with Fresnel's coefficient. If now it were legitimate to conclude from the present work that the aether is at rest with regard to the earth's surface, according to Lorentz there could not be a velocity potential, and his own theory also fails.

. . . .

STEFAN

Josef Stefan was born on March 24, 1835. In 1863 he became professor of physics at the University of Vienna and Director of the Physical Institute. He published numerous researches on light, sound and electricity. His death occurred in Vienna on January 7, 1893.

The paper which follows, "Ueber die Beziehung zwischen der Wärmestrahlung und der Temperatur," is taken from *Sitzungsberichte der Akademie der Wissenschaften in Wien*, Vol. 79, Part II, p. 391, 1879. The law given in it embodies the first successful attempt to connect radiation with absolute temperature, and has served as an important element in all subsequent researches in the theory of radiation.

TEMPERATURE RADIATION

1. On the Investigations of Dulong and Petit.

From their observations of the rate of cooling of a large mercury thermometer, the bulb of which in some experiments was bare and in others silvered, Dulong and Petit concluded that the quantity of heat radiated from a body increases in geometrical progression if its temperature increases uniformly. The quantity of heat radiated in unit time by a body at the temperature u can be represented by the formula ma^u in which m denotes a constant which depends on the magnitude and the nature of the surface of the body and a is the number 1.0077, which is the same for all bodies.

This law Dulong and Petit found to be in agreement with their observations between the temperatures of 0° and 280°. It was then assumed to hold also outside these limits and was first applied by Pouillet to the determination of the temperature of the sun.

The remarkably small number which Pouillet found for this quantity suggested that Dulong and Petit's law might not be applicable at higher temperatures, and that it cannot be used in such cases has been proved by the researches of Ericsson and Soret.

Dulong and Petit's formula is merely an empirical formula which exactly represents the loss of heat ascribed to radiation by the thermometer used in the experiment. Other formulas also would give the same result, but the formula of Dulong and Petit is distinguished by its extraordinary simplicity. I can however present another formula of equal, we may even say of greater, simplicity, which also represents the observations very well and in its theoretical relations has a certain preference.

We obtain numbers which come very close to the rates of cooling given by Dulong and Petit, if we assume that the heat radiated by a body is proportional to the fourth power of its absolute temperature. In the following table the second column contains the rates of cooling which Dulong and Petit found for the temperatures of their thermometer, which are given in the first column, when the spherical container in the middle of which the thermometer was placed was kept at 0°. The numbers given in the third column are obtained by dividing by 6 the differences $(273 + 80)^4 - 273^4$, $(273 + 100)^4 - 273^4$, etc. and placing the decimal point generally after the first digit, but in the last quotient after the second.

		Calculated	Difference
80°	1.74	1.66	− 0.08
100	2.30	2.30	0
120	3.02	3.05	− 3
140	3.88	3.92	− 4
160	4.89	4.93	− 4
180	6.10	6.09	+ 1
200	7.40	7.42	− 2
220	8.81	8.92	−11
240	10.69	10.62	+ 7

The values calculated from Dulong and Petit's formula differ from the observed values in the last place by

$$+2, -3, -3, -1, +2, +7, +6, -8, +1$$

Their formula agrees more closely with the observations and yet the deviations of the former one are not very great.

. . . .

The next part of the paper contains a criticism of the way in which Dulong and Petit undertook to correct their observations to take account of the effect of the air which remained in the radiating chamber. Stefan then proceeds as follows.

The rates of cooling calculated by Dulong and Petit, from the foregoing discussion, even if they are corrected for the heat conduction of the air, do not furnish an accurate measure of the radiation of heat by the thermometer. Only a very few of the experiments of de la Provostaye and Desains were carried out at such low pressures that we can assume that the action of the air is limited to its conduction, and in those cases in which they used a cylindrical thermometer the calculation of the heat conduction is not possible.

We can however, without knowing the effect of the influence of the air on the exchange of heat, apply these experiments on cooling to a test of the law of radiation, if observations have been made on the cooling of one and the same body with two different surfaces but under otherwise similar conditions.

The differences of the rates of cooling of the thermometer with a bare bulb and with a silvered bulb are the measure of the differences between the coefficients of radiation of glass and of silver. These numbers are not affected by the stem or by heat conduction, and at least with close approximation are not affected by air currents.

In this manner I will employ two series of observations given by Dulong and Petit of the cooling of a bare and of a silvered thermometer at the same pressure of 720 mm. The following table contains in the first column the differences of temperature between the thermometer and the container, which was at 20°.

The meaning of the numbers in the other columns is given above them.

	Rate of Cooling		
	Glass	Silver	Difference
100°	4.99	2.80	2.19
120	6.46	3.50	2.96
140	8.05	4.32	3.73
160	9.85	5.19	4.66
180	11.76	6.02	5.74
200	14.04	6.93	7.11

If Dulong and Petit's formula for the radiation of heat is correct for glass and silver, the differences of the rates of cooling must also be in accord with this formula. If the radiation of glass is given by ma^u, and that of silver by $m'a^u$, then their difference is given by $(m - m')a^u$ and if δ represents the difference of temperature between the thermometer and the container the difference of the rates of cooling is proportional to $(m - m')(a^\delta - 1)$.

If we divide the successive differences by the corresponding values of $a^\delta - 1$, we obtain the quotients

$$1.911, \ 1.977, \ 1.950, \ 1.947, \ 1.944, \ 1.972,$$

which differ only a little from the mean value 1.95. This agreement is a better proof of the applicability of the formula which has been used, than the agreement which the rates of cooling calculated by Dulong and Petit for vacuum show with the observed values.

The differences of which we are speaking agree equally well with the formula according to which the radiation is proportional to the fourth power of the absolute temperature. If T_1 represents the absolute temperature of the thermometer, and T_2 that of the container and if we divide the numbers given in the last column of the foregoing table by the corresponding values of $T_1{}^4 - T_2{}^4$ and pay no attention to the order of the numbers we shall obtain the quotients

$$1329, 1363, 1343, 1341, 1345, 1375,$$

which agree with each other as well as the former set and with respect to the departures from the mean are related in a similar way. The mean of these numbers is 1350 and its true value is $135 \cdot 10^{-12}$.

Similar comparisons are made of the results of the two formulas applied to observations of de la Provostaye and Desains. The final conclusion is then given.

We can therefore conclude from the comparison which we have now made that both formulas represent equally well the experiments of Dulong and Petit, which were made at higher temperatures, but that the formula given by Dulong and Petit does not fit the observations of de la Provostaye and Desains, some of which were carried out at lower temperatures, as well as the new formula of the fourth powers.

CHRISTIANSEN

Christian Christiansen was born at Lonborg, Jutland, on October 9, 1843. From 1886 until his death he was professor of physics at the University of Copenhagen. He died in Copenhagen on November 28, 1917.

In the short paper which follows, "Ueber die Brechungsverhältnisse einer weingeistigen Lösung des Fuchsin," from the *Annalen der Physik und Chemie*, Series 5, Vol. 141, p. 479, 1870, Christiansen describes his discovery of anomalous dispersion. A similar phenomenon had already been observed by Le Roux in iodine vapor.

Anomalous Dispersion

Copenhagen, Nov. 1870.

Permit me to inform you that I have been for some time occupied in the investigation of the refraction of red concentrated aniline (fuchsine) and that I have obtained very remarkable results. I propose to publish later fuller details on this matter, and I now only give the indices of refraction for an alcoholic solution which contains 18.8% aniline.

Fraunhofer's Lines	Index of Refraction
B	1.450
C	1.502
D	1.561
F	1.312
G	1.285
H	1.312

The index of refraction increases from B to D and a little further on, then very quickly falls to G, and increases from then on again.

It is easy to convince oneself of this by constructing a prism with a very small refracting angle from the fluid; by observing through it an illuminated slit one sees the colors in the following order: violet, red, yellow, where the last is most deviated. The consequences of this show themselves in the simplest and most beautiful way if we illuminate the hypothenuse of a right angled prism of the solution and examine the reflected light. We then have, instead of the colors at the limits of total reflection, colored light, rose red, violet, blue, green, for all incidences.

KUNDT

August Adolph Eduard Eberhard Kundt was born in Schwerin on November 18, 1839. He studied at Leipzig and Berlin. In 1868 he became professor of physics in the Polytechnic School at Zürich and four years later was called to the newly established university in Strassburg. In 1888 he succeeded Helmholtz at Berlin. He died in Israeldorf near Lübeck on May 21, 1894.

The extract which follows is from his paper "Ueber die anomale Dispersion der Körper mit Oberflächenfarben," taken from the *Annalen der Physik und Chemie*, 5th Series, Vol. 142, p. 163, 1871.

ANOMALOUS DISPERSION

.   .   .   .

Intermediate between transparent bodies and metals there is found a peculiar class of media which for some rays of light are transparent and for others are more or less like the metals, so that they reflect the rays of light with a metallic lustre.

These media have been classed as bodies with surface color. Most of them, though not all, are strong dye stuffs, transparent in solution as well as in small and even microscopic fragments,

like the most of the aniline dyes, indigo, carthamine, permanganate
of potassium, etc.

The optical peculiarities of these bodies have been investigated
by Brewster, Haidinger, Stokes and others. The principal result
is the law first announced by Haidinger, that the light which is
transmitted through these media is altogether or almost complementary to that reflected at the surface, that is, to the surface
color.

We can therefore also say that these bodies transmit only a
very little of the rays which they reflect strongly or possess for
these rays a very strong absorption. Stokes noticed that the
surface color of permanganate of potassium shows five light
maxima in the green, which correspond exactly to the five dark
bands which appear in the absorption spectrum of permanganate
of potassium in dilute solution.

. . . .

The attempts which I made in the course of last year to connect
anomalous dispersion with the surface colors of bodies or their
solutions, by means of interference phenomena, led to no result
on account of imperfect apparatus. A note in the last number,
No. 11, 1870 of Poggendorff's Annalen by Christiansen of Copenhagen gave me the first direct proof of my expectations and induced
me to take up my investigation again, and as Christiansen had
done, to investigate dispersion directly.

Christiansen reports in the place referred to that he investigated
the dispersion of a concentrated alcoholic solution of fuchsine
(a red aniline dye) and obtained the remarkable result that the
index of refraction of the solution increased from B to D, then
quickly sank to G and from there on increased again. Christiansen
simply states the fact without referring to the other optical
peculiarities of fuchsine and without mentioning any other similar
body.

Also the observation of Le Roux, who found in 1862 that iodine
vapor (iodine is a body with surface color) refracted the red rays
more than it did the blue, has remained an isolated fact.

My researches have in fact shown the general connection of
the anomalous dispersion of bodies with their surface color, for
the present only when they are in solution. Practically all the
bodies which in the solid condition show evident surface color
and which I could examine in the proper way in very concentrated
solutions give an anomalous dispersion.

By an anomalous dispersion I mean a succession of colors in the spectrum formed by dispersion which does not correspond with the order of the colors in the diffraction spectrum or in the dispersion spectrum of ordinary bodies.

A list of the substances with which Kundt's experiments were made has been omitted.

By the use of the method of observation which will be briefly described later on, it was only possible to detect complete anomalous dispersion, i.e. a complete exchange of the principal colors; we may therefore assume with confidence that in the case of many other bodies, whose surface color is not very conspicuous or which do not dissolve easily, smaller anomalies will be detected when finer methods of observation are used.

All the above named bodies refract the red light more than the blue, so that in the case of those bodies in which green is a principal part of the surface color and can also be clearly recognized in the spectrum, the green is the least deviated.

·   ·   ·   ·   ·   ·   ·   ·   ·   ·   ·   ·   ·   ·   ·   ·   ·   ·

ZEEMAN

Pieter Zeeman was born on May 25, 1865, at Zonnemaire, Zeeland. From 1885 to 1893 he studied at Leyden. In 1890 he became connected with the University as assistant and remained there until 1900 when he was made professor of physics at the University of Amsterdam.

The paper which follows contains Zeeman's account of the discovery of the effect known by his name. It is entitled "Ueber einen Einfluss der Magnetisirung auf die Natur des von einer Substanz emittirten Lichtes." It appeared in the *Verhandlungen der Physikalischen Gesellschaft zu Berlin*, No. 7, p. 128, 1896. A similar communication had been made shortly before to the Academy of Sciences in Amsterdam. The translation, by Arthur Stanton, appeared in *Nature*, February 11, 1897. It has been slightly changed in a few places.

THE ZEEMAN EFFECT

In consequence of my measurements of Kerr's magneto-optical phenomena, the thought occurred to me whether the period of the light emitted by a flame might be altered when the flame was acted upon by magnetic force. It has turned out that such an action really occurs. I introduced into an oxyhydrogen flame, placed between the poles of a Ruhmkorff's electromagnet, a filament of asbestos soaked in common salt. The light of the flame was examined with a Rowland's grating. Whenever the circuit was closed both D lines were seen to widen.

Since one might attribute the widening to the known effects of the magnetic field upon the flame, which would cause an alteration in the density and temperature of the sodium vapour, I had resort to a method of experimentation which is much more free from objection.

Sodium was strongly heated in a tube of biscuit porcelain, such as Pringsheim used in his interesting investigations upon the radiation of gases. The tube was closed at both ends by plane parallel glass plates, whose effective area was 1 cm. The tube was placed horizontally between the poles, at right angles to the lines of force. The light of an arc lamp was sent through. The absorption spectrum showed both D lines. The tube was continuously rotated round its axis to avoid temperature variations. Excitation of the magnet caused immediate widening of the lines. It thus appears very probable that the period of sodium light is altered in the magnetic field. It is remarkable that Faraday, as early as 1862, had made the first recorded experiment in this direction, with the incomplete resources of that period, but with a negative result (Maxwell, "Collected Works," vol. II. p. 790).

It has been already stated what, in general, was the origin of my own research on the magnetisation of the lines in the spectrum. The possibility of an alteration of period was first suggested to me by the consideration of the accelerating and retarding forces between the atoms and the Maxwell's molecular vortices; later came an example suggested by Lord Kelvin, of the combination of a quickly rotating system and a double pendulum. However, a true explanation appears to me to be afforded by the theory of electric phenomena propounded by Prof. Lorentz.

In this theory, it is considered that, in all bodies, there occur small molecular elements charged with electricity, that all electrical processes are to be referred to the equilibrium or motion of these "ions" and that the undulations of light are vibrations of the ions. It seems to me that in the magnetic field the forces directly acting on the ions suffice for the explanation of the phenomena.

Prof. Lorentz, to whom I communicated my idea, was good enough to show me how the motion of the ions might be calculated, and further suggested that if my application of the theory be correct there would follow these further consequences; that the light from the edges of the widened lines should be circularly polarised when the direction of vision lay along the lines of force; further, that the magnitude of the effect would lead to the determination of the ratio of the electric charge the ion bears to its mass. We may designate the ratio e/m. I have since found by means of a quarter-wave-length plate and an analyser, that the edges of the magnetically-widened lines are really circularly polarised when the line of sight coincides in direction with the lines of force. An altogether rough measurement gives 10^7 as the order of magnitude of the ratio e/m when e is expressed in electromagnetic units.

On the contrary, if one looks at the flame in a direction at right angles to the lines of force, then the edges of the broadened sodium lines appear plane polarised, in accordance with theory. Thus there is here direct evidence of the existence of ions.

This investigation was conducted in the Physical Institute of Leyden University, and will shortly appear in the "Communications of the Leyden University."

I return my best thanks to Prof. Kamerlingh Onnes for the interest he has shown in my work.

Amsterdam.   P. Zeeman.

MAGNETISM AND ELECTRICITY

GILBERT

William Gilbert was born in Colchester in 1540. From 1573 on he lived in London as a practising physician. He was physician to Queen Elizabeth and afterwards to her successor, James the First. He died in London in 1603.

In 1600 Gilbert published a book entitled *De Magnete Magneticisque Corporibus et de Magno Magnete Tellure Physiologia Nova,* in which he described the properties of magnets and presented his theory that the earth behaves like a great magnet. He also described experiments on the development of electricity by friction. Portions of the book dealing with these subjects are presented. The translation was made by E. Fleury Mottelay, and was published in 1893.

Magnetism and Electricity

The Loadstone Possesses Parts Differing In Their Natural Powers, And Has Poles Conspicuous For Their Properties.

The many qualities exhibited by the loadstone itself, qualities hitherto recognized yet not well investigated, are to be pointed out in the first place, to the end the student may understand the powers of the loadstone and of iron, and not be confused through want of knowledge at the threshold of the arguments and demonstrations. In the heavens, astronomers give to each moving sphere two poles; thus do we find two natural poles of excelling importance even in our terrestrial globe, constant points related to the movement of its daily revolution, to wit, one pole pointing to Arctos (Ursa) and the north; the other looking toward the opposite part of the heavens. In like manner the loadstone has from nature its two poles, a northern and a southern; fixed, definite points in the stone, which are the primary termini of the movements and effects, and the limits and regulators of the several actions and properties. It is to be understood, however, that not from a mathematical point does the force of the stone emanate, but from the parts themselves; and all these parts in the whole— while they belong to the whole—the nearer they are to the poles of the stone the stronger virtues do they acquire and pour out on other bodies. These poles look toward the poles of the earth,

387

and move toward them, and are subject to them. The magnetic poles may be found in every loadstone, whether strong and powerful (male, as the term was in antiquity) or faint, weak, and female; whether its shape is due to design or to chance, and whether it be long, or flat, or four-square, or three-cornered, or polished; whether it be rough, broken-off, or unpolished: the loadstone ever has and ever shows its poles.

. . . .

One Loadstone Appears To Attract Another In The Natural Position; But In The Opposite Position Repels It And Brings It To Rights.

First we have to describe in popular language the potent and familiar properties of the stone; afterward, very many subtle properties, as yet recondite and unknown, being involved in obscurities, are to be unfolded; and the causes of all these (nature's secrets being unlocked) are in their place to be demonstrated in fitting words and with the aid of apparatus. The fact is trite and familiar, that the loadstone attracts iron; in the same way, too, one loadstone attracts another. Take the stone on which you have designated the poles, *N.* and *S.,* and put it in its vessel so that it may float; let the poles lie just in the plane of the horizon, or at least in a plane not very oblique to it; take in your hand another stone the poles of which are also known, and hold it so that its south pole shall lie toward the north pole of the floating stone, and near it alongside; the floating loadstone will straightway follow the other (provided it be within the range and dominion of its powers), nor does it cease to move nor does it quit the other till it clings to it, unless by moving your hand away, you manage skilfully to prevent the conjunction. In like manner, if you oppose the north pole of the stone in your hand to the south pole of the floating one, they come together and follow each other. For opposite poles attract opposite poles. But, now, if in the same way you present *N.* to *N.* or *S.* to *S.,* one stone repels the other; and as though a helmsman were bearing on the rudder it is off like a vessel making all sail, nor stands nor stays as long as the other stone pursues. One stone also will range the other, turn the other around, bring it to right about and make it come to agreement with itself. But when the two come together and are conjoined in nature's order, they cohere firmly. For example, if you present the north pole of the stone in your hand to the Tropic of Capricorn (for so we may distinguish with mathematical circles the round

stone or terrella, just as we do the globe itself) or to any point
between the equator and the south pole: immediately the floating
stone turns round and so places itself that its south pole touches
the north pole of the other and is most closely joined to it.   In
the same way you will get like effect at the other side of the equator
by presenting pole to pole; and thus by art and contrivance we
exhibit attraction and repulsion, and motion in a circle toward
the concordant position, and the same movements to avoid hostile
meetings.   Furthermore, in one same stone we are thus able to
demonstrate all this: but also we are able to show how the self-same
part of one stone may by division become either north or south.
Take the oblong stone *ad* (Fig. 78) in which *a* is the north pole and

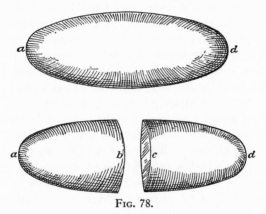

Fig. 78.

d the south.   Cut the stone in two equal parts, and put part *a* in
a vessel and let it float in water.

You will find that *a* the north point, will turn to the south
as before; and in like manner the point *d* will move to the north,
in the divided stone, as before division.   But *b* and *c*, before
connected, now separated from each other, are not what they were
before.   *b* is now south while *c* is north.   *b* attracts *c*, longing
for union and for restoration of the original continuity.   They
are two stones made out of one, and on that account the *c* of one
turning toward the *b* of the other, they are mutually attracted,
and being freed from all impediments and from their own weight,
borne as they are on the surface of the water, they come together
and into conjunction.   But if you bring the part or point *a* up to
c of the other, they repel one another and turn away; for by such a
position of the parts nature is crossed and the form of the stone is

perverted: but nature observes strictly the laws it has imposed
upon bodies: hence the flight of one part from the undue position
of the other, and hence the discord unless everything is arranged
exactly according to nature.   And nature will not suffer an unjust
and inequitable peace, or agreement, but makes war and employs
force to make bodies acquiesce fairly and justly.   Hence, when
rightly arranged the parts attract each other, i.e., both stones, the
weaker and the stronger, come together and with all their might tend
to union: a fact manifest in all loadstones, and not, as Pliny supposed, only in those from Ethiopia. The Ethiopic stones if strong,
and those brought from China, which are all powerful stones, show
the effect most quickly and most plainly, attract with most force
in the parts nighest the pole, and keep turning till pole looks
straight on pole. The pole of a stone has strongest attraction
for that part of another stone which answers to it (the *adverse* as it
is called); e.g., the north pole of one has strongest attraction for,
has the most vigorous pull on, the south part of another; so too it
attracts iron more powerfully, and iron clings to it more firmly,
whether previously magnetized or not. Thus it has been settled
by nature, not without reason, that the parts nigher the pole shall
have the greatest attractive force; and that in the pole itself shall
be the seat, the throne as it were, of a high and splendid power;
and that magnetic bodies brought near thereto shall be attracted
most powerfully and relinquished with most reluctance. So, too,
the poles are readiest to spurn and drive away what is presented
to them amiss, and what is inconformable and foreign.

. . . .

*Of Magnetic Coition; And, First, Of The Attraction Exerted By
Amber, or More Properly The Attachment of Bodies to Amber.*

Great has ever been the fame of the loadstone and of amber
in the writings of the learned: many philosophers cite the loadstone
and also amber whenever, in explaining mysteries, their minds
become obfuscated and reason can no farther go. Over-inquisitive
theologians, too, seek to light up God's mysteries and things
beyond man's understanding by means of the loadstone as a sort
of Delphic sword and as an illustration of all sorts of things.
Medical men also (at the bidding of Galen), in proving that
purgative medicines exercise attraction through likeness of substance and kinships of juices (a silly error and gratuitous!), bring
in as a witness the loadstone, a substance of great authority and

of noteworthy efficiency, and a body of no common order. Thus in very many affairs persons who plead for a cause the merits of which they cannot set forth, bring in as masked advocates the loadstone and amber. But all these, besides sharing the general misapprehension, are ignorant that the causes of the loadstone's movements are very different from those which give to amber its properties; hence they easily fall into errors, and by their own imaginings are led farther and farther astray. For in other bodies is seen a considerable power of attraction, differing from that of the loadstone,—in amber, for example. Of this substance a few words must be said, to show the nature of the attachment of bodies to it, and to point out the vast difference between this and the magnetic actions; for men still continue in ignorance, and deem that inclination of bodies to amber to be an attraction, and comparable to the magnetic coition. The Greeks call this substance ἤλεκτρον, because, when heated by rubbing, it attracts to itself chaff; whence it is also called ἄρπαξ and from its golden color, χρυσοφόρον. But the Moors call it *carabe*, because they used to offer it in sacrifices and in the worship of the gods; for in Arabic *carab* means oblation, not *rapiens paleas* (snatching chaff), as Scaliger would have it, quoting from the Arabic or Persian of Abohali (Hali Abbas). Many call this substance *ambra* (amber), especially that which is brought from India and Ethiopia. The Latin name *succinum* appears to be formed from *succus*, juice. The Sudavienses or Sudini call the substance *geniter*, as though *genitum terra* (produced by the earth). The erroneous opinion of the ancients as to its nature and source being exploded, it is certain that amber comes for the most part from the sea: it is gathered on the coast after heavy storms, in nets and through other means, by peasants, as by the Sudini of Prussia; it is also sometime found on the coast of our own Britain. But it seems to be produced in the earth and at considerable depth below its surface, like the rest of the bitumens; then to be washed out by the sea-waves, and to gain consistency under the action of the sea and the saltness of its waters. For at first it was a soft and viscous matter, and hence contains, buried in its mass forevermore (*aeternis sepulchris relucentes*), but still (shining) visible, flies, grubs, midges, and ants. The ancients as well as moderns tell (and their report is confirmed by experience) that amber attracts straws and chaff. The same is done by jet, a stone taken out of the earth in Britain, Germany, and many other regions: it is a hard concretion of black bitumen,—

a sort of transformation of bitumen to stone. Many modern
authors have written about amber and jet as attracting chaff and
about other facts unknown to the generality, or have copied from
other writers: with the results of their labors booksellers' shops
are crammed full. Our generation has produced many volumes
about recondite, abstruse, and occult causes and wonders, and in
all of them amber and jet are represented as attracting chaff; but
never a proof from experiments, never a demonstration do you find
in them. The writers deal only in words that involve in thicker
darkness subject-matter; they treat the subject esoterically,
miracle-mongeringly, abstrusely, reconditely, mystically. Hence
such philosophy bears no fruit; for it rests simply on a few Greek
or unusual terms—just as our barbers toss off a few Latin words
in the hearing of the ignorant rabble in token of their learning, and
thus win reputation—bears no fruit, because few of the philosophers themselves are investigators, or have any first-hand acquaintance with things; most of them are indolent and untrained, add
nothing to knowledge by their writings, and are blind to the things
that might throw a light upon their reasonings. For not only do
amber and (gagates or) jet, as they suppose attract light corpuscles
(substances): the same is done by diamond, sapphire, carbuncle,
iris stone, opal, amethyst, vincentina, English gem (Bristol stone,
bristola), beryl, rock crystal. Like powers of attracting are
possessed by glass, especially clear, brilliant glass; by artificial
gems made of (paste) glass or rock crystal, antimony glass, many
fluor-spars, and belemnites. Sulphur also attracts, and likewise
mastich, and sealing-wax (of lac), hard resin, orpiment (weakly).
Feeble power of attraction is also possessed in favoring dry atmosphere by sal gemma [native chloride of sodium], mica, rock alum.
This we may observe when in mid-winter the atmosphere is very
cold, clear, and thin; when the electrical effluvia of the earth offers
less impediment, and electric bodies are harder: of all this later.
These several bodies (electric) not only draw to themselves straws
and chaff, but all metals, wood, leaves, stones, earths, even water
and oil; in short, whatever things appeal to our senses or are
solid: yet we are told that it attracts nothing but chaff and twigs.
Hence Alexander Aphrodiseus incorrectly declares the question of
amber to be unsolvable, because that amber does attract chaff,
yet not the leaves of basil; but such stories are false, disgracefully
inaccurate. Now in order clearly to understand by experience
how such attraction takes place, and what those substances may

be that so attract other bodies (and in the case of many of these electrical substances, though the bodies influenced by them lean toward them, yet because of the feebleness of the attraction they are not drawn clean up to them, but are easily made to rise), make yourself a rotating-needle (electroscope—*versorium*) (Fig. 79), of any sort of metal, three or four fingers long, pretty light, and poised on a sharp point after the manner of a magnetic pointer. Bring near to one end of it a piece of amber or a gem, lightly rubbed, polished and shining: at once the instrument revolves. Several objects are seen to attract not only natural objects, but things artificially prepared, or manufactured, or formed by mixture. Nor is this a rare property possessed by one object or two (as is commonly supposed), but evidently belongs to a multitude of objects, both simple and compound, e.g., sealing-wax and other unctuous mixtures. But why this inclination and what these forces,—on which points a few writers have given a very small amount of information, while the common run of philosophers give us nothing,— these questions must be considered fully.

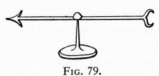

Fig. 79.

.   .   .   .

VON GUERICKE

The extract which follows is from von Guericke's (p. 80) *Works, Fourth Book, De Virtutibus Mundanis*, Chap. XV. In this chapter von Guericke describes the various virtues or powers exhibited by a globe of sulphur which has been electrified by friction. Many of these are the ordinary properties of all matter and are of no special interest. The extracts given contain some of his observations of the effects produced by the electrification of the globe.

Electric Attractions and Repulsions

If you please, take a sphere of glass which is called a phial, as large as a child's head; fill it with sulphur that has been pounded in a mortar and melt it sufficiently over a fire. When it is cooled again break the sphere and take out the globe and keep it in a dry place. If you think it best, bore a hole through it so that it can be turned around an iron rod or axle. In this way the globe is prepared.

.   .   .   .

To show the conservative power in this globe mount it by its axle on two fulcra in a frame so that it stands a palm high from the base, and under it place little fragments of all sorts, gold or silver foil, paper, leaves, or other such things. Now touch it gently with the dry hand so that it is rubbed or stroked two or three times; then it attracts these fragments: and when it is turned about its axle it takes them with it.

. . . .

The expulsive power is plainly to be seen if the globe is taken from the frame and is rubbed with a dry hand as has been described. For it not only attracts the small bodies in the same way, but it repels them from it again, and will not allow them to touch it until they have touched some other body.

. . . .

If a linen thread is so fastened at one end that it hangs down nearly to the globe and if the attempt is made to touch it with the finger or anything else, the thread moves away and cannot be touched by the finger. If a linen thread is fastened to a pointed piece of wood which is thrust into a table or bench and if the thread hangs down for more than an ell, so that it can touch something below it, being distant from it about an inch, whenever the excited globe is moved over the upper end of the wooden rod the lower end of the thread moves to the body which is near it: by which we have ocular demonstration that this power extends in the linen thread to its lowest parts, since it either attracts or is attracted by the body.

GRAY

Stephen Gray was born in England in the latter part of the seventeenth century and died in London on February 15, 1736. He was a member of the Royal Society, and an active experimental student of electricity. The extract which follows describes his discovery of the difference between electric conductors and non-conductors. It is taken from the *Philosophical Transactions*, Vol. 6, Part II, p. 6, 1731 (published in 1733).

ELECTRIC CONDUCTORS AND NONCONDUCTORS

Before I proceed to the experiments it may be necessary to give a description of the tube: Its length is three feet five inches, and near one inch two tenths in diameter: I give the mean dimensions, the tube being larger at each end than in the middle, the bore about one inch. To each end I fitted a cork, to keep the dust out when the tube was not in use.

The first experiment I made was to see if I could find any difference in its attraction, when the tube was stopped at both ends by the corks, or when left open; but could perceive none: for upon holding a down-feather over-against the upper end of the tube, it was attracted and repelled by the cork, as by the tube when it had been excited by rubbing. I then held the feather over-against the flat end of the cork, which attracted and repelled many times together; at which I was much surprised, and concluded that there was certainly an electric virtue communicated to it by the excited tube.

Having by me an ivory ball of about one inch three tenths diameter, with a hole through it, I fixed it upon a fir-stick about four inches long, thrusting the other end into the cork, and upon rubbing the tube found that the ball attracted and repelled the feather with more vigour than the cork had done, repeating its attractions and repulsions for many times together. I then fixed the ball on longer sticks, first upon one of eight inches, and afterwards upon one of twenty-four inches long, and found the same effect. Then I used first iron and next brass wire to fix the ball on, inserting the other end of the wire in the cork as before, and found the attraction the same as when the fir-sticks were made use of, and that when the feather was held over-against any part of the wire it was attracted by it; but though it was then nearer the tube, yet the attraction was not so strong as that of the ball. When a wire of two or three foot long was used, its vibrations caused by rubbing the tube made it somewhat troublesome to manage. This put me upon thinking, whether if the ball were hung by a pack-thread and suspended by a loop on the tube, the electricity would not be carried down the line to the ball: accordingly upon suspending the ball on the tube by a packthread about three foot long, when the tube had been excited by rubbing, the ivory ball attracted and repelled the leaf-brass over which it was held, as freely as it had done when it was suspended on sticks or wire: a ball of cork and another of lead weighing one pound and a quarter did the same.

. . . .

Some time after I made several attempts to carry the electric vertue in a line horizontally, since I had not an opportunity of carrying it from greater heights perpendicularly; but without success, occasioned by my using improper materials, as will appear from what follows. The first method I tried was by making a loop at each end of a line, one of which I hung on a nail drove into a beam; the other hung downwards, and through it I put the line with the ivory ball, the other end of it being hung perpendicular, the rest of it lay horizontal: then the leaf-brass being placed under the ball and the tube rubbed, yet not the least sign of attraction was perceivable. Upon this I concluded that when the electric vertue came to the loop that suspended on the beam, it went up the line to the beam; so that none, or very little of it at least, came down to the ball, which was afterwards verified, as will appear by the experiments mentioned hereafter. Upon this I gave over any farther attempts to carry the electricity horizontally, designing at my return to London, if I could get assistance, to have tried the experiment from the top of the cupola of St. Paul's, not doubting but the electric attraction would be carried perpendicularly down from thence to the ground.

June 30, 1729. I went to Otterden-Place to wait on Mr. Wheler, carrying with me a small solid glass cane of about eleven inches long, and ⅞ of an inch in diameter, with some other requisite materials, designing only to give a specimen of my experiments. The first was from the window in the long gallery that opened into the hall, about sixteen foot high; the next from the battlements of the house down into the fore court, twenty-nine foot; then from the clock-turret to the ground, which was thirty-four foot, this being the greatest height we could come at: and notwithstanding the smallness of the cane, the leaf-brass was attracted and repelled beyond what I expected. As we had no great heights here Mr. Wheler was desirous to try whether we could not carry the electric vertue horizontally. I told him of the unsuccessful attempt I had made, and the method and materials I had used. He then proposed a silk line to support that by which the electric vertue was to pass. I told him it might do better upon the account of its smallness; so that there would be less vertue carried from the line of communication.

Accordingly an experiment was made July 2, 1729, at ten in the morning. About four foot from the end of a gallery there was a cross line fixed by its ends to each side of the gallery with two nails; the middle part of the line was silk, the rest at each end packthread. The line to which the ivory ball was hung, and by which the electric vertue was to be conveyed to it from the tube, being 80½ foot in length, was laid on the cross silk line so that the ball hung about nine foot below it. The other end of the line was by a loop suspended on the glass cane, and the leaf-brass held under the ball on a piece of white paper; when the tube being rubbed the ball attracted the leaf-brass and kept it suspended on it for some time.

This experiment succeeding so well, and the gallery not permitting us to go any farther in one length, Mr. Wheler thought of an expedient for encreasing the length of our line, by putting up another cross line near the other end of the gallery; over the silk part of both the lines we laid a line of communication long enough to be returned to the other end where the ball hung: and though now both ends of the line were at the same end of the gallery, yet care was taken that the tube was far enough off from having any influence upon the leaf-brass, except what passed by the line of communication. Then the cane being rubbed, and the leaf-brass held under the ivory ball, the electric vertue passed by the line of communication to the other end of the gallery, and returned back again to the ivory ball which attracted the leaf-brass and suspended it as before. The whole length of the line was 147 foot.

We then thought of trying whether the attraction would not be stronger without doubling or returning the line, which we found means of doing in the barn, where we had one of 124 foot long, fourteen foot of which hung perpendicular from the silk line; and now the attraction was stronger than when the line was returned, as in the gallery.

July 3. Having now brought with me the great glass tube, between ten and eleven in the morning, we went again into the barn, and repeated the last mentioned experiment with both the tube and cane; but the attraction was not so strong as in the evening before, nor was there so great a difference in the attraction communicated by the solid cane and glass tube as one would have expected, considering the difference of their lengths and diameters.

We then proceeded farther by adding so much more line as would make a return to the other end of the barn, the whole length of it being now 293 foot; and though it was so much lengthened we found no sensible difference in the attraction. This encouraged us to add another return; but upon beginning to rub the tube, our silk lines broke, being too weak to bear the weight of the line, when shaken with the motion given it by rubbing the tube. Upon this, having brought with me both brass and iron wire, instead of the silk we put up small iron wire; but that also was too weak to bear the weight of the line. We then took brass wire of a somewhat larger size than that of iron. This supported our line of communication; but though the tube was well rubbed yet there was not the least attraction even with the great tube. By this we were now convinced that our former success depended upon the lines that supported the line of communication being silk, and not upon their being small, as before trial I imagined; the same effect happening here as when the line that is to convey the electric vertue is supported by packthread; viz. that when the effluvia come to the wires or packthread that support the line, they pass by them to the timber to which each end of them is fixed, and so go no farther forward in the line that is to carry them to the ivory ball.

. . . .

DU FAY

Charles Francois de Cisternay du Fay was born on September 14, 1698, in Paris and died there on July 16, 1739. While still very young he served as an officer in the army, but left it after the war was over and because of feeble health to devote himself to science. He was a member of the French Academy. His principal work was in electricity and he was the first who clearly stated the two fluid theory of electricity.

The following extract is from a paper which appeared in the *Philosophical Transactions*, Vol. 38, p. 258, 1734.

THE TWO FLUID THEORY OF ELECTRICITY

On making the experiment related by *Otho de Guerik*, in his Collection of experiments *de Spatio Vacuo*, which consists in making a ball of sulphur render'd electrical, to repel a down-feather, I perceived that the same effects were produced not only by the tube, but by all electrick bodies whatsoever; and I discovered a very simple principle, which accounts for a great part of the irregularities, and if I may use the term, of the caprices that seem to accompany most of the experiments on electricity. This principle is, that electrick bodies attract all those that are not so, and repel them as soon as they are become electrick, by the vicinity or contact of the electrick body. Thus leaf-gold is first attracted by the tube; and acquires an electricity by approaching it; and of consequence is immediately repell'd by it. Nor is it re-attracted, while it retains its electrick quality. But if, while it is thus sustain'd in the air, it chance to light on some other body, it straightways loses its electricity; and consequently is re-attracted by the tube, which, after having given it a new electricity, repels it a second time; which continues as long as the tube keeps its electricity. Upon applying this principle to the various experiments of electricity, one will be surprised at the number of obscure and puzzling facts it clears up. For Mr. *Hauksbee's* famous experiments of the glass globe, in which silk threads are put, is a necessary consequence of it. When these threads are ranged in form of rays by the electricity of the sides of the globe, if the finger be put near the outside of the globe, the threads within fly from it, as is well known; which happens only because the finger, or any other body applied near the glass globe, is thereby rendered electrical, and consequently repels the silk threads, which are endowed with the like quality. With a little reflection one may in the same manner account for most of the other *phaenomena*, and which seem inexplicable, without attending to this principle.

Chance has thrown in my way another principle, more universal and remarkable than the preceding one, and which casts a new light on the subject of electricity. This principle is, that there are two distinct electricities, very different from one another; one of which I call *vitreous electricity*, and the other *resinous electricity*. The first is that of glass, rock-crystal, precious stones, hair of animals, wool, and many other bodies; the second is that of amber, copal, gum-lack, silk, thread, paper, and a vast number of other substances. The characteristick of these two electricities is, that a body of the *vitreous electricity*, for example, repels all such as are of the same electricity; and on the contrary, attracts all those of the *resinous electricity;* so that the tube, made electrical, will repel glass, crystal, hair of animals, etc., when render'd electrick and will attract silk, thread, paper, etc., though render'd electrical likewise. Amber on the contrary will attract electrick glass, and other substances of the same class, and will repel gum-lac, copal, silk, thread, etc. Two silk ribbons rendered electrical, will repel each other; two woolen threads will do the like; but a woolen thread and a silk thread will mutually attract one another.   This principle very naturally explains why the ends of threads, of silk, or wool, recede from one another in form of a pencil or broom, when they have acquired an electrick quality. From this principle one may with the same ease deduce the explanation of a great number of other *phaenomena*.   And it is probable, that this truth will lead us to the further discovery of many other things.

In order to know immediately, to which of the two classes of electricity belongs any body whatsoever, one need only render electrical a silk thread, which is known to be of the *resinous electricity*, and see whether that body, rendered electrical, attracts or repels it.   If it attracts, it is certainly of that kind of electricity which I call *vitreous;* if on the contrary it repels, it is of the same kind of electricity with the silk, that is, of the *resinous*.   I have likewise observed that communicated electricity retains the same properties: for if a ball of ivory, or wood, be set on a glass stand, and this ball be rendered electrick by the tube, it will repel all such substances as the tube repels; but if it be rendered electrick by applying a cylinder of gum-lac near it, it will produce quite contrary effects, viz. precisely the same as gum-lac would produce. In order to succeed in these experiments, it is requisite that the two bodies, which are put near one another, to find out the nature of their electricity, be rendered as electrical as possible; for if one of them was not at all, or but weakly electrical, it would be attracted by the other, though it be of that sort, that should naturally be repelled by it.   But the experiment will always succeed perfectly well, if both the bodies are sufficiently electrical.

FRANKLIN

Benjamin Franklin was born in Boston, Massachusetts, on January 17, 1706.   His father was a soap and candle maker.   He apprenticed Benjamin to an older brother to learn the printing trade.   Dissensions arose between the brothers and the younger was finally freed from his indentures.   He went to Philadelphia, where he supported himself by his trade.   By industry and frugality he soon acquired a competence, so that in his later years he was able to engage in public business.   He represented the province of Pennsylvania in England for many years and after his return to America he played a prominent part in the Revolution.   He represented the United States in France during the war and was one of the commissioners who concluded the peace between Great Britain and the United States.   He died in Philadelphia on April 17, 1790.

While he lived in Philadelphia he became interested in the study of electricity. He proposed the one fluid theory of electricity, to which we owe the names positive and negative electricity, which have displaced those of vitreous and resinous electricity proposed by Du Fay. The extract which follows, taken from a letter dated June 1, 1747, which appeared in the *Philosophical Transactions*, Vol. 45, p. 98, 1750, contains his statement of the one fluid theory. The next extract is from a letter to Peter Collinson of London dated from Philadelphia, October 16, 1752. It contains the account of Franklin's famous experiment with the electric kite.

THE ONE FLUID THEORY OF ELECTRICITY

1. A person standing on wax, and rubbing a tube, and another person on wax drawing the fire; they will both of them, provided they do not stand so as to touch one another, appear to be electrified to a person standing on the floor; that is, he will perceive a spark on approaching each of them with his knuckle.

2. But if the persons on wax touch one another during the exciting of the tube, neither of them will appear to be electrified.

3. If they touch one another after the exciting the tube and drawing the fire as aforesaid, there will be a stronger spark between them than was between either of them and the person on the floor.

4. After such a strong spark neither of them discover any electricity.

These appearances we attempt to account for thus:

We suppose, as aforesaid, that electrical fire is a common element, of which every one of these three persons has his equal share before any operation is begun with the tube. *A*, who stands upon wax, and rubs the tube, collects the electrical fire from himself into the glass; and his communication with the common stock being cut off by the wax, his body is not again immediately supplied. *B*, who stands upon wax likewise, passing his knuckle along near the tube, receives the fire which was collected by the glass from *A*; and his communication with the common stock being cut off, he retains the additional quantity received. To *C* standing on the floor, both appear to be electrified; for he, having only the middle quantity of electrical fire, receives a spark upon approaching *B*, who has an over quantity, but gives one to *A*, who has an under quantity. If *A* and *B* approach to touch each other, the spark is stronger; because the difference between them is greater. After such touch, there is no spark between either of them and *C*, because the electrical fire in all is reduced to the original equality. If they touch while electrising, the equality is never destroyed, the fire only circulating. Hence have arisen some new terms among us. We say, *B* (and bodies alike circumstanced) is electrised positively; *A*, negatively; or rather, *B* is electrised *plus*, *A*, *minus*. And we daily in our experiments electrise *plus* or *minus*, as we think proper. To electrise *plus* or *minus*, no more needs be known than this; that the parts of the tube or sphere that are rubbed, do in the instant of the friction attract the electrical fire, and therefore take it from the thing rubbing. The same parts immediately, as the friction upon them ceases, are disposed to give the fire, they have received, to any body that has less. Thus you may circulate it, as *Mr. Watson* has shown; you may also accumulate or subtract it upon or from any body, as you connect that body with the rubber, or with the receiver, the communication with the common stock being cut off.

THE ELECTRIC KITE

To Peter Collinson, London

Philadelphia, Oct. 16, 1752

As frequent mention is made in public papers from Europe of the success of the Philadelphia experiment for drawing the electric fire from clouds by means of pointed rods of iron erected on high buildings, &c. it may be agreeable to the curious to be informed that the same experiment has succeeded in Philadelphia, though made in a different and more easy manner, which is as follows:

Make a small cross of two light strips of cedar, the arms so long as to reach to the four corners of a larger thin silk handkerchief when extended; tie the corners of the handkerchief to the extremities of the cross, so you have the body of a kite; which being properly accommodated with a tail, loop, and string, will rise in the air, like those made of paper; but this being of silk is fitter to bear the wet and wind of a thunder gust without tearing. To the top of the upright stick of the cross is to be fixed a very sharp pointed wire, rising a foot or more above the wood. To the end of the twine, next the hand, is to be tied a silk ribbon, and where the silk and twine join, a key may be fastened. This kite is to be raised when a thunder-gust appears to be coming on, and the person who holds the string must stand within a door or window, or under some cover, so that the silk ribbon may not be wet; and care must be taken that the twine does not touch the frame of the door or window. As soon as any of the thunder clouds come over the kite, the pointed wire will draw the electric fire from them, and the kite, with all the twine, will be electrified, and the loose filaments of the twine will stand out every way, and be attracted by an approaching finger. And when the rain has wetted the kite and twine, so that it can conduct the electric fire freely, you will find it stream out plentifully from the key on the approach of your knuckle. At this key the phial may be charged; and from electric fire thus obtained, spirits may be kindled, and all the other electric experiments be performed, which are usually done by the help of a rubbed glass globe or tube, and thereby the sameness of the electric matter with that of lightning completely demonstrated.

B. FRANKLIN.

NOLLET

Jean Antoine Nollet was born on November 19, 1700, in Pimpre. He was in orders and generally known as Abbé Nollet. He was professor of physics at Turin and later at Paris, a tutor of the royal children and a member of the French Academy. He died in Paris on April 24, 1770.

The extract which follows is from the *Mémoires de l'Académie Royale des Sciences*, Paris, p. 1, 1746. It is from a memoir entitled "Observations sur quelques nouveaux phénomènes d'Électricité," and contains an account of the discovery of the Leyden jar. The author of the letter quoted was Pieter van Musschenbroek, who was born in Leyden on March 14, 1692 and died there on September 19, 1761. He was professor of mathematics and physics successively at the Universities of Duisburg, Utrecht and Leyden, and the author of one of the first textbooks of physics.

THE LEYDEN JAR

In the month of January of the present year M. de Reaumur showed me a letter from M. Musschenbroek, Professor of Philosophy and Mathematics in the University of Leyden: among several matters which this letter contained there was one which particularly attracted the attention of the Academy. I give here its content translated from the Latin. "I am going to tell you about a new but terrible experiment which I advise you not to try yourself. . . . I was making some investigations on the force of electricity (this is the account of M. Musschenbroek); for this purpose I had suspended by two threads of blue silk a gun-barrel

AB (Fig. 80) which received by communication the electricity of a glass globe which was turned rapidly on its axis while it was rubbed by the hands placed against it; at the other end *B* there

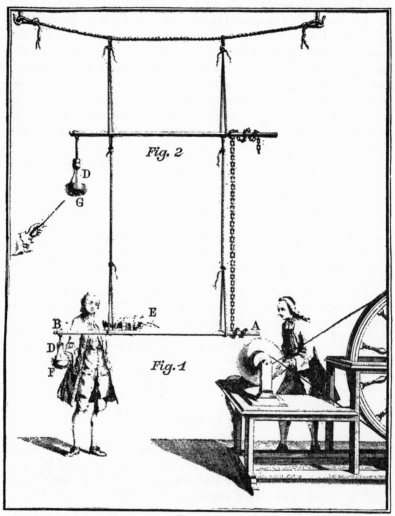

Fig. 80.

hung freely a brass wire, the end of which passed into a round glass flask *D*, partly filled with water, which I held in my right hand *F*, and with the other hand *E*, I tried to draw sparks from the electrified gun-barrel: all at once my right hand *F* was struck so violently that all my body was affected as if it had been struck by lightning; the flask, although made of thin glass, ordinarily does not break, and the hand is not displaced by this disturbance; but the arm and all the body are affected in a terrible way that I cannot describe: in a word I thought that it was all up with me. But it is very remarkable (adds the author of the letter) that when the experiment is tried with an English glass there is no effect or almost none; the glass must be German, it does not even do to have it from Holland; it is all the same whether it is rounded in the form of a sphere or of any other figure: we may use an ordinary goblet, large or small, thick or thin, deep or not; but what is absolutely necessary is that it be made of German or Bohemian glass; the one which I thought had killed me was made of a white thin glass and five inches in diameter. The person who tries the experiment may simply stand on the floor, but it is important that the same man holds the flask D in one hand and tries to draw the spark with the other; the effect is very slight if these actions are performed by two different persons: if the flask D is placed on a metal support standing on a wooden table, then the one who touches this metal even with the end of his finger and draws the spark with his other hand receives a great shock, etc."

Some days after I had read this letter, a letter reached me from M. Allaman, who also lives in Leyden, and who has devoted himself for a long time to experimental physics; he reports the following on the subject of electricity.

"You have learned by a letter from M. Musschenbroek to M. de Reaumur of a new experiment which we have made here which is very remarkable. It is as follows." (He then describes the procedure as it has been reported, and adds) "You will receive a prodigious shock which will affect all your arm and even all your body; it is a stroke of lightning; the first time that I received it I was affected to such a degree that for a few moments I could not breathe: two days afterward M. Musschenbroek, who tried the experiment with a hollow glass bowl, was so greatly affected that when he came to see me a few hours later he was still disturbed by it, and said that nothing in the world would tempt him to try the thing over again, etc."

This is the origin of this famous experiment which has excited admiration for the last three months, so that everybody flocks to see it, and which has resulted in making electricity so celebrated that it has been made a show of for the world.

In a note attached to this paper at this point it is stated that the first observer of this phenomenon was Cunaeus, who was a resident of Leyden and who came upon the result while he was amusing himself with the apparatus of von Musschenbroek. The same phenomenon had been previously observed by von Kleist of Camin in Pomerania. He communicated his discovery by letter to several friends but it was not published until later.

The memoir by Abbé Nollet deals with the repetition of the experiment. It was found that any kind of glass would serve, so long as it was well dried, and that probably the reason why von Musschenbroek succeeded only with German glass was because he was ignorant of this necessary condition.

AEPINUS

Franz Ulrich Theodor Aepinus was born on December 13, 1724, in Rostock. He was for a while privatdozent in the University of Rostock, then professor of astronomy at Berlin, and later professor of physics at St. Petersburg. He retired from this position and lived without any public function in Dorpat until his death on August 21, 1802. Aepinus is known in the history of electricity for his attempt to develop the one fluid theory of Franklin. His theory was for a while generally adopted, but was gradually displaced by the two fluid theory, in consequence chiefly of the necessity of ascribing to uncharged matter repulsions of the same force as those which were ascribed to electric charges. His theory exhibits interesting similarities to the present theory of the constitution of matter.

The extract which follows is from a memoir "Mémoire concernant quelques nouvelles expériences électriques remarquables," taken from the *Mémoires de l'Académie Royale des Sciences*, Berlin, Vol. 12, p. 105, 1756.

Pyroelectricity

The memoir begins after a few general remarks with a description of the mineral tourmaline. The author then proceeds as follows.

This stone has a property which distinguishes it from all others that are known at present, which is that when it is heated on hot coals it alternately attracts and repels the ashes which are around it. It does the same with metallic oxides, and in general with all other light bodies of whatever sort they are. The jewelers who put it in the fire to test its durability first noticed this property, and for this reason have given it the name of "the stone which attracts ashes." The authors whom I have cited also report this phenomenon but it has not hitherto been the object of any more particular investigations.

As soon as I heard of this remarkable peculiarity I conjectured at once that it owed its origin to electricity. I am under an obligation to our worthy colleague Mr. Lehmann, who first informed me of this property and who provided me with the means of making exact experiments on it. For this purpose he not only lent me a tourmaline stone which belonged to him, but procured for me also another one three times as heavy, which I acquired for myself. If I had not had this latter stone I should hardly have been in position clearly to recognize by way of experiments the astonishing properties of this stone, because it would have been very difficult, on account of the smallness of Mr. Lehmann's stone, to discriminate exactly among the different phenomena.

. . . .

Laws of the Electricity of Tourmaline

I. Tourmaline always has at the same time both positive and negative electricity; that is to say, when one of its faces is positive, the other is without exception negative, and reciprocally.

This rule is easy to verify by experiment. For if we have examined the electricity which is on one of the faces of the stone it is only necessary to turn it around, when the other side will never fail to show distinctly the opposite electricity. But although this law is incontestably right the stone sometimes, as I shall show in the sequel, is in a sort of mean state in which the truth of this law cannot be clearly perceived.

. . . .

II. With a small pair of tweezers or in any other way immerse the tourmaline in boiling water or in some other hot fluid and draw it out after it has been there a few minutes. You will always find that in this experiment, however often you think it proper to repeat it, one face of the stone is positively electrified and the other negatively. The face of the stone which always comes out positive I shall call in the future the positive face and that which comes out in the other state the negative face.

We should specially notice that a strong electric state is produced in a body surrounded by water, which in all other cases seems to be the substance which is most harmful to the electric virtue. There is no real necessity to have the water actually boiling. A less degree of heat also excites the electricity of tourmaline but to a less extent. When the water is heated only to 108 or

110 degrees of Fahrenheit's thermometer we can hardly discover any signs of electricity. The heat of boiling water seems to me to be in general that which renders the electricity of tourmaline most lively. If we heat this stone to a much higher degree on hot metal it shows only a feeble charge and does not become active until the stone has cooled down a little. The electricity that tourmaline acquires in boiling water still remains after the stone has entirely cooled down and I have even found it easily distinguishable in experiments made after six hours.

COULOMB

The extracts from Coulomb's (p. 97) work which follow describe his torsion balance and the way in which it was used to determine the law of electric force, also other methods of determining this law and the determination of a similar law for magnetic force. They are from two memoirs presented to the French Academy of Sciences, which appeared in the volume for 1785, published in 1788, of the *Mémoires de l'Académie Royale des Sciences* the first on p. 569, the second on p. 578.

LAW OF ELECTRIC FORCE

Construction and use of an electric balance based on the properties of metallic wires of having a force of reaction of torsion proportional to the angle of torsion.

Experimental determination of the law according to which the elements of bodies electrified with the same kind of electricity repel each other.

In a memoir presented to the Academy in 1784, I determined by experiment the laws of the force of torsion of a metallic wire, and I found that this force was in a ratio compounded of the angle of torsion, of the fourth power of the diameter of the suspended wire, and of the reciprocal of its length, all being multiplied by a constant coefficient which depends on the nature of the metal and which is easy to determine by experiment.

I showed in the same memoir that by using this force of torsion it was possible to measure with precision very small forces, as for example, a ten thousandth of a grain. I gave in the same memoir an application of this theory, by attempting to measure the constant force attributed to adhesion in the formula which expresses the friction of the surface of a solid body in motion in a fluid.

I submit today to the Academy an electric balance constructed on the same principle; it measures very exactly the state and the electric force of a body however slightly it is charged.

Construction of Balance

Although I have learned by experience that to carry out several electric experiments in a convenient way I should correct some defects in the first balance of this sort which I have made; nevertheless as it is so far the only one that I have used I shall give its description, simply remarking that its form and size may be and should be changed according to the nature of the experiments that one is planning to make. The first figure represents this balance in perspective and the details of it are as follows:

On a glass cylinder *ABCD* (Fig. 81) 12 inches in diameter and 12 inches high is placed a glass plate 13 inches in diameter, which entirely covers the glass vessel; this plate is pierced with two holes of about twenty lines in diameter, one of them in the middle, at *f*, above which is placed a glass tube 24 inches high; this tube is cemented over the hole *f* with the cement ordinarily used in electrical apparatus: at the upper end of the tube at *b* is placed a torsion micrometer which is seen in detail in figure 2. The upper part, No. 1, carries the milled head *b*, the index *io*, and the clamp *q*; this piece fits into the hole *G* of the piece No. 2; this piece No. 2 is made up of a circle *ab* divided on its edge into 360 degrees and of a copper tube Φ which fits into the tube *H*, No. 3, sealed to the interior of the upper end of the glass tube or column *fb* of figure 1. The clamp *q* (Fig. 81, 2, No. 1), is shaped much like the end of a solid crayon holder, which is closed by means of the ring *q*. In this holder is clamped the end of a very fine silver wire; the other end of the silver wire (Fig. 81, 3) is held at *P* in a clamp made of a cylinder *Po* of copper or iron with a diameter of not more than a line, whose upper end *P* is split so as to form a clamp which is closed by means of the sliding piece Φ. This small cylinder is enlarged at *C* and a hole bored through it, in which can be inserted (Fig. 81, 1) the needle *ag*: the weight of this little cylinder should be sufficiently great to keep the silver wire stretched without breaking it. The needle that is shown (Fig. 81, 1) at *ag* suspended horizontally about half way up in the large vessel which encloses it, is formed either of a silk thread soaked in Spanish wax or of a straw likewise soaked in Spanish wax and finished off from *q* to *a* for eighteen lines of its length by a cylindrical rod of shellac; at the end *a* of this needle is carried a little pith ball two or three lines in diameter; at *g* there is a little vertical piece of paper soaked in

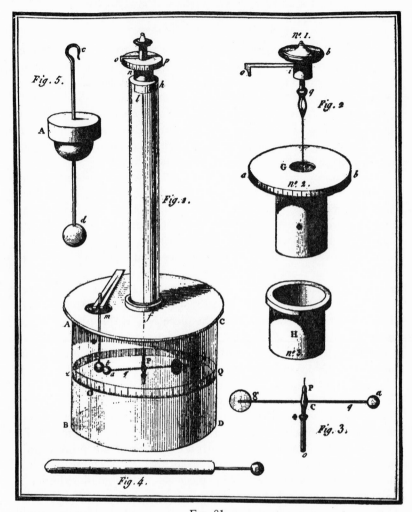

FIG. 81.

terebinth, which serves as a counterweight for the ball *a* and which slows down the oscillations.

We have said that the cover *AC* was pierced by a second hole at *m*. In this second hole there is introduced a small cylinder *mΦt*, the lower part of which *Φt* is made of shellac; at *t* is another pith ball; about the vessel, at the height of the needle, is described a circle *zQ* divided into 360 degrees: for greater simplicity I use a strip of paper divided into 360 degrees which is pasted around the vessel at the height of the needle.

To arrange this instrument for use I set on the cover so that the hole *m* practically corresponds to the first division of the circle *zoQ* traced on the vessel. I place the index *oi* of the micrometer on the point *o* or the first division of this micrometer; I then turn the micrometer in the vertical tube *fb* until, by looking past the vertical wire which suspends the needle and the center of the ball, the needle *ag* corresponds to the first division of the circle *zoQ*. I then introduce through the hole *m* the other ball *t* suspended by the rod *mΦt*, in such a way that it touches the ball *a* and that by looking past the suspension wire and the ball *t* we encounter the first division *o* of the circle *zoQ*. The balance is then in condition to be used for all our operations; as an example we go on to give the method which we have used to determine the fundamental law according to which electrified bodies repel each other.

FUNDAMENTAL LAW OF ELECTRICITY

The repulsive force between two small spheres charged with the same sort of electricity is in the inverse ratio of the squares of the distances between the centers of the two spheres.

Experiment

We electrify a small conductor, (Fig. 81, 4) which is simply a pin with a large head insulated by sinking its point into the end of a rod of Spanish wax; we introduce this pin through the hole *m* and with it touch the ball *t*, which is in contact with the ball *a*; on withdrawing the pin the two balls are electrified with electricity of the same sort and they repel each other to a distance which is measured by looking past the suspension wire and the center of the ball *a* to the corresponding division of the circle *zoQ*; then by turning the index of the micrometer in the sense *pno* we twist the suspension wire *lp* and exert a force proportional to the angle of torsion, which tends to bring the ball *a* nearer to the ball *t*. We observe in this way the distance through which different angles of torsion bring the ball *a* toward the ball *t*, and by comparing the forces of torsion with the corresponding distances of the two balls we determine the law of repulsion. I shall here only present some trials which are easy to repeat and which will at once make evident the law of repulsion.

First Trial. Having electrified the two balls by means of the pin head while the index of the micrometer points to *o*, the ball *a* of the needle is separated from the ball *t* by 36 degrees.

Second Trial. By twisting the suspension wire through 126 degrees as shown by the pointer *o* of the micrometer, the two balls approach each other and stand 18 degrees apart.

Third Trial. By twisting the suspension wire through 567 degrees the two balls approach to a distance of 8 degrees and a half.

Explanation and Result of This Experiment

Before the balls have been electrified they touch, and the center of the ball *a* suspended by the needle is not separated from the point where the torsion of the suspension wire is zero by more than half the diameters of the two balls. It must be mentioned that the silver wire *lp* which formed this suspension was twenty-eight inches long and was so fine that a foot of it weighed only $\frac{1}{16}$ grain. By calculating the force which is needed to twist this wire by acting on the point *a* four inches away from the wire *lp* or from the center of suspension, I have found by using the formulas explained in a memoir on the laws of the force of torsion of metallic wires, printed in the Volume of the Academy for 1784, that to twist this wire through 360 degrees the force that was needed when applied at the point *a* so as to act on the lever *an* four inches long was only $\frac{1}{340}$ grains: so that since the forces of torsion, as is proved in that memoir, are as the angles of torsion, the least repulsive force between the two balls would separate them sensibly from each other.

We found in our first experiment, in which the index of the micrometer is set on the point *o*, that the balls are separated by 36 degrees, which produces a force of torsion of $36° = \frac{1}{3400}$ of a grain; in the second trial the distance between the balls is 18 degrees, but as the micrometer has been turned through 126 degrees it results that at a distance of 18 degrees the repulsive force was equivalent to 144 degrees; so at half the first distance the repulsion of the balls is quadruple.

In the third trial the suspension wire was twisted through 567 degrees and the two balls are separated by only 8 degrees and a half. The total torsion was consequently 576 degrees, four times that of the second trial, and the distance of the two balls in this third trial lacked only one-half degree of being reduced to half of that at which it stood in the second trial. It results then from these three trials that the repulsive action which the two balls exert on each other when they are electrified similarly is in the inverse ratio of the square of the distances.

In the remarks which Coulomb adds to this account he points out first that when the very fine wire which he describes is used there is some uncertainty about the natural position of the zero. This uncertainty can be corrected by a suitable modification of the method of observation but he suggests that it is generally better to use a thicker wire. He also calls attention to the possible loss of electricity during the experiment, and suggests a way by which this can be observed and allowed for. He also points out that the repulsive action actually takes place along the chord of the arc by which the distances are measured, but shows that the errors introduced by measuring the distance by the degrees at least partly compensate each other, and that the errors are unimportant when the deflections do not exceed 25 to 30 degrees. He also shows how the instrument can be used to detect exceedingly small quantities of electricity.

SECOND MEMOIR ON ELECTRICITY AND MAGNETISM in which there are determined the laws according to which the magnetic fluid as also the electric fluid act either by repulsion or by attraction.

The first section of this memoir calls attention to the difficulties encountered in the use of the torsion balance when the electrical force is attractive, and describes the precautions which it was found necessary to take in order to obtain satisfactory results. The law of inverse squares was found to hold also in the case of attraction.

Second experimental method to determine the law with which a sphere one or two feet in diameter attracts a small body charged with electricity of a different sort from its own.

The method which we shall follow is analogous to that which we have used in the seventh volume of the *Savans Étrangers* to determine the magnetic force of a steel plate in relation to its length, its thickness, and its width. It consists in suspending a needle horizontally, of which the end only is electrified and which, when brought to a certain distance from a sphere, electrified with the other sort of electricity, is attracted and oscillates because of the action of the sphere: we determine then by calculation from the number of oscillations in a given time the attractive force at different distances, just as we determine the force of gravity by the oscillations of an ordinary pendulum.

We shall first consider some observations which have guided us in the experiments which are to follow. A silk fiber taken from a cocoon which can sustain 80 grains without breaking yields so readily to torsion that if we suspend horizontally to such a fiber three inches long in vacuum a small circular plate of which the weight and diameter are known we shall find from the period of oscillation of this little plate, using the formulas explained in a memoir on the force of torsion printed in the Volume of the Academy for 1784, that when we use a lever of 7 or 8 lines long to twist

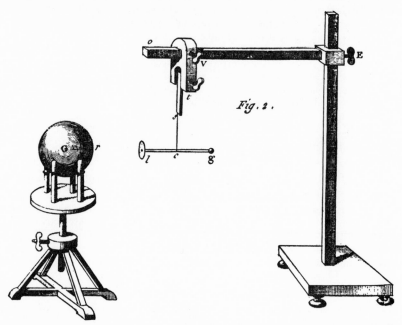

Fig. 2.

FIG. 82.

the fiber about its axis of suspension we shall need for a complete rotation to use usually not more than the force of a sixty thousandth of a grain; and if the suspended fiber is twice as long there will be needed only a hundred and twenty-thousandth of a grain. Therefore if we suspend a needle horizontally on this fiber, when the needle has come to rest and the fiber is entirely untwisted, and if by means of any force we set the needle in oscillations whose amplitude does not depart from the line in which the torsion is zero by more than 20 or 30 degrees, the force of torsion will have no sensible effect on the period of the oscillation, even when the force that produces the oscillations is not more than a hundredth of a grain. Premising this much, let us see how we proceed to determine the law of electrical attraction.

We suspend, (Fig. 82, 2) a needle *lg* made of shellac by a silk thread *sc* 7 to 8 inches long of a single fiber such as is drawn from the cocoon; at the end *l* we fix perpendicularly to the needle a little disc 8 or 10 lines in diameter, made very light and cut from a sheet of gilt paper; the silk thread is attached at *s* to the lower end of a little rod *st* dried in a furnace and coated with shellac or with Spanish wax; this rod is held at *t* by a clamp which slides along a ruled rod *oE* and can be placed anywhere we desire by means of the screw *V*.

G is a globe of copper or of cardboard covered with tin. It is carried on four uprights of glass coated with Spanish wax, and terminated, in order to make the insulation more perfect, by four rods of Spanish wax three or four inches long. The lower ends of these four uprights are set in a base which is placed on a little movable table that, as the figure shows, can be set at the height which is most convenient for the experiment; the rod *Eo* may also, by means of the screw *E*, be set at a convenient height.

When everything is ready we adjust the globe *G* in such a way that its horizontal diameter *Gr* is opposite the center of the plate *l*, which is some inches away from it. We give an electric spark to the sphere from a Leyden jar, we then touch the plate *l* with a conductor and the action of the electrified sphere on the electric fluid of the unelectrified plate gives to the plate a charge of the other sort from that of the sphere; so that when the conducting body is removed the sphere and the plate act on each other by attraction.

Experiment

The sphere *G* was a foot in diameter; the plate *l* was 7 lines in diameter; the shellac needle *lg* was 15 lines long; the suspension fiber *sc* was a silk fiber taken from the cocoon and 8 lines long: when the slider was at the point *o* the plate *l* touched the sphere at *r*, and as the slider was moved toward *E* the plate was removed from the center of the sphere by the quantity given by the divisions 0, 3, 6, 9, 12 inches, and when the sphere was electrified with what is called positive electricity and the plate with negative electricity by the method which has been described, we had:

Trial 1—The plate *l* being at......... 3 inches from the surface of the
    sphere or 9 inches from its center gave...15 oscillations in 20''.
Trial 2—The plate *l* distant by......... 18 inches from the center of the
    sphere gave...........................15 oscillations in 40''.
Trial 3—The plate *l* distant by......... 24 inches from the center of the
    sphere gave...........................15 oscillations in 60''.

Explanation of This Experiment and Its Result

When all the points of a spherical surface act by an attractive
or repulsive force which varies inversely as the square of the
distance on a point placed at any distance from this surface, it is
known that the action is the same as if all the spherical surface
were concentrated at the center of the sphere.

As in our experiment the plate *l* was only 7 lines in diameter
and as in the trials its least distance from the center of the sphere
was 9 inches, we may, without sensible error, suppose that all
the lines which are drawn from the center of the sphere to a point
of the plate are parallel and equal; and in consequence the total
action of the plate can be supposed to be united at its center just
as in the case of the sphere; so that for the small oscillations of
the needle, the action which makes the needle oscillate will be a
constant quantity for a given distance and will act along the line
which joins the two centers. Therefore if we call the force φ and
the time of a certain number of oscillations T we shall have T
proportional to $1/\sqrt{\varphi}$, but if d is the distance Gl from the center
of the sphere to the center of the plate and if the attractive forces
are proportional to the reciprocal of the square of the distances
or to $1/d^2$, it follows that T will be proportional to d or to the
distance; so that when we make our trials and change the distance,
the time of the same number of oscillations ought to be proportional to the distance from the center of the plate to the center
of the sphere: let us compare this theory with experiment.

Trial 1—Distance between centers...... 9 inches....15 oscillations in 20''—
Trial 2—...........................18 inches.................in 41''—
Trial 3—...........................24 inches.................in 60''—
The distances are as the numbers............ 3, 6, 8,
The times of the same number of oscillations. 20, 41, 60,
By theory they ought to have been.......... 20, 40, 54,

Thus in these three trials the difference between theory and
experiment is $\frac{1}{10}$ for the last trial compared with the first, and
almost nothing for the second trial compared with the first; but
it should be remarked that it took almost four minutes to make the

three trials; that although the electricity held pretty well on the day this experiment was tried, it nevertheless lost $\frac{1}{40}$ of its amount each minute. We shall see, in a memoir which will follow the one which I am presenting today, that when the electric density is not very great, the electric action of two electrified bodies diminishes in a given time exactly as the electric density or as the intensity of the action; therefore, since our trials lasted four minutes and since the electric action lost $\frac{1}{40}$ each minute from the first to the last trial, the action arising from the intensity of the electric density independently of the distance should be diminished by almost a tenth; consequently, to have the corrected time of the 15 oscillations in the last trial, we must set $\sqrt{(10)}:\sqrt{(9)}::60''$: the quantity required, which will be found to be 57 seconds, which differs only by $\frac{1}{20}$ from the 60 seconds found by experiment.

We have thus come, by a method absolutely different from the first, to a similar result; we may therefore conclude that the mutual attraction of the electric fluid which is called positive on the electric fluid which is ordinarily called negative is in the inverse ratio of the square of the distances; just as we have found in our first memoir, that the mutual action of the electric fluid of the same sort is in the inverse ratio of the square of the distance.

Law of Magnetic Force

Experiments to determine the law according to which the magnetic fluid acts whether by attraction or by repulsion.

Several experiments were tried by which it was proved that the center of action in each half of a magnetized wire is very near the end of the wire; so that in a steel wire 25 inches long we may, without sensible error, suppose that all the magnetic fluid is condensed near the end of the wire in 2 or 3 inches of its length.

The magnetic fluid acts by attraction or repulsion in a ratio compounded directly of the density of the fluid and inversely of the square of the distance of its molecules.

The first part of this proposition does not need to be proved; let us pass to the second.

We have seen that the magnetic fluid in our steel wire 25 inches long was concentrated at its ends in a length of 2 or 3 inches; that the center of action of each half of this needle was about 10 lines from its ends: therefore, by setting up some inches away from our steel wire a very short needle, in which, as we shall see in the sequel, the magnetic fluid may be supposed to be concentrated in 1 or 2 lines at its ends, we may calculate the mutual action of the wire on the needle and of the needle on the wire by supposing the magnetic fluid in the wire concentrated at a point 10 lines from its ends and in a needle an inch long at a point 1 or 2 lines from its end. These reflections have directed us in the experiment which follows:

Fourth Experiment

We suspended a steel wire weighing 70 grains and an inch in length, magnetized by the method of double touch, by a silk thread 3 lines long made of a single fiber taken from a cocoon; we allowed it to come to rest in the magnetic meridian, we then placed vertically in the meridian at different distances a steel wire 25 inches long, in such a way that its end was always ten lines below the level of the suspended needle; in each trial we changed the distance, and then by oscillating the suspended needle we counted the number of oscillations which it made in the same number of seconds. The following is the result of these experiments:

Trial 1—The free needle oscillating because of the action of the earth makes..............................15 oscillations in 60''.
Trial 2—The wire placed at 4 inches from the center of the needle........
41...........in 60''.
Trial 3—The wire placed 8 inches from the center of the needle..........
24...........in 60''.
Trial 4—The wire placed 16 inches from the center of the needle.........
17...........in 60''.

Explanation of This Experiment and Its Result.

When a pendulum is freely suspended and acted on by forces in a given direction, which make it oscillate, the forces are measured by the inverse ratio of the square of the time of the same number of oscillations, or, what comes to the same thing, by the direct ratio of the square of the number of oscillations made in the same time.

In the preceding experiment, the needle oscillates because of two different forces; the one is the magnetic force of the earth, the other is the action of all the points of the wire on the points of the needle. In our experiment all the forces are in the plane of the magnetic meridian, and since the needle is suspended horizontally the true force which makes it oscillate depends on those parts of all these forces which are resolved in the horizontal direction. Now we have seen from the three preceding experiments that since the magnetic fluid is concentrated in the ends of our wire, it may be supposed to be all brought together at a point 10 lines from the end of the wire. And, since the suspended needle is an inch long, that the boreal end is attracted at a distance of 3 inches and a half and that the austral end is repelled by the lower pole of the needle, which is distant from it $4\frac{1}{2}$ inches; it may be supposed without sensible error that the mean distance at which the lower end of the steel wire exerts its action on the two poles of the needle is 4 inches. Consequently, if the action of the magnetic fluid was in the inverse ratio of the square of the distances, the action of the lower pole of the steel wire on the needle should be proportional to $\frac{1}{4^2}, \frac{1}{8^2}, \frac{1}{16^2}$; or to $1, \frac{1}{4}, \frac{1}{16}$.

Now since the horizontal forces which make the needle oscillate are proportional to the square of the number of oscillations made in the same time, and since because of the magnetic force of the earth alone, the free needle makes 15 oscillations in 60″, this force can be measured by the square of these 15 oscillations or by 15^2. In the second trial the combined forces of the earth and of the steel wire make the needle make 41 oscillations in 60″; therefore, these forces combined are measured by 41^2, and the force resulting from the action of the magnetized steel wire alone is consequently measured by the difference of these two squares; it is thus proportional to $41^2 - 15^2$. We shall then have for the action of the wire on the needle:

	Force depending on the magnetic
Distance	*action of the steel wire.*
In *Trial 2*—At 4 inches..................	$41^2 - 15^2 = 1456.$
Trial 3— " 8 inches..................	$24^2 - 15^2 =\ \ 351.$
Trial 4— " 16 inches..................	$17^2 - 15^2 =\ \ \ \ 64.$

The second and third trials, in which the distances are as $1:2$, give very approximately for the forces the inverse ratio of the square of the distance. The fourth trial gives a number which is a little too small; but it may be remarked that in this fourth trial the distance of the lower pole of the steel wire from the center of the needle is 16 inches; and that the distance of the upper pole from the center of the needle is about $\sqrt{(16^2 + 23^2)}$: thus if we represent the action of the lower pole by $\frac{1}{16^2}$ the horizontal action of the upper pole will be $\dfrac{16}{(16^2 + 23^2)^{\frac{1}{2}}}$; so that the action of the lower pole is to that of the upper pole about as 100:19; from which it follows that since the oscillations of the needle are caused by the action of these two poles, and since the action of the upper pole is opposed in sense to that of the lower pole, the square of the oscillations which the action of the lower pole of the magnetized wire alone would produce is diminished by $^{19}\!/_{100}$ by the opposite action of the upper end of the same wire; and so to have the action of the lower part of the wire alone, we must, if we represent the true value of this force by x, set $(x - {}^{19}\!/_{100}x) = 64$, from which $x = 79$. If we substitute this quantity in the result of the fourth trial we shall find

> *Trial 2*— 4 inches of distance, the force................ 1456.
> *Trial 3*— 8 inches of distance, the force................ 351.
> *Trial 4*—16 inches of distance, the force................ 79.

These forces are very approximately as the numbers 16, 4, 1, or are in the inverse ratio of the square of the distance.

I have repeated this experiment several times by suspending needles two or three inches long and I have always found that when I have made the necessary corrections which I have just explained, the action of the magnetic fluid, whether repulsive or attractive, was inversely as the square of the distances.

The same law of attraction or repulsion was determined by a method in which the torsion balance was used.

GALVANI

Luigi Galvani was born on September 9, 1737 in Bologna and died there on December 4, 1798. He studied medicine in the University of Bologna and was called to a professorship in that institution. Apparently by accident he was led to notice the convulsions set up in frogs' legs, in certain circumstances, by the passage of an electric spark in the neighborhood. Galvani pursued this discovery with great diligence and eagerness, but he considered it from the first as produced by animal electricity and his later researches and theories had no influence on the development of physics. His fundamental discovery was taken up by Volta and led to the invention of the Voltaic pile and battery.

The extracts which follow are from Galvani's paper entitled "De viribus electricitatis in motu musculari commentarius," taken from *De Bononiensi Scientiarum et Artium Instituto atque Academia Commentarii*, Vol. 7, 1791. The translation was made from the German version, in Ostwald's *Klassiker der Exakten Wissenschaften*, No. 52.

The Electric Current

. . . .

The discovery was made in this way. I had dissected and prepared a frog as represented in (Fig. 83, 2) and while I was

Fig. 83.

attending to something else, I laid it on a table on which stood an electrical machine at some distance from its conductor and separated from it by a considerable space. Now when one of the persons who were present touched accidentally and lightly the inner crural nerves *DD* of the frog with the point of a scalpel all the muscles of the legs seemed to contract again and again as if they were affected by powerful cramps. Another one who was there, who was helping us in electrical researches, thought that he had noticed that the action was excited when a spark was discharged from the conductor of the machine. Being astonished by this new phenomenon he called my attention to it, who at that time had something else in mind and was deep in thought. Whereupon I was inflamed with an incredible zeal and eagerness to test the same and to bring to light what was concealed in it. I therefore myself touched one or the other nerve with the point of the knife and at the same time one of those present drew a spark. The phenomenon was always the same. Without fail there occurred lively contractions in every muscle of the leg at the same instant as that in which the spark jumped, as if the prepared animal was affected by tetanus.

With the thought that these motions might arise from the contact with the point of the knife, which perhaps caused the excited condition, rather than by the spark, I touched the same nerves again in the same way in other frogs with the point of the knife, and indeed with greater pressure, yet so that no one during this time drew off a spark. Now no motions could be detected. I therefore came to the conclusion that perhaps to excite the phenomenon there were needed both the contact of a body and the electric spark.

Therefore I again pressed the blade of the knife on the nerve and kept it there at rest while the spark passed and while the machine was not in motion. The phenomenon only occurred while the sparks were passing.

We repeated the experiment, always using the same knife. But it was remarkable that when the spark passed the motions observed sometimes occurred and sometimes not.

Excited by the novelty of the phenomenon, we undertook to investigate the thing in one way and another and to follow it up experimentally, while still using one and the same scalpel, so that if possible we might discover the causes of this unexpected difference. And this new effort was not without results. We discovered that the whole phenomenon was to be ascribed to the different parts of the scalpel by which it was held by the fingers.

The scalpel had a bone handle, and if this handle was held in the hand no contractions occurred when the spark passed; but they did occur if the finger rested on the metallic blade or on the iron rivet by which the blade was held in the handle.

Now since fairly dry bones have an electric nature but the metal blade and the iron rivet have a conducting or so called non-electric nature, we were led to assume that conditions were such that if we held the bone handle in the fingers the electric fluid which in some way or other was active in the frog would be kept from entering, but that it could enter if we touched the blade or the rivet which was connected with it.

Now to put the thing beyond all doubt we used instead of the scalpel sometimes a slender glass rod *H*, (Fig. 83, 2) which had been wiped clean from dampness and dust, and sometimes an iron rod *G*. With the glass rod we not only touched the nerves of the leg but rubbed them hard while the sparks were passing. But in vain; in spite of all our trouble the phenomenon never appeared, even when a number of powerful sparks were drawn from the conductor of the machine at a small distance from the animal. The phenomenon occurred however if we even lightly touched the same nerve with the iron rod and only little sparks passed.

THE FORCE OF ANIMAL ELECTRICITY IN MUSCULAR MOTION

After we had investigated the forces of atmospheric electricity during thunder storms our hearts burned with desire to investigate also the force of electricity in quiet times during the day.

Therefore, as I had casually noticed that the prepared frogs, which were hung by a brass hook passing through the spinal cord to the iron grating which enclosed a hanging garden of our house, showed the usual contractions not only when there was lightning but also when the sky was clear and fair, I thought that the origin of these contractions might be found in the changes which nevertheless were going on in the atmospheric electricity. Therefore I began, not without hope, carefully to investigate the action of these changes in the muscular motion and to set up experiments in one way and another. Thus I observed at different hours and indeed for days at a time suitably arranged animals, but scarcely ever did a motion of the muscles occur. Finally, tired of this useless waiting, I began to squeeze and press the hooks which were fastened in the spinal cord against the iron grating, in order to see whether such an artifice might excite the contraction of the muscles and whether instead of its depending on the condition of the atmosphere and its electricity any other change and alteration might have an influence.   I quite often observed contractions, but none which depended upon the different conditions of the atmosphere and its electricity.

As I had observed these contractions only in the open air and as hitherto no researches had been undertaken in other places, there seemed to be little lacking to my argument and I might have referred such contractions to the atmospheric electricity which enters the animal and accumulates there and suddenly leaves it when the hook is brought in contact with the iron grating.   So easy is it to deceive oneself in experimenting, and to think that we have seen and found that which we wish to see and find.

But when I transferred the animal to a closed room, had laid it on an iron plate, and begun to press the hook which was in the spinal cord against the plate, behold, the same contractions, the same motions!   I repeated the experiment by using other metals at other places and on other hours and days; with the same result, only that the contractions were different when different metals were used, being more lively for some and more sluggish for the others.   At last it occurred to us to use other bodies which conduct electricity only a little or not at all, made of glass, rubber, resin, stone or wood and always dried, and with these nothing similar occurred, no muscular contractions and motions could be seen. Naturally such a result excited in us no slight astonishment and caused us to think that possibly the electricity was present in the animal itself.   We were confirmed in this view by the assumption of a very fine nervous fluid which during the occurrence of the phenomenon flows from the nerves to the muscle like the electric current in the Leyden jar.

.   .   .   .

To make the thing plainer I had with the greatest success laid the frog on a nonconducting plate, as on glass or on resin, and then using a curved rod, sometimes conducting and sometimes either altogether or partly non-conducting, I touched with one end of it the hook which entered the spinal cord and with the other end either the muscles of the leg or the feet.   In this experiment, when we used the conducting rod, (Fig. 84), we saw the contractions occur, but when we used the rod which was non-conducting, as in Fig. 84, 10, there were no contractions.   The conducting rod

was an iron wire, the hook was a brass wire. After these last discoveries it appeared to us that the contractions which, as has been
said, occur in a frog laid on a metallic plate if the hook in the spinal

Fig. 84.

cord is pressed against the plate, must come from a similar circuit in
place of which, in a way, the metallic plate acts, and therefore it
happens that the contractions will not occur in frogs which are laid

on a non-conducting plate even when the same procedure is employed.

This opinion of ours has clearly explained a not unwelcome and accidentally observed phenomenon, if my judgment is correct. If the frog is held up in the fingers by one leg so that the hook that is fastened in the spinal cord touches a silver plate and the other free leg can come in contact with the plate, (Fig. 84, 11), then if this leg touches the plate the muscles are repeatedly contracted, so that the leg is lifted, but soon, when it becomes quiet and falls down to the plate again, it again comes in contact with it and therefore is lifted again and in short it so proceeds to rise and fall that in a certain sense it seems like an electrical pendulum, to the greatest astonishment and delight of the observer.

· · · · · · · · · · · · · · · · · · ·

Before we end our discussion of the use of the curved rod and its forces we should not neglect to make clear its significance and, if I may so say, its necessity for such muscular contractions. These occur more clearly and more quickly not only with one but with two curved rods, if they are so applied and arranged that the end of one of them touches the muscle and the end of the other the nerves in a similar way, and the two other ends are brought in contact with each other or if necessary are rubbed together, (Fig. 84, 12). In this connection it is noticeable that the electricity which causes the contractions is not conducted away and dissipated by the contact of the hands with the two rods or by the repeated contacts of the rods with the parts of the animal's body. Furthermore we were fortunate enough to observe this peculiar and remarkable phenomenon, that the use of more than one metallic substance and the differences between them contribute much to the excitation, as also especially to the increase of the muscular contraction, far more indeed than when one and the same metal is used. Thus for example, if the whole rod was iron or the hook was iron and the conducting plate also, the contractions either did not occur or were very small. But if one of them was iron and the other brass, or better if it was silver (silver seems to us the best of all the metals for conducting animal electricity) there occur repeated and much greater and more prolonged contractions. The same thing occurs with one and the same arrangement of the non-conducting plate. If strips of different metals are laid in the same way on two points separate from each other, as for example, if you put a zinc strip at one place and a brass strip at the other, the contractions are usually much greater than when metal strips of the same sort are used, even when both places are brought in contact with silver.

.   .   .   .

VOLTA

Alessandro Volta was born in Como on February 18, 1745 and died there on March 5, 1827. In 1774 he became professor of physics at Como and in 1779 he was called to the University of Pavia. He subsequently made several long journeys through Europe, on which he met many of the distinguished scientific men of the time. He was greatly honored by Napoleon. In 1804 he resigned his professorship. He returned to academic work for a short time as director of the Philosophical Faculty at the University of Padua but withdrew from this position in 1819 and retired to private life.

Volta contributed much to the science of electricity. He invented the condensing electroscope and the electrophorus. His most important work was his study of the electrifications brought about by the contact of metals, which led him to the invention of the voltaic pile and the voltaic battery. The extract which follows is from a letter from Como of March 20, 1800, to Sir Joseph Banks, the president of the Royal Society. It appeared in the *Philosophical Transactions*, for 1800, p. 403, with the title "On the Electricity excited by the mere contact of conducting substances of different kinds."

THE VOLTAIC PILE

Como, in the Milanais
March 20th, 1800

After a long silence, which I do not attempt to excuse, I have the pleasure of communicating to you, Sir, and through you to the Royal Society, some striking results to which I have come in carrying out my experiments on electricity excited by the simple mutual contact of metals of different sorts, and even by the contact of other conductors, also different among themselves, whether liquids or containing some liquid, to which property they owe their conducting power. The most important of these results, which includes practically all the others, is the construction of an apparatus which, in the effects which it produces, that is, in the disturbances which it produces in the arms etc., resembles Leyden jars, or better still electric batteries feebly charged, which act unceasingly or so that their charge after each discharge reestablishes itself; which in a word provides an unlimited charge or imposes a

perpetual action or impulsion on the electric fluid; but which otherwise is essentially different from these, both because of this continued action which is its property and because, instead of being
made, as are the ordinary jars and electric batteries, of one or more
insulating plates in thin layers of those bodies which are thought
to be the only electric bodies, coated with conductors or bodies
called non-electrics, this new apparatus is formed altogether of
several of these latter bodies, chosen even among the best conductors and therefore the most remote, according to what has
always been believed, from the electric nature.   Yes! the apparatus of which I speak, and which will doubtless astonish you, is
only an assemblage of a number of good conductors of different
sorts arranged in a certain way.   30, 40, 60, pieces or more of
copper, or better of silver, each in contact with a piece of tin, or
what is much better, of zinc and an equal number of layers of
water or some other liquid which is a better conductor than pure
water, such as salt-water or lye and so forth, or pieces of cardboard
or of leather, etc. well soaked with these liquids; when such layers
are interposed between each couple or combination of the two
different metals, such an alternative series of these three sorts
of conductors always in the same order, constitutes my new instrument; which imitates, as I have said, the effects of Leyden jars or
of electric batteries by giving the same disturbances as they; which
in truth, are much inferior to these batteries when highly charged
in the force and noise of their explosions, in the spark, in the
distance through which the charge can pass, etc., and equal in
effect only to a battery very feebly charged, but a battery nevertheless of an immense capacity; but which further infinitely
surpasses the power of these batteries in that it does not need,
as they do, to be charged in advance by means of an outside source;
and in that it can give the disturbance every time that it is properly
touched, no matter how often.

A short section on the electric organ of the torpedo is omitted.

I proceed to give a more detailed description of this apparatus
and of some other analogous ones, as well as the most remarkable
experiments made with them.
I provided myself with several dozen small round plates or
discs of copper, of brass, or better of silver, an inch in diameter
more or less (for example, coins) and an equal number of plates of

tin, or which is much better, of zinc, approximately of the same shape and size; I say approximately because precision is not necessary, and in general the size as well as the shape of the metallic pieces is arbitrary: all that is necessary is that they may be arranged easily one above the other in a column. I further provided a sufficiently large number of discs of cardboard, of leather, or of some other spongy matter which can take up and retain much water, or the liquid with which they must be well soaked if the experiment is to succeed. These pieces, which I will call the moistened discs, I make a little smaller than the metallic discs or plates, so that when placed between them in the way that I shall soon describe, they do not protrude.

Now having in hand all these pieces in good condition, that is to say, the metallic discs clean and dry, and the other non-metallic ones well soaked in water or which is much better, in brine, and afterwards slightly wiped so that the liquid does not come out in drops, I have only to arrange them in the proper way; and this arrangement is simple and easy.

I place horizontally on a table or base one of the metallic plates, for example, one of the silver ones, and on this first plate I place a second plate of zinc; on this second plate I lay one of the moistened discs; then another plate of silver, followed immediately by another of zinc, on which I place again a moistened disc. I thus continue in the same way coupling a plate of silver with one of zinc, always in the same sense, that is to say, always silver below and zinc above or *vice versa*, according as I began, and inserting between these couples a moistened disc; I continue, I say, to form from several of these steps a column as high as can hold itself up without falling (Fig. 85, 2, 3, 4).

An account of the shocks that can be obtained from this instrument and details of the best method of experimenting with it are omitted.

Coming back to the mechanical construction of my apparatus, which admits of several variations, I proceed to describe here not all those which I have thought out and constructed either on a large or small scale, but some only which are either more curious or more useful or which present some real advantage, such as being easier or quicker to construct, more certain in their effects or keeping in good condition longer. To begin with one of these which unites almost all of these advantages, which in its form

differs the most from the columnar apparatus described before
but which has the disadvantage of being a much larger apparatus,
I present this new apparatus which I call the crown of cups in
the next figure.   (Fig. 85, 1.)

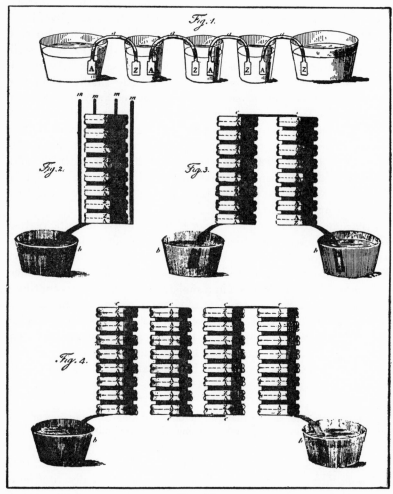

Fig. 85.

We set up a row of several cups or bowls made of any material
whatever except the metals, cups of wood, of shell, of clay, or
better of crystal (small drinking glasses or goblets are very suitable)
half-full of pure water, or better of brine or of lye; and we join them

all together in a sort of chain by means of metallic arcs of which one arm Aa or only the end A which is placed in one of the goblets is of red or yellow copper, or better of silvered copper, and the other Z, which is placed in the next goblet is of tin or better of zinc. I may observe here in passing that lye and the other alkaline liquids are preferable when one of the metals which is immersed in them is tin; brine is preferable when it is zinc. The two metals of which each arc is composed are soldered together somewhere above the part which is immersed in the liquid and which ought to touch it with a sufficiently large surface: for this purpose it is suitable that this part should be an inch square or very little less; the rest of the arc may be as much narrower as we please, and may even be a simple metallic wire. It may also be made of a third metal different from the two which are immersed in the liquid of the goblets; since the action on the electric fluid which results from all the contacts of several metals which are in immediate contact, the force with which this fluid is driven at the end, is the same absolutely or nearly as that which it would have received by the immediate contact of the first metal with the last without any of the intermediate metals, as I have verified by direct experiment, of which I shall have occasion to speak elsewhere.

Now then a train of 30, 40, 60 of these goblets joined up in this mannner and arranged either in a straight line or in a curve or set round in any way forms the whole of this new apparatus, which fundamentally and in substance is the same as the other one of the column tried before; the essential feature, which consists in the immediate connection of the different metals which form each pair and the mediate connection of one couple with another by the intermediary of a damp conductor, appears in this apparatus as well as in the other.

NICHOLSON

William Nicholson was born in London in 1753, and died there on May 21, 1815. He was occupied with business and as a civil engineer, but devoted himself also to science. He founded and published a journal of natural philosophy, chemistry and the arts, of which forty-one volumes were published. He was allowed by Sir Joseph Banks to see the letter in which Volta describes his invention of the pile. In collaboration with Sir Anthony Carlisle he constructed a voltaic pile and with it discovered the evolution of gases by the passage of the electric current through water.

The extract which follows is from a paper published in *Nicholson's Journal of Natural Philosophy, Chemistry and the Arts*, July, 1800, entitled, "Account of the New Electrical or Galvanic Apparatus of Sig. Alex. Volta, and Experiments performed with the same." It contains the account of the decomposition of water.

THE DECOMPOSITION OF WATER

The first part of this paper contains a description of the construction of the voltaic pile and of the combination which Volta called the crown of cups.

Thus far I have followed this able philosopher; who, to his former researches into the nature and laws of electricity, has now added a discovery which must for ever remove the doubt whether galvanism be an electrical phenomenon. But I cannot here look back without some surprise, and observe that the chemical phenomena of galvanism, which had been much so insisted on by Fabbroni, more especially the rapid oxidation of the zinc, should constitute no part of his numerous observations.

On the 30th of April, Mr. Carlisle had provided a pile consisting of 17 half crowns, with a like number of pieces of zinc, and of pasteboard, soaked in salt water. These were arranged in the order of silver, zinc, card, etc. which order I shall denote by saying, that the silver was undermost, that is to say, under the zinc; and I make this remark because some philosophers have used the expression that the silver was undermost when they used the order of silver, card, zinc, etc. which, as the reader will easily perceive, is contrary to the order here spoken of. This is of no consequence to the effect, though it is material to a clear understanding of the terms we use. This pile gave us the shock as before described, and a very acute sensation wherever the skin was broken. Our first research was directed to ascertain that the shock we felt was really an electrical phenomenon. For this purpose the pile was placed upon Bennett's gold leaf electrometer, and a wire was then made to communicate from the top of the pile to the metallic stand or foot of the instrument, so that the circuit of the shock would have been through the leaves, if they had diverged. But no signs of electricity appeared. Recourse was then had to the revolving doubler, described at page 95 of our present volume. The plate *A* was connected with the top of the electrometer and the silver end of the pile; and the plate *B* and ball were made to touch the top of the system by an uninsulated brass wire. The doubler had been previously cleared of electricity by twenty turns in connection with the earth. The negative divergence was produced in the electrometer. Repeated experiments of this kind showed that the silver end was in the minus, and the zinc end in the plus state.

In all these experiments it was observed, that the action of the instrument was freely transmitted through the usual conductors of electricity, but stopped by glass and other non-conductors. Very early in this course, the contacts being made sure by placing a drop of water upon the upper plate, Mr. Carlisle observed a disengagement of gas round the touching wire. This gas, though very minute, evidently seemed to me to have the smell afforded by hydrogen when the wire of communication was steel. This, with some other facts, led me to propose to break the circuit by the substitution of a tube of water between two wires. On the second of May we, therefore, inserted a brass wire through each of two corks inserted in a glass tube of half an inch internal diameter. The tube was filled with New river water, and the distance between the points of the wires in the water was one inch and three quarters. This compound discharger was applied so that the external ends of its wire were in contact with the two extreme plates of a pile of thirty-six half crowns with the correspondent pieces of zinc and pasteboard. A fine stream of minute bubbles immediately began to flow from the point of the lower wire in the tube, which communicated with the silver, and the opposite point of the upper wire became tarnished, first deep orange, and then black. On reversing the tube, the gas came from the other point, which was now lowest, while the upper in its turn became tarnished and black. Reversing the tube again, the phenomena again changed their order. In this state the whole was left for two hours and a half. The upper wire gradually emitted whitish filmy clouds, which, towards the end of the process, became of a pea green colour, and hung in perpendicular threads from the extreme half inch of the wire, the water being rendered semiopaque by what fell off, and in a great part lay, a pale green, on the lower surface of the tube, which, in this disposition of the apparatus, was inclined about forty degrees to the horizon. The lower wire of three quarters of an inch long, constantly emitted gas, except when another circuit, or complete wire, was applied to the apparatus; during which time the emission of gas was suspended. When this last mentioned wire was removed, the gas reappeared as before, not instantly, but after the lapse of four beats of a half second clock standing in the room. The product of gas, during the whole two hours and a half, was two-thirtieths of a cubic inch. It was then mixed with an equal quantity of common air, and exploded by the application of a lighted waxed thread.

It might seem almost unnecessary to have reversed the order of the pile in building up, as reversing the tube must have answered exactly the same purpose. We chose, however, to do this, and found that when the zinc was at the bottom, its effects were reversed, that is to say, the gas still came from the wire communicating with the silver, etc.

We had been led by our reasoning on the first appearance of hydrogen to expect a decomposition of the water; but it was with no little surprise that we found the hydrogen extricated at the contact with one wire, while the oxygen fixed itself in combination with the other wire at the distance of almost two inches. This new fact still remains to be explained and seems to point at some general law of the agency of electricity in chemical operations. As the distance between the wires formed a striking feature in this result, it became desirable to ascertain whether it would take place to greater distances. When a tube three quarters of an inch in diameter, and thirty-six inches long, was made use of, the effect failed, though the very same wires, inserted into a shorter tube, operated very briskly. The solicitation of other objects of enquiry prevented trial being made of all the various intermediate distances; but from the general tenor of experiments, it appears to be established, that this decomposition is more effectual the less the distance between the wires, but that it ceases altogether when the wires come into contact.

May 6—Mr. Carlisle repeated the experiment with copper wires and tincture of litmus. The oxidating wire, namely, from the zinc side, was the lowest in the tube; it changed the tincture red in about ten minutes as high as the upper extremity of the wire. The other portion remained blue. Hence it seems either an acid was formed, or that a portion of the oxigen combined with the litmus, so as produce the effect of an acid.

It may be here offered as a general remark, that the electric pile with card, or with woolen cloth, continues in order for about two days, or scarcely three; that from a series of glasses set up by Mr. Carlisle, as well as from the pile itself, it appears that the same process of decomposition of water is carried on between each pair of plates, the zinc being oxided on the wet face, and hydrogen given out; that the common salt is decomposed, and exhibits an efflorescence of soda round the edges of the pile, extruded, most probably, by the hydrogen: and that on account of the corrosion of the faces of the zinc, it is necessary to renew them previous to each construction of the pile. This may be done by scraping or grinding. I found it most convenient to lay the piece in a hole in a board, and give it a stroke with a float file, or file of which the teeth are not crossed. It might, perhaps be less troublesome to clean them with diluted muriatic acid; but this I have not tried.

As the ample field of physiological research to which Mr. Carlisle's attention is directed, and the multiplicity of my own avocations, rendered it less convenient for us to pursue our enquiries together, I constructed an apparatus for my own use.

. . . .

The decomposition of water, and oxidation of metallic wire, gave birth to a variety of speculations and projects of experiments. Among others it became a question, what would be the habitude of metals of difficult oxidation. Two wires of platina, one of which was round, and one fortieth of an inch in diameter, and the other nearly of the same mass, but flatted to the breadth of one twenty-fifth of an inch, were inserted into a short tube of $\frac{1}{4}$ of an inch inside diameter. When placed in the circuit, the silver side gave a plentiful stream of fine bubbles, and the zinc side also a stream less plentiful. No turbidness nor oxidation, nor tarnish appeared, during the course of four hours continuance of this operation. It was natural to conjecture, that the larger stream from the silver side was hydrogen, and the smaller oxigen. Thick gold leaf was tried with the same effects. A wire of brass was then substituted instead of one of the slips of gold. When the brass was on the minus, or silver side, the two gases were extricated for two hours, without oxidation as before; but when the brass was, by reversing the tube, brought to the plus side, it became oxided in the same manner as if both the wires had been brass. When the slips of gold were long subjected to this action, the extremity of the slip communicating with the zinc, acquired a coppery or purplish tinge, which was deepest near the end. Whether this arose from oxidation of the gold, or of the copper, of which gold leaf contains about a seventieth part, cannot from this experiment be decided.

The simple decomposition of water by platina wires without oxidation, offered a means of obtaining the gases separate from each other. With this intention, Mr. Carlisle's pile of thirty-six was combined with my two sets of sixteen repetitions. His pile was built with the zinc uppermost, and mine in the reverse order; so that by connecting the upper plates the whole constituted one range, and the communications could be made from the bottom of the one to the bottom of the other. The two platina wires were made to protrude out of two separate tubes, each containing a little water, and through the opposite corks of each were passed copper wires of communication. These tubes were slightly greased on the outside to prevent their becoming damp; and in this state the extremities, armed with the platina, were plunged in a shallow glass vessel of water, in which two small inverted vessels, quite full of water, were so disposed, that the platina of one tube was beneath one vessel, and the platina of the other tube was beneath the other, the distance between their extremities being about two inches. The copper wires of these tubes respectively were made to communicate with the extremities of the intire pile of sixty-eight sets. A cloud of gas arose from each wire, but most from the silver, or minus side. Bubbles were extricated from all parts of the water, and adhered to the whole internal surface of the vessels. The process was continued for thirteen hours, after which the wires were disengaged, and the gases decanted into separated bottles. On measuring the quantities, which was done by weighing the bottles, it was found, that the quantities of water displaced by the gases, were respectively, 72 grains by the gas from the zinc side, and 142 grains by the gas from the silver side; so that the whole volume of gas was 1.17 cubic inches, or near an inch and a quarter. These are nearly the proportions in bulk, of what are stated to be the component parts of water.

OERSTED

Hans Christian Oersted was born on August 14, 1777, in Iudkjobing, Denmark. His father was an apothecary. He studied in Copenhagen and from 1806 taught physics in the University of Copenhagen. In 1829 he became director of the newly founded Polytechnic Institute. He died on March 9, 1851, in Copenhagen.

The extract which follows contains the account of Oersted's famous discovery of the magnetic field accompanying an electric current. This discovery was announced under date of July 21, 1820, in a pamphlet entitled *Experimenta circa efficaciam conflictus electrici in acum magneticam.* The translation was made by the Rev. J. E. Kempe and appeared in the *Journal of Telegraph Engineers*, Vol. 5, p. 459, 1876 (1877).

THE ACTION OF CURRENTS ON MAGNETS

The first experiments on the subject which I undertake to illustrate were set on foot in the classes for electricity, galvanism, and magnetism, which were held by me in the winter just past. By these experiments it seemed to be shown that the magnetic needle was moved from its position by the help of a galvanic apparatus, and that, when the galvanic circuit was closed, but not when open, as certain very celebrated physicists in vain attempted several years ago. As, however, these experiments were conducted with somewhat defective apparatus, and, on that account, the phenomena which were produced did not seem clear enough for the importance of the subject, I got my friend Esmarch, the King's Minister of Justice, to join me, that the experiments might be repeated and extended with the great galvanic apparatus which we fitted up together. A distinguished man, Wleugel, Knight of the Danish Order, and President of our Pilot Board, was also present at our experiments as a partner and a witness. Besides these there were witnesses at these experiments that most excellent man, decorated by the King with the highest of honours,—Hauch, whose acquaintance with natural science has long been celebrated,—that most acute man Reinhardt, Professor of Natural History; Jacobsen, Professor of Medicine, a man of the utmost sagacity in conducting experiments; and the most experienced chemist, Zeise, Doctor of Philosophy. I have indeed somewhat frequently carried out by myself experiments relating to the matter proposed, but the phenomena which it thus befel me to disclose I repeated in the presence of these most learned men.

In reviewing my experiments I will pass over everything which, though they conduced to the discovery of the reason of the thing, yet, when this is discovered, cannot any further illustrate it. Those things, therefore, which clearly demonstrate the reason of the thing, let us take for granted.

The galvanic apparatus which we made use of consists of 20 rectangular copper receptacles, the length and height of which are alike 12 inches, the breadth, however, scarcely exceeding 2½ inches. Every receptacle is furnished with two copper plates, so inclined that they can carry a copper bar which supports a zinc plate in the water of the next receptacle. The water of the receptacles contains $\frac{1}{60}$ of its weight of sulphuric acid and likewise $\frac{1}{60}$ of its weight of nitric acid. The part of each plate which is immersed in the solution is square, the side being about 10 inches long. Even smaller apparatus may be used, provided they are able to make a metallic wire red hot.

Let the opposite poles of the galvanic apparatus be joined by a metallic wire, which, for brevity, we will call hereafter the joining conductor or else the joining wire. To the effect, however, which takes place in this conductor and surrounding space, we will give the name of electric conflict.

Let the rectilinear part of this wire be placed in a horizontal position over the magnetic needle duly suspended, and parallel to it. If necessary, the joining wire can be so bent that the suitable part of it may obtain the position necessary for the experiment. These things being thus arranged, the magnetic needle will be moved, and indeed, under that part of the joining wire which receives electricity most immediately from the negative end of the galvanic apparatus, will decline towards the west.

If the distance of the joining wire from the magnetic needle does not exceed $\frac{3}{4}$ of an inch, the declination of the needle makes an angle of about 45°. If the distance is increased the angles decrease as the distances increase. The declination, however, varies according to the efficiency of the apparatus.

The joining wire can change its place either eastward or westward, provided it keeps a position parallel to the needle, without any other change of effect than as respects magnitude; and thus the effect can by no means be attributed to attraction, for the same pole of the magnetic needle which approaches the joining wire while it is placed at the east side of it ought to recede from the same when it occupies a position at the west side of it if these declinations depended upon attractions or repulsions.

The joining conductor may consist of several metallic wires or bands connected together. The kind of metal does not alter the effects, except, perhaps, as regards quantity. We have employed with equal success wires of platinum, gold, silver, copper, iron, bands of lead and tin, a mass of mercury. A conductor is not wholly without effect when water interrupts, unless the interruption embraces a space of several inches in length.

The effects of the joining wire on the magnetic needle pass through glass, metal, wood, water, resin, earthenware, stones: for if a plate of glass, metal, or wood be interposed, they are by no means destroyed, nor do they disappear if plates of glass, metal, and wood be simultaneously interposed; indeed, they seem to be scarcely lessened. The result is the same if there is interposed a disc of amber, a plate of porphyry, an earthenware vessel, even if filled with water. Our experiments have also shown that the effects already mentioned are not changed if the magnetic needle is shut up in a copper box filled with water. It is unnecessary to state that the passing of the effects through all these materials in electricity and galvanism has never before been observed. The effects, therefore, which take place in electric conflict are as different as possible from the effects of one electric force or another.

If the joining wire is placed in a horizontal plane under the magnetic needle, all the effects are the same as in the plane over the needle, only in an inverse direction, for the pole of the magnetic needle under which is that part of the joining wire which receives electricity most immediately from the negative end of the galvanic apparatus will decline towards the east.

That these things may be more easily remembered let us use this formula: the pole *over* which negative electricity enters is turned towards the west, that *under* which it enters towards the east.

If the joining wire is so turned in a horizontal plane as to form with the magnetic meridian a gradually increasing angle, the declination of the magnetic needle is increased if the motion of the wire tends towards the place of the disturbed needle, but is lessened if the wire goes away from this place.

The joining wire placed in the horizontal plane in which the magnetic needle moves balanced by means of a counterpoise, and parallel to the needle, disturbs the same neither eastward nor westward but only makes it quiver in the plane of inclination, so that the pole near which the negative electric force enters the wire is depressed when it is situated at the west side and elevated when at the east.

If the joining wire is placed perpendicular to the plane of the magnetic meridian, either above or below the needle, the latter remains at rest, unless the wire is very near to the pole, for then the pole is elevated when the entrance is made from the western part of the wire and depressed when it is made from the eastern.

When the joining wire is placed perpendicular to the pole of the magnetic needle, and the upper end of the wire receives electricity from the negative end of the galvanic apparatus, the pole is moved towards the east; but when the wire is placed opposite to a point situated between the pole and the middle of the needle it is driven towards the west. When the upper end of the wire receives electricity from the positive end reverse phenomena will occur.

If the joining wire is so bent that it is made parallel to itself at both parts of the bend, or forms two parallel legs, it repels or attracts the magnetic poles according to the different conditions of the case. Let the wire be placed opposite to either pole of the needle so that the plane of the parallel legs is perpendicular to the magnetic meridian, and let the eastern leg be joined with the negative end of the galvanic apparatus, the western with the positive, and when this is so arranged the nearest pole will be repelled either eastward or westward according to the position of the plane of the legs. When the eastern leg is joined with the positive end, and the western with the negative, the nearest pole is attracted. When the plane of the legs is placed perpendicular to a spot between the pole and the middle of the needle the same effects occur, only inverted.

A needle of copper, suspended like a magnetic needle, is not moved by the effect of a joining wire. Also needles of glass, or of so-called gum-lac, subjected to the like experiments, remain at rest.

From all this it may be allowable to adduce some considerations in explanation of these phenomena.

Electric conflict can only act upon magnetic particles of matter. All non-magnetic bodies seem to be penetrable through electric conflict; but magnetic bodies, or rather their magnetic particles, seem to resist the passage of this conflict, whence it is that they can be moved by the impulse of contending forces.

That electric conflict is not inclosed in the conductor, but as we have already said is at the same time dispersed in the surrounding space, and that somewhat widely is clear enough from the observations already set forth.

In like manner it is allowable to gather from what has been observed that this conflict performs gyrations, for this seems to be a condition without which it is impossible that the same part of the joining wire, which, when placed beneath the magnetic pole, carries it eastward, drives it westward when placed above; for this is the nature of a gyration, that motions in opposite parts have an opposite direction. Moreover, motion by circuits combined with progressive motion, according to the length of the conductor, seems bound to form a cochlea or spiral line, which, however, if I am not mistaken, contributes nothing to the explanation of phenomena hitherto observed.

All the effects on the northern pole, here set forth, are easily understood by stating that negatively electric force or matter runs through a spiral line bending to the right, and propels the northern pole, but does not act at all upon the southern. The effects on the southern pole are similarly explained if we attribute to force or matter positively electrified a contrary motion and the power of acting on the southern pole but not on the northern. The agreement of this law with nature will be better seen by the repetition of experiments than by a long explanation. To judge of the experiments, however, will be made much easier if the course of the electric force on the joining wire is indicated by marks, either painted or incised.

I will add this only to what has been said: that I have demonstrated in a book, published seven years ago, that heat and light are in electric conflict. From observations lately brought to bear we may now conclude that motion by gyrations also occurs in these effects; and I think that this does very much to illustrate the phenomena which they call the polarity of light.

BIOT AND SAVART

Jean Baptiste Biot was born on April 21, 1774, in Paris. He became professor of physics at the Collège de France and of astronomy in the Faculty of Sciences at Paris. He was a very active investigator, particularly in optics, and was a violent opponent of the undulatory theory of light, for which he undertook to substitute an emission theory of his own. He died in Paris on February 3, 1862.

Félix Savart was born on June 30, 1791, in Mézières. He was for a time a practising physician at Strassburg and afterwards taught physics in Paris and was in charge of the Physical Cabinet at the Collège de France. He died in Paris on March, 16, 1841.

The extract which follows is a note on the magnetism of Volta's pile, taken from the *Annales de Chimie et de Physique*, Vol. 15, p. 222, 1820. It contains the account of the discovery of the law of force in a magnetic field around a long straight current.

BIOT AND SAVART'S LAW

In the session of October 30th, 1820, MM. Biot and Savart presented to the Academy of Sciences a memoir in which they undertake to determine by accurate measurement the physical laws according to which metallic wires connected with the two poles of the voltaic apparatus act on magnetized bodies. The experiments were made by suspending with silk fibers rectangular rods or cylindrical wires of tempered steel magnetized by the method of double touch, and observing their times of oscillation and also their positions of equilibrium when they were placed at different distances and in different directions from the metallic wire by which the two poles of the pile were joined. The action of the earth's magnetism was sometimes combined with that of the wire, sometimes compensated for and destroyed by the opposing action of an artificial magnet placed at a distance. The apparatus used was a trough which contained ten couples, each with a square decimeter of surface. The observations were combined by a method of alternations which corrected the progressive variations that might be experienced; the times were measured by an excellent chronometer beating half seconds and with double stop and made by MM. Bréguet.

By this method MM. Biot and Savart were led to the following result, which expresses accurately the actions exerted on a molecule of austral or boreal magnetism when placed at any distance from a thin cylindrical wire of indefinite length rendered magnetic by the voltaic current. From the point where the molecule is, draw a perpendicular to the axis of the wire; the force which acts on the molecule is perpendicular to this line and to the axis of the wire. Its intensity is inversely as the distance. The nature of its action is the same as that of a magnetized needle, which is placed on the contour of the wire in a direction determined and always fixed with respect to the direction of the voltaic current; so that a molecule of boreal magnetism and a molecule of austral magnetism will always be acted on in opposite senses, although always along the straight line determined by the preceding construction.

By the help of this law of force we can anticipate and calculate in numbers all the motions impressed upon magnets by the connecting wire, whatever may be the direction of this wire with respect to them. We may equally well deduce, by using the ordinary laws of magnetic action, the sense or the nature of the magnetization which it can impress on steel or iron wires which are exposed to its action in a permanent way in a direction given with respect to its length.

ARAGO

The two extracts which follow present discoveries made by Arago (p. 324) relating to magnetism. The first gives an account of the magnetization of iron and steel by the action of the voltaic current. It is taken from the *Annales de Chimie et de Physique*, Vol. 15, p. 93, 1820. The second describes the famous experiment which seemed to show that magnetism was developed by moving conductors and which for so many years presented an unsolved problem. The extract is from Arago's *Collected Works*, Vol. IV, p. 424.

MAGNETIZATION BY THE CURRENT

As is well known, M. Oersted has recently discovered an action exerted by the voltaic current on a steel needle previously magnetized. While I was repeating the experiments of the Danish physicist, I noticed that the same current produces a strong development of the magnetic condition in bars of iron or steel which before were entirely devoid of it.

I will report the experiments which established this result almost in the order in which they were made.

Having joined a fine cylindrical copper wire to one of the poles of the voltaic pile, I noticed that, as soon as this wire was put in connection with the opposite pole, it attracted iron filings as a magnet would have done.

The wire, dipped in the filings, took them up equally all around so that it acquired by this addition a diameter almost equal to that of a quill.

As soon as the connecting wire was no longer in communication with the two poles of the pile the filings detached themselves at once and fell.

These effects do not depend upon a previous magnetization of the filings, since wires of soft iron or steel do not attract them at all.

They cannot be explained by attributing them to ordinary electric actions; for when I repeated the experiment with filings of copper or brass, or with sawdust, I found that they did not attach themselves in any case perceptibly to the connecting wire.

This attraction that the connecting wire exerts on the iron filings diminishes very rapidly as the action of the pile grows more feeble. Perhaps we shall find some day in the weight of the amount of filings lifted by a given length of wire a means of measuring the energy of this instrument at different epochs of the same experiment.

The action of the connecting wire on the iron is exerted at a distance: it is in fact easy to see that the filings are lifted well before the wire is in contact with them.

Hitherto I have spoken only of a connecting wire of brass; but wires of silver, platinum, etc. give analogous results. It remains to examine whether with similarity of form, of mass and of diameter, the wires of different metals act exactly with the same intensity.

The connecting wire communicates to soft iron only temporary magnetization; if we use little morsels of steel it gives them a permanent magnetization. I have even succeeded in completely magnetizing a sewing needle.

M. Ampère, to whom I showed these experiments, had just made the important discovery that two straight and parallel wires, through which two electric currents pass, attract each other when the currents are in the same sense and repel when they are in opposite senses; he further deduced from this by analogy the consequence that the attractive and repulsive properties of magnets depend on electric currents which circulate about the molecules of iron and steel in a direction perpendicular to a line which joins the two poles. M. Ampère assumes that in a horizontal magnet pointing toward the North the current in the upper part of it moves from West to East. These theoretical views suggested to him at once the thought that one might obtain a stronger magnetization by substituting for the straight connecting wire which I had used a wire wound into a helix, in the middle of which a steel needle is placed; he hoped further that we would obtain in that way a constant position of the poles—a result which was not attained by my method. Let us see how we brought these conjectures to the test of experiment:

A copper wire bent into a helix was terminated by two straight portions which could be fitted as we chose to the opposite poles of a powerful voltaic pile lying horizontally; a steel needle wrapped in paper was introduced into the helix, but only after the connection between the two poles had been set up, so that the effect which we looked for could not be attributed to the electric discharge which shows itself at the instant that the connecting wire is joined to the two poles. During the experiment the portion of the wire in which the steel needle was enclosed remained constantly perpendicular to the magnetic meridian, so that there was nothing to fear from the action of the earth.

After remaining a few minutes in the helix, the steel needle received a considerable charge of magnetism; the position of the north and south poles conforms perfectly to the result that M. Ampère had deduced in advance from the direction of the elements of the helix and from the hypothesis that the electric current traverses the connecting wire by going from the zinc end of the pile to the copper end.

. . . .

Magnetism of Rotation

The first publication of this discovery which I made is thus described in the report of the sitting of November 22nd, 1824, of the Academy of Sciences.

"M. Arago reported verbally the results of some experiments which he has made on an influence which the metals and many other substances exert on a magnetic needle, of which the effect is to diminish rapidly the amplitude of the oscillations without sensibly changing their period."

While I was determining, in company with my friend Alexander Humboldt, the magnetic intensity on the hillside of Greenwich in 1822, I noticed that the dip needle, when set in vibration, came to rest sooner when it was in its box than when it was not near any other bodies. It seemed to me that this effect ought to lead to important consequences as indicating a wider distribution of magnetic phenomena, which, up to that time, were very limited and so to speak isolated among the sciences. I have never ceased considering this line of thought and even today when I can no longer see or observe, it seems to me that there are still many researches which should be undertaken in the field which I have opened up, in spite of the apparently satisfactory explanation which has been given by Faraday of a part of the phenomena which I have discovered.

On March 7th, 1825 I presented a second communication to the Academy of Sciences on this subject; it is given as follows in the Annales de Chimie et de Physique, Vol. XXVIII, p. 325, 1825:

"M. Arago exhibited to the Academy an apparatus which shows in a new way the action that magnetized bodies and those which are not magnetized exert on one another.

"In his first experiments M. Arago proved that a sheet of copper or any other substance, solid or liquid, placed below a magnetic needle, exerts on this needle an action whose immediate effect is to change the amplitude of the oscillation without sensibly changing its period. The phenomenon which he exhibited to the Academy, is, so to speak, the inverse of the preceding one. Since a needle in motion is arrested by a plate at rest M. Arago thought that it might follow that a needle at rest would be carried along by a plate in motion. In fact, if we revolve a plate of copper, for example, with a definite velocity, under a magnetic needle contained in a closed vessel, the needle will no longer stand in its ordinary position; it takes a position out of the magnetic meridian and so much further from this plane as the rotation is more rapid. If this motion of rotation is sufficiently great, the needle at some distance from the plate turns continually about the filament by which it is suspended.

AMPÈRE

André Marie Ampère was born in Lyons on January 20, 1775. His father was a well-to-do retired merchant. He devoted himself to the education of his son, who early showed a genius for mathematics, though his reading and study ranged over many other fields. When the Revolution broke out his father returned to Lyons where he fell a victim during the Reign of Terror. The young Ampère was so broken by this catastrophe that for a year he could do nothing. His health and spirits gradually returned, and he settled again in Lyons, where he gave lessons in mathematics. A paper which he published on the mathematical theory of games of chance attracted the attention of Delambre and led to his appointment as professor of mathematics at the Lyceum at Lyons. He then obtained a post at the École Polytechnique in Paris and in 1809 was made professor of analysis in that school. He became a member of the Institute in 1814. The discovery of Oersted (p. 436) led to Ampère's experimental investigation of the forces exerted by currents on currents and to the development of a mathematical theory to describe them. This work put Ampère in the first rank of the investigators in the field of electricity and magnetism. He occupied himself also with philosophical questions and prepared a work on the classification of the sciences, which appeared after his death. He died at Marseilles on June 10, 1836.

The first short extract is from a paper entitled "Experiments on the New Electrodynamical Phenomena," in the *Annales de Chimie et de Physique,* Series II, Vol. 20, p. 60, 1822. It is a note attached to the title and is inserted for the sake of the names which Ampère here gave for the first time.

The second extract is from a memoir presented to the Academy of Sciences, October 2, 1820, as it appeared in the *Annales de Chimie et de Physique* Series II, Vol. 15, p. 59, 1820. It contains the account of the experiments by which the forces between two currents were examined.

The remaining extracts are from Ampère's work entitled "Théorie des phénomènes électrodynamiques uniquement déduite de l'expérience," published in Paris in 1826. They are taken from those parts of the work in which the author states in words the results of his calculations. The calculations themselves are too extensive to permit of quotation. The first extract is given for the sake of the names which Ampère introduces.

NEW NAMES

The paragraph here given is a note attached to the title.

The word *'electromagnetic'* which is used to characterize the phenomena produced by the conducting wires of the voltaic pile, could not suitably describe them except during the period when the only phenomena which were known of this sort were those which M. Oersted discovered, exhibited by an electric current and a magnet. I have determined to use the word *electrodynamic* in order to unite under a common name all these phenomena, and particularly to designate those which I have observed between two voltaic conductors. It expresses their true character, that of being produced by electricity in motion: while the electric attractions and repulsions, which have been known for a long time, are *electrostatic* phenomena produced by the unequal distribution of electricity at rest in the bodies in which they are observed.

ACTIONS BETWEEN CURRENTS

I. On the Mutual Action of Two Electric Currents.

1. Electromotive action is manifested by two sorts of effects which I believe I should first distinguish by precise definitions.

I shall call the first *electric tension*, the second *electric current*.

The first is observed when two bodies, between which this action occurs, are separated from each other by non-conducting bodies at all the points of their surfaces except those where it is established; the second occurs when the bodies make a part of a circuit of conducting bodies, which are in contact at points on their surface different from those at which the electromotive action is produced. In the first case, the effect of the electromotive action is to put the two bodies, or the two systems of bodies, between which it exists, in two states of tension, of which the difference is constant when this action is constant, when, for example, it is produced by the contact of two substances of different sorts; this difference may be variable, on the contrary, with the cause which produces it, if it results from friction or from pressure.

The first case is the only one which can arise when the electromotive action develops between different parts of the same nonconducting body; tourmaline is an example of this when its temperature changes.

In the second case there is no longer any electric tension, light bodies are not sensibly attracted and the ordinary electrometer can no longer be of service to indicate what is going on in the body; nevertheless the electromotive action continues; for if, for example, water, or an acid or an alkali or a saline solution forms part of the circuit, these bodies are decomposed, especially when the electromotive action is constant, as has been known for some time; and furthermore as M. Oersted has recently discovered, when the electromotive action is produced by the contact of metals, the magnetic needle is turned from its direction when it is placed near any portion of the circuit; but these effects cease, water is no longer decomposed, and the needle comes back to its ordinary position as soon as the circuit is broken, when the tensions are reestablished and light bodies are again attracted. This proves that the tensions are not the cause of the decomposition of water, or of the changes of direction of the magnetic needle discovered by M. Oersted. This second case is evidently the only one which can occur if the electromotive action is developed between the different parts of the same conducting body. The consequences deduced in this memoir from the experiments of M. Oersted will lead us to recognize the existence of this condition in the only case where there is need as yet to admit it.

2. Let us see in what consists the difference of these entirely different orders of phenomena, one of which consists in the tension and attractions or repulsions which have been long known, and the other, in decomposition of water and a great many other substances, in the changes of direction of the needle, and in a sort of attractions and repulsions entirely different from the ordinary electric attractions and repulsions; which I believe I have first discovered and which I have named *voltaic attractions* and *repulsions* to distinguish them from the others. When there is not conducting continuity from one of the bodies, or systems of bodies, in which the electromotive action develops, to the other, and when these bodies are themselves conductors, as in Volta's pile, we can only conceive this action as constantly carrying positive electricity into the one body and negative electricity into the other: in the first moment, when there is nothing opposed to the effect that it

tends to produce, the two electricities accumulate, each in the part of the whole system to which it is carried, but this effect is checked as soon as the difference of electric tensions gives to their mutual attraction, which tends to reunite them, a force sufficient to make equilibrium with the electromotive action. Then everything remains in this state, except for the leakage of electricity, which may take place little by little across the non-conducting body, the air, for example, which interrupts the circuit; for it appears that there are no bodies which are perfect insulators. As this leakage takes place the tension diminishes, but since when it diminishes, the mutual attraction of the two electricities no longer makes equilibrium with the electromotive action, this last force, in case it is constant, carries anew positive electricity on one side and negative electricity on the other, and the tensions are reestablished. It is this state of a system of electromotive and conducting bodies that I called *electric tension*. We know that it exists in the two halves of this system when we separate them or even in case they remain in contact after the electromotive action has ceased, provided that then it arose by pressure or friction between bodies which are not both conductors. In these two cases the tension is gradually diminished because of the leakage of electricity of which we have recently spoken.

But when the two bodies or the two systems of bodies between which the electromotive action arises are also connected by conducting bodies in which there is no other electromotive action equal and opposite to the first, which would maintain the state of electrical equilibrium, and consequently the tensions which result from it, these tensions would disappear or at least would become very small and the phenomena occur which have been pointed out as characterizing this second case. But as nothing is otherwise changed in the arrangement of the bodies between which the electromotive action develops, it cannot be doubted that it continues to act, and as the mutual attraction of the two electricities, measured by the difference in the electric tensions, which has become nothing or has considerably diminished, can no longer make equilibrium with this action, it is generally admitted that it continues to carry the two electricities in the two senses in which it carried them before; in such a way that there results a double current, one of positive electricity, the other of negative electricity, starting out in opposite senses from the points where the electromotive action arises, and going out to reunite in the parts of the

circuits remote from these points. The currents of which I am
speaking are accelerated until the inertia of the electric fluids
and the resistance which they encounter because of the imperfec-
tion of even the best conductors make equilibrium with the
electromotive force, after which they continue indefinitely with
constant velocity so long as this force has the same intensity,
but they always cease on the instant that the circuit is broken.
It is this state of electricity in a series of electromotive and conducting bodies which I name, for brevity, the *electric current;*
and as I shall frequently have to speak of the two opposite senses
in which the two electricities move, I shall understand every time
that the question arises, to avoid tedious repetition, after the
words *"sense of the electric current"* these words, *of positive electricity;* so that if we are considering, for example, a voltaic pile, the
expression: *direction of the electric current in the pile*, will designate
the direction from the end where hydrogen is disengaged in the
decomposition of water to that end where oxygen is obtained; and
this expression, *direction of the electric current in the conductor which
makes connection between the two ends of the pile*, will designate
the direction which goes, on the contrary, from the end where
oxygen appears to that where the hydrogen develops.   To include
these two cases in a single definition we may say that what we may
call the direction of the electric current is that followed by hydrogen and the bases of the salts when water or some saline substance
is a part of the circuit, and is decomposed by the current, whether,
in the voltaic pile, these substances are a part of the conductor
or are interposed between the pairs of which the pile is constructed.

The paragraphs in which it is pointed out that the electric tensions cannot
be the cause of chemical or magnetic actions are omitted.

3. The ordinary electrometer indicates tension and the intensity
of the tension; there was lacking an instrument which would
enable us to recognize the presence of the electric current in a pile
or a conductor and which would indicate the energy and the
direction of it.   This instrument now exists; all that is needed
is that the pile, or any portion of the conductor, should be placed
horizontally, approximately in the direction of the magnetic
meridian, and that an apparatus similar to a compass, which, in
fact, differs from it only in the use that is made of it, should be
placed above the pile or either above or below a portion of the

conductor. So long as the circuit is interrupted, the magnetic needle remains in its ordinary position, but it departs from this position as soon as the current is established, so much the more as the energy is greater, and it determines the direction of the current from this general fact, that if one places oneself in thought in the direction of the current in such a way that it is directed from the feet to the head of the observer and that he has his face turned toward the needle; the action of the current will always throw toward the left that one of the ends of the needle which points toward the north and which I shall always call the austral pole of the magnetic needle, because it is the pole similar to the southern pole of the earth. I express this more briefly by saying, that the austral pole of the needle is carried to the left of the current which acts on the needle. I think that to distinguish this instrument from the ordinary electrometer we should give it the name of *galvanometer* and that it should be used in all experiments on electric currents, as we habitually use an electrometer on electric machines, so as to see at every instant if a current exists and what is its energy.

The first use that I have made of this instrument is to employ it to show that the current in the voltaic pile, from the negative end to the positive end, has the same effect on the magnetic needle as the current in the conductor which goes on the contrary from the positive end to the negative end.

It is well to have for this experiment two magnetic needles, one of them placed on the pile and the other above or below the conductor; we see the austral pole of each needle move to the left of the current near which it is placed; so that when the second is above the conductor it is turned to the opposite side from that toward which the needle turns which has been placed on the pile, because the currents have opposite directions in these two portions of the circuit; the two needles, on the contrary, are turned toward the same side, remaining nearly parallel with each other, when the one is above the pile and the other below the conductor. As soon as the circuit is broken they come back at once in both cases to their ordinary position.

4. Such are the differences already recognized in the effects produced by electricity in the two states which I have described, of which the one consists, if not in rest, at least in a movement which is slow and only produced because of the difficulty of completely insulating the bodies in which the electric tension exhibits

itself, the other, in a double current of positive and negative electricity along a continuous circuit of conducting bodies. In the ordinary theory of electricity, we suppose that the two fluids of which we consider it composed are unceasingly separated one from the other in a part of a circuit and carried rapidly in contrary senses into another part of the same circuit, where they are continually reunited. Although the electric current thus defined can be produced with an ordinary machine by arranging it in such a way as to develop the two electricities and by joining by a conductor the two parts of the apparatus where they are produced, we cannot, unless we use a very large machine, obtain the current with an appreciable energy except by the use of the voltaic pile, because the quantity of electricity produced by a frictional machine remains the same in a given time whatever may be the conducting power of the rest of the circuit, whereas that which the pile sets in motion during a given time increases indefinitely as we join the two extremities by a better conductor.

But the differences which I have recalled are not the only ones which distinguish these two states of electricity. I have discovered some more remarkable ones still by arranging in parallel directions two straight parts of two conducting wires joining the ends of two voltaic piles; the one was fixed and the other, suspended on points and made very sensitive to motion by a counterweight, could approach the first or move from it while keeping parallel with it. I then observed that when I passed a current of electricity in both of these wires at once they attracted each other when the two currents were in the same sense and repelled each other when they were in opposite senses. Now these attractions or repulsions of electric currents differ essentially from those that electricity produces in the state of repose; first, they cease, as chemical decompositions do, as soon as we break the circuit of the conducting bodies; secondly, in the ordinary electric attractions and repulsions the electricities of opposite sort attract and those of the same name repel; in the attractions and repulsions of electric currents we have precisely the contrary; it is when the two conducting wires are placed parallel in such a way that their ends of the same name are on the same side and very near each other that there is attraction, and there is repulsion when the two conductors, still always parallel, have currents in them in opposite senses, so that the ends of the same name are as far apart as possible. Thirdly, in the case of attraction, when it is sufficiently strong to bring the movable conductor into contact with the fixed conductor, they remain attached to one another like two magnets and do not separate after a while, as happens when two conducting bodies which attract each other because they are electrified, one positively and the other negatively, come to touch. Finally, and it appears that this last circumstance depends on the same cause as the preceding, two electric currents attract or repel in vacuum as in air, which is contrary to that which we observe in the mutual action of two conducting bodies charged with ordinary electricity. It is not the place here to explain these new phenomena; the attractions and repulsions which occur between two parallel currents, according as they are directed in the same sense or in opposite senses, are facts given by an experiment which is easy to repeat. It is necessary in this experiment, in order to prevent the motions which would be given to the movable conductor by agitation of the air, to place the apparatus under a glass cover within which we introduce, through the base which carries it, those parts of the conductors which can be joined to the two ends of the pile. The most convenient arrangement of these conductors is to place one of them on two supports in a horizontal position, (Fig. 86) in which it is fixed, and to hang up the other by two metallic wires, which are joined to it, on a glass rod which is above the first conductor and which rests on two other metal supports by very fine steel points; these points are soldered to the two ends of the metallic wires of which I have spoken, in such a way that connection is established through the supports by the aid of these points.

The two conductors are thus parallel and one beside the other in a horizontal plane; one of them is movable because of the oscillations which it can make about the horizontal line passing through the ends of the two steel points and when it thus moves it necessarily remains parallel to the fixed conductor.

There is introduced above and in the middle of the glass rod a counterweight, to increase the mobility of the oscillating part of the apparatus, by raising its center of gravity.

I first thought that it would be necessary to set up the electric current in the two conductors by means of two different piles; but this is not necessary. The conductors may both make parts of the same circuit; for the electric current exists everywhere with the same intensity. We may conclude from this observation that the electric tensions of the two ends of the pile have nothing to do with the phenomena with which we are concerned; for there is certainly no tension in the rest of the circuit. This view is confirmed by our being able to move the magnetic needle at a great distance away from the pile by means of a very long conductor, the middle of which is curved over in the direction of the magnetic meridian above or below the needle. This experiment was suggested to me by the illustrious savant to whom the physico-mathematical sciences owe so much of the great progress that they have made in our time: it has fully succeeded.

Designate by *A* and *B* the two ends of the fixed conductor, by *C* the end of the movable conductor which is on the side of *A*

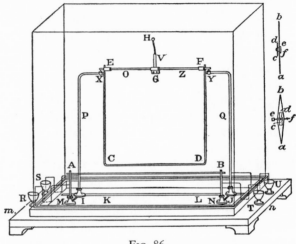

Fig. 86.

and by *D* that of the same conductor which is on the side of *B*; it is plain that if one of the ends of the pile is joined to *A*, *B* to *C*, and *D* to the other end of the pile, the electric current will be in the same sense in the two conductors; then we shall see them attract each other; if on the other hand, while *A* always is joined to one end of the pile, *B* is joined to *D* and *C* to the other end of the pile, the current will be in opposite senses in the two conductors and then they repel each other. Further, we may recognize that since the attractions and repulsions of electric currents act at all points in the circuit we may, with a single fixed conductor, attract and repel as many conductors and change the direction of as many magnetic needles as we please. I propose to have made two movable conductors within the same glass case so arranged that by making them parts of the same circuit, with a common fixed conductor, they may be alternately both attracted or both repelled, or one of them attracted and the other repelled at the same time, according to the way in which the connections are made. Following up the success of the experiment which was suggested to me by the Marquis de Laplace, by employing as many conducting wires and magnetized needles as there are letters, by fixing each letter on a different magnet, and by using a pile at a distance from these needles, which can be joined alternately by its own ends to the ends of each conductor, we may form a sort of telegraph, by which we can write all the matters which we may wish to transmit, across whatever obstacles there may be, to the person whose duty it is to observe the letters carried by the needles. By setting up above the pile a key-board of which the keys carry the same letters and by making connection by pressing them down, this method of correspondence could be managed easily and would take no more time than is necessary to touch the keys at one end and to read off each letter at the other.

If the movable conductor, instead of being adjusted so as to move parallel to the fixed conductor, can only turn in a plane parallel to the fixed conductor about a common perpendicular passing through their centers, it is clear, from the law that we have discovered of the attractions and repulsions of electric currents, that each half of the two conductors will attract or repel at the same time, according as the currents are in the same sense or in opposite senses; and consequently that the movable conductor will turn until it becomes parallel to the fixed conductor, in such a way that the currents are directed in the same sense: from which it follows that in the mutual action of two electric currents the directive action and the attractive or repulsive action depend on the same principle and are only different effects of one and the same action. It is no longer necessary, therefore, to set up between these two effects the distinction which it is so important to make, as we shall see very soon, when we are dealing with the mutual action of an electric current and of a magnet considered, as we ordinarily do, with respect to its axis, because in this action the two bodies tend to place themselves perpendicular to each other.

We now turn to the examination of this last action and of the action of two magnets on each other and we shall see that they both come under the law of the mutual action of two electric currents, if we conceive one of these currents as set up at every point of a line drawn on the surface of a magnet from one pole to the other, in planes perpendicular to the axis of the magnet, so that from the simple comparison of facts it seems to me impossible to doubt that there are really such currents about the axis of a magnet, or rather that magnetization consists in a process by which we give to the particles of steel the property of producing, in the sense of the currents of which we have spoken, the same electromotive action as is shown by the voltaic pile, by the oxidized zinc of the mineralogists, by heated tourmaline, and even in a pile made up of damp cardboard and discs of the same metal at two different temperatures.   However, since this electromotive action is set up in the case of a magnet between the different particles of the same body, which is a good conductor, it can never, as we have previously remarked, produce any electric tension, but only a continuous current similar to that which exists in a voltaic pile re-entering itself in a closed curve.   It is sufficiently evident from the preceding observations that such a pile cannot produce at any of its points either electric tensions or attractions or repulsions or chemical phenomena, since it is then impossible to insert a liquid in the circuit; but that the current which is immediately established in this pile will act to direct it or to attract or repel it either by another electric current or by a magnet, which, as we shall see, is only an assemblage of electric currents.

It is thus that we come to this unexpected result, that the phenomena of the magnet are produced by electricity and that there is no other difference between the two poles of a magnet than their positions with respect to the currents of which the magnet is composed, so that the austral pole is that which is to the right of these currents and the boreal pole that which is to the left.

The rest of the memoir in the same volume of the *Annales* contains descriptions of more elaborate experiments by which the discoveries announced in the first part of the memoir are confirmed.

The Solenoid

Now let us imagine in space any line *MmO* which is encircled by electric currents, making very small closed circuits about it, in planes infinitely near each other, which are perpendicular to it, in such a way that the areas contained by these circuits are all equal to one another, and so that their centers of gravity are on

the line and so placed that there is the same distance measured
on the line between two consecutive planes. If we call this
distance g, which we shall consider infinitely small, the number of
currents which will correspond to an element ds of the line MmO
will be ds/g; and we must multiply by this number the values of
A, B, C which we have found for a single circuit so as to get those
values which refer to the circuits of the element ds; and finally
by integrating from one end of the arc s to the other we shall obtain
the values for A, B, C which hold for the assemblage of all the
circuits which encircle the line.   To this assemblage I have given
the name *electrodynamic solenoid*, from the Greek word σωληνοειδής
which denotes that which has the form of a canal, that is to say,
the surface of this form in which all the circuits lie.

CIRCUITS AND MAGNETIC SHELLS

To justify the way in which I conceive the phenomena presented
by magnets, by considering them as if they were assemblages of
electric currents in very small circuits about their particles, it
must be demonstrated, by starting from the formula by which I
have represented the mutual action of two elements of electric
current, that there shall result from certain assemblages of these
small circuits forces which depend only on the positions of two
definite points of this system, and which, relative to these two
points, exhibit all the properties of the forces which we attribute
to what we call the molecules of austral fluid and of boreal fluid,
when we explain by means of these two fluids the phenomena
presented by magnets, either in their mutual action or in that
which they exert on the conducting wire.   Now we know that
the physicists who prefer the explanation in which we suppose
the existence of these molecules to that which I have deduced
from the properties of electric currents, assume that to each
molecule of austral fluid there corresponds in each particle of the
magnetized body a molecule of boreal fluid of the same intensity,
and that if we give the name magnetic element to the combination
of these two molecules, which we may consider to be the two poles
of this element, it is necessary for the explanation of the phenomena
which are presented by the two kinds of action which we are now
considering: (1) That the mutual action of two magnetic elements
is compounded of four forces, two attractive and two repulsive,
directed along the straight lines which join the two molecules of
one of these elements to the two molecules of the other, and of

which the intensity is in the inverse ratio of the squares of these lines; (2) That when one of these elements acts on an infinitely small portion of the conducting wire, there result two forces perpendicular to the planes which pass through the two molecules of the element and through the direction of the small portion of the wire, and which are proportional to the sines of the angles which this direction makes with the straight lines which measure the distances to the two molecules and in the inverse ratio of the squares of these distances. So long as one does not accept the explanation which I give of the action of magnets and so long as one attributes these two sorts of forces to molecules of an austral fluid and of a boreal fluid, it is impossible to refer them to a single principle; but if one adopts my way of looking at the constitution of magnets, one sees by the preceding calculations that both these types of action and the values of the forces which result from them are deduced immediately from my formula, and that to find these values it is sufficient to substitute for the combination of two molecules, one of them of austral fluid, and the other of boreal fluid, a solenoid whose ends, which are the two definite points on which depend the forces with which we are dealing, are situated exactly at the points where one supposes the molecules of the two fluids to be situated.

Thus two systems of very small solenoids will act on each other according to my formula like two magnets made up of as many magnetic elements as we suppose there are solenoids in these two systems; one of the same systems will act also on an element of an electric current as a magnet does; and consequently all the calculations and all the explanations based on the consideration of the attractive and repulsive forces of these molecules in the inverse ratio of the squares of their distances, and also on the consideration of the forces of rotation between one of these molecules and an element of the electric current, the law of which I have stated in the form which is adopted by the physicists who do not receive my theory, are necessarily the same whether one explains, as I do, by electric currents the phenomena which magnets present in these two cases or prefers the hypothesis of two fluids. It is therefore not in these calculations or explanations that one should seek either objections to my theory or proofs in its favor. The proof on which I rely follows altogether from this, that my theory reduces to a single principle three sorts of actions which the totality of the phenomena proves to result from a

common cause and which cannot be reduced to one principle in
any other way.

. . . .

We have now to concern ourselves with the actions that a closed
circuit, whatever may be its form, its size, and its position, exerts
either on a solenoid or on another circuit of any form, size and
position. The principal result of these researches consists in
the analogy which exists between the forces produced by this
circuit, whether it acts on another closed circuit or on a solenoid,
and the forces which would be exerted by points, the action of
which is precisely that which is attributed to the molecules of that
which we call the austral fluid and the boreal fluid; these points
being distributed in the way that I have explained on surfaces
bounded by these circuits and the ends of the solenoid being replaced
by two magnetic molecules of opposite sort. This analogy appears
at once so complete that all the electrodynamic phenomena seem
to be thus referred to the theory in which we assume these two
fluids; but we soon perceive that this does not hold except for
voltaic conductors which form rigid closed circuits, and that
only those phenomena which are produced by conductors forming
such circuits can be explained in this way, and that finally the
forces expressed by my formula are the only ones which agree with
all the facts.

. . . .

Instead of replacing each circuit by two neighboring surfaces,
of which one is covered with austral fluid and the other with
boreal fluid, these fluids being distributed as has been said before,
we can replace each circuit by a single surface on which are uniformly distributed magnetic elements, such as they have been
defined by M. Poisson in a memoir read to the Academy of Sciences,
February 2, 1824.

The author of this memoir, by calculating the formulas by which
he has brought within the range of analysis all questions relative
to the magnetization of bodies, whatever may be the cause which
is assigned for that, has given the values of the three forces exerted
by a magnetic element on a molecule of austral or boreal fluid.
These values are identical with those which I have deduced from
my formula for the three quantities A, B, C, in the case of a very
small closed plane circuit, when we suppose that the constant

coefficients are the same. It is easy to deduce from them a theorem from which we see immediately:

1. That the action of an electrodynamic solenoid, calculated from my formula, is in every case the same as that of a series of magnetic elements of the same intensity distributed uniformly along the straight or curved line which is encircled by all the small circuits of the solenoid, if we give to the axes of the elements at each point of the line the direction of the line;

2. That the action of a rigid closed voltaic circuit, calculated from my formula, is precisely that which magnetic elements of the same intensity will exert, if they are uniformly distributed on any surface bounded by this circuit with the axes of the magnetic elements everywhere normal to this surface.

The same theorem also leads us to this consequence, that if we conceive a surface, enclosing on all sides a very small volume, and if we suppose on the one hand molecules of austral fluid and of boreal fluid distributed on this small surface in equal quantities, as they ought to be if they are to constitute a magnetic element such as M. Poisson has conceived it, and on the other hand if we suppose the same surface covered with electric currents which form on this surface small closed circuits in parallel and equidistant planes, and if we calculate the action of these currents from my formula, the forces exerted in both cases, whether on an element of the conducting wire or on a magnetic molecule, are precisely the same, are independent of the form of the small surface, and are proportional to the volume which it encloses, when the axes of the magnetic elements coincide with the perpendicular to the planes of the circuits.

When the identity of these forces has been demonstrated we may consider as simple corollaries all the results which I have given in this treatise regarding the possibility of substituting for magnets, without changing the effects produced, assemblages of electric currents forming closed circuits about their particles. I think that it will be easy for the reader to deduce this consequence, and the theorem on which it rests, from the preceding calculations; I have also developed it in another paper where, at the same time, I discussed from this new point of view everything relative to the mutual action of a magnet and of a voltaic conductor.

SEEBECK

Thomas Johann Seebeck was born on April 9, 1770, in Reval. He held no public or academic position. He died in Berlin on December 10, 1831.

The extract which follows contains the account of the discovery of thermo-electricity. It appeared in the *Abhandlungen der Königlichen Akademie der Wissenschaften in Berlin*, 1822–1823, under the somewhat misleading title "Magnetische Polarisation der Metalle und Erze durch Temperatur-Differenz."

THERMOELECTRIC CURRENTS

. . . .

As I was carrying on investigations on the mutual relations of the electric, chemical and magnetic activities in galvanic batteries, I encountered phenomena which seemed to me to show that two metals by themselves, simply joined with each other in a circuit, may become magnetic without the additional action of a liquid conductor. Other reasons also seemed in favor of this view. For, from several facts, particularly from that noticed in a former paper, it seems to follow that the magnetic polarization of the whole closed circuit is determined not merely by the action at the point of contact of the metals with each other, but rather by the inequality of the actions at the two points of contact of the metals with the moist conductor; also it cannot be doubted that even if the action at the first named point of contact must be recognized as contributing a portion to the excitation of magnetism, yet surely the excess of action at one of the points of contact over that at both the others could produce a magnetic tension; and this I believe justifies the expectation that if there is any difference in the conditions at the points of contact of two metals joined in circuit with each other a magnetic polarization can occur.

For the first research undertaken from this point of view I chose two metals, bismuth and antimony, which I had found in many respects peculiar and variable when they were used as elements in conjunction with copper in the ordinary galvanic cell. With both these I found my expectation fulfilled and yet their action was different.

1. A plate of bismuth lying immediately on a plate of copper between the two ends of a spirally wound copper ribbon 40 feet long and $2\frac{1}{2}$ lines wide and lying in the magnetic meridian, showed, when the circuit was closed, a distinct divergence of the magnetic needle.

A number of such experiments are described, in which the plates of bismuth or antimony were used generally in connection with copper. The relative positions of these plates and of the spiral in which the current was set up were changed, and the deviations of the needle were recorded.

5. In all these researches I closed the circuit by laying the metallic plate which was to be investigated on the lower end of the spiral or of the simple strip and by pressing down the upper free end upon the plate with my fingers. In the early investigations, therefore, the question might arise whether the hand did not take the place of the moistened conductor, and whether bismuth and antimony did not give their opposite deflections because the one of them by reason of the dampness of the hand becomes with copper $+E$, while the other becomes $-E$.

Several experiments are then detailed which prove that the action discovered could not arise from the moisture of the hand.

11. In all these experiments the action was strongest when the metals and the ores were immediately in contact with the hand; it was weaker when the circuit was closed with a thin sheet of some body between them, (which, however, if it is non-metallic should not be placed between the spiral and the metal or ore which is to be investigated, but on them both), and all action on the magnetic needle failed if the ends of the spiral were pressed down on the plate of metal with rods of glass, wood or metal 2 feet long. However, a movement of the magnetic needle soon showed itself if the hand was placed on the lower end of the metallic rod near the place where it touched the circuit, and was kept there for a while. From these observations the thought arises that the heat which is communicated by the hand more strongly to one of the points of contact of the metals may be the cause of the magnetism in these circuits with two elements. Hence it was to be expected that a higher degree of temperature than that which could be communicated to the metals by the hand would excite a greater magnetic tension. The following investigation confirmed this.

12. A plate of bismuth was brought in contact with both ends of a copper spiral; a cold plate of copper was placed under the closed circuit and above it a plate of copper heated over a lamp. A deviation followed at once, which was much greater than those obtained in the former experiment. The magnetic needle within the spiral was deflected through 50° or 60° and remained stationary

at 17°. In other respects the deviation was similar to that of former experiments, that is, it was toward the West when the apparatus was arranged as in Fig. 87.

If the warmed plate of copper was placed under the bismuth, which was in contact with the copper strip, there followed, when

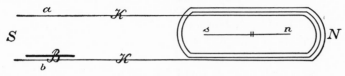

Fig. 87.

everything else remained unchanged, a deviation toward the East which was quite as great as the former one toward the West.

13. When the bismuth plate itself was warmed and laid immediately on the lower end of the spiral, there followed similarly a deviation toward the East when the upper end of the spiral touched the bismuth. In this case the lower end of the spiral was the warmer, since it remained always in contact with the plate of bismuth; on the other hand, the upper end, which only touched the plate for a short time, was colder, and so the same deviation should occur as in the last experiment of the preceding paragraph.

If both ends of the spiral were brought in contact for the same length of time with the heated plate of bismuth no deviation of the magnetic needle was shown.

A number of other experiments of the same general nature are then presented.

18. From all these experiments we can conclude that the first and most important condition for the appearance of free magnetism

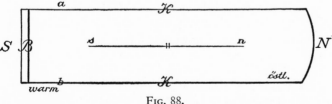

Fig. 88.

in these metallic circuits is the difference of temperature at the points of contact of the two elements.

Without doubt magnetism will also be excited if both points of contact of the metals or the ores are heated at the same time and to an equal degree; there will be no action on the magnetic needle, however, in this case, because by this procedure a double and opposite magnetic polarization will be excited in the circuit, and because this is everywhere of equal strength. By heating the upper point of contact (Fig. 88) we have the condition for a Westerly deviation, and by heating the lower point of contact that for an Easterly deviation. These keep each other in equilibrium and the needle must therefore remain at rest.

Magnetism must be excited by the mere contact of the half circuits with each other without any change of temperature, but it remains latent, because the actions of the two metals on each other are of equal strength at both points of contact and the magnetic polarizations excited by them have opposite directions.

19. Artificial cooling of one of the two points of contact will produce a magnetic tension in these metallic circuits of two elements, just as well as heating does, by breaking the magnetic equilibrium.

The rest of this section is omitted.

20. The magnetic tension in these metallic circuits is greater when the difference of temperature at the two points of contact is greater. Even if this tension does not increase uniformly in all cases as the temperature rises, and metallic alloys furnish many exceptions, as we shall find further on, yet this law seems to be valid for most combinations of metals and particularly for the purer metals. How the heating takes place is a matter of indifference, whether over a lamp, or on a hot bolt, or by means of a burning glass. The magnetic polarization of a circuit remains always the same, whichever one of these methods is used. It is also a matter of indifference in respect to this polarization whether only one of the two metals at one end is warmed, and which of them, or whether both of them are warmed at the same time; and yet when both are warmed at the same time at one point of contact the magnetic polarization is as a rule greater.

The rest of this long paper contains a detailed study of various combinations of metals. No further extracts from it need be made.

OHM

George Simon Ohm was born in Erlangen on March 16, 1789. His father was a master locksmith and a student of mathematics and philosophy. He taught his two sons mathematics and prepared them for the university. For want of means George's career at the university was interrupted and it was not until 1811 that he took his degree at Erlangen. He then taught mathematics and physics in several positions with great success. He soon turned his attention to the voltaic battery and published several papers on the laws of conduction in the circuit, the last of which was his mathematical treatment of the galvanic circuit, published in 1827. He was so disappointed by the reception given this work, in particular by the neglect of it by the Berlin Academy, that he resigned his position in Cologne and lived for several years in retirement. In 1833 he was called to be professor of physics at the Polytechnikum at Nuremberg. In the following years he gave himself largely to researches in acoustics. In 1849 he was called to Munich as a professor of physics and mathematics. He died there on July 6, 1854.

Ohm was the discoverer of the law which is known by his name. The theoretical treatise in which this law is developed has already been mentioned. The extract which follows is from a paper published in the *Journal für Chemie und Physik* (Schweigger's *Journal*) Vol. 46, p. 137, 1826, under the title "Bestimmung des Gesetzes, nach welchem Metalle die Contaktelektricität leiten, nebst einem Entwurfe zu einer Theorie des Voltaischen Apparates und des Schweiggerschen Multiplicators." In it is given an experimental proof of Ohm's law.

Ohm's Law

Ohm begins by referring to a paper in which he had already published a law governing the conduction of electricity in metallic conductors, and calls attention to differences between his results and those obtained by Barlow and Becquerel. He then describes experiments made with voltaic batteries, and concludes that these sources of current fluctuate so much while in use as to make them unsatisfactory for the purpose for which he used them. He then proceeds as follows.

I therefore turned to the use of the thermo-electric battery, the suitability of which for my purposes was suggested by Herr Poggendorf; and since the results obtained in this way give the law of conduction in a definite manner, I think it not superfluous to describe my apparatus at length, so that the degree of confidence which can be placed in the results obtained with it can be more easily estimated.

A piece of bismuth was cast in the form of a rectangular brace *abb'a'* (Fig. 89) whose longer side was 6½ inches long, and whose shorter legs *ab*, *a'b'* were each 3½ inches long. It was throughout

9 lines broad and 4 lines thick. On each of the two legs I fastened with two screws copper strips *abcd*, *a'b'c'd'*, which were 9 lines broad, 1 line thick, and together were 28 inches long. These were so bent, that their free ends *cd*, *c'd'* were immersed in mercury contained in two cups *m*, *m'* which stood on the wooden base *fghi*.

On the upper plate of the base was placed the torsion balance, in the description of which I shall be a little diffuse, since its con-

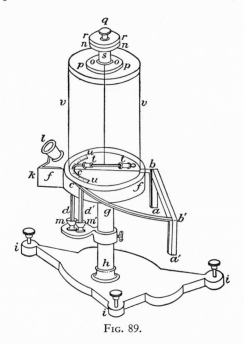

Fig. 89.

struction differs somewhat from the ordinary. The glass cylinder *vv*, on which it is set, is 6 inches high and 4½ inches wide. It itself consists of two parts, one of which *nop* is provided with a slightly conical socket, and is cemented fast to the upper plate of the glass cylinder; the other, *qrs*, with a conical projection 8 lines thick fitting closely in the socket, and with a plate *rr* 3 inches wide resting on the plate *nn* of the first part of the same width. The midpoint of the projection *qs* was marked with great care on the lathe by a slight conical depression, and the metal was then filed off for a half inch of its length until the plane surfaces which are thus developed show the conical depression as a complete triangle. By a special arrangement the thread by which the needle is suspended is fastened to the projection so that its midpoint falls exactly in the apex of the triangle.

The magnetic needle *tt* is made of steel wire 0.8 lines thick and is not quite 2 inches long. Its two ends are inserted in cylindrical pieces of ivory, to one of which is attached a brass wire, cut to a fine point and bent slightly downward. This brass pointer, which serves as an indicator, comes close to a brass arc *uu*, standing on the base and divided into degrees. At the start I made the magnet so long that its end moved immediately over the graduated scale; but the sluggishness of its motion, as shown by the small number of its vibrations, reminded me of the experiment lately made by Arago, and led me to choose the other arrangement.

The needle thus prepared is suspended by a strip of flattened gold wire 5 inches long, which is fastened to the torsion balance exactly in the axis of rotation. These ribbon-like strips of metal are in my judgment much better fitted for experiments with the torsion balance than cylindrical wires. The strip which I used in my torsion balance, not to mention its shortness, which in many respects is so desirable, possesses in so high a degree all the requirements for investigations with the torsion balance, that the needle, after the strip has experienced a strain of more than three complete revolutions, will again resume its old position after it is released from the strain. Nevertheless after each experiment I examined the needle in the position of rest, so as to be assured that the apparatus had undergone no change. Furthermore I think I should remark that experiments carried out with a similar needle of brass convinced me that small and great vibrations (I examined them from two whole revolutions down to a few degrees) are made in exactly the same time, so that in this respect also there is nothing to be feared.

The torsion balance was cemented to the upper plate of the base in such a way that a straight line drawn down the width of the copper strip *bc* in the direction of the midpoint of the divided arc *uu* and of a simple silk thread set perpendicularly before this arc, was in the magnetic meridian and the magnetic needle also, when its pointer marked zero on the scale. On a projection *k* of the base was carried a convex lens *l* of an inch focal length, set in such a position that the lower divided scale could be observed, and in order to avoid parallax, the eye was always so placed during an observation that the silk thread and the midpoint of the scale coincided. The observations were made in the following manner:

whenever the needle was deflected by the electric current in the apparatus, the strip was twisted in the opposite sense by the rotating part of the balance until the brass pointer of the needle stood behind the silk thread on the midpoint of the scale; then the torsion was read off in hundredths of a revolution on the upper scale, which number as is well known gives the force which acts on the needle.

The ends of the conductors which were used in the experiment were dipped in the mercury cups m, m' above which for greater security a simple arrangement secured that the ends of each of the conductors were always put in contact with the mercury in a similar manner. In addition the ends of the conductors, so far

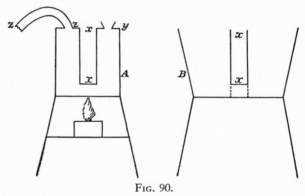

Fig. 90.

up as there was any reason to fear contact with the mercury, were coated with rosin, and the end surfaces were filed off clean with a fine file, and were renewed every time. A perfect metallic contact of the several parts is an indispensable condition in researches of this sort, since otherwise the observations will not agree with one another.

Finally, in order to bring the parts of the apparatus where the bismuth and copper are in contact to the proper difference of temperature, I had prepared two tin vessels, whose cross sections are represented on a larger scale in Fig. 90. Each of them had in its middle part a space xx, open at the top, and otherwise completely enclosed to receive the parts ab, $a'b'$. In one of them, marked A, water was kept constantly boiling; this had at y a hole that could be closed with a cork, through which water could be introduced into the vessel, and on the other side a tube zz through which the steam could escape. In the other vessel B snow or cracked ice was placed. The parts *ab*, *a'b'* were sewed up in thin but closely woven silk, were then pushed down into the spaces *xx*, and these were then filled up to the height of about an inch with small shot, and then packed to their tops with powdered glass. In this way all points of contact of the bismuth and the copper were in the space filled with lead, which conducted heat well, and the layer of glass protected this region from rapid changes of temperature from the surrounding air.

After this elaborate description of the apparatus I come to the experiments which I made with it. I had prepared 8 different conductors, which in the future I shall distinguish as 1, 2, 3, 4, 5, 6, 7, 8, which were respectively 2, 4, 6, 10, 18, 34, 66, 130, inches long, $\frac{7}{8}$ lines thick, cut from one specimen of so called plated copper wire, prepared in the way formerly described. After the water had been boiling for half an hour, these wires were introduced into the circuit one after the other. Between any two sets of experiments, which lasted from 3 to 4 hours, there was always a pause of one hour, while some fresh water already warmed was poured in, which soon began to boil, and then the conductors were again introduced into the circuit in series, but in the opposite order. I thus obtained the following results.

Time of the observations	Set of experiments	Conductors							
		1	2	3	4	5	6	7	8
Jan. 8	I	326¾	300¾	277¾	238¼	190¾	134½	83¼	48½
Jan. 11	II	311¼	287	267	230¼	183½	129¼	80	46
	III	307	284	263¾	226¼	181	128¾	79	44½
Jan. 15	IV	305¼	281½	259	224	178½	124¾	79	44½
	V	305	282	258¼	223½	178	124¾	78	44

It appears that the force fell off perceptibly from one day to another. Whether the reason for this is to be sought in a change in the surfaces of contact or perhaps in the fact that the 8th. and 11th. of January were very cold days, and the ice box stood by the window of a poorly heated and badly protected room, I do not dare to decide; I think I should add that from the 15th. on I observed no more such differences.

Special emphasis should be laid upon the fact that no trace of fluctuations can be detected, such as appeared when the hydro-

electric battery was in circuit. When the needle was brought
to rest, it remained in its place without further motion. I have
often watched it for half an hour after a set of observations was
concluded without perceiving the slightest change of position.
Indeed when the needle with conductor 1 was brought to equilibrium and was kept in the same position by a stop placed on one
side of it, and then the circuit was completed again with the same
conductor, which had been removed from the circuit for some time,
there was not the slightest motion toward the opposite side. This
justifies the conclusion that the fluctuations have their origin in
changes in the fluid, which are conditioned by the electric current
itself and rise and fall with it. It seems as if a separation of certain
constituents of the fluid was brought about by the moving electricity, which takes place in accordance with exactly the same laws
as those which have been determined for the action of electricity
at rest; an increase of the force results in an increased separation of
the constituents, a decrease of the force permits a partial reunion,
which becomes complete when the force vanishes. It is very
probable, and we shall later find support for this view, that this
separation of the fluid by the current produces a change not only
in the exciting force of the circuit, but also in the conductivity
of the fluid, and it is just this variability in the hydroelectric
circuit which makes the law of conduction in it so confused and
so hard to unravel. It appears clearly at once that when we are
trying to determine only the influence of the metals in the conduction of the electric current, the hydroelectric circuit is not
suited for the purpose, because it gives rise to so many irregularities; whereas the thermo-electric circuit is perfectly fitted for this
purpose. We shall now see what it gives.

The numbers already given can be represented very satisfactorily
by the equation

$$X = \frac{a}{b + x}$$

in which X is the strength of the magnetic action when the conductor is used whose length is x, and a and b are constants which
represent magnitudes depending on the exciting force and the
resistance of the rest of the circuit. If for example we set b equal
to $20\frac{1}{4}$ and a in the different series equal to 7285, 6965, 6885, 6800,
6800, we obtain by calculation the following results

Series	Conductors							
	1	2	3	4	5	6	7	8
I	328	300½	277½	240¾	190½	134½	84¼	48½
II	313	287¼	265⅓	230¼	182	128⅓	80¾	46⅓
III	309½	284	262⅓	228	180	127	79¾	45¾
IV	305½	280½	259	224¾	177¾	125¼	79	45
V	305½	280½	259	224¾	177¾	125¼	79	45

If we compare these numbers found by calculation with the former set found by experiment, it will appear that the differences are very small, and are of the order that one might expect in researches of this kind. I shall not delay over this point, but proceed to prove the correctness of the formula in extreme cases, a method which is most serviceable in establishing the general applicability of a law which has been deduced from a few instances.

To this end I made four conductors, a, b, c, d, in order 2, 4, 8, 16 inches long, from the brass wire 0.3 lines thick, which I had used in my previous researches with the hydroelectric circuit; these gave in the circuit the numbers 111½, 64¾, 37, 19¾, while the conductor 1 gave 305. From the above equation the lengths may be determined which correspond to these numbers. We find them to be 40¾, 84¾, 163½, 324, which numbers in general agreement show that an inch of the brass wire is equivalent to 20½ inches of the plated copper wire. After this preliminary work, I introduced into the circuit a conductor of the same brass wire, 23 feet long, which I designated as 5 in this set; it gave 1¼. And we actually get this number almost exactly if we use for x in the equation $23 \cdot 12 \cdot 20½ = 5658$. We see by this example that the equation fits with experiment very accurately nearly up to the extinction of the force by the resistance of the conductors.

Furthermore I kept one end of the copper-bismuth couple at the temperature of 0° by the use of ice, while the other end was exposed to the temperature of the room, which was shown to be steadily 7½°R. by a thermometer hanging near the apparatus during the observations. The conductors, brought into the circuit in the following order, 1, 2, 3, 4, 5, 6, 7, 8, 7, 6, 5, 4, 3, 2, 1, gave the numbers: 27, 25, 23⅓, 20, 15½, 10¾, 6½, 3⅔, 6½, 10¾, 15½, 20, 23½, 25¼, 27¾. If we set in our equation $b = 20¼$, and so determine a that $a/22¼ = 27⅜$, we obtain by calculation numbers which in no case differ from the above by more than half a division; from which it appears that the equation holds for any value of the exciting force. From this last investigation two additional important points are evident. First there is the noteworthy circumstance that the value of b remains unchanged, while the force is more than 10 times less, so that a appears to depend only on the exciting force, and b only on the unchanged part of the circuit. Secondly it seems to follow from these experiments that the force of the thermo-electric circuit is exactly proportional to the difference of temperature between its two ends.

I cannot avoid mentioning here, at the close of this research, an observation which in a more direct way confirms Davy's conclusion that the conductivity of metals is increased by lowering the temperature. I took a 4 inch brass conductor, and brought it into the circuit; it gave 159 divisions. When I heated it in the middle with an alcohol flame, the force gradually decreased by 20 or more divisions, and the action was the same if I moved the flame more toward one or the other end of the conductor; but when I placed on it a layer of snow the force increased by 2 divisions. The temperature of the room was $8\frac{1}{4}°$ Reaumur. This fact is not out of place here, because it may give rise to slight anomalies.

FARADAY

Michael Faraday was born at Newington, Surrey, on September 22, 1791 His father was a blacksmith and Faraday himself learned the trade of a bookbinder. His interest in scientific investigation led to his applying for a position under Sir Humphry Davy. In 1813 he was made assistant in the laboratory of the Royal Institution and for some years carried on investigations in chemistry and electricity under the direction of Davy. He accompanied Davy on an extensive tour on the Continent and met many of the distinguished scientific men of the time. In 1825 he was made director of the laboratory and in 1833 he was appointed Fullerian Professor of Chemistry, and to leave him more free for his investigations he was relieved of the obligation of giving a course of lectures. Faraday's whole life was devoted to experimental investigation, mostly in the field of electricity. In the later years of his life his memory failed him so that he was at last compelled to withdraw from active work. He died at Hampton Court on August 25, 1867.

On November 24, 1831, Faraday read to the Royal Society the first of his long series of "Experimental Researches in Electricity." These were published in the *Philosophical Transactions* between the years 1831 and 1854 and were subsequently collected in three volumes with the same title.

The paper of November 24, 1831, appeared in the *Philosophical Transactions* of 1832, p. 125, and may be found also in the *Experimental Researches*, Vol. I, p. 1. The extracts from it which follow contain the account of Faraday's discovery of electromagnetic induction and his statement of the law of the production of currents by this induction.

INDUCED CURRENTS

1. The power which electricity of tension possesses of causing an opposite electrical state in its vicinity has been expressed by the general term Induction; which, as it has been received into scientific language, may also, with propriety, be used in the same general sense to express the power which electrical currents may possess of inducing any particular state upon matter in their immediate neighbourhood, otherwise indifferent. It is with this meaning that I purpose using it in the present paper.

2. Certain effects of the induction of electrical currents have already been recognized and described: as those of magnetization; Ampere's experiments of bringing a copper disc near to a flat spiral; his repetition with electro-magnets of Arago's extraordinary experiments, and perhaps a few others. Still it appeared unlikely that these could be all the effects which induction by currents could produce; especially as, upon dispensing with iron, almost the whole of them disappear, whilst yet an infinity of bodies, exhibiting definite phenomena of induction with electricity of tension, still remain to be acted upon by the induction of electricity in motion.

3. Further: Whether Ampere's beautiful theory were adopted, or any other, or whatever reservation were mentally made, still it appeared very extraordinary, that as every electric current was accompanied by a corresponding intensity of magnetic action at right angles to the current, good conductors of electricity, when placed within the sphere of this action, should not have any current induced through them, or some sensible effect produced equivalent in force to such a current.

4. These considerations, with their consequence, the hope of obtaining electricity from ordinary magnetism, have stimulated me at various times to investigate experimentally the inductive effect of electric currents. I lately arrived at positive results, and not only had my hopes fulfilled, but obtained a key which appeared to me to open out a full explanation of Arago's magnetic phenomena, and also to discover a new state, which may probably

have great influence in some of the most important effects of electric currents.

5. These results I purpose describing, not as they were obtained, but in such a manner as to give the most concise view of the whole.

1. *Induction of Electric Currents.*

6. About twenty-six feet of copper wire one twentieth of an inch in diameter were wound round a cylinder of wood as a helix, the different spires of which were prevented from touching by a thin interposed twine. This helix was covered with calico, and then a second wire applied in the same manner. In this way twelve helices were superposed, each containing an average length of wire of twenty-seven feet, and all in the same direction. The first, third, fifth, seventh, ninth, and eleventh of these helices were connected at their extremities end to end, so as to form one helix; the others were connected in a similar manner; and thus two principal helices were produced, closely interposed, having the same direction, not touching anywhere, and each containing one hundred and fifty-five feet in length of wire.

7. One of these helices was connected with a galvanometer, the other with a voltaic battery of ten pairs of plates four inches square, with double coppers and well charged; yet not the slightest sensible deflection of the galvanometer needle could be observed.

8. A similar compound helix, consisting of six lengths of copper and six of soft iron wire, was constructed. The resulting iron helix contained two hundred and fourteen feet of wire, the resulting copper helix two hundred and eight feet; but whether the current from the trough was passed through the copper or the iron helix, no effect upon the other could be perceived at the galvanometer.

9. In these and many similar experiments no difference in action of any kind appeared between iron and other metals.

10. Two hundred and three feet of copper wire in one length were coiled round a large block of wood; other two hundred and three feet of similar wire were interposed as a spiral between the turns of the first coil, and metallic contact everywhere prevented by twine. One of these helices was connected with a galvanometer, and the other with a battery of one hundred pairs of plates four inches square, with double coppers, and well charged. When the contact was made, there was a sudden and very slight effect at the galvanometer, and there was also a similar slight effect when the contact with the battery was broken. But whilst

the voltaic current was continuing to pass through the one helix, no galvanometrical appearances nor any effect like induction upon the other helix could be perceived, although the active power of the battery was proved to be great, by its heating the whole of its own helix, and by the brilliancy of the discharge when made through charcoal.

11. Repetition of the experiments with a battery of one hundred and twenty pairs of plates produced no other effects; but it was ascertained, both at this and the former time, that the slight deflection of the needle occurring at the moment of completing the connexion, was always in one direction, and that the equally slight deflection produced when the contact was broken, was in the other direction; and also, that these effects occurred when the first helices were used (6, 8).

12. The results which I had by this time obtained with magnets led me to believe that the battery current through one wire, did, in reality, induce a similar current through the other wire, but that it continued for an instant only, and partook more of the nature of the electrical wave passed through from the shock of a common Leyden jar than of the current from a voltaic battery, and therefore might magnetise a steel needle, although it scarcely affected the galvanometer.

13. This expectation was confirmed: for on substituting a small hollow helix, formed round a glass tube, for the galvanometer, introducing a steel needle, making contact as before between the battery and the inducing wire (7, 10), and then removing the needle before the battery contact was broken, it was found magnetised.

14. When the battery contact was first made, then an unmagnetised needle introduced into the small indicating helix (13), and lastly the battery contact broken, the needle was found magnetised to an equal degree apparently as before; but the poles were of the contrary kind.

15. The same effects took place on using the large compound helices first described (6, 8).

16. When the unmagnetised needle was put into the indicating helix, before contact of the inducing wire with the battery, and remained there until the contact was broken, it exhibited little or no magnetism; the first effect having been nearly neutralised by the second (13, 14). The force of the induced current upon making contact was found always to exceed that of the induced current at breaking of contact; and if therefore the contact was

made and broken many times in succession, whilst the needle remained in the indicating helix, it at last came out not unmagnetised, but a needle magnetised as if the induced current upon making contact had acted alone on it. This effect may be due to the accumulation (as it is called) at the poles of the unconnected pile, rendering the current upon first making contact more powerful than what it is afterwards, at the moment of breaking contact.

17. If the circuit between the helix or wire under induction and the galvanometer or indicating spiral was not rendered complete *before* the connexion between the battery and the inducing wire was completed or broken, then no effects were perceived at the galvanometer. Thus, if the battery communications were first made, and then the wire under induction connected with the indicating helix, no magnetising power was there exhibited. But still retaining the latter communications, when those with the battery were broken, a magnet was formed in the helix, but of the second kind (14), i.e. with poles indicating a current in the same direction to that belonging to the battery current, or to that always induced by that current at its cessation.

18. In the preceding experiments the wires were placed near to each other, and the contact of the inducing one with the battery made when the inductive effect was required; but as the particular action might be supposed to be exerted only at the moments of making and breaking contact, the induction was produced another way. Several feet of copper wire were stretched in wide zigzag forms, representing the letter *W*, on one surface of a broad board; a second wire was stretched in precisely similar forms on a second board, so that when brought near the first, the wires should everywhere touch, except that a sheet of thick paper was interposed. One of these wires was connected with the galvanometer, and the other with a voltaic battery. The first wire was then moved towards the second, and as it approached, the needle was deflected. Being then removed, the needle was deflected in the opposite direction. By first making the wires approach and then recede, simultaneously with the vibrations of the needle, the latter soon became very extensive; but when the wires ceased to move from or towards each other, the galvanometer-needle soon came to its usual position.

19. As the wires approximated, the induced current was in the *contrary* direction to the inducing current. As the wires receded, the induced current was in the *same* direction as the inducing

current. When the wires remained stationary, there was no induced current (54).

20. When a small voltaic arrangement was introduced into the circuit between the galvanometer (10) and its helix or wire, so as to cause a permanent deflection of 30° or 40°, and then the battery of one hundred pairs of plates connected with the inducing wire, there was an instantaneous action as before (11); but the galvanometer-needle immediately resumed and retained its place unaltered, notwithstanding the continued contact of the inducing wire with the trough: such was the case in whichever way the contacts were made (33).

21. Hence it would appear that collateral currents, either in the same or in opposite directions, exert no permanent inducing power on each other, affecting their quantity or tension.

22. I could obtain no evidence by the tongue, by spark, or by heating fine wire or charcoal, of the electricity passing through the wire under induction; neither could I obtain any chemical effects, though the contacts with metallic and other solutions were made and broken alternately with those of the battery, so that the second effect of induction should not oppose or neutralize the first (13. 16).

23. This deficiency of effect is not because the induced current of electricity cannot pass fluids, but probably because of its brief duration and feeble intensity; for on introducing two large copper plates into the circuit on the induced side (20), the plates being immersed in brine, but prevented from touching each other by an interposed cloth, the effect at the indicating galvanometer or helix occurred as before. The induced electricity could also pass through a voltaic trough (20). When, however, the quantity of interposed fluid was reduced to a drop, the galvanometer gave no indication.

24. Attempts to obtain similar effects by the use of wires conveying ordinary electricity were doubtful in the results. A compound helix similar to that already described, containing eight elementary helices (6), was used. Four of the helices had their similar ends bound together by wire, and the two general terminations thus produced connected with the small magnetising helix containing an unmagnetised needle (13). The other four helices were similarly arranged, but their ends connected with a Leyden jar. On passing the discharge, the needle was found to be a magnet; but it appeared probable that a part of the electricity

of the jar had passed off to the small helix, and so magnetised the needle. There was indeed no reason to expect that the electricity of a jar possessing as it does great tension, would not diffuse itself through all the metallic matter interposed between the coatings.

25. Still it does not follow that the discharge of ordinary electricity through a wire does not produce analogous phenomena to those arising from voltaic electricity; but as it appears impossible to separate the effects produced at the moment when the discharge begins to pass, from the equal and contrary effects produced when it ceases to pass (16), inasmuch as with ordinary electricity these periods are simultaneous, so there can be scarcely any hope that in this form of the experiment they can be perceived.

26. Hence it is evident that currents of voltaic electricity present phenomena of induction somewhat analogous to those produced by electricity of tension, although, as will be seen hereafter, many differences exist between them. The result is the production of other currents (but which are only momentary), parallel, or tending to parallelism, with the inducing current. By reference to the poles of the needle formed in the indicating helix (13, 14) and to the deflections of the galvanometer-needle (11), it was found in all cases that the induced current, produced by the first action of the inducing current, was in the contrary direction to the latter, but that the current produced by the cessation of the inducing current was in the same direction (19). For the purpose of avoiding periphrasis, I propose to call this action of the current from the voltaic battery *volta-electric induction*. The properties of the second wire, after induction has developed the first current, and whilst the electricity from the battery continues to flow through its inducing neighbor (10. 18), constitute a peculiar electric condition, the consideration of which will be resumed hereafter (60). All these results have been obtained with a voltaic apparatus consisting of a single pair of plates.

2. *Evolution of Electricity from Magnetism.*

27. A welded ring was made of soft round bar-iron, the metal being seven eighths of an inch in thickness, and the ring six inches in external diameter. Three helices were put round one part of this ring, each containing about twenty-four feet of copper wire one twentieth of an inch thick; they were insulated from the iron and each other, and superposed in the manner before described (6), occupying about nine inches in length upon the ring. They could

be used separately or conjointly; the group may be distinguished by the letter A (Fig. 91, 1). On the other part of the ring about sixty feet of similar copper wire in two pieces were applied in the same manner, forming a helix B, which had the same common direction with the helices of A, but being separated from it at each extremity by about half an inch of the uncovered iron.

28. The helix B was connected by copper wires with a galvanometer three feet from the ring. The helices of A were connected end to end so as to form one common helix, the extremities of which were connected with a battery of ten pairs of plates four inches square. The galvanometer was immediately affected, and to a degree far beyond what has been described when with a battery of tenfold power helices *without iron* were used (10); but though the contact was continued, the effect was not permanent, for the needle soon came to rest in its natural position, as if quite indifferent to the attached electro-magnetic

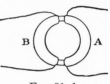

Fig. 91, 1.

arrangement. Upon breaking the contact with the battery, the needle was again powerfully deflected, but in the contrary direction to that induced in the first instance.

29. Upon arranging the apparatus so that B should be out of use, the galvanometer be connected with one of the three wires of A (27), and the other two made into a helix through which the current from the trough (28) was passed, similar but rather more powerful effects were produced.

30. When the battery contact was made in one direction, the galvanometer-needle was deflected on the one side; if made in the other direction, the deflection was on the other side. The deflection on breaking the battery contact was always the reverse of that produced by completing it. The deflection on making a battery contact always indicated an induced current in the opposite direction to that from the battery; but on breaking the contact the deflection indicated an induced current in the same direction as that of the battery. No making or breaking of the contact at B side, or in any part of the galvanometer circuit, produced any effect at the galvanometer. No continuance of the battery current caused any deflection of the galvanometer-needle. As the above results are common to all these experiments, and to similar ones with ordinary magnets to be hereafter detailed, they need not be again particularly described.

31. Upon using the power of one hundred pairs of plates (10), with this ring, the impulse at the galvanometer, when contact was completed or broken, was so great as to make the needle spin round rapidly four or five times, before the air and terrestrial magnetism could reduce its motion to mere oscillation.

32. By using charcoal at the ends of the *B* helix, a minute *spark* could be perceived when the contact of the battery with *A* was completed. This spark could not be due to any diversion of a part of the current of the battery through the iron to the helix *B*; for when the battery contact was continued, the galvanometer still resumed its perfectly indifferent state (28). The spark was rarely seen on breaking contact. A small platina wire could not be ignited by this induced current; but there seems every reason to believe that the effect would be obtained by using a stronger original current or a more powerful arrangement of helices.

33. A feeble voltaic current was sent through the helix *B* and the galvanometer, so as to deflect the needle of the latter 30° or 40°, and then the battery of one hundred pairs of plates connected with *A*; but after the first effect was over, the galvanometer-needle resumed exactly the position due to the feeble current transmitted by its own wire. This took place in whichever way the battery contacts were made, and shows that here again (20) no permanent influence of the currents upon each other, as to their quantity and tension, exists.

34. Another arrangement was then employed connecting the former experiments on volta-electric induction (6–26) with the present. A combination of helices like that already described (6) was constructed upon a hollow cylinder of pasteboard: there were eight lengths of copper wire, containing altogether 220 feet; four of these helices were connected end to end, and then with the galvanometer (7); the other intervening four were also connected end to end, and the battery of one hundred pairs discharged through them. In this form the effect on the galvanometer was hardly sensible (11), though magnets could be made by the induced current (13). But when a soft iron cylinder seven eighths of an inch thick, and twelve inches long, was introduced into the pasteboard tube, surrounded by the helices, then the induced current affected the galvanometer powerfully, and with all the phenomena just described (30). It possessed also the power of making magnets with more energy, apparently, than when no iron cylinder was present.

35. When the iron cylinder was replaced by an equal cylinder of copper, no effect beyond that of the helices alone was produced. The iron cylinder arrangement was not so powerful as the ring arrangement already described (27).

36. Similar effects were then produced by *ordinary magnets:* thus the hollow helix just described (34) had all its elementary helices connected with the galvanometer by two copper wires, each five feet in length; the soft iron cylinder was introduced into its

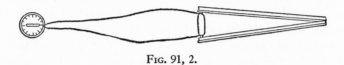

Fig. 91, 2.

axis; a couple of bar magnets, each twenty-four inches long, were arranged with their opposite poles at one end in contact, so as to resemble a horse-shoe magnet, and then contact made between the other poles and the ends of the iron cylinder, so as to convert it for the time into a magnet (Fig. 91, 2): by breaking the magnetic contacts, or reversing them, the magnetism of the iron cylinder could be destroyed or reversed at pleasure.

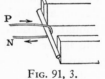

Fig. 91, 3.

37. Upon making magnetic contact, the needle was deflected; continuing the contact, the needle became indifferent, and resumed its first position; on breaking the contact, it was again deflected, but in the opposite direction to the first effect, and then it again became indifferent. When the magnetic contacts were reversed the deflections were reversed.

38. When the magnetic contact was made, the deflection was such as to indicate an induced current of electricity in the opposite direction to that fitted to form a magnet, having the same polarity as that really produced by contact with the bar magnets. Thus when the marked and unmarked poles were placed as in (Fig. 91, 3), the current in the helix was in the direction represented, *P* being supposed to be the end of the wire going to the positive pole of the battery, or that end towards which the zinc plates face, and *N* the negative wire. Such a current would have converted the cylinder into a magnet of the opposite kind to that formed by contact with the poles *A* and *B*; and such a current moves in the opposite

direction to the currents which in M. Ampere's beautiful theory are considered as constituting a magnet in the position figured.

39. But as it might be supposed that in all the preceding experiments of this section, it was by some peculiar effect taking place during the formation of the magnet, and not by its mere virtual approximation, that the momentary induced current was excited, the following experiment was made. All the similar ends of the compound hollow helix (34) were bound together by copper wire, forming two general terminations, and these were connected with the galvanometer. The soft iron cylinder (34) was removed, and a cylindrical magnet, three quarters of an inch in diameter and eight inches and a half in length, used instead. One end of this magnet was introduced into the axis of the helix (Fig. 91, 4), and then, the galvanometer-needle being stationary, the magnet was suddenly thrust in; immediately the needle was deflected in the same direction as if the magnet had been formed

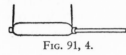

FIG. 91, 4.

by either of the two preceding processes (34. 36). Being left in, the needle resumed its first position, and then the magnet being withdrawn the needle was deflected in the opposite direction. These effects were not great; but by introducing and withdrawing the magnet, so that the impulse each time should be added to those previously communicated to the needle, the latter could be made to vibrate through an arc of 180° or more.

40. In this experiment the magnet must not be passed entirely through the helix, for then a second action occurs. When the magnet is introduced, the needle at the galvanometer is deflected in a certain direction; but being in, whether it be pushed quite through or withdrawn, the needle is deflected in a direction the reverse of that previously produced. When the magnet is passed in and through at one continuous motion, the needle moves one way, is then suddenly stopped, and finally moves the other way.

41. If such a hollow helix as that described (34) be laid east and west (or in any other constant position), and a magnet be retained east and west, its marked pole always being one way; then whichever end of the helix the magnet goes in at, and consequently whichever pole of the magnet enters first, still the needle is deflected the same way: on the other hand, whichever direction is followed in withdrawing the magnet, the deflection is constant, but contrary to that due to its entrance.

42. These effects are simple consequences of the *law* hereafter to be described (114).

. . . .

114. The relation which holds between the magnetic pole, the moving wire or metal, and the direction of the current evolved, i.e. *the law* which governs the evolution of electricity by magneto-electric induction, is very simple, although rather difficult to express. If in (Fig. 91, 5) *PN* represent a horizontal wire passing by a marked magnetic pole, so that the direction of its motion shall coincide with the curved line proceeding from below upwards; or if its motion parallel to itself be in a line tangential to the curved line, but in the general direction of the arrows; or if it pass the pole in other directions, but so as to cut the magnetic curves in the same general direction, or on the same side as they would be cut by the wire if moving along the dotted curved line;—then the current of electricity in the wire is from *P* to *N*. If it be carried in the reverse directions, the electric current will be from *N* to *P*. Or if the wire be in the vertical position, figured *P'N'*, and it be carried in similar directions, coinciding with the dotted horizontal curve so far as to

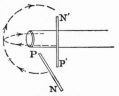

Fig. 91, 5.

cut the magnetic curves on the same side with it, the current will be from *P'* to *N'*. If the wire be considered a tangent to the curved surface of the cylindrical magnet, and it be carried round that surface into any other position, or if the magnet itself be revolved on its axis, so as to bring any part opposite to the tangential wire,—still, if afterwards the wire be moved in the directions indicated, the current of electricity will be from *P* to *N*; or if it be moved in the opposite direction, from *N* to *P*; so that as regards the motions of the wire past the pole, they may be reduced to two, directly opposite to each other, one of which produces a current from *P* to *N*, and the other from *N* to *P*.

115. The same holds true of the unmarked pole of the magnet, except that if it be substituted for the one in the figure, then, as the wires are moved in the direction of the arrows, the current of electricity would be from *N* to *P*, and when they move in the reverse direction, from *P* to *N*.

116. Hence the current of electricity which is excited in metal when moving in the neighbourhood of a magnet, depends for

its direction altogether upon the relation of the metal to the resultant of magnetic action, or to the magnetic curves, and may be expressed in a popular way thus: Let *AB* (Fig. 91, 6) represent a cylinder magnet, *A* being the marked pole, and *B* the unmarked pole; let *PN* be a silver knife-blade resting across the magnet with its edge upward, and with its marked or notched side towards the pole *A*; then in whatever direction or position this knife be moved edge foremost, either about the marked or the unmarked pole, the current of electricity produced will be from *P* to *N*, provided the intersected curves proceeding from *A* abut upon the notched surface of the knife, and those from *B* upon the unnotched side.   Or if the knife be moved with its back foremost, the current will be from *N* to *P* in every possible position and direction, pro-

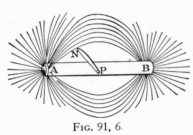

vided the intersected curves abut on the same surfaces as before. A little model is easily construct-ed, by using a cylinder of wood for a magnet, a flat piece for the blade, and a piece of thread connecting one end of the cylin-der with the other, and passing through a hole in the blade, for

FIG. 91, 6

the magnetic curves: this readily gives the result of any possible direction.

117. When the wire under induction is passing by an electro-magnetic pole, as for instance one end of a copper helix traversed by the electric current (34), the direction of the current in the approaching wire is the same with that of the current in the parts or sides of the spirals nearest to it, and in the receding wire the reverse of that in the parts nearest to it.

118. All these results show that the power of inducing electric currents is circumferentially exerted by a magnetic resultant or axis of power, just as circumferential magnetism is dependent upon and is exhibited by an electric current.

119. The experiments described combine to prove that when a piece of metal (and the same may be true of all conducting matter (213)) is passed either before a single pole, or between the opposite poles of a magnet, or near electro-magnetic poles, whether fer-ruginous or not, electrical currents are produced across the metal transverse to the direction of motion; and which therefore, in Arago's experiments, will approximate towards the direction of

radii. If a single wire be moved like the spoke of a wheel near a magnetic pole, a current of electricity is determined through it from one end towards the other. If a wheel be imagined, constructed of a great number of these radii, and this revolved near the pole, in the manner of the copper disc (85), each radius will have a current produced in it as it passes by the pole. If the radii be supposed to be in contact laterally, a copper disc results, in which the directions of the currents will be generally the same, being modified only by the coaction which can take place between the particles, now that they are in metallic contact.

The extract which follows contains Faraday's discovery of self-induction. It is taken from the *Philosophical Transactions* of 1835, p. 41, and may be found also in the *Experimental Researches*, Vol. I, p. 322. In the discovery of self-induction Faraday had been anticipated by Joseph Henry (p. 513).

SELF-INDUCTION

. . . .

1048. The following investigations relate to a very remarkable inductive action of electric currents, or of the different parts of the same current (74), and indicate an immediate connexion between such inductive action and the direct transmission of electricity through conducting bodies, or even that exhibited in the form of a spark.

1049. The inquiry arose out of a fact communicated to me by Mr. Jenkin, which is as follows. If an ordinary wire of short length be used as the medium of communication between the two plates of an electromotor consisting of a single pair of metals, no management will enable the experimenter to obtain an electric shock from this wire; but if the wire which surrounds an electro-magnet be used, a shock is felt each time the contact with the electromotor is broken, provided the ends of the wire be grasped one in each hand.

1050. Another effect is observed at the same time, which has long been known to philosophers, namely, that a bright electric spark occurs at the place of disjunction.

1051. A brief account of these results, with some of a corresponding character which I had observed in using long wires, was published in the Philosophical Magazine for 1834; and I added to them some observations on their nature. Further investigations led me to perceive the inaccuracy of my first notions and ended in identifying these effects with the phenomena of induction which

I had been fortunate enough to develop in the First Series of these Experimental Researches (1.–59.). Notwithstanding this identity, the extension and the peculiarity of the views respecting electric currents which the results supply, lead me to believe that they will be found worthy of the attention of the Royal Society.

1052. The *electromotor* used consisted of a cylinder of zinc introduced between the two parts of a double cylinder of copper, and preserved from metallic contact in the usual way by corks. The zinc cylinder was eight inches high and four inches in diameter. Both it and the copper cylinder were supplied with stiff wires, surmounted by cups containing mercury; and it was at these cups that the contacts of wires, helices, or electro-magnets, used to complete the circuit, were made or broken. These cups I will call G and E throughout the rest of this paper (1079).

1053. Certain *helices* were constructed, some of which it will be necessary to describe. A pasteboard tube had four copper wires, one twenty-fourth of an inch in thickness, wound round it, each forming a helix in the same direction from end to end: the convolutions of each wire were separated by string, and the superposed helices prevented from touching by intervening calico. The lengths of the wires forming the helices were 48, 49.5, 48, and 45 feet. The first and third wires were united together so as to form one consistent helix of 96 feet in length; and the second and fourth wires were similarly united to form a second helix, closely interwoven with the first, and 94.5 feet in length. These helices may be distinguished by the numbers i and ii. They were carefully examined by a powerful current of electricity and a galvanometer, and found to have no communication with each other.

1054. Another helix was constructed upon a similar pasteboard tube, two lengths of the same copper wire being used, each forty-six feet long. These were united into one consistent helix of ninety-two feet, which therefore was nearly equal in value to either of the former helices, but was not in close inductive association with them. It may be distinguished by the number iii.

1055. A fourth helix was constructed of very thick copper wire, being one fifth of an inch in diameter; the length of wire used was seventy-nine feet, independent of the straight terminal portions.

1056. The principal *electro-magnet* employed consisted of a cylindrical bar of soft iron twenty-five inches long, and one inch and three quarters in diameter, bent into a ring, so that the ends nearly touched, and surrounded by three coils of thick copper wire, the similar ends of which were fastened together; each of these terminations was soldered to a copper rod, serving as a conducting continuation of the wire. Hence any electric current sent through the rods was divided in the helices surrounding the ring, into three parts, all of which, however, moved in the same direction. The three wires may therefore be considered as representing one wire, of thrice the thickness of the wire really used.

1057. Other electro-magnets could be made at pleasure by introducing a soft iron rod into any of the helices described (1053, &c.).

1058. The *galvanometer* which I had occasion to use was rough in its construction, having but one magnetic needle, and not at all delicate in its indications.

1059. The effects to be considered *depend on the conductor* employed to complete the communication between the zinc and copper plates of the electromotor; and I shall have to consider this conductor under four different forms: as the helix of an electro-magnet (1056); as an ordinary helix (1053, &c.); as a *long* extended wire, having its course such that the parts can exert little or no mutual influence; and as a *short* wire. In all cases the conductor was of copper.

1060. The peculiar effects are best shown by the *electro-magnet* (1056). When it was used to complete the communication at the electromotor, there was no sensible spark on *making* contact, but on *breaking* contact there was a very large and bright spark, with considerable combustion of the mercury. Then, again, with respect to the shock: if the hands were moistened in salt and water, and good contact between them and the wires retained, no shock could be felt upon *making* contact at the electromotor, but a powerful one on *breaking* contact.

1061. When the *helix* i or iii (1053, &c.) was used as the connecting conductor, there was also a good spark on breaking contact, but none (sensibly) on making contact. On trying to obtain the shock from these helices, I could not succeed at first. By joining the similar ends of i and ii so as to make the two helices equivalent to one helix, having wire of double thickness, I could just obtain the sensation. Using the helix of thick wire (1055) the shock was distinctly obtained. On placing the tongue between two plates of silver connected by wires with the parts which the hands had heretofore touched (1064), there was a powerful shock on *breaking* contact, but none on *making* contact.

1062. The power of producing these phenomena exists therefore in the simple helix, as in the electro-magnet, although by no means in the same high degree.

1063. On putting a bar of soft iron into the helix, it became an electro-magnet (1057), and its power was instantly and greatly raised. On putting a bar of copper into the helix, no change was produced, the action being that of the helix alone. The two helices i and ii, made into one helix of twofold length of wire, produced a greater effect than either i or ii alone.

1064. On descending from the helix to the mere *long wire*, the following effects were obtained. A copper wire, 0.18 of an inch in diameter, and 132 feet in length, was laid out upon the floor of the laboratory, and used as the connecting conductor (1059); it gave no sensible spark on making contact, but produced a bright one on breaking contact, yet not so bright as that from the helix (1061). On endeavouring to obtain the electric shock at the moment contact was broken, I could not succeed so as to make it pass through the hands; but by using two silver plates fastened by small wires to the extremity of the principal wire used, and introducing the tongue between those plates, I succeeded in obtaining powerful shocks upon the tongue and gums, and could easily convulse a flounder, an eel, or a frog. None of these effects could be obtained directly from the electromotor, i.e. when the tongue, frog, or fish was in a similar, and therefore comparative manner, interposed in the course of the communication between the zinc and copper plates, separated everywhere else by the acid used to excite the combination, or by air. The bright spark and the shock, produced only on breaking contact, are therefore effects of the same kind as those produced in a higher degree by the helix, and in a still higher degree by the electro-magnet.

1065. In order to compare an extended wire with a helix, the helix i, containing ninety-six feet, and ninety-six feet of the same-sized wire lying on the floor of the laboratory, were used alternately as conductors: the former gave a much brighter spark at the moment of disjunction than the latter. Again, twenty-eight feet of copper wire were made up into a helix, and being used gave a good spark on disjunction at the electromotor; being then suddenly pulled out and again employed, it gave a much smaller spark than before, although nothing but its spiral arrangement had been changed.

1066. As the superiority of a helix over a wire is important to the philosophy of the effect, I took particular pains to ascertain the fact with certainty. A wire of copper sixty-seven feet long was bent in the middle so as to form a double termination which could be communicated with the electromotor; one of the halves of this wire was made into a helix and the other remained in its extended condition. When these were used alternately as the connecting wire, the helix half gave by much the strongest spark. It even gave a stronger spark than when it and the extended wire were used conjointly as a double conductor.

1067. When a *short wire* is used, *all* these effects disappear. If it be only two or three inches long, a spark can scarcely be perceived on breaking the junction. If it be ten or twelve inches long and moderately thick, a small spark may be more easily obtained. As the length is increased, the spark becomes proportionately brighter, until from extreme length the resistance offered by the metal as a conductor begins to interfere with the principal result.

The following letter on "Static Electrical Inductive Action" appeared in the *Philosophical Magazine*, Vol. 22, p. 200, 1843, and also in the *Experimental Researches*, Vol. II, p. 279. It contains the account of Faraday's well-known ice pail experiment.

STATIC ELECTRIC INDUCTION

To R. Phillips, Esq., F.R.S.
Dear Phillips,
Perhaps you may think the following experiments worth notice; their value consists in their power to give a very precise and decided idea to the mind respecting certain principles of inductive electrical action, which I find are by many accepted with a degree of doubt or obscurity that takes away much of their importance: they are the expression and proof of certain parts of my view of induction. Let A in the diagram (Fig. 92) represent an insulated pewter ice-pail ten and a half inches high and seven inches diameter, connected by a wire with a delicate gold-leaf electrometer E, and let C be a round brass ball insulated by a dry thread of white silk, three or four feet in length, so as to remove the influence of the hand holding it from the ice-pail below. Let A be perfectly discharged, then let C be charged at a distance by a machine or Leyden jar, and introduced into A as in the figure. If C be posi-

tive, E also will diverge positively; if C be taken away, E will collapse perfectly, the apparatus being in good order.   As C enters the vessel A the divergence of E will increase until C is about three inches below the edge of the vessel, and will remain quite steady and unchanged for any greater depression.   This shows that at that distance the inductive action of C is entirely exerted upon the interior of A, and not in any degree directly upon external objects.   If C be made to touch the bottom of A, all its charge is communicated to A; there is no longer any inductive action between C and A, and C, upon being withdrawn and examined, is found perfectly discharged.

These are all well-known and recognised actions, but being a little varied, the following conclusions may be drawn from them.

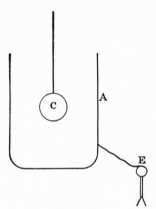

FIG. 92.

If C be merely suspended in A it acts upon it by induction, evolving electricity of its own kind on the outside of A; but if C touch A its electricity is then communicated to it, and the electricity that is afterwards upon the outside of A may be considered as that which was originally upon the carrier C.   As this change, however, produces no effect upon the leaves of the electrometer, it proves that the electricity induced by C and the electricity in C are accurately equal in amount and power.

Again, if C charged be held equidistant from the bottom and sides of A at one moment, and at another be held as close to the bottom as possible without discharging to A, still the divergence remains absolutely unchanged, showing that whether C acts at a considerable distance or at the very smallest distance, the amount of its force is the same.   So also if it be held excentric and near to the side of the ice-pail in one place, so as to make the inductive action take place in lines expressing almost every degree of force in different directions, still the sum of their forces is the same constant quantity as that obtained before; for the leaves alter not.   Nothing like expansion or coercion of the electric force appears under these varying circumstances.

I can now describe experiments with many concentric metallic vessels arranged as in the diagram (Fig. 93), where four ice-pails are represented insulated from each other by plates of shell-lac on which they respectively stand. With this system the charged carrier *C* acts precisely as with the single vessel, so that the intervention of many conduction plates causes no difference in the amount of inductive effect. If *C* touch the inside of vessel 4, still the leaves are unchanged. If 4 be taken out by a silked thread, the leaves perfectly collapse; if it be introduced again, they open out to the same degree as before. If 4 and 3 be connected by a wire let down between them by a silk thread, the leaves remain the same, and so they still remain if 3 and 2 be connected by a similar wire; yet all the electricity originally on the carrier and acting at a considerable distance, is now on the outside of 2, and acting through only a small non-conducting space. If at last it be communicated to the outside of 1, still the leaves remain unchanged.

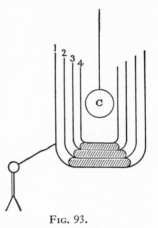

Fig. 93.

Again, consider the charged carrier *C* in the centre of the system, the divergence of the electrometer measures its inductive influence; this divergence remains the same whether 1 be there alone, or whether all four vessels be there; whether these vessels be separate as to insulation, or whether 2, 3 and 4 be connected so as to represent a very thick metallic vessel, or whether all four vessels be connected.

Again, if in place of the metallic vessels 2, 3, 4, a thick vessel of shell-lac or of sulphur be introduced, or if any other variation in the character of the substance within the vessel 1 be made, still not the slightest change is by that caused upon the divergence of the leaves.

If in place of one carrier many carriers in different positions are within the inner vessel, there is no interference of one with the other; they act with the same amount of force outwardly as if the electricity were spread uniformly over one carrier, however much the distribution on each carrier may be disturbed by its neighbours. If the charge of one carrier be by contact given to vessel 4 and distributed over it, still the others act through and across it with the same final amount of force; and no state of charge given to any of the vessels 1, 2, 3, or 4, prevents a charged carrier introduced within 4 acting with precisely the same amount of force as if they were uncharged. If pieces of shell-lac, slung by white thread and excited, be introduced into the vessel, they act exactly as the metallic carriers, except that their charge cannot be communicated by contact to the metallic vessels.

Thus a certain amount of electricity acting within the centre of the vessel *A* exerts exactly the same power externally, whether it act by induction through the space between it and *A*, or whether it be transferred by conduction to *A*, so as absolutely to destroy the previous induction within. Also, as to the inductive action, whether the space between *C* and *A* be filled with air, or with shell-lac or sulphur, having above twice the specific inductive capacity of air; or contain many concentric shells of conducting matter; or be nine-tenths filled with conducting matter, or be metal on one side and shell-lac on the other; or whatever other means be taken to vary the forces either by variation of distance or substance, or actual charge of the matter in this space, still the amount of action is precisely the same.

. . . .

I am, my dear Phillips, ever yours,
M. Faraday

The extract which follows is taken from a paper by Faraday on "Electrochemical Decomposition." It contains the nomenclature which Faraday introduced and an account of his investigation of the laws of electrolysis. The paper was published in the *Philosophical Transactions* of 1834, p. 77, and can be found in the *Experimental Researches* Vol. I, p. 195.

Laws of Electrolysis

Preliminary

661. The theory which I believe to be a true expression of the facts of electrochemical decomposition, and which I have therefore detailed in a former series of these Researches, is so much at variance with those previously advanced that I find the greatest difficulty in stating results, as I think, correctly, whilst limited to the use of terms which are current with a certain accepted meaning. Of this kind is the term *pole*, with its prefixes of positive and negative, and the attached ideas of attraction and repulsion. The general phraseology is that the positive pole *attracts* oxygen,

acids, etc., or more cautiously, that it *determines* their evolution upon its surface; and that the negative pole acts in an equal manner upon hydrogen, combustibles, metals, and bases. According to my view, the determining force is *not* at the poles, but *within* the body under decomposition; and the oxygen and acids are rendered at the *negative* extremity of that body, whilst hydrogen, metals, etc., are evolved at the *positive* extremity.

662. To avoid, therefore, confusion and circumlocution, and for the sake of greater precision of expression than I can otherwise obtain, I have deliberately considered the subject with two friends, and with their assistance and concurrence in framing them I purpose henceforward using certain other terms, which I will now define. The *poles*, as they are usually called, are only the doors or ways by which the electric current passes into and out of the decomposing body; and they of course, when in contact with that body, are the limits of its extent in the direction of the current. The term has been generally applied to the metal surfaces in contact with the decomposing substance; but whether philosophers generally would also apply it to the surfaces of air and water, against which I have effected electro-chemical decomposition, is subject to doubt. In place of the term pole, I propose using that of *electrode*, and I mean thereby that substance, or rather surface, whether of air, water, metal, or any other body, which bounds the extent of the decomposing matter in the direction of the electric current.

663. The surfaces at which, according to common phraseology. the electric current enters and leaves a decomposing body are most important places of action, and require to be distinguished apart from the poles, with which they are mostly, and the electrodes, with which they are always in contact. Wishing for a natural standard of electric direction to which I might refer these, expressive of their difference and at the same time free from all theory, I have thought it might be found in the earth. If the magnetism of the earth be due to electric currents passing round it, the latter must be in constant direction, which, according to the present usage of speech, would be from east to west, or, which will strengthen this to help the memory, that in which the sun appears to move. If in any case of electro-decomposition we consider the decomposing body as placed so that the current passing through it shall be in the same direction, and parallel to that supposed to exist in the earth, then the surfaces at which the electricity is passing into

and out of the substance would have an invariable reference, and exhibit constantly the same relations of powers.   Upon this notion we purpose calling that towards the east the *anode,* and that towards the west the *cathode;* and whatever changes may take place in our views of the nature of electricity and electrical action, as they must affect the *natural standard* referred to, in the same direction, and to an equal amount with any decomposing substances to which these terms may at any time be applied, there seems no reason to expect that they will lead to confusion or tend in any way to support false views.   The *anode* is therefore that surface at which the electric current according to our present expression, enters; it is the *negative* extremity of the decomposing body; is where oxygen, chlorine, acids, etc., are evolved; and is against or opposite the positive electrode.   The *cathode* is that surface at which the current leaves the decomposing body, and is its *positive* extremity; the combustible bodies, metals, alkalies, and bases, are evolved there, and it is in contact with the negative electrode.

664.   I shall have occasion in these Researches, also, to class bodies together according to certain relations derived from their electrical actions; and wishing to express those relations without at the same time involving the expression of any hypothetical views, I intend using the following names and terms.   Many bodies are decomposed directly by the electric current, their elements being set free; these I propose to call *electrolytes.* Water, therefore, is an electrolyte.   The bodies which, like nitric or sulphuric acids, are decomposed in a secondary manner are not included under this term.   Then for *electrochemically decomposed,* I shall often use the term *electrolyzed,* derived in the same way, and implying that the body spoken of is separated into its components under the influence of electricity; it is analogous in its sense and sound to analyze, which is derived in a similar manner. The term *electrolytical* will be understood at once: muriatic acid is electrolytical, boracic acid is not.

665.   Finally, I require a term to express those bodies which can pass to the *electrodes,* or, as they are usually called, the poles. Substances are frequently spoken of as being *electro-negative* or *electro-positive,* according as they go under the supposed influence of a direct attraction to the positive or negative pole.   But these terms are much too significant for the use to which I should have to put them; for, though the meanings are perhaps right, they are

only hypothetical, and may be wrong; and then, through a very imperceptible, but still very dangerous, because continual, influence, they do great injury to science by contracting and limiting the habitual views of those engaged in pursuing it. I propose to distinguish such bodies by calling those *anions* which go to the *anode* of the decomposing body; and those passing to the *cathode, cations;* and when I have occasion to speak of these together, I shall call them *ions*. Thus, the chloride of lead is an electrolyte, and when electrolyzed evolves the two *ions*, chlorine and lead, the former being an *anion*, and the latter a *cation*.

666. These terms, being once well defined, will, I hope, in their use enable me to avoid much periphrasis and ambiguity of expression. I do not mean to press them into service more frequently than will be required, for I am fully aware that names are one thing and science another.

667. It will be well understood that I am giving no opinion respecting the nature of the electric current now, beyond what I have done on former occasions; and that though I speak of the current as proceeding from the parts which are positive to those which are negative, it is merely in accordance with the conventional, though in some degree tacit, agreement entered into by scientific men, that they may have a constant, certain, and definite means of referring to the direction of the forces of that current.

. . . .

729. Although not necessary for the practical use of the instrument I am describing, yet as connected with the important point of constant electrochemical action upon water, I now investigated the effects produced by an electric current passing through aqueous solutions of acids, salts, and compounds, exceedingly different from each other in their nature, and found them to yield astonishingly uniform results. But many of them which are connected with a secondary action will be more usefully described hereafter.

730. When solutions of caustic potassa or soda, or sulphate of magnesia, or sulphate of soda, were acted upon by the electric current, just as much oxygen and hydrogen was evolved from them as from the diluted sulphuric acid, with which they were compared. When a solution of ammonia, rendered a better conductor by sulphate of ammonia, or a solution of subcarbonate of potassa, was experimented with, the *hydrogen* evolved was in the same quantity as that set free from the diluted sulphuric acid with which

they were compared.   Hence *changes in the nature of the solution do not alter the constancy of electrolytic action upon water.*

731. I have already said, respecting large and small electrodes, that change of order caused no change in the general effect.   The same was the case with different solutions, or with different intensities; and however the circumstances of an experiment might be varied, the results came forth exceedingly consistent, and proved that the electrochemical action was still the same.

732. I consider the foregoing investigation as sufficient to prove the very extraordinary and important principle with respect to WATER, *that when subjected to the influence of the electric current, a quantity of it is decomposed exactly proportionate to the quantity of electricity which has passed,* notwithstanding the thousand variations in the conditions and circumstances under which it may at the time be placed; and further, that when the interference of certain secondary effects, together with the solution or recombination of the gas and the evolution of air, are guarded against, *the products of the decomposition may be collected with such accuracy as to afford a very excellent and valuable measurer of the electricity concerned in their evolution.*

.   .   .   .

821. All these facts combine into, I think, an irresistible mass of evidence, proving the truth of the important proposition which I at first laid down—namely, *that the chemical power of a current of electricity is in direct proportion to the absolute quantity of electricity which passes.*   They prove, too, that this is not merely true of one substance, as water, but generally with all electrolytic bodies; and, further, that the results obtained with any *one substance* do not merely agree amongst themselves, but also with those obtained from *other substances*, the whole combining together into *one series of definite electrochemical actions.*   I do not mean to say that no exceptions will appear; perhaps some may arise, especially amongst substances existing only by weak affinity; but I do not expect that any will seriously disturb the result announced.   If, in the well considered, well examined, and, I may surely say, well-ascertained doctrines of the definite nature of ordinary chemical affinity, such exceptions occur, as they do in abundance, yet, without being allowed to disturb our minds as to the general conclusion, they ought also to be allowed if they should present themselves at this, the opening of a new view of electrochemical action; not being

held up as obstructions to those who may be engaged in rendering that view more and more perfect, but laid aside for a while, in hopes that their perfect and consistent explanation will ultimately appear.

. . . .

825. A summary of certain points already ascertained respecting *electrolytes, ions,* and *electrochemical equivalents* may be given in the following general form of propositions, without, I hope, including any serious error.

826. I. A single *ion,* i.e., one not in combination with another, will have no tendency to pass to either of the electrodes, and will be perfectly indifferent to the passing current, unless it be itself a compound of more elementary *ions,* and so subject to actual decomposition. Upon this fact is founded much of the proof adduced in favor of the new theory of electrochemical decomposition, which I put forth in a former series of these Researches.

827. II. If one *ion* be combined in right proportions with another strongly opposed to it in its ordinary chemical relations, i.e., if an *anion* be combined with a *cation,* then both will travel, the one to the *anode,* the other to the *cathode,* of the decomposing body.

828. III. If, therefore, an *ion* pass towards one of the electrodes, another *ion* must also be passing simultaneously to the other electrode, although, from secondary action, it may not make its appearance.

829. IV. A body decomposable directly by the electric current, i.e., an *electrolyte,* must consist of two *ions,* and must also render them up during the act of decomposition.

830. V. There is but one *electrolyte* composed of the same two elementary *ions;* at least such appears to be the fact, dependent upon a law, that *only single electrochemical equivalents of elementary ions can go to the electrodes, and not multiples.*

831. VI. A body not decomposable when alone, as boracic acid, is not directly decomposable by the electric current when in combination. It may act as an *ion* going wholly to the *anode* or *cathode,* but does not yield up its elements, except occasionally by a secondary action. Perhaps it is superfluous for me to point out that this proposition has *no relation* to such cases as that of water, which, by the presence of other bodies, is rendered a better conductor of electricity, and *therefore* is more freely decomposed.

832. VII. The nature of the substance of which the electrode is formed, provided it be a conductor, causes no difference in the electro-decomposition, either in kind or degree; but it seriously influences, by secondary action, the state in which the *ions* finally appear. Advantage may be taken of this principle in combining and collecting such *ions* as, if evolved in their free state, would be unmanageable.

833. VIII. A substance which, being used as the electrode, can combine with the *ion* evolved against it, is also, I believe, an *ion*, and combines, in such cases, in the quantity represented by its *electrochemical equivalent*. All the experiments I have made agree with this view; and it seems to me, at present, to result as a necessary consequence. Whether, in the secondary actions that take place, where the *ion* acts not upon the matter of the electrode, but on that which is around it in the liquid, the same consequence follows, will require more extended investigation to determine.

834. IX. Compound *ions* are not necessarily composed of electrochemical equivalents of simple *ions*. For instance, sulphuric acid, boracic acid, phosphoric acid, are *ions*, but not *electrolytes*, i.e., not composed of electrochemical equivalents of *ions*.

835. X. Electrochemical equivalents are always consistent, i.e., the same number which represents the equivalent of a substance, *A*, when it is separating from a substance, *B*, will also represent *A* when separating from a third substance *C*. Thus, 8 is the electrochemical equivalent of oxygen, whether separating from hydrogen, or tin, or lead; and 103.5 is the electrochemical equivalent of lead, whether separating from oxygen, or chlorine, or iodine.

836. XI. Electrochemical equivalents coincide, and are the same, with ordinary chemical equivalents.

837. By means of experiment and the preceding propositions, a knowledge of *ions* and their electrochemical equivalents may be obtained in various ways.

The following short extract from the *Philosophical Transactions* of 1838, p. 1, contains an account of Faraday's experiment to prove that the electricity of a conductor resides on its surface. This conclusion had already been drawn by Franklin and by Coulomb but Faraday felt that their experiments were not conclusive. The large scale upon which some of the experiments were tried made them very striking.

The paper also appears in the *Experimental Researches*, Vol. I, p. 365.

THE CHARGE RESIDES ON SURFACE

1169. Can matter, either conducting or non-conducting, be charged with one electric force independently of the other, in any degree, either in a sensible or latent state?

1170. The beautiful experiments of Coulomb upon the equality of action of *conductors*, whatever their substance, and the residence of *all* the electricity upon their surfaces, are sufficient, if properly viewed, to prove that *conductors cannot be bodily charged;* and as yet no means of communicating electricity to a conductor so as to place its particles in relation to one electricity, and not at the same time to the other in exactly equal amount, has been discovered.

1171. With regard to electrics or non-conductors, the conclusion does not at first seem so clear. They may easily be electrified bodily, either by communication or excitement; but being so charged, every case in succession, when examined, came out to be a case of induction, and not of absolute charge. Thus, glass within conductors could easily have parts not in contact with the conductor brought into an excited state; but it was always found that a portion of the inner surface of the conductor was in an opposite and equivalent state, or that another part of the glass itself was in an equally opposite state, an *inductive* charge and not an *absolute* charge having been acquired.

1172. Well-purified oil of turpentine, which I find to be an excellent liquid insulator for most purposes, was put into a metallic vessel, and, being insulated, an endeavour was made to charge its particles, sometimes by contact of the metal with the electrical machine, and at others by a wire dipping into the fluid within; but whatever the mode of communication, no electricity of one kind only was retained by the arrangement, except what appeared on the exterior surface of the metal, that portion being present there only by an inductive action through the air to the surrounding conductors. When the oil of turpentine was confined in glass vessels, there were at first some appearances as if the fluid did receive an absolute charge of electricity from the charging wire, but these were quickly reduced to cases of common induction jointly through the fluid, the glass, and the surrounding air.

1173. I carried these experiments on with air to a very great extent. I had a chamber built, being a cube of twelve feet. A slight cubical wooden frame was constructed, and copper wire passed along and across it in various directions, so as to make the sides a large net-work, and then all was covered in with paper, placed in close connexion with the wires, and supplied in every direction with bands of tin foil, that the whole might be brought into good metallic communication, and rendered a free conductor in every part. This chamber was insulated in the lecture-room of the Royal Institution; a glass tube about six feet in length was passed through its side, leaving about four feet within and two feet on the outside, and through this a wire passed from the large electrical machine to the air within. By working the machine, the air in this chamber could be brought into what is considered a highly electrified state (being, in fact, the same state as that of the air of a room in which a powerful machine is in operation), and at the same time the outside of the insulated cube was everywhere strongly charged. But putting the chamber in communication with the perfect discharging train described in a former series, and working the machine so as to bring the air within to its utmost degree of charge if I quickly cut off the connexion with the machine, and at the same moment or instantly after insulated the cube, the air within had not the least power to communicate a further charge to it. If any portion of the air was electrified, as glass or other insulators may be charged, it was accompanied by a corresponding opposite action *within* the cube, the whole effect being merely a case of induction. Every attempt to charge air bodily and independently with the least portion of either electricity failed.

1174. I put a delicate gold-leaf electrometer within the cube, and then charged the whole by an *outside* communication, very strongly, for some time together; but neither during the charge or after the discharge did the electrometer or air within show the least signs of electricity. I charged and discharged the whole arrangement in various ways, but in no case could I obtain the least indication of an absolute charge; or of one by induction in which the electricity of one kind had the smallest superiority in quantity over the other. I went into the cube and lived in it, and using lighted candles, electrometers, and all other tests of electrical states, I could not find the least influence upon them, or indication of anything particular given by them, though all the time the outside of the cube was powerfully charged, and large sparks and brushes were darting off from every part of its outer surface. The conclusion I have come to is, that non-conductors, as well as conductors, have never yet had an absolute and independent charge of one electricity communicated to them, and that to all appearance such a state of matter is impossible.

Faraday's paper on Specific Inductive Capacity is so extensive and so detailed that it is difficult to extract from it any part which will clearly represent the whole. ı have preferred to introduce a supplementary note to the large paper which appeared in the *Philosophical Transactions* of 1838, p. 79. It is found also in the *Experimental Researches*, Vol. I, p. 413. In it is described a simple experiment by which the features of specific inductive capacity can be illustrated.

SPECIFIC INDUCTIVE CAPACITY

1307. I have recently put into an experimental form that general statement of the question of *specific inductive capacity* which is given at No. 1253 of Series XI., and the result is such as to lead me to hope the Council of the Royal Society will authorize its addition to the paper in the form of a supplementary note. Three circular brass plates, about five inches in diameter, were mounted side by side upon insulating pillars; the middle one, *A*, was a fixture, but the outer plates *B* and *C* were movable on slides, so that all three could be brought with their sides almost into contact, or separated to any required distance. Two gold leaves were suspended in a glass jar from insulated wires; one of the outer plates *B* was connected with one of the gold leaves, and the other outer plate with the other leaf. The outer plates *B* and *C* were adjusted at the distance of an inch and a quarter from the middle plate *A*, and the gold leaves were fixed at two inches apart; *A* was then slightly charged with electricity, and the plates *B* and *C*, with their gold leaves, thrown out of insulation *at the same time*, and then left insulated. In this state of things *A* was charged positive inductrically, and *B* and *C* negative inducteously; the same dielectric, air, being in the two intervals, and the gold leaves hanging, of course, parallel to each other in a relatively unelectrified state.

1308. A plate of shell-lac three quarters of an inch in thickness, and four inches square, suspended by clean white silk thread, was very carefully deprived of all charge (so that it produced no effect on the gold leaves if *A* were uncharged) and then introduced between plates *A* and *B*; the electric relation of the three plates was immediately altered, and the gold leaves attracted each other. On removing the shell-lac this attraction ceased; on introducing it between A and C it was renewed; on removing it the attraction again ceased; and the shell-lac when examined by a delicate Coulomb electrometer was still without charge.

1309. As A was positive, B and C were of course negative; but as the specific inductive capacity of shell-lac is about twice that of air, it was expected that when the lac was introduced between A and B, A would induce more towards B than towards C; that therefore B would become more negative than before towards A, and, consequently, because of its insulated condition, be positive externally, as at its back or at the gold leaves; whilst C would be less negative towards A, and therefore negative outwards or at the gold leaves. This was found to be the case; for on whichever side of A the shell-lac was introduced the external plate at that side was positive, and the external plate on the other side negative towards each other, and also to uninsulated external bodies.

1310. On employing a plate of sulphur instead of shell-lac, the same results were obtained; consistent with the conclusions drawn regarding the high specific inductive capacity of that body already given.

1311. These effects of specific inductive capacity can be exalted in various ways, and it is this capability which makes the great value of the apparatus. Thus I introduced the shell-lac between A and B, and then for a moment connected B and C, uninsulated them, and finally left them in the insulated state; the gold leaves were of course hanging parallel to each other. On removing the shell-lac the gold leaves attracted each other; on introducing the shell-lac between A and C this attraction was *increased*, (as had been anticipated from theory), and the leaves came together, though not more than four inches long, and hanging three inches apart.

1312. By simply bringing the gold leaves nearer to each other I was able to show the difference of specific inductive capacity when only thin plates of shell-lac were used, the rest of the dielectric space being filled with air. By bringing B and C nearer to A another great increase of sensibility was made. By enlarging the size of the plates still further power was gained. By diminishing the extent of the wires, &c. connected with the gold leaves, another improvement resulted. So that in fact the gold leaves became, in this manner, as delicate a test of *specific inductive action* as they are, in Bennet's and Singer's electrometers, of ordinary electrical charge.

The extract which follows is taken from the *Philosophical Transactions* of 1846, p. 21, or from Faraday's *Experimental Researches* Vol. III, p. 30. The title of the paper is "On New Magnetic Actions and On the Magnetic Condition of All Matter." In it is described what is now known as diamagnetism. Faraday called those bodies diamagnetic through which magnetic lines of force pass without producing the magnetic phenomena exhibited by iron. Such bodies were found to be affected by the magnetic field and the action thus exhibited was called diamagnetic action.

Faraday first describes the powerful magnets which were needed for this investigation and refers particularly to the importance of avoiding spurious effects which may arise from the presence of small quantities of iron or from other causes. The action which he describes in the following paragraphs was found to be exhibited by almost all substances, particularly by bismuth and some of the other metals.

DIAMAGNETISM

2253. The bar of silicated borate of lead, or heavy glass already described as the substance in which magnetic forces were first made effectually to bear on a ray of light, and which is 2 inches long, and about 0.5 of an inch wide and thick, was suspended centrally between the magnetic poles, and left until the effect of torsion was over. The magnet was then thrown into action by making contact at the voltaic battery: immediately the bar moved, turning round its point of suspension, into a position across the magnetic curve or line of force, and after a few vibrations took up its place of rest there. On being displaced by hand from this position, it returned to it, and this occurred many times in succession.

2254. Either end of the bar indifferently went to either side of the axial line. The determining circumstance was simply inclination of the bar one way or the other to the axial line, at the beginning of the experiment. If a particular or marked end of the bar were on one side of the magnetic, or axial line, when the magnet was rendered active that end went further outwards, until the bar had taken up the equatorial position.

2255. Neither did any change in the magnetism of the poles, by change in the direction of the electric current, cause any difference in this respect. The bar went by the shortest course to the equatorial position.

2256. The power which urged the bar into this position was so thoroughly under command, that if the bar were swinging it could easily be hastened in its course into this position, or arrested as it was passing from it, by seasonable contacts at the voltaic battery.

2257. There are two positions of equilibrium for the bar; one stable, the other unstable. When in the direction of the axis or magnetic line of force, the completion of the electric communication causes no change of place; but if it be the least oblique to this position, then the obliquity increases until the bar arrives at the equatorial position; or if the bar be originally in the equatorial position, then the magnetism causes no further changes, but retains it there.

2258. Here then we have a magnetic bar which points east and west, in relation to north and south poles, i.e. points perpendicularly to the lines of magnetic force.

2259. If the bar be adjusted so that its point of suspension, being in the axial line, is not equidistant from the poles, but near to one of them, then the magnetism again makes the bar take up a position perpendicular to the magnetic lines of force; either end of the bar being on the one side of the axial line, or the other, at pleasure. But at the same time there is another effect, for at the moment of completing the electric contact, the centre of gravity of the bar recedes from the pole and remains repelled from it as long as the magnet is retained excited. On allowing the magnetism to pass away, the bar returns to the place due to it by its gravity.

2260. Precisely the same effect takes place at the other pole of the magnet. Either of them is able to repel the bar, whatever its position may be, and at the same time the bar is made to assume a position, at right angles, to the line of magnetic force.

2261. If the bar be equidistant from the two poles, and in the axial line, then no repulsive effect is or can be observed.

2262. But preserving the point of suspension in the equatorial line, i.e. equidistant from the two poles, and removing it a little on one side or the other of the axial line, then another effect is brought forth. The bar points as before across the magnetic line of force, but at the same time it recedes from the axial line, increasing its distance from it, and this new position is retained as long as the magnetism continues, and is quitted with its cessation.

2263. Instead of two magnetic poles, a single pole may be used, and that either in a vertical or a horizontal position. The effects are in perfect accordance with those described above; for the bar, when near the pole, is repelled from it in the direction of the line of magnetic force, and at the same time it moves into a position perpendicular to the direction of the magnetic lines passing through it. When the magnet is vertical and the bar by its side, this action makes the bar a tangent to the curve of its surface.

2264. To produce these effects, of pointing across the magnetic curves, the form of the heavy glass must be long; a cube, or a fragment approaching roundness in form, will not point, but a long piece will. Two or three rounded pieces or cubes, placed side by side in a paper tray, so as to form an oblong accumulation, will also point.

2265. Portions, however, of any form, are repelled: so if two pieces be hung up at once in the axial line, one near each pole, they are repelled by their respective poles, and approach, seeming to attract each other. Or if two pieces be hung up in the equatorial line, one on each side of the axis, then they both recede from the axis, seeming to repel each other.

2266. From the little that has been said, it is evident that the bar presents in its motion a complicated result of the force exerted by the magnetic power over the heavy glass, and that, when cubes or spheres are employed, a much simpler indication of the effect may be obtained. Accordingly, when a cube was thus used with the two poles, the effect was repulsion or recession from either pole, and also recession from the magnetic axis on either side.

2267. So, the indicating particle would move, either along the magnetic curves, or across them; and it would do this either in one direction or the other; the only constant point being, that its tendency was to move from stronger to weaker places of magnetic force.

2268. This appeared much more simply in the case of a single magnetic pole, for then the tendency of the indicating cube or sphere was to move outwards, in the direction of the magnetic lines of force. The appearance was remarkably like a case of weak electric repulsion.

2269. The cause of the pointing of the bar, on any oblong arrangement of the heavy glass, is now evident. It is merely a result of the tendency of the particles to move outwards, or into the positions of weakest magnetic action. The joint exertion of the action of all the particles brings the mass into the position, which, by experiment, is found to belong to it.

The use of the idea of lines of force, not only in their geometrical relations, but as representing a peculiar physical state in the region between the bodies to which they belong, occurs very frequently in Faraday's writings, not only in connection with magnetism but also in connection with electricity. The following short paper is selected for reproduction, partly because it gives a succinct description of these lines of force as they were conceived by Faraday and partly because it exhibits Faraday's mind in its more philosophical aspects. It appeared under the title, "On the Physical Lines of Magnetic Force," in the *Proceedings of the Royal Institution* for June 11, 1851. It also may be found in the *Experimental Researches*, Vol. III, p. 438.

PHYSICAL LINES OF MAGNETIC FORCE

On a former occasion certain lines about a bar-magnet were described and defined (being those which are depicted to the eye by the use of iron filings sprinkled in the neighbourhood of the magnet), and were recommended as expressing accurately the nature, condition, direction, and amount of the force in any given region either within or outside of the bar. At that time the lines were considered in the abstract. Without departing from or unsettling anything said, the inquiry is now entered upon of the possible and probable physical existence of such lines. Those who wish to reconsider the different points belonging to these parts of magnetic science may refer to two papers in the first part of the Phil. Trans. for 1852 for data concerning the representative lines of force, and to a paper in the Phil. Mag. 4th. Series, 1852, vol. III, p. 401, for the argument respecting the physical lines of force.

Many powers act manifestly at a distance; their physical nature is incomprehensible to us; still we may learn much that is real and positive about them, and amongst other things something of the condition of the space between the body acting and that acted upon, or between the two mutually acting bodies. Such powers are presented to us by the phaenomena of gravity, light, electricity, magnetism, etc. These when examined will be found to present remarkable differences in relation to their respective lines of force; and at the same time that they establish the existence of real physical lines in some cases, will facilitate the consideration of the question as applied especially to magnetism.

When two bodies, *a*, *b*, gravitate towards each other, the line in which they act is a straight line, for such is the line which either would follow if free to move. The attractive force is not altered, either in *direction* or *amount*, if a third body is made to act by gravitation or otherwise upon either or both of the two first. A balanced cylinder of brass gravitates to the earth with a weight exactly the same, whether it is left like a pendulum freely to hang towards it, or whether it is drawn aside by other attraction or by tension, whatever the amount of the latter may be. A new gravitating force may be exerted upon *a*, but that does not in the least affect the amount of power which it exerts towards *b*. We have no evidence that *time* enters in any way into the exercise of this power, whatever the distance between the acting bodies, as that from the sun to the earth, or from star to star. We can hardly conceive of this force in one particle by itself; it is when two or more are present that we comprehend it: yet in gaining this idea we perceive no difference in the character of the power in the different particles; all of the same kind are *equal, mutual,* and *alike.* In the case of gravitation, no effect which sustains the idea of an independent or physical line of force is presented to us; and as far as we at present know, the line of gravitation is merely an ideal line representing the direction in which the power is exerted.

Take the Sun in relation to another force which it exerts upon the earth, namely its illuminating or warming power. In this case rays (which are lines of force) pass across the intermediate space; but then we may affect these lines by different media applied to them in their course. We may alter their direction either by reflection or refraction; we may make them pursue curved or angular courses. We may cut them off at their origin and then search for and find them before they have attained their object. They have a relation to *time*, and occupy 8 minutes in coming from the sun to the earth: so that they may exist independently either of their source or their final home, and have in fact a clear distinct physical existence. They are in extreme contrast with the lines of gravitating power in this respect; as they are also in respect of their condition at their terminations. The two bodies terminating a line of gravitating force are alike in their actions in every respect, and so the line joining them has like relations in both directions. The two bodies at the terminals of a ray are utterly unlike in action; one is a source, the other a destroyer of the line and the line itself has the relation of a stream flowing in one direction. In these two cases of gravity and radiation, the difference between an abstract and a physical line of force is immediately manifest.

Turning to the case of Static Electricity we find here attractions (and other actions) at a distance as in the former cases; but when we come to compare the attraction with that of gravity, very striking distinctions are presented which immediately affect the question of a physical line of force.   In the first place, when we examine the bodies bounding or terminating the lines of attraction, we find them as before, mutually and equally concerned in the action; but they are not alike: on the contrary, though each is endued with a force which speaking generally is of the like nature, still they are in such contrast that their actions on a third body in a state like either of them are precisely the reverse of each other,—what the one attracts the other repels; and the force makes itself evident as one of those manifestations of power endued with a dual and antithetical condition.   Now with all such dual powers, attraction cannot occur unless the two conditions of force are present and in face of each other through the lines of force. Another essential limitation is that these two conditions must be exactly equal in amount, not merely to produce the effects of attraction, but in every other case; for it is impossible so to arrange things that there shall be present or be evolved more electric power of the one kind than of the other.   Another limitation is that they must be in physical relation to each other; and that when a positive and a negative electrified surface are thus associated, we cannot cut off this relation except by transferring the forces of these surfaces to equal amounts of the contrary forces provided elsewhere.   Another limitation is that the power is definite in amount.   If a ball *a* be charged with 10 of positive electricity, it may be made to act with that amount of power on another ball *b* charged with 10 of negative electricity; but if 5 of its power be taken up by a third ball *c* charged with negative electricity, then it can only act with 5 of power on ball *a*, and that ball must find or evolve 5 of positive power elsewhere: this is quite unlike what occurs with gravity, a power that presents us with nothing dual in its character.   Finally, the electric force acts in curved lines. If a ball be electrified positively and insulated in the air, and a round metallic plate be placed about 12 or 15 inches off, facing it and uninsulated, the latter will be found, by the necessity mentioned above, in a negative condition; but it is not negative only on the side facing the ball, but on the other or outer face also, as may be shown by a carrier applied there, or by a strip of gold or silver leaf hung against that outer face.   Now the power affecting this face does not pass through the uninsulated plate, for the thinnest gold leaf is able to stop the inductive action, but round the edges of the face, and therefore acts in curved lines.   All these points indicate the existence of physical lines of electric force:—the absolutely essential relation of positive and negative surfaces to each other, and their dependence on each other contrasted with the known mobility of the forces, admit of no other conclusion. The action also in curved lines must depend upon a physical line of force. And there is a third important character of the force leading to the same result, namely its affection by media having different specific inductive capacities.

When we pass to Dynamic Electricity the evidence of physical lines of force is far more patent. A voltaic battery having its extremities connected by a conducting medium, has what has been expressively called a current of force running round the circuit, but this current is an axis of power having equal and contrary forces in opposite directions. It consists of lines of force which are compressed or expanded according to the transverse action of the conductor, which changes in direction with the form of the conductor, which are found in every part of the conductor, and can be taken out from any place by channels properly appointed for the purpose; and nobody doubts that they are physical lines of force.

Finally as regards a Magnet, which is the object of the present discourse. A magnet presents a system of forces perfect in itself, and able, therefore, to exist by its own mutual relations. It has the dual and antithetic character belonging to both static and dynamic electricity; and this is made manifest by what are called its polarities, i.e. by the opposite powers of like kind found at and towards its extremities. These powers are found to be absolutely equal to each other; one cannot be changed in any degree as to amount without an equal change of the other; and this is true when the opposite polarities of a magnet are not related to each other, but to the polarities of other magnets. The polarities, or the *northness* and *southness* of a magnet are not only related to each other, through or within the magnet itself, but they are also related externally to opposite polarities (in the manner of static electric induction), or they cannot exist; and this external relation involves and necessitates an exactly equal amount of the new opposite polarities to which those of the magnet are related. So that if the force of a magnet *a* is related to that of another magnet *b*, it cannot act on a third magnet *c* without being taken off from *b*, to an amount proportional to its action on *c*. The lines of magnetic force are shown by the moving wire to exist both within and outside of the magnet; also they are shown to be closed curves passing in one part of their course through the magnet; and the amount of those within the magnet at its equator is exactly equal in force to the amount in any section including the whole of those on the outside. The lines of force outside a magnet can be affected in their direction by the use of various media placed in their course. A magnet can in no way be procured having only one magnetism, or even the smallest excess of northness or southness one over the other. When the polarities of a magnet are not related externally to the forces of other magnets, then they are related to each other: i.e. the northness and southness of an isolated magnet are externally dependent on and sustained by each other.

Now all these facts, and many more, point to the existence of physical lines of force external to the magnets as well as within. They exist in curved as well as in straight lines; for if we conceive of an isolated straight bar-magnet, or more especially of a round disc of steel magnetized regularly, so that its magnetic axis shall be in one diameter, it is evident that the polarities must be related to each other externally by curved lines of force; for no straight line can at the same time touch two points having northness and southness. Curved lines of force can, as I think, only consist with physical lines of force.

The phaenomena exhibited by the moving wire confirm the same conclusion. As the wire moves across the lines of force, a current of electricity passes or tends to pass through it, there being no such current before the wire is moved. The wire when quiescent has no such current, and when it moves it need not pass into places where the magnetic force is greater or less. It may travel in such a course that if a magnetic needle were carried through the same course it would be entirely unaffected magnetically, i.e. it would be a matter of absolute indifference to the needle whether it were moving or still. Matters may be so arranged that the wire when still shall have the same diamagnetic force as the medium surrounding the magnet, and so in no way cause disturbance of the lines of force passing through both; and yet when the wire moves, a current of electricity shall be generated in it. The mere fact of motion cannot have produced this current: there must have been a state or condition around the magnet and sustained by it, within the range of which the wire was placed: and this state shows the physical constitution of the lines of magnetic force.

What this state is, or upon what it depends, cannot as yet be declared. It may depend upon the aether, as a ray of light does, and an association has already been shown between light and magnetism. It may depend upon a state of tension, or a state of vibration, or perhaps some other state analogous to the electric current, to which the magnetic forces are so intimately related. Whether it of necessity requires matter for its sustentation will depend upon what is understood by the term matter. If that is to be confined to ponderable or gravitating substances, then matter is not essential to the physical lines of magnetic force any more than to a ray of light or heat; but if in the assumption of an aether we admit it to be a species of matter, then the lines of force may depend upon some function of it. Experimentally mere space is magnetic; but then the idea of such mere space must include that of the aether, when one is talking on that belief; or if hereafter any other conception of the state or condition of space rise up, it must be admitted into the view of that, which just now in relation to experiment is called mere space. On the other hand it is, I think, an ascertained fact, that ponderable matter is not essential to the existence of physical lines of magnetic force.

LENZ

Heinrich Friedrich Emil Lenz was born on February 12, 1804 in Dorpat, and died in Rome on February 10, 1865. He became professor of physics at the University of St. Petersburg. He investigated the conductivity of many materials for electricity and the effect of temperature on conductivity. He also studied the heat produced by the current and discovered the law which is known by the name of Joule.

The extract which follows is from a paper published in the *Annalen der Physik und Chemie*, Vol. 31, p. 483, 1834, entitled "Ueber die Bestimmung der Richtung der durch elektrodynamische Vertheilung erregten galvanischen Ströme." It contains the statement of the law which is known by his name.

Lenz' Law

In his experimental researches in electricity which contain the discovery of the so-called electrodynamic induction, Faraday determines the direction of the galvanic currents produced by it in the following manner: 1. A galvanic current excites in a parallel wire, which is moved nearer to it, a current in a sense opposite to its own, but in a wire which is removed from it a current in the same direction as its own, and 2. A magnet produces a current in a conductor which is moved in its neighborhood, which depends on the direction in which the conductor in its motion cuts the magnetic curves. Leaving out of account the fact that two entirely different rules are given for one and the same phenomenon (since according to Ampère's beautiful theory, a magnet may be considered as a system of circular galvanic currents) the rule is not at once, at least immediately, sufficient, in that there are many cases which it does not include, for example, the case in which a conductor which is placed perpendicular to a current moves along it; and finally in the second statement, in my opinion, it has not the simplicity which is desirable, so that it can be easily applied to special cases, and I believe that other readers of this remarkable memoir will agree with me in this, if they recall §116, where Faraday tries to make the above rule clear by the motion of a knife blade along a magnet; yes, Faraday himself calls attention to the difficulty of giving a clear statement for the direction of the current.

Nobili: (in his memoir, Poggendorff's Annalen 1833 No. 3) starts from Faraday's first law that when a conductor is brought nearer a galvanic current parallel to it, a current in the opposite direction is produced, and when it is carried away, a current in the same direction, and he endeavors from this alone to explain all the phenomena and the directions of the currents which are produced by electrodynamic induction. I must say that this work, which is so worthy of appreciation in other respects, does not in many points give me the degree of evidence which we are justified in expecting in physical memoirs, in particular in the explanation of those currents which arise in a conductor which is placed perpendicular to a galvanic current and is moved along it. Faraday is certainly right when he makes the criticism of the theory of the Italian physicist in general, that in the case of a rotation of a magnet about its own axis, with a suitable arrangement of the wire, a galvanic current can be excited, although in this case there occurs neither an approach nor a departure of the currents from the magnets, while on the contrary all parts of the arrangement retain their mutual distances. When I read Faraday's memoir, it seemed to me that all the experiments in electrodynamic induction might very easily be referred to the laws of electrodynamic motion, so that if we assume that these are known the others are therefore determined, and since I was able to confirm this view by many experiments, I will present it in what follows and prove it partly by well known observations and partly by some which I have arranged for the purpose.

The law, according to which the magnetoelectric phenomena are reduced to electromagnetic phenomena, is the following:

If a metallic conductor moves in the neighborhood of a galvanic current or of a magnet, a galvanic current will be produced in it which will have such a direction that it would have occasioned in the wire, if it were at rest, a motion which is exactly opposite to that here given to the wire, provided that the wire when at rest is movable only in the direction of the motion and in the opposite direction.

In order therefore to represent the sense of the direction of the current excited in the moving wire by electrodynamic induction, we consider in what sense the current must be directed, according to the electromagnetic laws, if it were to produce the motion; the current in the wire will be excited in the opposite direction. As an example, we may consider the well known Faraday's rotation experiment in which a movable conductor hanging vertically downward is traversed by a galvanic current from above downward, and in consequence travels around the North pole of a magnet placed directly under it in the direction from *N* through *E* to *S*; now, if we do not allow the current to traverse the conductor, but give it the motion which has just been described by mechanical means, then by our law a current will be excited in it, which, being opposite to the former one, passes through the moving wire from below upward and can be exhibited in it if the lower and the upper ends of it are connected with a galvanometer.

The remainder of the paper contains several examples taken from Ampère and others of electrodynamic attractions or repulsions, and the comparison of these with corresponding cases of the production of induced currents.

HENRY

Joseph Henry was born on December 17, 1797, in Albany, New York. His father was a day laborer. He studied at the Albany Academy and in 1826 he was elected to the professorship of mathematics and natural philosophy in that institution. In 1832 he was called to the chair of natural philosophy at the College of New Jersey in Princeton, New Jersey. During these academic years he devoted himself to researches in electricity. He improved the electromagnet and used it in the construction of an electromagnetic telegraph. He studied induced currents, at first independently of Faraday, and discovered the current of self induction. He also studied induced currents of higher orders than the first and those produced by the discharge of a Leyden jar. From the behavior of these currents he recognized the oscillatory character of this discharge.

In 1846 Henry was appointed the first secretary of the newly established Smithsonian Institution. He gave himself entirely to the work of developing the functions of the Institution and its present important position as a source of scientific information is the result of his labors. He died in Washington, D. C., on May 13, 1878.

The first extract which follows is from a paper entitled "On the Production of Currents and Sparks of Electricity from Magnetism," which appeared in the *American Journal of Science*, Vol. 22, p. 403, 1832. The second extract on the same subject was published in the *Journal of the Franklin Institute*, March, 1835, Vol. 15, p. 169. It describes the discovery of self-induction. The third extract is from the *Proceedings of the American Philosophical Society*, Vol. 2, p. 193, 1842. It contains a discussion of the oscillatory discharge.

SELF-INDUCTION

I

I have made several other experiments in relation to the same subject, but which more important duties will not permit me to verify in time for this paper. I may however mention one fact which I have not seen noticed in any work, and which appears to me to belong to the same class of phenomena as those before described; it is this: when a small battery is moderately excited by diluted acid, and its poles, which should be terminated by cups of mercury, are connected by a copper wire not more than a foot in length, no spark is perceived when the connection is either formed or broken; but if a wire thirty or forty feet long be used instead of the short wire, though no spark will be perceptible when the connection is made, yet when it is broken by drawing one end of the wire from its cup of mercury, a vivid spark is produced. If the action of the battery be very intense, a spark will be given by the short wire; in this case it is only necessary to wait a few minutes until the action partially subsides, and until no more sparks are given from the short wire; if the long wire be now substituted a spark will again be obtained. The effect appears somewhat increased by coiling the wire into a helix; it seems also to depend in some measure on the length and thickness of the wire. I can account for these phenomena only by supposing the long wire to become charged with electricity, which by its reaction on itself projects a spark when the connection is broken.

II

The following facts in reference to the spark, shock, etc. from a galvanic battery of a single pair when the poles are united by a long conductor, were communicated by Prof. Joseph Henry, and those relating to the spark were illustrated experimentally:

1. A long wire gives a more intense spark than a short one. There is however with a given surface of zinc a length beyond which the effect is not increased; a wire of one hundred and twenty feet gave about the same intensity of spark as one of two hundred and forty feet.

2. A thick wire gives a larger spark than a smaller one of the same length.

3. A wire coiled into a helix gives a more vivid spark than the same wire when uncoiled.

4. A ribbon of copper, coiled into a flat spiral, gives a more intense spark than any other arrangement yet tried.

5. The effect is increased, by using a longer and wider ribbon, to an extent not yet determined. The greatest effect has been produced by a coil ninety-six feet long and weighing 15 lbs; a larger conductor has not been received.

6. A ribbon of copper, first doubled into two strands and then coiled into a flat spiral, gives no spark, or a very feeble one.

7. Large copper handles, soldered to the ends of a coil of ninety-six feet, and these both grasped, one by each hand, a shock is felt at the elbows, when the contact is broken in a battery of a single pair with one and a half feet of zinc surface.

8. A shock is also felt when the copper of the battery is grasped with one hand and one of the handles with the other; the intensity however is not as great as in the last case. This method of receiving the shock may be called the direct method; the other, the lateral one.

9. The decomposition of a liquid is effected by the use of the coil with a battery of a single pair, by interrupting the current and introducing a pair of decomposing wires.

10. A mixture of oxygen and hydrogen is also exploded by means of the coil, and breaking the contact, in a bladder containing the mixture.

11. The property of producing an intense spark is induced, on a short wire, by introducing at any point of a compound galvanic current a large flat spiral, and joining the poles by the short wire.

12. A spark is produced when the plates of a single battery are separated by a foot or more of diluted acid.

13. Little or no increase in the effect is produced by inserting a piece of soft iron into the centre of a flat spiral.

14. The effect produced by an electro-magnet, in giving the shock, is due principally to the coiling of the long wire which surrounds the soft iron.

THE OSCILLATORY DISCHARGE

Professor Henry presented the record of a series of experiments on induction from ordinary electricity, as the fifth number of his Contributions to Electricity and Magnetism. Of these experiments he gave an oral account, of which the following is the substance.

In the third number of his Contributions he had shown on this subject: 1. That the discharge of a Leyden battery through a conductor, developed in an adjoining parallel conductor an induced current, analogous to that which, under similar circumstances, is produced by a galvanic current. 2. That the direction of the induced current, as indicated by the polarity given to a steel needle, changes its sign with a change of distance of the two conductors, and also with a change in the quantity of the discharge of electricity. 3. That when the induced current is made to act on a third conductor, a second induced current is developed, which can again develop another, and so on through a series of successive inductions. 4. That when a plate of metal is interposed between any two of the consecutive conductors, the induced current is neutralized by the adverse action of a current in the plate.

The direction of the induced currents in all the author's experiments was indicated by the polarity given to steel needles enclosed in a spiral, the wire of which formed a part of the circuit. But some doubts were reasonably entertained of the true indications of the direction of a current by this means, since M. Savary had announced in 1826, that when several needles are placed at different distances above a wire through which the discharge of a Leyden battery is passed, they are magnetized in different directions, and that by constantly increasing the discharge through a spiral, several reversions of the polarity of the contained needles are obtained.

It was therefore very important before attempting further advances in the discovery of the laws of the phenomena, that the results obtained by M. Savary should be carefully studied; and accordingly the first experiments of the new series relate to the repetition of them. The author first attempted to obtain them by using needles of a larger size, Nos. 3, and 4, such as he had generally employed in all his previous experiments; but although nearly a thousand needles were magnetized in the course of the experiments, he did not succeed in getting a single change in the polarity. The needles were always magnetized in a direction conformable to the direction of the electrical discharge. When however very fine needles were employed he did obtain several changes in the polarity in the case of the spiral, by merely increasing the quantity of the electricity, while the direction of the discharge remained the same.

This anomaly which has remained so long unexplained, and which at first sight appears at variance with all our theoretical ideas of the connection of electricity and magnetism, was after considerable study satisfactorily referred by the author to an action of the discharge of the Leyden jar which had never before been recognized. The discharge, whatever may be its nature, is not correctly represented (employing for simplicity the theory of Franklin) by the single transfer of an imponderable fluid from one side of the jar to the other; the phenomena require us to admit *the existence of a principal discharge in one direction, and then several reflex actions backward and forward, each more feeble than the preceding, until the equilibrium is obtained.* All the facts are shown to be in accordance with this hypothesis, and a ready explanation is afforded by it of a number of phenomena which are to be found in the older works on electricity, but which have until this time remained unexplained.

The same action is evidently connected with the induction of a current on its own conductor, in the case of an open circuit, such as that of the Leyden jar, in which the two ends of the conductor are separated by the thickness of the glass. And hence, if an induced current could be produced in this case, one should also be obtained in that of a second conductor, the ends of which are separated; and this was detected by attaching to the ends of the open circuit a quantity of insulated metal, or by connecting one end with the earth.

The next part of the research relates to a new examination of the phenomena of the change in the direction of the induced currents, with a change of distance, etc. These are shown to be due to the fact that the discharge from a jar does not produce a single induced current in one direction, but several successive currents in opposite directions. The effect on the needle is principally produced by two of these: the first is the more powerful, and in the adverse direction with that of the jar; the second is less powerful, and in the same direction with that of the jar. To explain the change of polarity, let us suppose the capacity of the needle to receive magnetism to be represented by ± 10, while the power of the first induced current to produce magnetism is represented by -15, and that of the second by $+12$; then the needle will be magnetized to saturation or to -10, by the first induced current, and immediately afterwards all this magnetism will be neutralized by the adverse second induction, and a power of $+2$ will remain; so that the polarity of the needle in this case will indicate an induced current in the same direction as that of the jar. Next, let the conductors be so far separated, or the charge so much diminished, that the power of the first current to develop magnetism may be reduced to -8, while that of the second current is reduced to $+6$, the magnetic capacity of the needle remaining the same. It is evident then that the first current will magnetize the needle to -8, and that the second current will immediately afterwards neutralize 6 of this, and consequently the needle will retain a magnetism of -2, or will indicate an induced current in an opposite direction to that of the jar.

In extending the researches relative to this part of the investigations, a remarkable result was obtained in regard to the distance at which inductive effects are produced by a very small quantity of electricity; a single spark from the prime conductor of the machine, of about an inch long, thrown on the end of a circuit of wire, in an upper room, produced an induction sufficiently powerful to magnetize needles in a parallel circuit of wire placed in the cellar beneath, at a perpendicular distance of thirty feet with two floors and ceilings, each fourteen inches thick, intervening. The author is disposed to adopt the hypothesis of an electrical *plenum*, and from the foregoing experiment it would appear that the transfer of a single spark is sufficient to disturb perceptibly the electricity of space throughout at least a cube of 400,000 feet of capacity; and when it is considered that the magnetism of the needle is the result of the difference of two actions, it may be further inferred that the diffusion of motion in this case is almost comparable with that of a spark from a flint and steel in the case of light.

The author next alludes to a proposition which he advanced in the second number of his Contributions, namely, that the phenomena of dynamic induction may be referred to the known electrical laws, as given by the common theories of electricity; and he gives a number of experiments to illustrate the connection between statical and dynamical induction.

The last part of the series of experiments relates to induced currents from atmospheric electricity. By a very simple arrangement, needles are strongly magnetized in the author's study, even when the flash is at the distance of seven or eight miles, and when the thunder is scarcely audible. On this principle he proposes a simple self-registering electrometer, connected with an elevated exploring rod.

GAUSS

Karl Friedrich Gauss was born on April 30, 1777, near Braunschweig. His family was very poor and could do nothing toward his education. His mathematical ability showed itself so plainly to his masters in the school he attended that he was recommended to the Duke as one worthy of special support. He was thus enabled to attend the University of Göttingen. He received his doctor's degree at Helmstadt *in absentia*. For several years his attention was turned to astronomy and an effort was made to build an observatory for his use in Braunschweig. This effort failed perhaps because of the disturbance of the country by the Napoleonic wars. In 1807 he was called to the University of Göttingen where a new observatory was built for him, in which he took up his residence in 1816. In 1831 Wilhelm Weber, whom Gauss had met a few years before, was called to the professorship of physics in Göttingen and the two friends collaborated in an extensive study of the earth's magnetism. Gauss's health began to fail in 1852 and he died on February 23, 1855, in Göttingen.

The extract which follows is from Gauss's paper, "Intensitas vis Magneticae terrestris ad mensuram absolutam revocata." It appeared in the *Commentationes Societatis regiae Scientiarum Gottingensis recentiores*, Vol. 8, 1841. It contains the first example of the measurement of magnetic and electric quantities in absolute units.

THE ABSOLUTE MEASURE OF MAGNETIC FORCE

For the complete determination of the magnetic force of the earth in a given place three elements are required: the declination, or the angle between the plane in which the magnet lies and a meridian; the inclination of its direction to the horizontal plane; and in the third place, the intensity. The declination, which may be considered the primary element in respect to all the applications to nautical and geodetic purposes, has from the beginning occupied observers and physicists, who also now for a century have given considerable attention to the inclination. On the other hand the third element, the intensity, which is an object equally worthy of scientific attention, has up to recent times been almost entirely neglected. Praise is due to the illustrious Humboldt for this reason among many, that he very early gave his attention to this subject, and in his journeys collected a great number of observations of the relative intensity of terrestrial magnetism, from which he determined the continual increase of this intensity, as we go from the magnetic equator toward the pole. Very many physicists, following in the footsteps of this observer of nature, have now collected such a number of measurements, that Hansteen, a most distinguished man and especially deserving of praise for his work in terrestrial magnetism, was able to issue a universal map showing the isodynamic lines.

The method which was used in this investigation consists in the observation of the time in which the same magnetic needle makes the same number of oscillations in different places, or of the number of oscillations made in the same interval of time. The intensity is set proportional to the square of the number of oscillations in a given time. In this way the total intensities are compared with one another, when the dipping needle suspended at its center of gravity oscillates about a horizontal axis perpendicular to the magnetic meridian, or the intensities of the horizontal force, when the horizontal needle oscillates about a vertical axis: the latter method admits of greater precision in the observations, and the results which are obtained, when the inclinations are known, are easily reduced to the total intensities.

Plainly the validity of this method depends on the supposition that the distribution of free magnetism in the particles of the magnet which is used for such a comparison remains unchanged in the several experiments: for if the magnetic force of the needle were to become weaker with lapse of time, the needle would oscillate more slowly and the observer who was ignorant of this change would attribute too small a value to the intensity of the earth's magnetism for those stations in which the later observations were made. If the experiments are completed within a short time, and if the needle which is used is made of hardened steel and carefully magnetized, no considerable loss of its power need be feared; the uncertainty is further diminished if several needles are used in the comparison; and one may have greater confidence in this supposition if, when the journey is finished, the needle when examined at the starting point is found not to have changed its time of vibration. But, whatever care is taken, some slight weakening of the force of the needle can scarcely be avoided, and so such an agreement can hardly be expected after a long absence; and so when we compare the intensities in widely separated stations on the earth it will not be possible to attain the precision and certitude which is desirable.

Another inconvenience of this method is not so serious, so long as we are dealing with the comparison of intensities which are simultaneous or which are taken at times not very far apart. But since experiment shows that both the declination and the inclination at a given place experience continual changes which after many years become considerable, it cannot be doubted that the intensity also is subject to similar secular changes; and it is plain that as regards this question the method described is not effective. And yet it is highly desirable for the increase of natural knowledge, that this most important question should be brought into the fullest light, which certainly cannot be done, unless we substitute for this purely comparative method another one which is independent of the accidental changes of the needle and reduces the intensity of terrestrial magnetism to fixed units and absolute measurements.

It is not difficult to establish the theoretical principles on which such a method, so long desired, should be based. The number of oscillations which the needle makes in a given time depends partly on the intensity of the earth's magnetism, partly on the constitution of the needle, that is, on the static moment of the elements of free magnetism contained in it, and on its moment of inertia. Since this moment of inertia can be easily determined, it appears that the observation of the oscillations will furnish us with the product of the intensity of terrestrial magnetism and the static moment of the magnetism of the needle; but these two quantities cannot be separated unless observations of another sort are made, which furnish a different combination of them. For this purpose a second needle may be used, which is subjected to the action both of the earth's magnetism and of the magnetism of the first magnet, so that we may determine the ratio between these two actions. Both of these actions will depend on the distribution of free magnetism in the second magnet: and further on the constitution of the first magnet and the distance between the centers, the position of the line joining the centers of the two magnets with respect to their magnetic axes, and finally on the law which the magnetic actions and repulsions obey. The immortal Tobias Mayer was the first who conjectured that this law was similar to the law of gravitation, in that these actions also decrease according to the inverse square of the distances: the experiments of the distinguished Coulomb and Hansteen have shown that this conjecture is highly plausible, and recent experiments have set it beyond a doubt. But it should be carefully noticed that this law applies to the individual elements of free magnetism: the effect of the whole magnetized body will be very different, and at considerable distances, as can be deduced from the law itself, it nearly approaches the inverse third power of the distances, so that the action of the needle multiplied by the cube of the distance, as the distance, other things being equal, continually increases, converges asymptotically to a constant value, which, since the distances are expressed by numbers, when an arbitrary line is taken as a unit, will be homogeneous and comparable to the action of the terrestrial force. By a suitable arrangement and execution of the experiments the limit of this ratio may be found. Since this involves the static moment of magnetism of the first magnet there will be obtained a quotient arising from the division of this moment by the intensity of the terrestrial force, and this, compared with the product of these quantities which has been previously obtained, will serve for the elimination of this moment and will give the value of the intensity of terrestrial magnetism.

As respects the method by which we can submit to experiment the actions of terrestrial magnetism and of the first magnet on the second magnet two ways are possible, since we can observe the second magnet either in motion or in equilibrium. The first method consists in this, that the oscillations of this magnet are observed when the action of terrestrial magnetism is joined with the action of the first magnet placed at suitable distances and so that its axis is in the magnetic meridian passing through the center of the oscillating magnet: with this arrangement the oscillations will be either accelerated or retarded according as attracting poles or repelling poles confront each other, and the comparison either of the times of vibration for each position of the first magnet or of any one of these times with the time of vibration under the action of terrestrial magnetism alone, with the first magnet removed, will give the ratio of this force to the action of the first magnet. In the other method the first magnet is so placed that the direction of the force which it exerts in the region occupied by the second magnet when freely suspended, makes an angle (e.g. a right angle) with the magnetic meridian. By this arrangement the second magnet is deflected from the magnetic meridian, and from the amount of this deflection the ratio between the terrestrial magnetic force and the action of the first magnet is determined.

The former method essentially agrees with that which the illustrious Poisson proposed several years ago. The experiments tried on this plan by several physicists, so far as I know them, were either entirely unsuccessful or furnished only a rough approximation.

The difficulty of the task depends particularly on this, that a certain limit must be computed from the actions of the magnets observed at moderate distances which holds for an infinitely great distance, and that the eliminations necessary to this end are rendered difficult and even erroneous by very small errors of observation, particularly because several unknowns are to be eliminated which depend on the individual conditions of the magnets: the problem can only be reduced to a small number of unknowns when the actions are observed at distances which are sufficiently great in comparison with the length of the needles, so that these actions become very small. But for the accurate measurement of such small actions the arrangements hitherto used in practice are not satisfactory.

First of all therefore I saw that I had to prepare new arrangements by which I might observe and measure not only the times of the oscillations but also the directions of the magnets with much greater precision than had hitherto been possible. My labors undertaken to this end and continued through several months, in which I was assisted in many ways by the distinguished Weber, brought me finally to the desired results, so that they not only met my expectation but far surpassed it, and now nothing remains to be desired to attain an accuracy in the experiments comparable with that of astronomical observations except a place protected from the influence of neighboring iron and from agitation of the air. I have now made two pieces of apparatus which are not less distinguished by their simplicity than by their accuracy, the description of which I ought to reserve for another opportunity, while I report in this paper the experiments instituted to determine the intensity of terrestrial magnetism in our observatory of physics.

JOULE

The following extract is taken from a paper published in the *Philosophical Magazine*, Vol. 19, p. 260, 1841, "On the Heat Evolved by Metallic Conductors of Electricity and in the Cells of a Battery During Electrolysis." It contains the statement of Joule's (p. 203) law connecting the heat developed in a circuit with the strength of the current and the resistance.

JOULE'S LAW

There are few facts in science more interesting than those which establish a connexion between heat and electricity. Their value, indeed, cannot be estimated rightly, until we obtain a complete knowledge of the grand agents upon which they shed so much light. I hope, therefore, that the results of my careful investigation on the heat produced by voltaic action are of sufficient interest to justify me in laying them before the Royal Society.

Chap. I.   Heat Evolved by Metallic Conductors.

It is well known that the facility with which a metallic wire is heated by the voltaic current is in inverse proportion to its conducting-power; and it is generally believed that this proportion is exact; nevertheless I wished to ascertain the fact for my own satisfaction, and especially as it was of the utmost importance to know whether resistance to conduction is the *sole* cause of the heating effects. The detail, therefore, of some experiments confirmatory of the law, in addition to those already recorded in the pages of science, will not, I hope, be deemed superfluous.

It was absolutely essential to work with a *galvanometer* the indications of which could be depended upon as marking definite quantities of electricity. I bent a rod of copper into the shape of a rectangle AB, (Fig. 94), 12 inches long and 6 inches broad. This I secured in a vertical position by means of the block of wood C; N is the magnetic needle, 3¾ inches long, pointed at its extremities, and suspended upon a fine steel pivot over a graduated card placed a little before the centre of the instrument.

On account of the large relative size of the rectangular conductor of my galvanometer, the tangents of the deviations of the needle are very nearly proportional to the quantities of current electricity. The small correction which it is necessary to apply to the tangents, I obtained by means of the rigorous experimental process which I have some time ago described in the 'Annals of Electricity'.

I have expressed my quantities of electricity on the basis of Faraday's great discovery of definite electrolysis; and I venture

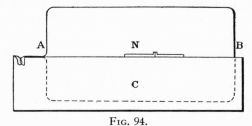

FIG. 94.

to suggest that that quantity of current electricity which is able to electrolyze a chemical equivalent expressed in grains in one hour of time, be called a *degree*. Now, by a number of experiments I found that the needle of my galvanometer deviated 33°.5 of the graduated card when a current was pass-ing in sufficient quantity to decompose nine grains of water per hour; that deviation, therefore, indicates one *degree of current electricity* on the scale that I propose to be adopted. We shall see in the sequel some of the practical advantages which I have had by using this measure.

The thermometer which I used had its scale grad-uated on the glass stem. The divisions were wide, and accurate. In taking temperatures with it, I stir the liquid gently with a feather; and then, suspending the thermometer by the top of its stem, so as to cause it to assume a vertical position, I bring my eye to a level with the top of the mercury. In this way a little practice has enabled me to estimate tempera-ture to the tenth part of Fahrenheit's degree with certainty.

FIG. 95.

In order to ascertain the heating-power of a given metallic wire, it was passed through a thin glass tube, and then closely coiled upon it. The extremities of the coil thus formed were then drawn asunder, so as to leave a small space between each convolution; and if this could not be well done, a piece of cotton thread was interposed. The apparatus thus prepared, when placed in a glass jar containing a given quantity of water, was ready for experiment. Fig. 95 will explain the dispositions: A is the coil of wire; B the glass jar, partly filled with water; T represents the thermometer. When the voltaic electricity is transmitted through the wire, no appreciable quantity passes from it to take the shorter course through the water. No trace of such a current could be detected, either by the evolution of hydrogen, or the oxidation of metal.

Previous to each of the experiments, the necessary precaution was taken of bringing the water in the glass jar and the air of the room to the same temperature. When this is accurately done, the results of the experiments bear the same proportions to one another as if no extraneous cooling agents, such as radiation, were present; for their effects in a given time are proportional to the difference of the temperatures of the cooling and cooled bodies; and hence, although towards the conclusion of some experiments this cooling effect is very considerable, the *absolute quantities* alone of heat are affected, not the *proportions* that are generated in the same time. (See the table of heats produced during half an hour and one hour.)

Exp. 1.—I took two copper wires, each two yards long, one of them $\frac{1}{28}$th of an inch, the other $\frac{1}{50}$th of an inch thick, and arranged them in coils in the manner that I have described. These were immersed in two glass jars, each of which contained nine ounces avoirdupois of water. A current of the mean quantity $1°.1Q$ (I place Q at the end of my *degrees*, to distinguish them from those of the graduated card) was then passed consecutively through both coils; and at the close of one hour I observed that the water in which the thin wire was immersed had gained $3°.4$, whilst the thick wire had produced only $1°.3$.

Now, by direct experiment, I found that three feet of thin wire could conduct exactly as well as eight feet of the thick wire; and hence it is evident that the resistances of two yards of each were in the ratio of 3.4 to 1.27, which approximates very closely to the ratio of the heating effects exhibited by the experiment.

Exp. 2.—I now substituted a piece of iron wire $\frac{1}{27}$ of an inch thick, and two yards long, for the thick copper wire used in Exp. 1, and placed each coil in half a pound of water. A current of $1°.25Q$ was passed through both during one hour, when the augmentation of temperature caused by the iron was $6°$, whilst that produced by the copper wire was 5°.5. In this case the resistances of the iron and copper wires were found to be in the ratio of 6 to 5.51.

Exp. 3.—A coil of copper wire was then compared with one of mercury, which was accomplished by enclosing the latter in a bent glass tube. In this way I had immersed, each in half a pound of water, $11\frac{1}{4}$ feet of copper wire $\frac{1}{50}$th of an inch thick and $22\frac{3}{4}$ inches of mercury 0.065 of an inch in diameter. At the close of one hour, during which the same current of electricity was passed through both, the former had caused a rise of temperature of 4°.4, the latter of 2°.9. The resistances were found by a careful experiment to be in the ratio of 4.4 to 3.

Other trials were made, with results of precisely the same character: they all conspire to confirm the fact, that when a given quantity of voltaic electricity is passed through a metallic conductor for a given length of time, the quantity of heat evolved by it is always proportional to the resistance which it presents, whatever may be the length, thickness, shape, or kind of that metallic conductor.

On considering the above law, I thought that the effect produced by the increase of the intensity of the electric current would be as the square of that element; for it is evident that in that case the resistance would be augmented in a double ratio, arising from the increase of the *quantity* of electricity passed in a given time, and also from the increase of the *velocity* of the same. We shall immediately see that this view is actually sustained by experiment.

I took the coil of copper wire used in Exp. 3, and have found the different quantities of heat gained by half a pound of water in which it was immersed, by the passage of electricities of different degrees of tension. My results are arranged in the table shown on p. 528.

The differences between the numbers in columns three and four, and in columns five and six, are very inconsiderable, taking into account the nature of the experiments, and are principally owing to the difficulty which exists in keeping the air of the room in the same state of quiet, of hygrometry, &c. during the different days on which the experiments were made. They are much less when a larger quantity of water is used, so as to reduce the cooling effects.

We see, therefore, that when a current of voltaic electricity is propagated along a metallic conductor, the heat evolved in a

Mean Deviations of the Needle of the Galvanometer.	Quantities of Current Electricity expressed in Degrees (5).	Quantities of Heat produced in half an hour by the Intensities in Column 2.	Proportional to the Squares of the Intensities in Column 2.	Quantities of Heat produced in one hour by the Intensities in Column 2.	Proportional to the Squares of the Intensities in Column 2.
16°	0.43Q	....	....	1.2	1
31½	0.92Q	3	2.9	4.7	4.55
55	2.35Q	19.4	18.8		
57⅔	2.61Q	23	23.2		
58½	2.73Q	25	25.4	39.6	40

given time is proportional to the resistance of the conductor multiplied by the square of the electric intensity.

The above law is of great importance. It teaches us the right use of those instruments which are intended to measure electric currents by the quantities of heat which they evolve. If such instruments be employed (though in their present state they are far inferior in point of accuracy to many other forms of the galvanometer), it is obvious that the *square roots* of their indications are alone proportional to the intensities which they are intended to measure.

MAXWELL

The following extract is from Maxwell's (p. 257) paper, "A Dynamical Theory of the Electromagnetic Field," published in the *Philosophical Transactions*, Vol. 155, p. 459, 1865. In this paper he gives the mathematical theory, based on Faraday's ideas of the transmission of electric and magnetic force through a medium, which he afterwards elaborated in his *Treatise on Electricity and Magnetism*. The mathematical treatment itself is too long for insertion and extracts from it would be useless. What is here given is the introduction, in which Maxwell describes the results of his mathematical investigation and announces the electromagnetic theory of light.

A DYNAMICAL THEORY OF THE ELECTROMAGNETIC FIELD

(1) The most obvious mechanical phenomenon in electrical and magnetical experiments is the mutual action by which bodies in certain states set each other in motion while still at a sensible

distance from each other. The first step, therefore, in reducing these phenomena into scientific form, is to ascertain the magnitude and direction of the force acting between the bodies, and when it is found that this force depends in a certain way upon the relative position of the bodies and on their electric or magnetic condition, it seems at first sight natural to explain the facts by assuming the existence of something either at rest or in motion in each body, constituting its electric or magnetic state, and capable of acting at a distance according to mathematical laws.

In this way mathematical theories of statical electricity, of magnetism, of the mechanical action between conductors carrying currents, and of the induction of currents have been formed. In these theories the force acting between the two bodies is treated with reference only to the condition of the bodies and their relative position, and without any express consideration of the surrounding medium.

These theories assume, more or less explicitly, the existence of substances the particles of which have the property of acting on one another at a distance by attraction or repulsion. The most complete development of a theory of this kind is that of M. W. Weber, who has made the same theory include electrostatic and electromagnetic phenomena.

In doing so, however, he has found it necessary to assume that the force between two electric particles depends on their relative velocity, as well as on their distance.

This theory, as developed by MM. W. Weber and C. Neumann, is exceedingly ingenious, and wonderfully comprehensive in its application to the phenomena of statical electricity, electromagnetic attractions, induction of currents and diamagnetic phenomena; and it comes to us with the more authority, as it has served to guide the speculations of one who has made so great an advance in the practical part of electric science, both by introducing a consistent system of units in electrical measurement, and by actually determining electrical quantities with an accuracy hitherto unknown.

(2) The mechanical difficulties, however, which are involved in the assumption of particles acting at a distance with forces which depend on their velocities are such as to prevent me from considering this theory as an ultimate one, though it may have been, and may yet be useful in leading to the coordination of phenomena.

I have therefore preferred to seek an explanation of the facts in another direction, by supposing them to be produced by actions which go on in the surrounding medium as well as in the excited bodies, and endeavouring to explain the action between distant bodies without assuming the existence of forces capable of acting directly at sensible distances.

(3) The theory I propose may therefore be called a theory of the *Electromagnetic Field*, because it has to do with the space in the neighbourhood of the electric or magnetic bodies, and it may be called a *Dynamical* Theory, because it assumes that in that space there is matter in motion, by which the observed electromagnetic phenomena are produced.

(4) The electromagnetic field is that part of space which contains and surrounds bodies in electric or magnetic conditions.

It may be filled with any kind of matter, or we may endeavour to render it empty of all gross matter, as in the case of Geissler's tubes and other so-called vacua.

There is always, however, enough of matter left to receive and transmit the undulations of light and heat, and it is because the transmission of these radiations is not greatly altered when transparent bodies of measurable density are substituted for the so-called vacuum, that we are obliged to admit that the undulations are those of an aethereal substance, and not of the gross matter, the presence of which merely modifies in some way the motion of the aether.

We have therefore some reason to believe, from the phenomena of light and heat, that there is an aethereal medium filling space and permeating bodies, capable of being set in motion and of transmitting that motion from one part to another, and of communicating that motion to gross matter so as to heat it and affect it in various ways.

(5) Now the energy communicated to the body in heating it must have formerly existed in the moving medium, for the undulations had left the source of heat some time before they reached the body, and during that time the energy must have been half in the form of motion of the medium and half in the form of elastic resilience. From these considerations Professor W. Thomson has argued, that the medium must have a density capable of comparison with that of gross matter, and has even assigned an inferior limit to that density.

(6) We may therefore receive, as a datum derived from a branch of science independent of that with which we have to deal, the

existence of a pervading medium, of small but real density, capable of being set in motion, and of transmitting motion from one part to another with great, but not infinite, velocity.

Hence the parts of this medium must be so connected that the motion of one part depends in some way on the motion of the rest; and at the same time these connexions must be capable of a certain kind of elastic yielding, since the communication of motion is not instantaneous, but occupies time.

The medium is therefore capable of receiving and storing up two kinds of energy, namely, the "actual" energy depending on the motion of its parts, and "potential" energy, consisting of the work which the medium will do in recovering from displacement in virtue of its elasticity.

The propagation of undulations consists in the continual transformation of one of these forms of energy into the other alternately, and at any instant the amount of energy in the whole medium is equally divided, so that half is energy of motion, and half is elastic resilience.

(7) A medium having such a constitution may be capable of other kinds of motion and displacement than those which produce the phenomena of light and heat, and some of these may be of such a kind that they may be evidenced to our senses by the phenomena they produce.

(8) Now we know that the luminiferous medium is in certain cases acted on by magnetism; for Faraday discovered that when a plane polarized ray traverses a transparent diamagnetic medium in the direction of the lines of magnetic force produced by magnets or currents in the neighbourhood, the plane of polarization is caused to rotate.

This rotation is always in the direction in which positive electricity must be carried round the diamagnetic body in order to produce the actual magnetization of the field.

M. Verdet has since discovered that if a paramagnetic body, such as solution of perchloride of iron in ether, be substituted for the diamagnetic body, the rotation is in the opposite direction.

Now Professor W. Thomson has pointed out that no distribution of forces acting between the parts of a medium whose only motion is that of the luminous vibrations, is sufficient to account for the phenomena, but that we must admit the existence of a motion in the medium depending on the magnetization, in addition to the vibratory motion which constitutes light.

It is true that the rotation by magnetism of the plane of polarization has been observed only in media of considerable density;
but the properties of the magnetic field are not so much altered
by the substitution of one medium for another, or for a vacuum,
as to allow us to suppose that the dense medium does anything
more than merely modify the motion of the ether. We have
therefore warrantable grounds for inquiring whether there may
not be a motion of the ethereal medium going on wherever magnetic
effects are observed, and we have some reason to suppose that this
motion is one of rotation, having the direction of the magnetic
force as its axis.

(9) We may now consider another phenomenon observed in
the electromagnetic field. When a body is moved across the
lines of magnetic force it experiences what is called an electromotive force; the two extremities of the body tend to become
oppositely electrified, and an electric current tends to flow through
the body. When the electromotive force is sufficiently powerful,
and is made to act on certain compound bodies, it decomposes
them, and causes one of their components to pass towards one
extremity of the body, and the other in the opposite direction.

Here we have evidence of a force causing an electric current in
spite of resistance; electrifying the extremities of a body in opposite
ways, a condition which is sustained only by the action of the
electromotive force, and which, as soon as that force is removed,
tends, with an equal and opposite force, to produce a counter
current through the body and to restore the original electrical
state of the body; and finally, if strong enough, tearing to pieces
chemical compounds and carrying their components in opposite
directions, while their natural tendency is to combine, and to
combine with a force which can generate an electromotive force
in the reverse direction.

This, then, is a force acting on a body caused by its motion
through the electromagnetic field, or by changes occurring in that
field itself; and the effect of the force is either to produce a current
and heat the body, or to decompose the body, or, when it can do
neither, to put the body in a state of electric polarization,—a state
of constraint in which opposite extremities are oppositely electrified, and from which the body tends to relieve itself as soon as
the disturbing force is removed.

(10) According to the theory which I propose to explain, this
"electromotive force" is the force called into play during the

communication of motion from one part of the medium to another, and it is by means of this force that the motion of one part causes motion in another part. When electromotive force acts on a conducting circuit, it produces a current, which, as it meets with resistance, occasions a continual transformation of electrical energy into heat, which is incapable of being restored again to the form of electrical energy by any reversal of the process.

(11) But when electromotive force acts on a dielectric it produces a state of polarization of its parts similar in distribution to the polarity of the parts of a mass of iron under the influence of a magnet, and like the magnetic polarization, capable of being described as a state in which every particle has its opposite poles in opposite conditions.

In a dielectric under the action of electromotive force, we may conceive that the electricity in each molecule is so displaced that one side is rendered positively and the other negatively electrical, but that the electricity remains entirely connected with the molecule, and does not pass from one molecule to another. The effect of this action on the whole dielectric mass is to produce a general displacement of electricity in a certain direction. This displacement does not amount to a current, because when it has attained to a certain value it remains constant, but it is the commencement of a current, and its variations constitute currents in the positive or the negative direction according as the displacement is increasing or decreasing. In the interior of the dielectric there is no indication of electrification, because the electrification of the surface of any molecule is neutralized by the opposite electrification of the surface of the molecules in contact with it; but at the bounding surface of the dielectric, where the electrification is not neutralized, we find the phenomena which indicate positive or negative electrification.

The relation between the electromotive force and the amount of electric displacement it produces depends on the nature of the dielectric, the same electromotive force producing generally a greater electric displacement in solid dielectrics, such as glass or sulphur, than in air.

(12) Here, then, we perceive another effect of electromotive force, namely, electric displacement, which according to our theory is a kind of elastic yielding to the action of the force, similar to that which takes place in structures and machines owing to the want of perfect rigidity of the connexions.

(13) The practical investigation of the inductive capacity of dielectrics is rendered difficult on account of two disturbing phenomena. The first is the conductivity of the dielectric, which, though in many cases exceedingly small, is not altogether insensible. The second is the phenomenon called electric absorption, in virtue of which, when the dielectric is exposed to electromotive force, the electric displacement gradually increases, and when the electromotive force is removed, the dielectric does not instantly return to its primitive state, but only discharges a portion of its electrification, and when left to itself gradually acquires electrification on its surface, as the interior gradually becomes depolarized Almost all solid dielectrics exhibit this phenomenon, which gives rise to the residual charge in the Leyden jar, and to several phenomena of electric cables described by Mr. F. Jenkin.

(14) We have here two other kinds of yielding besides the yielding of the perfect dielectric, which we have compared to a perfectly elastic body. The yielding due to conductivity may be compared to that of a viscous fluid (that is to say, a fluid having great internal friction), or a soft solid on which the smallest force produces a permanent alteration of figure increasing with the time during which the force acts. The yielding due to electric absorption may be compared to that of a cellular elastic body containing a thick fluid in its cavities. Such a body, when subjected to pressure, is compressed by degrees on account of the gradual yielding of the thick fluid; and when the pressure is removed it does not at once recover its figure, because the elasticity of the substance of the body has gradually to overcome the tenacity of the fluid before it can regain complete equilibrium.

Several solid bodies in which no such structure as we have supposed can be found, seem to possess a mechanical property of this kind; and it seems probable that the same substances, if dielectrics, may possess the analogous electrical property, and if magnetic, may have corresponding properties relating to the acquisition, retention, and loss of magnetic polarity.

(15) It appears therefore that certain phenomena in electricity and magnetism lead to the same conclusion as those of optics, namely, that there is an aethereal medium pervading all bodies, and modified only in degree by their presence; that the parts of this medium are capable of being set in motion by electric currents and magnets; that this motion is communicated from one part of the medium to another by forces arising from the connexions

of those parts; that under the action of these forces there is a certain yielding depending on the elasticity of these connexions; and that therefore energy in two different forms may exist in the medium, the one form being the actual energy of motion of its parts, and the other being the potential energy stored up in the connexions, in virtue of their elasticity.

(16) Thus, then, we are led to the conception of a complicated mechanism capable of a vast variety of motion, but at the same time so connected that the motion of one part depends, according to definite relations, on the motion of other parts, these motions being communicated by forces arising from the relative displacement of the connected parts, in virtue of their elasticity. Such a mechanism must be subject to the general laws of Dynamics, and we ought to be able to work out all the consequences of its motion, provided we know the form of the relation between the motions of the parts.

(17) We know that when an electric current is established in a conducting circuit, the neighbouring part of the field is characterized by certain magnetic properties, and that if two circuits are in the field, the magnetic properties of the field due to the two currents are combined. Thus each part of the field is in connexion with both currents, and the two currents are put in connexion with each other in virtue of their connexion with the magnetization of the field. The first result of this connexion that I propose to examine, is the induction of one current by another, and by the motion of conductors in the field.

The second result, which is deduced from this, is the mechanical action between conductors carrying currents. The phenomenon of the induction of currents has been deduced from their mechanical action by Helmholtz and Thomson. I have followed the reverse order, and deduced the mechanical action from the laws of induction. I have then described experimental methods of determining the quantities L, M, N, on which these phenomena depend.

(18) I then apply the phenomena of induction and attraction of currents to the exploration of the electromagnetic field, and the laying down systems of lines of magnetic force which indicate its magnetic properties. By exploring the same field with a magnet, I shew the distribution of its equipotential magnetic surfaces, cutting the lines of force at right angles.

In order to bring these results within the power of symbolical calculation, I then express them in the form of the General Equations of the Electromagnetic Field. These equations express—

(A) The relation between electric displacement, true conduction, and the total current, compounded of both.

(B) The relation between the lines of magnetic force and the inductive coefficients of a circuit, as already deduced from the laws of induction.

(C) The relation between the strength of a current and its magnetic effects, according to the electromagnetic system of measurement.

(D) The value of the electromotive force in a body, as arising from the motion of the body in the field, the alteration of the field itself, and the variation of electric potential from one part of the field to another.

(E) The relation between electric displacement, and the electromotive force which produces it.

(F) The relation between an electric current, and the electromotive force which produces it.

(G) The relation between the amount of free electricity at any point, and the electric displacements in the neighbourhood.

(H) The relation between the increase or diminution of free electricity and the electric currents in the neighbourhood.

There are twenty of these equations in all, involving twenty variable quantities.

(19) I then express in terms of these quantities the intrinsic energy of the Electromagnetic Field as depending partly on its magnetic and partly on its electric polarization at every point.

From this I determine the mechanical force acting, 1st, on a moveable conductor carrying an electric current; 2ndly, on a magnetic pole; 3rdly, on an electrified body.

The last result, namely, the mechanical force acting on an electrified body, gives rise to an independent method of electrical measurement founded on its electrostatic effects. The relation between the units employed in the two methods is shewn to depend on what I have called the "electric elasticity" of the medium, and to be a velocity, which has been experimentally determined by MM. Weber and Kohlrausch.

I then shew how to calculate the electrostatic capacity of a condenser, and the specific inductive capacity of a dielectric.

The case of a condenser composed of parallel layers of substances of different electric resistances and inductive capacities is next examined, and it is shewn that the phenomenon called electric absorption will generally occur, that is, the condenser, when suddenly discharged, will after a short time shew signs of a *residual* charge.

(20) The general equations are next applied to the case of a magnetic disturbance propagated through a non-conducting field, and it is shewn that the only disturbances which can be so propagated are those which are transverse to the direction of propagation, and that the velocity of propagation is the velocity v, found from experiments such as those of Weber, which expresses the number of electrostatic units of electricity which are contained in one electromagnetic unit.

This velocity is so nearly that of light, that it seems we have strong reason to conclude that light itself (including radiant heat, and other radiations if any) is an electromagnetic disturbance in the form of waves propagated through the electromagnetic field according to electromagnetic laws. If so, the agreement between the elasticity of the medium as calculated from the rapid alternations of luminous vibrations, and as found by the slow processes of electrical experiments, shews how perfect and regular the elastic properties of the medium must be when not encumbered with any matter denser than air. If the same character of the elasticity is retained in dense transparent bodies, it appears that the square of the index of refraction is equal to the product of the specific dielectric capacity and the specific magnetic capacity. Conducting media are shewn to absorb such radiations rapidly, and therefore to be generally opaque.

The conception of the propagation of transverse magnetic disturbances to the exclusion of normal ones is distinctly set forth by Professor Faraday in his "Thoughts on Ray Vibrations." The electromagnetic theory of light, as proposed by him, is the same in substance as that which I have begun to develope in this paper, except that in 1846 there were no data to calculate the velocity of propagation.

(21) The general equations are then applied to the calculation of the coefficients of mutual induction of two circular currents and the coefficient of self-induction in a coil. The want of uniformity of the current in the different parts of the section of a wire at the commencement of the current is investigated, I believe for the first time, and the consequent correction of the coefficient of self-induction is found.

These results are applied to the calculation of the self-induction of the coil used in the experiments of the Committee of the British

Association on Standards of Electric Resistance, and the value compared with that deduced from the experiments.

ROWLAND

The following extract is from a paper "On the Magnetic Effect of Electric Convection," which appeared in the *American Journal of Science*, Series 3, Vol. 15, p. 30, 1878. It is Rowland's (p. 365) first important contribution to physics.

THE MAGNETIC EFFECT OF ELECTRIC CONVECTION

The experiments described in this paper were made with a view of determining whether or not an electrified body in motion produces magnetic effects. There seems to be no theoretical ground upon which we can settle the question, seeing that the magnetic action of a conducted electric current may be ascribed to some mutual action between the conductor and the current. Hence an experiment is of value. Professor Maxwell, in his "Treatise on Electricity," Art. 770, has computed the magnetic action of a moving electrified surface, but that the action exists has not yet been proved experimentally or theoretically.

The apparatus employed consisted of a vulcanite disc 21.1 centimetres in diameter and .5 centimetre thick which could be made to revolve around a vertical axis with a velocity of 61 turns per second. On either side of the disc at a distance of .6 cm. were fixed glass plates having a diameter of 38.9 cm. and a hole in the centre of 7.8 cm. The vulcanite disc was gilded on both sides and the glass plates had an annular ring of gilt on one side, the outside and inside diameters being 24.0 cm. and 8.9 cm. respectively. The gilt sides could be turned toward or from the revolving disc but were usually turned toward it so that the problem might be calculated more readily and there should be no uncertainty as to the electrification. The outside plates were usually connected with the earth; and the inside disc with an electric battery, by means of a point which approached within one-third of a millimetre of the edge and turned toward it. As the edge was broad, the point would not discharge unless there was a difference of potential between it and the edge. Between the electric battery and the disc, a commutator was placed, so that the potential of the latter could be made plus or minus at will. All parts of the apparatus were of non-magnetic material.

Over the surface of the disc was suspended, from a bracket in the wall, an extremely delicate astatic needle, protected from electric action and currents of air by a brass tube. The two needles were 1.5 cm. long and their centres 17.98 cm. distant from each other. The readings were by a telescope and scale. The opening in the tube for observing the mirror was protected from electrical action by a metallic cone, the mirror being at its vertex. So perfectly was this accomplished that no effect of electrical action was apparent either on charging the battery or reversing the electrification of the disc. The needles were so far apart that any action of the disc would be many fold greater on the lower needle than the upper. The direction of the needles was that of the motion of the disc directly below them, that is, perpendicular to the radius drawn from the axis to the needle. As the support of the needle was the wall of the laboratory and the revolving disc was on a table beneath it, the needle was reasonably free from vibration.

In the first experiments with this apparatus no effect was observed other than a constant deflection which was reversed with the direction of the motion. This was finally traced to the magnetism of rotation of the axis and was afterward greatly reduced by turning down the axis to .9 cm. diameter. On now rendering the needle more sensitive and taking several other precautions a distinct effect was observed of several millimetres on reversing the electrification and it was separated from the effect of magnetism of rotation by keeping the motion constant and reversing the electrification. As the effect of the magnetism of rotation was several times that of the moving electricity, and the needle was so extremely sensitive, numerical results were extremely hard to be obtained, and it is only after weeks of trial that reasonably accurate results have been obtained. But the qualitative effect, after once being obtained, never failed. In hundreds of observations extending over many weeks, the needle always answered to a change of electrification of the disc. Also on raising the potential above zero the action was the reverse of that when it was lowered below. The *swing* of the needle on reversing the electrification was about 10 or 15 millimetres and therefore the point of equilibrium was altered 5 or 7½ millimetres. This quantity varied with the electrification, the velocity of motion, the sensitiveness of the needle, etc.

The direction of the action may be thus defined. Calling the motion of the disc + when it moved like the hands of a watch laid on the table with its face up, we have the following, the needles being over one side of the disc, with the north pole pointing in the direction of positive motion. The motion being +, on electrifying the disc + the north pole moved toward the axis, and on changing the electrification, the north pole moved away from the axis. With − motion and + electrification, the north pole moved away from the axis, and with − electrification, it moved toward the axis. The direction is therefore that in which we should expect it to be.

To prevent any suspicion of currents in the gilded surfaces, the latter, in many experiments, were divided into small portions by radial scratches, so that no tangential currents could take place without sufficient difference of potential to produce sparks. But to be perfectly certain, the gilded disc was replaced by a plane thin glass plate which could be electrified by points on one side, a gilded induction plate at zero potential being on the other. With this arrangement, effects in the same direction as before were obtained, but smaller in quantity, seeing that only one side of the plate could be electrified.

The inductor plates were now removed, leaving the disc perfectly free, and the latter was once more gilded with a continuous gold surface, having only an opening around the axis of 3.5 cm. The gilding of the disc was connected with the axis and so was at a potential of zero. On one side of the plate, two small inductors formed of pieces of tinfoil on glass plates, were supported, having the disc between them. On electrifying these, the disc at the points opposite them was electrified by induction but there could be no electrification except at points near the inductors. On now revolving the disc, if the inductors were very small, the electricity would remain nearly at rest and the plate would as it were revolve through it. Hence in this case we should have conduction without motion of electricity, while in the first experiment we had motion without conduction. I have used the term "nearly at rest" in the above, for the following reasons. As the disc revolves the electricity is being constantly conducted in the plate so as to retain its position. Now the function which expresses the potential producing these currents and its differential coefficients must be continuous throughout the disc, and so these currents must pervade the whole disc.

To calculate these currents we have two ways. Either we can consider the electricity at rest and the motion of the disc through it to produce an electromotive force in the direction of motion and proportional to the velocity of motion, to the electrification, and to the surface resistance; or, as Professor Helmholtz has suggested, we can consider the electricity to move with the disc and as it comes to the edge of the inductor to be set free to return by conduction currents to the other edge of the inductor so as to supply the loss there. The problem is capable of solution in the case of a disc without a hole in the centre but the results are too complicated to be of much use. Hence scratches were made on the disc in concentric circles about .6 cm. apart by which the radial component of the currents was destroyed and the problem became easily calculable.

For, let the inductor cover $\frac{1}{n}$th part of the circumference of any one of the conducting circles; then, if C is a constant, the current in the circle outside the inductor will be $+C/n$, and inside the area of the inductor $-C\frac{(n-1)}{n}$. On the latter is superposed the convection current equal to $+C$. Hence the motion of electricity throughout the whole circle is $1/n$ what it would have been had the inductor covered the whole circle.

In one experiment n was about 8. By comparison with the other experiments we know that had electric conduction alone produced effect we should have observed at the telescope -5. mm. Had electric convection alone produced magnetic effect we should have had $+5.7$ mm. And if they both had effect it would have been $+.7$ mm., which is practically zero in the presence of so many disturbing causes. No effect was discovered or at least no *certain* effect, though every care was used. Hence we may conclude with reasonable certainty that electricity produces nearly if not quite the same magnetic effect in the case of convection as of conduction, provided the same quantity of electricity passes a given point in the convection stream as in the conduction stream.

. . . .

HALL

Edwin Herbert Hall was born on November 7, 1855, in Gorham, Maine. He studied at Bowdoin College and at Johns Hopkins University, where he

received the doctor's degree in 1880.   In the next year he became connected
with Harvard University as an instructor in physics.   He advanced until in
1895 he became professor of physics.   He was made professor emeritus in 1921.

The extract which follows is from a paper entitled "On a New Action of the
Magnet on Electric Currents." This paper appeared in the *American Journal
of Mathematics*, Vol. 2, p. 287, 1879, and also in the *Philosophical Magazine*,
Vol. 9, Series 5, p. 225, 1880. The phenomenon which is described in it is
known as the Hall effect.

THE HALL EFFECT

Sometime during the last University year, while I was reading
Maxwell's "Electricity and Magnetism" in connexion with Professor Rowland's lectures, my attention was particularly attracted
by the following passage in vol. ii. p. 144:—

"It must be carefully remembered, that the mechanical force
which urges a conductor carrying a current across the lines of
magnetic force, acts, not on the electric current, but on the conductor which carries it.   If the conductor be a rotating disk or a
fluid, it will move in obedience to this force; and this motion may
or may not be accompanied with a change of position of the electric
current which it carries.   But if the current itself be free to choose
any path through a fixed solid conductor or a network of wires,
then, when a constant magnetic force is made to act on the system,
the path of the current through the conductors is not permanently
altered, but after certain transient phenomena, called induction-
currents, have subsided, the distribution of the current will be
found to be the same as if no magnetic force were in action.   The
only force which acts on electric currents is electromotive force,
which must be distinguished from the mechanical force which
is the subject of this chapter."

This statement seemed to me to be contrary to the most natural
supposition in the case considered, taking into account the fact
that a wire not bearing a current is in general not affected by a
magnet, and that a wire bearing a current is affected exactly in
proportion to the strength of the current, while the size and, in
general, the material of the wire are matters of indifference.
Moreover, in explaining the phenomena of statical electricity,
it is customary to say that charged bodies are attracted toward
each other or the contrary solely by the attraction or repulsion
of the charges for each other.

Soon after reading the above statement in Maxwell I read an
article by Prof. Edlund, entitled "Unipolar Induction" (Phil. Mag.

Oct. 1878, or *Annales de Chimie et de Physique,* Jan. 1879), in which the author evidently assumes that a magnet acts upon a current in a fixed conductor just as it acts upon the conductor itself when free to move.

Finding these two authorities at variance, I brought the question to Prof. Rowland. He told me he doubted the truth of Maxwell's statement, and had some time before made a hasty experiment for the purpose of detecting, if possible, some action of the magnet on the current itself, though without success. Being very busy with other matters however, he had no immediate intention of carrying the investigation further.

I now began to give the matter more attention, and hit upon a method that seemed to promise a solution of the problem. I laid my plan before Prof. Rowland, and asked whether he had any objection to my making the experiment. He approved of my method in the main, though suggesting some very important changes in the proposed form and arrangement of the apparatus. The experiment proposed was suggested by the following reflection:—If the current of electricity in a fixed conductor is itself attracted by a magnet, the current should be drawn to one side of the wire, and therefore the resistance experienced should be increased.

To test this theory, a flat spiral of German-silver wire was enclosed between two thin disks of hard rubber, and the whole placed between the poles of an electromagnet in such a position that the lines of magnetic force would pass through the spiral at right angles to the current of electricity. The wire of the spiral was about $\frac{1}{2}$ millim. in diameter, and the resistance of the spiral was about two ohms. The magnet was worked by a battery of twenty Bunsen cells joined four in series and five abreast. The strength of the magnetic field in which the coil was placed was probably fifteen or twenty thousand times H, the horizontal intensity of the earth's magnetism.

Making the spiral one arm of a Wheatstone's bridge, and using a low-resistance Thomson galvanometer, so delicately adjusted as to betray a change of about one part in a million in the resistance of the spiral, I made, from October 7th to October 11th inclusive, thirteen series of observations, each of forty readings. A reading would first be made with the magnet active in a certain direction, then a reading with the magnet inactive, then one with the magnet active in the direction opposite to the first, then with the magnet inactive, and so on till the series of forty readings was completed.

Some of the series seemed to show a slight increase of resistance due to the action of the magnet, some a slight decrease, the greatest change indicated by any complete series being a decrease of about one part in a hundred and fifty thousand. Nearly all the other series indicated a very much smaller change, the average change shown by the thirteen series being a decrease of about one part in five millions.

Apparently, then, the magnet's action caused no change in the resistance of the coil.

But though conclusive, apparently, in respect to any change of resistance, the above experiments are not sufficient to prove that a magnet cannot affect an electric current. If electricity is assumed to be an incompressible fluid, as some suspect it to be, we may conceive that the current of electricity flowing in a wire cannot be forced into one side of the wire or made to flow in any but a symmetrical manner. The magnet may *tend* to deflect the current without being able to do so. It is evident, however, that in this case there would exist a state of stress in the conductor, the electricity pressing, as it were, toward one side of the wire. Reasoning thus, I thought it necessary, in order to make a thorough investigation of the matter, to test for a difference of potential between points on opposite sides of the conductor.

This could be done by repeating the experiment formerly made by Prof. Rowland, and which was the following:—A disk or strip of metal, forming part of an electric circuit, was placed between the poles of an electromagnet, the disk cutting across the lines of force. The two poles of a sensitive galvanometer were then placed in connexion with different parts of the disk, through which an electric current was passing, until two nearly equipotential points were found. The magnet-current was then turned on and the galvanometer was observed, in order to detect any indication of a change in the relative potential of the two poles.

Owing probably to the fact that the metal disk used had considerable thickness, the experiment at that time failed to give any positive result. Prof. Rowland now advised me, in repeating this experiment, to use gold-leaf mounted on a plate of glass as my metal strip. I did so, and, experimenting as indicated above, succeeded on the 28th of October in obtaining, as the effect of the magnet's action, a decided deflexion of the galvanometer-needle.

This deflexion was much too large to be attributed to the direct action of the magnet on the galvanometer-needle, or to any similar cause. It was, moreover, a permanent deflexion, and therefore not to be accounted for by induction. The effect was reversed when the magnet was reversed. It was not reversed by transferring the poles of the galvanometer from one end of the strip to the other. In short, the phenomena observed were just such as we should expect to see if the electric current were pressed, but not moved, toward one side of the conductor.

In regard to the direction of this pressure or tendency, as dependent on the direction of the current in the gold-leaf and the direction of the lines of magnetic force, the following statement may be made:—If we regard an electric current as a single stream flowing from the positive to the negative pole, i.e. from the carbon pole of the battery through the circuit to the zinc pole, in this case the phenomena observed indicate that two *currents*, parallel and in the same direction, tend to repel each other. If, on the other hand, we regard the electric current as a stream flowing from the negative to the positive pole, in this case the phenomena observed indicate that two *currents* parallel and in the same direction tend to attract each other.

It is, of course, perfectly well known that two *conductors*, bearing currents parallel and in the same direction, are drawn toward each other. Whether this fact, taken in connexion with what has been said above, has any bearing upon the question of the absolute direction of the electric current, it is perhaps too early to decide.

In order to make some rough quantitative experiments, a new plate was prepared, consisting of a strip of gold-leaf about 2 centims. wide and 9 centims. long mounted on plate-glass. Good contact was ensured by pressing firmly down on each strip of gold-leaf a thick piece of brass polished on the underside. To these pieces of brass the wires from a single Bunsen cell were soldered. The portion of the gold-leaf strip not covered by the pieces of brass was about $5\frac{1}{2}$ centims. in length, and had a resistance of about 2 ohms. The poles of a high-resistance Thomson galvanometer were placed in connexion with points opposite each other on the edges of the strip of gold-leaf, and midway between the pieces of brass. The glass plate bearing the gold-leaf was fastened, as the first one had been, by a soft cement to the flat end of one pole of the magnet, the other pole of the magnet being brought to within about 6 millims. of the strip of gold-leaf.

The apparatus being arranged as above described, on the 12th of November a series of observations was made for the purpose of determining the variations of the observed effect with known variations of the magnetic force and the strength of current through the gold-leaf.

The experiments were hastily and roughly made, but are sufficiently accurate, it is thought, to determine the law of variation above mentioned as well as the order of magnitude of the current through the Thomson galvanometer compared with the current through the gold-leaf and the intensity of the magnetic field.

The results obtained are as follows:—

Current through gold-leaf strip C.	Strength of magnetic field M.	Current through Thomson galvanometer c.	$\dfrac{C \times M}{c}$
.0616	11420H	.00000000232	303000000000.
.0249	11240"	. 085	329 .
.0389	11060"	. 135	319 .
.0598	7670"	. 147	312 .
.0595	5700"	. 104	326 .

H is the horizontal intensity of the earth's magnetism = .19 approximately.

Though the greatest difference in the last column above amounts to about 8 per cent. of the mean quotient, yet it seems safe to conclude that, with a given form and arrangement of apparatus, the action on the Thomson galvanometer is proportional to the product of the magnetic force by the current through the gold-leaf. This is not the same as saying that the effect on the Thomson galvanometer is under all circumstances proportional to the current which is passing between the poles of the magnet. If a strip of copper of the same length and breadth as the gold-leaf, but ¼ millim. in thickness, is substituted for the latter, the galvanometer fails to detect any current arising from the action of the magnet, except an induction-current at the moment of making or breaking the magnet circuit.

It has been stated above that in the experiments thus far tried the current apparently tends to move, without actually moving, toward the side of the conductor. I have in mind a form of apparatus which will, I think, allow the current to follow this tendency, and move across the lines of magnetic force. If this experiment succeeds, one or two others immediately suggest themselves.

To make a more complete and accurate study of the phenomena described in the preceding pages, availing myself of the advice and assistance of Prof. Rowland, will probably occupy me for some months to come.

Baltimore, Nov. 19th, 1879.

It is perhaps allowable to speak of the action of the magnet as setting up in the strip of gold-leaf a new electromotive force at right angles to the primary electromotive force.

This new electromotive force cannot under ordinary conditions manifest itself, the circuit in which it might work being incomplete. When the circuit is completed by means of the Thomson galvanometer, a current flows.

The actual current through this galvanometer depends of course upon the resistance of the galvanometer and its connexions, as well as upon the distance between the two points of the gold-leaf at which the ends of the wires from the galvanometer are applied. We cannot, therefore, take the ratio of C and c above as the ratio of the primary and the transverse electromotive forces just mentioned.

If we represent by E' the difference of potential of two points a centimetre apart on the transverse diameter of the strip of gold-leaf, and by E the difference of potential of two points a centimetre apart on the longitudinal diameter of the same, a rough and hasty calculation for the experiments already made shows the ratio E/E' to have varied from about 3000 to about 6500.

The transverse electromotive force E' seems to be, under ordinary circumstances, proportional to Mv, where M is the intensity of the magnetic field and v is the *velocity* of the electricity in the gold-leaf. Writing for v the equivalent expression C/s, where C is the primary current through a strip of the gold-leaf 1 centim. wide, and s the area of section of the same, we have $E' \propto \dfrac{MC}{c}$.

November 22nd, 1879.

P. CURIE

The following short paper is the first of several on the same subject. It deals with the development by pressure of electric polarization in hemihedral crystals with inclined faces.   The work was done in collaboration with Jacques Curie and appeared in the *Comptes Rendus*, Vol. 91, p. 294, 1880.

It was soon recognized by both Friedel and Curie that the pyroelectric phenomena observed by Friedel's method were not those of pyroelectricity as generally understood and that the effects produced by the heat in Friedel's method of observation really resulted from pressures introduced by irregular heating.

Piezoelectricity

Crystals which have one or more axes whose ends are unlike, that is to say, hemihedral crystals with inclined faces, have a special physical property, that they exhibit two electric poles of opposite names at the ends of those axes, when they undergo a change of temperature: this is the phenomenon known as pyroelectricity.

We have found a new way to develop electric polarisation in crystals of this sort, which consists in subjecting them to different pressures along their hemihedral axes.

The effects produced are analogous to those caused by heat: during a compression, the ends of the axis along which we are acting are charged with opposite electricities; when the crystal is brought back to the neutral state and the compression is relieved, the phenomenon occurs again, but with the signs reversed; the end which was positively charged by compression becomes negative when the compression is removed and reciprocally.

To make an experiment we cut two faces parallel to each other, and perpendicular to a hemihedral axis, in the substance which we wish to study; we cover these faces with two sheets of tin which are insulated on their outer sides by two sheets of hard rubber; when the whole thing is placed between the jaws of a vise, for example, we can exert pressure on the two cut surfaces, that is to say, along the hemihedral axis itself.   To perceive the electrification we used a Thomson electrometer.   We may show the difference of potential between the ends by connecting each sheet of tin with two of the sectors of the instrument while the needle is charged with a known sort of electricity.   We may also recognize each of the electricities separately; to do this we connect one of the tin sheets with the earth, the other with the needle, and we charge the two pairs of sectors from a battery.

Although we have not yet undertaken the study of the laws of this phenomenon, we are able to say that the characteristics which it exhibits are identical with those of pyroelectricity, as they have been described by Gaugain in his beautiful work on tourmaline.

We have made a comparative study of the two ways of developing electric polarisation in a series of non-conducting substances, hemihedral with inclined faces, which includes almost all those which are known as pyroelectric.

The action of heat has been studied by the process indicated by M. Friedel, a process which is very convenient.

Our experiments have been made on blende, sodium chlorate, boracite, tourmaline, quartz, calamine, topaz, tartaric acid (right handed), sugar, and Seignette's salt.

In all these crystals the effects produced by compression are in the same sense as those produced by cooling; those which result from relieving the pressure are in the same sense as those which come from heating.

There is here an evident relation which allows us to refer the phenomena in both cases to the same cause and to bring them under the following statement:

Whatever may be the determining cause, whenever a hemihedral crystal with inclined faces, which is also a non-conductor, contracts, electric poles are formed in a certain sense; whenever the crystal expands the electricities are separated in the opposite sense.

If this way of looking at the matter is correct the effects arising from compression ought to be in the same sense as those resulting from heating in a substance which has a negative coefficient of dilatation along the hemihedral axis.

HERTZ

Heinrich Rudolph Hertz was born on February 22, 1857, in Hamburg. He studied engineering and physics at Munich and Berlin. He was graduated in 1880 from the University of Berlin and served for a few years as Helmholtz' assistant. He was privatdocent for theoretical physics at Kiel and professor of physics in the technical high school at Karlsruhe. It was while there that he carried out the researches on electric waves which made him famous. They were universally recognized as an experimental confirmation of Maxwell's theory. In 1889 he became professor of physics at the University of Bonn. His health soon failed and he died on January 1, 1894.

The first paper published by Hertz on this subject appeared in the *Annalen der Physik*, Vol. 31, 1887. The last one of the series appeared in 1890. These papers were collected and published in 1892 and an English version by D. E. Jones was published in 1893.

The particular paper chosen for reproduction is the one entitled "Ueber Strahlen elektrischer Kraft," originally published in the *Sitzungsberichte der Berliner Akademie der Wissenschaften*, December 13, 1888, and in the *Annalen der Physik*, Vol. 36, p. 769, 1889.   It contains an account of experiments by which some of the properties of electric waves were investigated.   The translation is that of Jones.

ELECTRIC RADIATION

As soon as I had succeeded in proving that the action of an electric oscillation spreads out as a wave into space, I planned experiments with the object of concentrating this action and making it perceptible at greater distances by putting the primary conductor in the focal line of a large concave parabolic mirror. These experiments did not lead to the desired result, and I felt certain that the want of success was a necessary consequence of the disproportion between the length (4–5 metres) of the waves used and the dimensions which I was able, under the most favourable circumstances, to give to the mirror.   Recently I have observed that the experiments which I have described can be carried out quite well with oscillations of more than ten times the frequency, and with waves less than one-tenth the length of those which were first discovered.   I have, therefore, returned to the use of concave mirrors, and have obtained better results than I had ventured to hope for.   I have succeeded in producing distinct rays of electric force, and in carrying out with them the elementary experiments which are commonly performed with light and radiant heat.   The following is an account of these experiments:—

The Apparatus.

The short waves were excited by the same method which we used for producing the longer waves.   The primary conductor used may be most simply described as follows:—Imagine a cylindrical brass body, 3 cm. in diameter and 26 cm. long, interrupted midway along its length by a spark-gap whose poles on either side are formed by spheres of 2 cm. radius.   The length of the conductor is approximately equal to the half wave-length of the corresponding oscillation in straight wires; from this we are at once able to estimate approximately the period of oscillation. It is essential that the pole-surfaces of the spark-gap should be frequently repolished, and also that during the experiments they should be carefully protected from illumination by simultaneous side-discharges; otherwise the oscillations are not excited. Whether the spark-gap is in a satisfactory state can always be recognized by the appearance and sound of the sparks. The discharge is led to the two halves of the conductor by means of two gutta-percha-covered wires which are connected near the spark-gap on either side. I no longer made use of the large Ruhmkorff, but found it better to use a small induction-coil by Keiser and Schmidt; the longest sparks, between points, given by this were 4.5 cm. long. It was supplied with current from three accumulators, and gave sparks 1–2 cm. long between the spherical knobs of the primary conductor. For the purpose of the experiments the spark-gap was reduced to 3 mm.

Here, again, the small sparks induced in a secondary conductor were the means used for detecting the electric forces in space. As before, I used partly a circle which could be rotated within itself and which had about the same period of oscillation as the primary conductor. It was made of copper wire 1 mm. thick, and had in the present instance a diameter of only 7.5 cm. One end of the wire carried a polished brass sphere a few millimetres in diameter; the other end was pointed and could be brought up, by means of a fine screw insulated from the wire, to within an exceedingly short distance from the brass sphere. As will be readily understood, we have here to deal only with minute sparks of a few hundredths of a millimetre in length; and after a little practice one judges more according to the brilliancy than the length of the sparks.

The circular conductor gives only a differential effect, and is not adapted for use in the focal line of a concave mirror. Most of the work was therefore done with another conductor arranged as follows:—Two straight pieces of wire, each 50 cm. long and 5 mm. in diameter, were adjusted in a straight line so that their near ends were 5 cm. apart. From these ends two wires, 15 cm. long and 1 mm. in diameter, were carried parallel to one another and perpendicular to the wires first mentioned to a spark-gap arranged just as in the circular conductor. In this conductor the resonance-action was given up, and indeed it only comes slightly into play in this case. It would have been simpler to put the spark-gap directly in the middle of the straight wire; but the observer could not then have handled and observed the spark-gap in the focus of the mirror without obstructing the aperture. For this reason the arrangement above described was chosen in preference to the other which would in itself have been more advantageous.

The Production of the Ray.

If the primary oscillator is now set up in a fairly large free space, one can, with the aid of the circular conductor, detect in its neighbourhood on a smaller scale all those phenomena which I have already observed and described as occurring in the neighbourhood of a larger oscillation. The greatest distance at which sparks could be perceived in the secondary conductor was 1.5 metre, or, when the primary spark-gap was in very good order, as much as 2 metres. When a plane reflecting plate is set up at a suitable distance on one side of the primary oscillator, and parallel to it, the action on the opposite side is strengthened. To be more precise:—If the distance chosen is either very small, or somewhat greater than 30 cm., the plate weakens the effect; it strengthens the effect greatly at distances of 8–15 cm., slightly at a distance of 45 cm., and exerts no influence at greater distances. We have drawn attention to this phenomenon in an earlier paper, and we conclude from it that the wave in air corresponding to the primary oscillation has a half wave-length of about 30 cm. We may expect to find a still further reinforcement if we replace the plane surface by a concave mirror having the form of a parabolic cylinder, in the focal line of which the axis of the primary oscillation lies. The focal length of the mirror should be chosen as small as possible, if it is properly to concentrate the action. But if the direct wave is not to annul immediately the action of the reflected wave, the focal length must not be much smaller than a quarter wave-length. I therefore fixed on 12½ cm. as the focal length, and constructed the mirror by bending a zinc sheet 2 metres long, 2 metres broad, and ½ mm. thick into the desired shape over a wooden frame of the exact curvature. The height of the mirror was thus 2 metres, the breadth of its aperture 1.2 metre, and its depth 0.7 metre. The primary oscillator was fixed in the middle of the focal line. The wires which conducted the discharge were led through the mirror; the induction-coil and the cells were accordingly placed behind the mirror so as to be out of the way. If we now investigate the neighbourhood of the oscillator with our conductors, we find that there is no action behind the mirror or at either side of it; but in the direction of the optical axis of the mirror the sparks can be perceived up to a distance of 5–6 metres. When a plane conducting surface was set up so as to oppose the advancing waves at right angles, the sparks could be detected in its neighbourhood at even greater distances—up to about 9–10 metres. The waves reflected from the conducting surface reinforce the advancing waves at certain points. At other points again the two sets of waves weaken one another. In front of the plane wall one can recognize with the rectilinear conductor very distinct maxima and minima, and with the circular conductor the characteristic interference-phenomena of stationary waves which I have described in an earlier paper. I was able to distinguish four nodal points, which were situated at the wall and at 33, 65, and 98 cm. distance from it. We thus get 33 cm. as a closer approximation to the half wave-length of the waves used, and 1.1 thousand-millionth of a second as their period of oscillation, assuming that they travel with the velocity of light. In wires the oscillation gave a wave-length of 29 cm. Hence it appears that these short waves also have a somewhat lower velocity in wires than in air; but the ratio of the two velocities comes very near to the theoretical value—unity—and does not differ from it so much as appeared to be probable from our experiments on longer waves. This remarkable phenomenon still needs elucidation. Inasmuch as the phenomena are only exhibited in the neighbourhood of the optic axis of the mirror, we may speak of the result produced as an electric ray proceeding from the concave mirror.

I now constructed a second mirror, exactly similar to the first, and attached the rectilinear secondary conductor to it in such a way that the two wires of 50 cm. length lay in the focal line, and the two wires connected to the spark-gap passed directly through the walls of the mirror without touching it. The spark-gap was thus situated directly behind the mirror, and the observer could adjust and examine it without obstructing the course of the waves. I expected to find that, on intercepting the ray with this apparatus, I should be able to observe it at even greater distances; and the event proved that I was not mistaken. In the rooms at my disposal I could now perceive the sparks from one end to the other. The greatest distance to which I was able, by availing myself of a doorway, to follow the ray was 16 metres; but according to the results of the reflection-experiments (to be presently described), there can be no doubt that sparks could be obtained at any rate up to 20 metres in open spaces. For the remaining experiments such great distances are not necessary, and it is convenient that the sparking in the secondary conductor should not be too feeble; for most of the experiments a distance of 6–10 metres is most suitable. We shall now describe the simple phenomena which can be exhibited with the ray without difficulty. When the contrary is not expressly stated, it is to be assumed that the focal lines of both mirrors are vertical.

Rectilinear Propagation.

If a screen of sheet zinc 2 metres high and 1 metre broad is placed on the straight line joining both mirrors, and at right angles to the direction of the ray, the secondary sparks disappear completely. An equally complete shadow is thrown by a screen of tinfoil or gold-paper. If an assistant walks across the path of the ray, the secondary spark-gap becomes dark as soon as he intercepts the ray, and again lights up when he leaves the path clear. Insulators do not stop the ray—it passes right through a wooden partition or door; and it is not without astonishment that one sees the sparks appear inside a closed room. If two conducting screens, 2 metres high and 1 metre broad, are set up symmetrically on the right and left of the ray, and perpendicular to it, they do not interfere at all with the secondary spark so long as the width of the opening between them is not less than the aperture of the mirrors, viz., 1.2 metre. If the opening is made narrower the sparks become weaker, and disappear when the width of the opening is reduced below 0.5 metre. The sparks also disappear if the opening is left with a breadth of 1.2 metre, but is shifted to one side of the straight line joining the mirrors. If the optical axis of the mirror containing the oscillator is rotated to the right or left about 10° out of the proper position, the secondary sparks become weak, and a rotation through 15° causes them to disappear.

There is no sharp geometrical limit to either the ray or the shadows; it is easy to produce phenomena corresponding to diffraction. As yet, however, I have not succeeded in observing maxima and minima at the edge of the shadows.

Polarisation.

From the mode in which our ray was produced we can have no doubt whatever that it consists of transverse vibrations and is plane-polarised in the optical sense. We can also prove by experiment that this is the case. If the receiving mirror be rotated about the ray as axis until its focal line, and therefore the secondary conductor also, lies in a horizontal plane, the secondary sparks become more and more feeble, and when the two focal lines are at right angles, no sparks whatever are obtained even if the mirrors are moved close up to one another. The two mirrors behave like the polariser and analyser of a polarisation apparatus.

I next had made an octagonal frame, 2 metres high and 2 metres broad; across this were stretched copper wires 1 mm. thick, the wires being parallel to each other and 3 cm. apart. If the two mirrors were now set up with their focal lines parallel, and the wire screen was interposed perpendicularly to the ray and so that the direction of the wires was perpendicular to the direction of the focal lines, the screen practically did not interfere at all with the secondary sparks. But if the screen was set up in such a way that its wires were parallel to the focal lines, it stopped the ray completely. With regard then, to transmitted energy the screen behaves towards our ray just as a tourmaline plate behaves towards a plane-polarised ray of light. The receiving mirror was now placed once more so that its focal line was horizontal; under these circumstances, as already mentioned, no sparks appeared. Nor were any sparks produced when the screen was interposed in the path of the ray, so long as the wires in the screen were either horizontal or vertical. But if the frame was set up in such a position that the wires were inclined at 45° to the horizontal on either side, then the interposition of the screen immediately produced sparks in the secondary spark-gap. Clearly the screen resolves the advancing oscillation into two components and transmits only that component which is perpendicular to the direction of its wires. This component is inclined at 45° to the focal line of the second mirror, and may thus, after being again resolved by the mirror, act upon the secondary conductor. The phenomenon is exactly analogous to the brightening of the dark field of two crossed Nicols by the interposition of a crystalline plate in a suitable position.

With regard to the polarisation it may be further observed that, with the means employed in the present investigation, we are only able to recognise the electric force. When the primary oscillator is in a vertical position the oscillations of this force undoubtedly take place in the vertical plane through the ray, and are absent in the horizontal plane. But the results of experiments with slowly alternating currents leave no room for doubt that the electric oscillations are accompanied by oscillations of magnetic force which take place in the horizontal plane through the ray and are zero in the vertical plane. Hence the polarisation of the ray does not so much consist in the occurrence of oscillations in the vertical plane, but rather in the fact that the oscillations in the vertical plane are of an electrical nature, while those in the horizontal plane are of a magnetic nature. Obviously, then, the question, in which of the two planes the oscillation in our ray occurs, cannot be answered unless one specifies whether the question relates to the electric or the magnetic oscillation. It was Herr Koláček who first pointed out clearly that this consideration is the reason why an old optical dispute has never been decided.

Reflection.

We have already proved the reflection of the waves from conducting surfaces by the interference between the reflected and the advancing waves, and have also made use of the reflection in the construction of our concave mirrors. But now we are able to go further and to separate the two systems of waves from one another. I first placed both mirrors in a large room side by side, with their apertures facing in the same direction, and their axes converging to a point about 3 metres off. The spark-gap of the receiving mirror naturally remained dark. I next set up a plane vertical wall made of thin sheet zinc, 2 metres high and 2 metres broad, at the point of intersection of the axes, and adjusted it so that it was equally inclined to both. I obtained a vigorous stream of sparks arising from the reflection of the ray by the wall. The sparking ceased as soon as the wall was rotated around a vertical axis through about 15° on either side of the correct position; from this it follows that the reflection is regular, not diffuse. When the wall was moved away from the mirrors, the axes of the latter being still kept converging towards the wall, the sparking diminished very slowly. I could still recognise sparks when the wall was 10 metres away from the mirrors, i.e. when the waves had to traverse a distance of 20 metres. This arrangement might be adopted with advantage for the purpose of comparing the rate of propagation, e.g. through cables.

In order to produce reflection of the ray at angles of incidence greater than zero, I allowed the ray to pass parallel to the wall of the room in which there was a doorway. In the neighbouring room to which this door led I set up the receiving mirror so that its optic axis passed centrally through the door and intersected the direction of the ray at right angles. If the plane conducting surface was now set up vertically at the point of intersection, and adjusted so as to make angles of 45° with the ray and also with the axis of the receiving mirror, there appeared in the secondary conductor a stream of sparks which was not interrupted by closing the door. When I turned the reflecting surface about 10° out of the correct position the sparks disappeared. Thus the reflection is regular, and the angles of incidence and reflection are equal. That the action proceeded from the source of disturbance to the plane mirror, and hence to the secondary conductor, could also be shown by placing shadow-giving screens at different points of this path. The secondary sparks then always ceased immediately; whereas no effect was produced when the screen was placed anywhere else in the room. With the aid of the circular secondary conductor it is possible to determine the position of the wave-front in the ray; this was found to be at right angles to the ray before and after reflection, so that in the reflection it was turned through 90°.

Hitherto the focal lines of the concave mirrors were vertical, and the plane of oscillation was therefore perpendicular to the plane of incidence. In order to produce reflection with the oscillations in the plane of incidence, I placed both mirrors with their focal lines horizontal. I observed the same phenomena as in the previous position; and, moreover, I was not able to recognise any difference in the intensity of the reflected ray in the two cases. On the other hand, if the focal line of the one mirror is vertical, and of the other horizontal, no secondary sparks can be observed. The inclination of the plane of oscillation to the plane of incidence is therefore not altered by reflection provided this inclination has one of the two special values referred to; but in general this statement cannot hold good. It is even questionable whether the ray after reflection continues to be plane-polarised. The interferences which are produced in front of the mirror by the intersecting wave-systems, and which, as I have remarked, give rise to characteristic phenomena in the circular conductor, are most likely to throw light upon all problems relating to the change of phase and amplitude produced by reflection.

One further experiment on reflection from an electrically eolotropic surface may be mentioned. The two concave mirrors were again placed side by side, as in the reflection-experiment first described; but now there was placed opposite to them, as a reflecting surface, the screen of parallel copper wires which has already been referred to. It was found that the secondary spark-gap remained dark when the wires intersected the direction of the oscillations at right angles, but that sparking began as soon as the wires coincided with the direction of the oscillations. Hence the analogy between the tourmaline plate and our surface which conducts in one direction is confined to the transmitted part of the ray. The tourmaline plate absorbs the part which is not transmitted; our surface reflects it. If in the experiment last described the two mirrors are placed with their focal lines at right angles, no sparks can be excited in the secondary conductor by reflection from an isotropic screen; but I proved to my satisfaction that sparks are produced when the reflection takes place from the eolotropic wire grating, provided this is adjusted so that the wires are inclined at 45° to the focal lines. The explanation of this follows naturally from what has been already stated.

Refraction.

In order to find out whether any refraction of the ray takes place in passing from air into another insulating medium, I had a large prism made of so-called hard pitch, a material like asphalt. The base was an isosceles triangle 1.2 metres in the side, and with a refracting angle of nearly 30°. The refracting edge was placed vertical, and the height of the whole prism was 1.5 metres. But since the prism weighed about 12 cwt., and would have been too heavy to move as a whole, it was built up of three pieces, each 0.5 metre high, placed one above the other. The material was cast in wooden boxes which were left around it, as they did not appear to interfere with its use. The prism was mounted on a support of such height that the middle of its refracting edge was at the same height as the primary and secondary spark-gaps. When I was satisfied that refraction did take place, and had obtained some idea of its amount, I arranged the experiment in the following manner:—The producing mirror was set up at a distance of 2.6 metres from the prism and facing one of the refracting surfaces, so that the axis of the beam was directed as nearly as possible towards the centre of mass of the prism, and met the refracting surface at an angle of incidence of 25° (on the side of the normal towards the base). Near the refracting edge and also at the opposite side of the prism were placed two conducting screens which prevented the ray from passing by any other path than that through the prism. On the side of the emerging ray there was marked upon the floor a circle of 2.5 metres radius, having as its centre the centre of mass of the lower end of the prism. Along this the receiving mirror was now moved about, its aperture being always directed towards the centre of the circle. No sparks were obtained when the mirror was placed in the direction of the incident ray produced; in this direction the prism threw a complete shadow. But sparks appeared when the mirror was moved towards the base of the prism, beginning when the angular deviation from the first position was about 11°. The sparking increased in intensity until the deviation amounted to about 22°, and then again decreased. The last sparks were observed with a deviation of about 34°. When the mirror was placed in a position of maximum effect, and then moved away from the prism along the radius of the circle, the sparks could be traced up to a distance of 5–6 metres. When an assistant stood either in front of the prism or behind it the sparking invariably ceased, which shows that the action reaches the secondary conductor through the prism and not in any other way. The experiments were repeated after placing both mirrors with their focal lines horizontal, but without altering the position of the prism. This made no difference in the phenomena observed. A refracting angle of 30° and a deviation of 22° in the neighbourhood of the minimum deviation corresponds to a refractive index of 1.69. The refractive index of pitch-like materials for light is given as being between 1.5 and 1.6. We must not attribute any importance to the magnitude or even the sense of this difference, seeing that our method was not an accurate one, and that the material used was impure.

We have applied the term rays of electric force to the phenomena which we have investigated. We may perhaps further designate them as rays of light of very great wave-length. The experiments described appear to me, at any rate, eminently adapted to remove any doubt as to the identity of light, radiant heat, and electro-magnetic wave-motion. I believe that from now on we shall have greater confidence in making use of the advantages which this identity enables us to derive both in the study of optics and of electricity.

.   .   .   .

Explanation of the Figures. In order to facilitate the repetition and extension of these experiments, I append in the accompanying Figs. 96, 97, *a*, and 97, *b*, illustrations of the apparatus which I used, although these were constructed simply for the purpose of experimenting at the time and without any regard to durability. Fig. 96 shows in plan and elevation (section) the producing mirror.

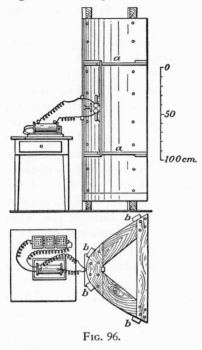

It will be seen that the framework of it consists of two horizontal frames (*a, a*) of parabolic form, and four vertical supports (*b, b*) which are screwed to each of the frames so as to support and connect them. The sheet metal reflector is clamped between the frames and the supports project above and below beyond the sheet metal so that they can be used as handles in handling the mirror. Fig. 97, *a* represents the primary conductor on a somewhat larger scale. The two metal parts slide with friction in two sleeves of strong paper which are held together by indiarubber bands. The sleeves themselves are fastened by four rods of sealing-wax to a board which again is tied by indiarubber bands to a strip of wood forming part of the frame which can be seen in Fig. 96. The two leading wires (covered with gutta-

FIG. 96.

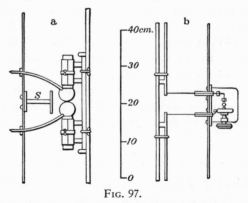

FIG. 97.

percha) terminate in two holes bored in the knobs of the primary conductor. This arrangement allows of all necessary motion and adjustment of the various parts of the conductor; it can be taken to pieces and put together again in a few minutes, and this is essential in order that the knobs may be frequently repolished. Just at the points where the leading wires pass through the mirror, they are surrounded during the discharge by a bluish light. The smooth wooden screen is introduced for the purpose of shielding the spark-gap from this light, which otherwise would interfere seriously with the production of the oscillations. Lastly, Fig. 97, *b* represents the secondary spark-gap. Both parts of the secondary conductor are again attached by sealing-wax rods and indiarubber bands to a slip forming part of the wooden framework. From the inner ends of these parts the leading wires, surrounded by glass tubes, can be seen proceeding through the mirror and bending towards one another. The upper wire carries at its pole a small brass knob. To the lower wire is soldered a piece of watch-spring which carries the second pole, consisting of a fine copper point. The point is intentionally chosen of softer metal than the knob; unless this precaution is taken the point easily penetrates into the knob, and the minute sparks disappear from sight in the small hole thus produced. The figure shows how the point is adjusted by a screw which presses against the spring that is insulated from it by a glass plate. The spring is bent in a particular way in order to secure finer motion of the point than would be possible if the screw alone were used.

No doubt the apparatus here described can be considerably modified without interfering with the success of the experiments. Acting upon friendly advice, I have tried to replace the spark-gap in the secondary conductor by a frog's leg prepared for detecting currents; but this arrangement which is so delicate under other conditions does not seem to be adapted for these purposes.

HITTORF

Johann Wilhelm Hittorf was born on March 27, 1824, in Bonn. After studying at Bonn and Berlin he took his doctor's degree at Bonn in 1846. In 1852 he became professor of physics and chemistry at Münster, where he died on November 28, 1914.

The extract here given is from a paper entitled "Ueber die Elektricitäts-leitung der Gase," which appeared in the *Annalen der Physik und Chemie*, Vol. 136, p. 1, 1869. It contains a more explicit statement than had appeared before of the rectilinear character of the radiation from the cathode.

The Cathode Discharge

In the following description of the conditions which the electric discharge occasions, we shall assume that the medium is air, although the same conditions occur in all gases.

The three parts, the positive light, the dark space, and the negative glow, occur with any density of the medium if the electric discharge lasts for a perceptible time. At ordinary pressures, in the sheath of light which follows the momentary flash of the induction coil, the blue glow on the surface of the cathode covers only a small spot whose magnitude is little different from the cross section of this sheath and, with Neef and DuMoncel, we must bring in magnification to perceive the dark space between the different colored lights. The further the exhaustion of the gas proceeds, over so much the greater surface of the cathode will the same discharge extend its glow. It increases especially fast if the pressure is lowered below 2 mm. When in the axis of a glass tube 60 mm. wide, a wire 82 cm. long was fastened as cathode, the glow, which always begins on the end nearest the anode, covered the whole length of it, for an air pressure which still amounted to $\frac{1}{3}$ mm. One would see the surface of any wire, as long as one pleases, sheathed with it, if the evacuation had gone far enough and the diameter of the tube were taken large enough.

As the density diminishes the glow extends not only over the surface of the cathode, but also out into the surrounding space. The thickness of the blue sheath which surrounds the negative pole increases with special rapidity if the pressure is under 2 mm., and with extreme exhaustion it fills the largest tubes.

For the same density of the medium, the glow stretches out further from the cathode when its surface is smaller. At the same time the tensions increase which excite the discharge at the electrodes and the fluorescence of the glass becomes more vivid. In order to examine this relation we fix on the pump two similar cylindrical tubes which have straight wire cathodes of different lengths in their axes, and we introduce in them, one after the other, the same current from the induction machine, at such exhaustions that both cathodes are entirely covered with the glow. This experiment is especially informing if we reduce the free surface of one of the wires to its last cross-section by drawing it into its capillary and thus produce a cathode which is approximately a point. There then appears around this a beautiful blue hemi-

sphere of glow, whose radius when the pressure of the air is such
that the sheath around an extended cathode shows a scarcely
measurable thickness, is greater than 1 cm. and increases very fast
as the density diminishes.   So long as it is still small the negative
light is separated from the positive by a dark space.   This disappears, however, and the two lights meet each other if the first
is very much extended.

Any solid or fluid body, whether an insulator or a conductor,
which is placed in front of the cathode cuts off the glow, which lies
between it and the cathode; no bendings out of straight lines occur.

If we take a tube bent at right angles, whose two legs are very
different in length, and fix at the ends two similar point electrodes,
the negative light with the proper exhaustion fills the whole long
leg if b is the cathode.   On the other hand, it remains confined to
the short leg and does not follow the bending of the tube if a is the
negative pole.   We see most clearly the rectilinear transmission
of the glow if it goes out from a point cathode and, through a great
length of the tube, brings the surface of the glass to fluorescence.
If, in such conditions, any object is set in the space filled with the
glow, it casts a sharp shadow on the fluorescing wall by cutting
off from it the cone of light which goes out from the cathode as an
origin.

We will therefore, in what follows, speak of rectilinear paths or
rays of glow, and consider any point of the cathode as the source
of a cone of rays.

In any long cylindrical tube which contains in its axis a point
cathode, the more the rays deviate from the direction of the axis
so much the sooner will they be limited by the walls.   Therefore,
at considerable distances from the cathode, with suitable exhaustion, there remains a beam of nearly parallel rays of glow, which
later will be of use to us.

If the glow spreads out from the points of the cathode in straight
lines, it must be independent of the direction of the positive light.
This we can see very clearly if the point cathode is turned away
from the anode.

CROOKES

Sir William Crookes was born on June 17, 1832, and died on April 4, 1919.
He studied at the Royal College of Chemistry, London, and served there as an
assistant to Hoffmann.   In 1859 he founded the *Chemical News* and remained

its proprietor and editor until his death. He early attracted attention by his discovery of the element thallium by spectroscopic methods. He was an active investigator in many fields of physics and contributed greatly to the advance of knowledge by his study of the radiometer and of the electric discharge in rarefied gases.

The extract which follows is from his Bakerian Lecture "On the Illumination of Lines of Electrical Pressure, and the Trajectory of Molecules," published in the *Philosophical Transactions*, Part I, p. 135, 1879.

THE CATHODE DISCHARGE

Several sections are omitted.

Green Phosphorescent Light of Molecular Impact.

510. When the exhaustion approaches 30 M, (M represents a millionth of a millimetre) a new phenomenon makes its appearance. The dark space has spread out so much that it nearly fills the bulb; the violet light by which the focus was rendered visible has become so faint as to be difficultly traced, but with care it can be seen converging to a focus beyond the focal point noticed at lower exhaustions. At the part of the bulb on which the rays impinge, a faint spot of greenish-yellow light is observed, sharp in outline. On exhausting to 14 M, and making the cup the negative pole of the coil, the projection from the cup is represented by a brilliant green spot of light about 7 millims, diameter, and the focus can scarcely be traced. The rest of the bulb is nearly dark, but at those parts furthest removed from the negative pole the faintly luminous boundary of the dark space can still be seen. A little blue light is seen round the positive pole extending somewhat into the bulb. On reversing the poles and making the cup positive, the bulb becomes beautifully illuminated with greenish-yellow light.

511. This phosphorescent light only appears in its full intensity when the dark space surrounding the negative pole extends to the surface of the bulb. At lower exhaustions, it can be detected when specially sought for outside the luminous boundary of the dark space, but it is faint and not easily noticeable. The colour depends on the kind of glass used. Most of my apparatus are made of soft German glass, and this gives a phosphorescent light of a greenish-yellow colour. English glass phosphoresces of a blue colour; uranium glass becomes green; a diamond became brilliantly blue.

Projection of Molecular Shadows.

524. In ordinary vacuum tubes, illuminated by the induction current, the luminous phenomena follow the tube through any amount of curves and angles; a hollow spiral becomes illuminated just as well as if the tube were in a straight line. Not so, however, the phenomena of green phosphorescence observed at these high exhaustions. The molecular ray which gives birth to green light absolutely refuses to turn a corner, and radiates from the negative

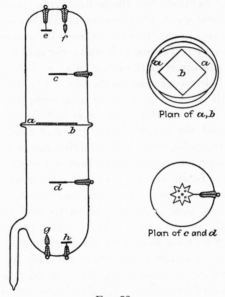

Plan of a, b

Plan of c and d

FIG. 98.

pole in straight lines, casting strong and sharply-defined shadows of anything which happens to be in its path. In a U tube with poles at each end, one leg will be bright green and the other almost dark, the light being cut off sharply by the bend of the glass, a shadow being projected on the curvature. I can detect no trace of polarisation in the green phosphorescent light on the surface of the glass, except, of course, when it emerges at an angle through the side of the glass tube.

525. The projection from the negative pole of a shadow rendered visible by a sharply-defined image on the side of the glass, seemed worthy of more close examination. A tube was accordingly made, as shown in Fig. 98. In the centre, dividing the tube into nearly

equal parts, is a screen of thin mica, a a, loosely fitting into a groove blown round the tube. A flat plate of uranium glass, b, about half a millimetre thick, is rivetted to the mica on one side. c is a star shaped piece of aluminum foil attached to a platinum terminal, and d is a similar star made of mica. At each end of the tube are two terminals, e and b, being flat aluminum disks, and f and g aluminum points.

With this apparatus experiments were carried on during exhaustion. When the exhaustion is moderate (say 1 or 2 millims.) so as to show stratifications and the ordinary phenomena of vacuum tubes, the luminosity extends from one pole to the other. Thus, if e and g are the two poles, the light extends the whole length of the tube; if, however, e or f is one pole and c the other, the luminosity only occupies the upper part of the tube, and if e and f are the two poles, the light keeps close to the top. The whole appearance shows that both poles are at work in producing the phenomena.

526. When, however, the vacuum is sufficiently high for the dark space round the negative pole to have swollen out to the dimensions of the tube, there is little difference in the phenomena of green phosphorescence and projection of the shadow of c on b, whichever is the positive pole, provided e be made the negative. The appearances are almost the same, and the shadows projected from the negative pole e are just as sharp and intense whether I make f or b the positive pole. The positive pole, in fact, seems to have little or nothing to do with the phenomena.

527. The best and sharpest shadows are cast by the flat disks e and b. The shadows thrown by the pointed poles f and g are faint and undecided in outline. An aluminum ring scarcely makes any shadow; a spherical pole, owing to the rays from it diverging more, gives faint and broad shadows; a square pole acts the same as a disk. Using the upper flat pole e as the negative, the shadow of the star c is thrown distinctly on the uranium plate b, where it is seen magnified about two diameters, but perfectly sharp in outline; either f, g, or b, and even the star itself may be made the positive pole without affecting the appearance of its shadow on b.

528. The whole upper part of the tube which is in the line of direct projection from the negative pole, glows with an intense yellowish-green fluorescent light. The uranium plate is still more brilliant, and of a greenish colour. Where the shadow of the star falls on it, no phosphorescence whatever is visible. The mica plate a, where uncovered at the side of the uranium plate, gives no

phosphorescence, and no shadow is therefore seen on it.   When the lower pole, *b*, is made negative, so as to project the shadow of the mica star *d*, no shadow is seen on the mica plate, neither is any seen on the uranium plate above the mica.   The thin film of mica entirely prevents the uranium glass from becoming fluorescent under the influence of the negative pole.   Other experiments have, however, shown that the mica star gives just as sharp and intense a shadow as the aluminum star, provided a suitable screen is used to receive it on.

529. If the aluminum star is made the positive pole, any one of the others being the negative pole, it casts an enlarged and somewhat distorted image of itself all over the upper part of the tube.   This image is not sharply defined.

The sharpness of the shadows cast by the negative pole is slightly affected by the intensity of the current; when the spark is very strong, the shadow widens out a little.

530. I have already advanced the theory that the thickness of the dark space surrounding the negative pole is the measure of the mean length of the path of the gaseous molecules between successive collisions.   The electrified molecules are projected from the negative pole with enormous velocity, varying, however, with the degree of exhaustion and intensity of the induction current.

In the dark space they are few in number in comparison to what they are at the luminous boundary.   When the exhaustion is so high that the mean path of the molecules stretches right across the tube, their velocity is suddenly checked by the glass walls, and the production of light is the consequence of this sudden arrest of velocity.   The light actually proceeds from the glass, and is caused by fluorescence or phosphorescence in or on its surface, and not by an evolution of light by the molecules themselves, crowding together and striking each other on the surface of the glass. Had this been the case—had the molecules themselves been the lamps—they would shine equally well whatever were the arresting surface, and their light would have shown the spectral characteristics of the gas whose residue they constituted.   But no light is caused by a mica or quartz screen, however near it may be brought to the negative pole; and generally speaking the more fluorescent the material of the screen, the better the luminosity.

531. The theory best supported by experiment, and the one which although new is not at all improbable in the present state

of our knowledge respecting molecules, is that the greenish-yellow phosphorescence of the glass is caused by the direct impact of the molecules on the surface of the glass. The shadows are not optical but molecular shadows, only they are revealed by an ordinary illuminating effect. The sharpness of the shadow, when projected from a wide pole, proves them to be molecular. Had the projection from the negative pole radiated in all directions, after the manner of light radiating from a luminous disk, the shadows would not be perfectly sharp, but be surrounded by a penumbra. Being, however, projected material molecules in the same electrical state, they do not cross each other, but travel on in slightly divergent paths, giving perfectly sharp shadows with no penumbrae.

.   .   .   .

Mechanical Action of Projected Molecules.

541. It was noticed that when the coil was first turned on, the thin glass film was driven back at the moment of becoming phosphorescent. This seemed to point to an actual material blow being given by the molecular impact, and the following experiment was devised to render this mechanical action more evident.

A large somewhat egg-shaped bulb (Fig. 99, elevation) is furnished at each end with flat aluminum poles, a and b; a pointed aluminum pole is inserted at c. At d, a little indicator is suspended from jointed glass fibres, so as to admit of being brought into any position near the middle of the bulb, by tilting the apparatus. The indicator consists of a small radiometer fly 8 millims, in diameter, furnished with clear mica vanes 2 millims, across, and delicately supported on a glass cup and needle point. A screen cut out of a flat aluminum plate 12 millims. wide and 30 millims. high, is supported upright in the bulb at e, a little on one side of its axis, being attached to the bulb by a platinum wire passing through the glass, so that if needed the screen e can be used as a pole.

542. This apparatus was designed with a double object. The indicator fly is not blacked on one side or favourably presented, therefore if immersed in a full stream of projected molecules, there will be no tendency for it to turn one way rather than the other. If, however, I tilt the bulb so as to bring the indicator half in and half out of the molecular shadow cast by the screen, I should expect to see the fly driven round to the right or to the left by the molecules striking one side only, thus confirming the observa-

tion on the movement of the thin glass film under the molecular impact. (541)

543. The other subject which I had in view was the following. It is well known that a movable conductor carrying a current of

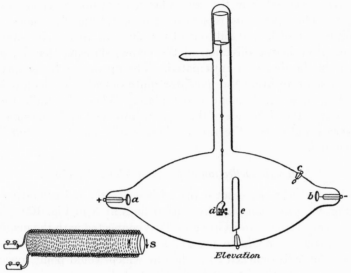

Elevation

FIG. 99.

electricity is deflected under the influence of magnetic force and experiments tried very early in this research, and repeated with the apparatus already described, showed that the stream of molecules projected from the negative pole obeyed in a very marked manner the power of a magnet. It was hoped that the

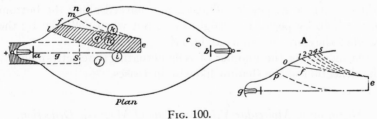

Plan

FIG. 100.

form of apparatus now under experiment would throw some light on this action of magnetism on molecules.

544. The bulb being exhausted to the necessary high degree the pole *b* is made negative, so as to cast the shadow of the screen *e* across the tube, where it can be traced as a broad band along the

lower part of the glass. Fig. 100, plan, shows the appearance, the shadow of e projected by the pole b being enclosed within the lines fe, ge, shaded diagonally.

The indicator fly is first brought into position h, where it is entirely screened from the molecular stream; no movement takes place. The apparatus is slightly tilted, till the fly comes into position i, half in and half out of the shadow; very rapid rotation takes place in the direction of the arrow, showing that impacts occur in the direction anticipated. The apparatus being further tilted, so as to bring the indicator quite outside the shadow into position j, no movement takes place. When the indicator is brought to the other side of the shadow into position k, the rotation is very rapid in the direction of the arrow, opposite to what it is when at i.

Magnetic Deflection of Lines of Molecular Force.

545. An electro-magnet is placed beneath the bulb, shown at S in Fig. 99, elevation, and by dotted lines at S in Fig. 100, plan. A battery of from 1 to 5 Grove's cells is connected with the magnet. The current is made to pass in such a direction that the pole under the bulb (marked S) is the one which would point towards the south were the magnet freely suspended.

546. The induction current being turned on, the shadow of e is projected straight along the tube in the position fe, ge; the edge of the shadow f is called the zero position, marked O on Fig. 100, A. The electro-magnet is now excited by 1 cell. The shadow is deflected sideways to the position shown by 1, p, the edge fe now moving to 1, and the edge ge moving to pe. The lines fe and ge, when under no magnetic influence, are marked along the bottom of the bulb by perfectly straight lines, but when deflected by the magnet they are curved as at 1e, pe.

On increasing the number of cells actuating the electro-magnet, the deflection of the shadow likewise increases.

. . . .

Alteration of Molecular Velocity—Law of Magnetic Deflection.

555. It has been shown (545) that the position of the edge of the shadow is affected by varying the magnetic power used to deflect it. It became of interest to see if by keeping the magnetic power constant, the position of the edge of the shadow could be altered by any circumstance affecting the intensity of the spark,

such as intercalating a Leyden jar in the circuit, or screwing the contact breaker one way or the other. If the molecules are projected from the negative pole with different velocities we might expect that under a constant magnetic deflection the higher velocities would show the flatter trajectories. With the apparatus shown at Figs. 99 and 100, these variations in the trajectory of the molecules were not obtained in a decided manner, although indications of an alteration of curve by intensifying the spark were apparent.

556. Another apparatus was accordingly made in order to test this point, and also to obtain a more definite relation between the dimensions of the dark space round the negative pole, the

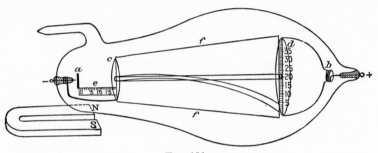

FIG. 101.

commencement of the green phosphorescence, and the magnetic deflection under varying conditions of pressure in different gases.

I have spoken of shadows being deflected by the magnet as a convenient way of describing the phenomena observed; but it will be understood that what is really deflected is the path of the molecules driven from the negative pole and whose impact on the phosphorescent surface causes light. The shadows are the effect of a material obstacle in the way of the molecules.

In the apparatus now about to be described, a ray of light was used instead of a shadow. Figure 101 shows the arrangement.

The poles are at *a* and *b*. The negative pole *a* is a flat aluminum disk with a notch cut in it. The pole *b* is a ring of aluminum; *c* is a mica screen with a small hole in the middle about 1 mil im. in diameter; *d* is a flat plate of German glass with a millimetre scale engraved on it vertically; *e* is a mica scale of millimetres. The scale *e* is to measure the thickness of the dark space as the exhaustion proceeds; the hole in *c* is to enable a spot of light to be thrown on the scale d from the pole a; the notch in the pole is to enable me to see if the spot of light projected on d is an image of the pole a, or of the hole in c; the scale on d is to enable me to measure the deflection of the ray proceeding from a, through c, to d, when bent by the magnet; ff is a vertical screen of mica in the plane of the movement of the ray, covered with a phosphorescent powder. On this the path of the ray traces itself in a straight line when the magnet is absent, and curved when the magnet is present.

. . . .

576. These experiments prove several important points. In par. 559, when working with an air vacuum, it is recorded that the spot of green light is visible on the screen at a pressure of 102.6 M when the thickness of the dark space is only 12 millims. from the pole. Assuming, as I do, that this is a measure of the free path of the molecules before collision, it follows that some of the molecules sufficient to cause green phosphorescence on the screen, are projected the whole distance from the pole to the screen, or 102 millims., without being stopped by collisions. It is probable that this would have occurred at a still lower exhaustion, for on reference to par. 566, it is seen that the green spot was detected on the screen when the mean path in carbonic acid was 8.5 millims., and it was seen with hydrogen (570) when the mean path was only 7 millims.

577. If we suppose the magnet permanently in position, and thus exerting a uniform downward pull on the molecules, it is seen that their trajectory is much curved at low exhaustions, and gets flatter as the exhaustion increases. A flatter trajectory corresponding to a higher velocity of the molecules, it follows that the molecules move quicker the better the exhaustion. This may arise from one of two causes: either the initial impulse given by the negative pole is stronger, or the collisions are less frequent. I consider the latter to be the true cause. The molecules which produce the green phosphorescence must be looked upon as in a different state from those which are arrested by frequent collisions. These impede the velocity of the free molecules and allow longer time for the magnetism to act on them; and although the deflecting force of magnetism might be expected to increase with the velocity of the molecules, Professor Stokes has pointed out that it would have to increase as the square of the velocity in order that the deflection should be as great at high as at low velocities.

Comparing the free molecules to cannon balls, the magnetic pull to the earth's gravitation, and the electrical excitation of the negative pole to the explosion of the powder in the gun, the trajectory will be quite flat when no gravitation acts, and gets curved under the influence of gravitation; it is also much curved when the ball passes through a dense resisting medium; it is less curved when the resisting medium gets rarer, and, from par. 562, it is seen that intensifying the induction spark, equivalent to increasing the charge of powder, gives greater initial velocity, and therefore flattens the trajectory. The parallelism is still closer if we compare the evolution of light seen when the shot strikes the target, with the phosphorescence produced in the glass screen by molecular impacts.

578. In carbonic acid the mean free path of the molecules is shorter for the same degree of exhaustion than in air, and the velocity of the molecules measured by the flatness of the trajectory under magnetic influence is also lower than what it is in air at similar pressures.

579. Although in many cases, especially at a moderate exhaustion and using a feeble spark, the image of the screen was a perfectly well-defined circle, I could detect no image of the notch cut in the negative pole. The hole in the mica was quite small enough to have given a good image of the negative pole inverted on the screen had it been shining by ordinary light, but the rays being corpuscular and the particles not crossing, no image of the pole is formed, but only the image of the hole in the mica.

580. Attempts to obtain continuous rotation of the ray of molecular light by means of a magnet have hitherto failed. The stream of molecules does not obey *Ampère's Law* as it would were it a perfectly flexible conductor joining the negative and positive pole. The molecules are projected from the negative pole, but the position of the positive pole, whether in front, at the side, or even behind the negative pole, has no influence on their subsequent behaviour, either in causing phosphorescence, producing mechanical effects, or in their magnetic deflection (519, 526, 527, 549). The magnet seems to give them a spiral twist, greater or less, according to its power, but diminishing as the molecules get further off, and independent of their direction.

581. In the experiment described at pars. 536 to 540, the heating effect of the molecular bombardment is assumed to be the cause of certain phenomena. It was thought that by concentrating

the molecular impacts to one point the heat produced might be rendered apparent. The experiments tried in the apparatus shown in Fig. 102 prove that this supposition is correct. A polished aluminum cup a is made the negative pole in a properly exhausted tube. The focus is seen very sharp and distinct, as at A, and of a dark blue colour. The light, although blue in the centre of the tube, when it spreads out and strikes the tube at the end, illuminates it beautifully with the yellowish-green light. By means of a magnet the focus was deflected to the side of the tube, as shown at B, the path of the rays being beautifully curved. On the tube

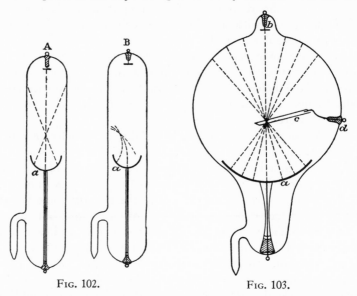

Fig. 102.   Fig. 103.

the appearance was that of a sharply-defined oval of a yellowish-green colour with a dark spot in the centre. To ascertain if heat were developed here I touched it with my finger and immediately raised a blister. The spot where the focus fell was nearly red hot.

582. Another apparatus was now made, as shown in Fig. 103. A nearly hemispherical cup of polished aluminum a is made one pole in a bulb, and a small disk of aluminum b is made the other pole. At c a strip of platinum is held by a wire passing through the glass, and forming another pole at d. The tip of the platinum strip is brought to the centre of curvature, and the whole is exhausted to a very high point. On first turning on the induction

current, the cup being made the negative pole, the platinum strip entered into a very rapid vibration. This soon stopped and the platinum quickly rose to a white heat, and would have melted had I not stopped the action of the coil.

The same phenomena of ignition take place if the platinum strip itself is made the positive pole.

583. Experiments in a similar tube in which a piece of charcoal is the body to be ignited, show that the ignition takes place at a less high exhaustion in hydrogen gas than it does in air.

I have great pleasure in expressing my continued obligation to the great skill in glass-blowing and manipulation possessed by my friend and assistant, Mr. *C. H. Gimingham*, whose dexterity in executing complicated forms of apparatus has rendered easy a research which otherwise would have been full of difficulties.

584. I hope I may be allowed to record some theoretical speculations which have gradually formed in my mind during the progress of these experiments. I put them forward only as working hypotheses, useful, perhaps necessary, in the first dawn of new knowledge, but only to be retained as long as they are of assistance; for experimental research is necessarily and slowly progressive, and one's early provisional hypotheses have to be modified, adjusted, perhaps altogether abandoned in deference to later observations.

An Ultra-Gaseous State of Matter.

585. The modern idea of the gaseous state of matter is based upon the supposition that a given space of the capacity of, say, a cubic centimetre, contains millions of millions of molecules in rapid motion in all directions, each having millions of encounters in a second. In such a case the length of the mean free path of the molecules is excessively small as compared with the dimensions of the vessel, and properties are observed which constitute the ordinary gaseous state of matter, and which depend upon constant collisions. But by great rarefaction the free path may be made so long that the hits in a given time are negligible in comparison to the misses, in which case the average molecule is allowed to obey its own motions or laws without interference; and if the mean free path is comparable to the dimensions of the vessel, the properties which constitute gaseity are reduced to a minimum, and the matter becomes exalted to an ultra-gaseous or

molecular state, in which the very decided but hitherto masked properties now under investigation come into play.

The phenomena in these exhausted tubes reveal to physical science a new world—a world where matter may exist in a fourth state, where the corpuscular theory of light may be true, and where light does not always move in straight lines, but where we can never enter, and with which we must be content to observe and experiment from the outside.

GOLDSTEIN

Eugen Goldstein was born on September 5, 1850, in Gleiwitz. He took his doctor's degree at Berlin in 1881. In 1888 he became connected as physicist with the astronomical observatory of the University of Berlin.

In the extract which follows Goldstein describes his discovery of the Canal Rays. The title of the paper is "Ueber eine noch nicht untersuchte Strahlungsform an der Kathode inducirter Entladungen." It appeared in the *Sitzungsberichte der Königlichen Akademie der Wissenschaften zu Berlin,* July 29, p. 691, 1886. In the first part of this paper the author describes the apparatus by which he made a fortuitous observation of the appearances in the high vacuum tube which were subsequently named Canal Rays. The rays seemed to be connected with a yellow discharge which appeared in the neighborhood of the cathode. He then proceeds as follows.

THE CANAL RAYS

In order to investigate this phenomenon as far as possible under simpler conditions than those under which it first appeared, it was necessary first to investigate the general condition for its occurrence. The result was that the appearance of the yellow light without any perceptible admixture of the blue cathode light occurs whenever the cathode divides the space within the discharge vessel into two parts, in such a way that one part contains the anode and the two parts communicate only by narrow openings which are made through the substance of the cathode itself. For further investigation, therefore, in most cases, plane plates with holes bored through them were used instead of the cylindrical wire netting as cathode.

I will here give schematic diagrams of the two arrangements which were most often used. Fig. 104, 1 represents a tube which/ has been pressed out flat like a plate at *x*, and carries there the plane parallel cathode plate *K* in which holes are bored. Electric connection of the cathode with the inductorium is made through the thin wire *d* which is fastened at one end to a hook on the edge of the plate, and at the other end to the ring *b*. The anode is formed of the wire *a*.

In the tube (Fig. 104, 2) the cathode plate is not laid loosely on a surface, but forms the bottom of a little box which fits tightly over a glass tube *r*, while again a wire *d* carries the negative electricity, and *a* forms the anode. The surface of the cathode which in such vessels is turned toward the anode may, in the future, be called its front surface or front side, and the side that looks away from the anode, its back side. The hole through it, if nothing else is said, may be thought of as cylindrical, as bored perpendicularly to the plane of the plate, and as about ⅔ mm. in diameter.

The front side of such cathodes shows the customary cathode light, the most of which is formed by the blue rays. The first yellow layer is again narrow. On the back side, on the other hand, the yellow light develops into long columns of light, like fire. With sufficient exhaustion the blue light does not appear on this side of the cathode at all. The conducting wire *d* has no light around it, so that in the part of the tube which lies behind the cathode the yellow light which is sent out from the cathode plate is all the light there is.

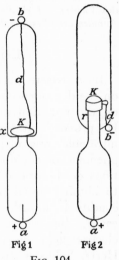

FIG. 104.

By means of such arrangements, the following properties of the yellow light can be observed:

The yellow light consists of regular rays which travel in straight lines. From every opening in the cathode there arises a straight, bright, slightly divergent yellow beam of rays. The separate bright beams are surrounded with a widely distributed cloud which is very weak in light, but in general has the same color as the beams.

Some paragraphs dealing with the peculiarities of these rays and apparent contrasts between them and the cathode rays, are omitted.

Until a suitable name has been found, these rays, which we now cannot distinguish any longer by their color, which changes from gas to gas, may be known as "canal rays".

Other properties of these rays are investigated, in particular the effect of a magnetic field upon them. No deflection in the magnetic field could be detected.

HALLWACHS

Wilhelm Hallwachs was born in 1859. He received his doctor's degree from the University of Strassburg in 1883. In 1900 he became professor of physics in the Technical High School at Dresden, where he remained until his death on June 20, 1922.

The extract which follows is from a paper entitled "Ueber den Einfluss des Lichtes auf elektrostatisch geladene Körper," which appeared in the *Annalen der Physik und Chemie*, n.f., Vol. 33, p. 301, 1888.

Electric Discharge by Light

In a recent publication Hertz has described investigations on the dependence of the maximum length of an induction spark on the radiation received by it from another induction spark. He proved that the phenomenon observed is an action of the ultraviolet rays of light. No further light on the nature of the phenomenon could be obtained, because of the complicated conditions of the research in which it appeared. I have endeavored to obtain related phenomena which would occur under simple conditions, in order to make the explanation of the phenomenon easier. Success was obtained by investigating the action of the electric light on electrically charged bodies. Already before Hertz, Schuster and after him Arrhenius have described phenomena which seem to be closely connected with that of Hertz. Hertz' observations go further than theirs in that in them there is given a hint of the first cause of the phenomenon.

The investigations which I have undertaken on the influence of the electric light on electrically charged bodies, were mostly carried out in the following way. A thoroughly cleaned circular plate of zinc about 8 cm. in diameter was set on an insulating stand and connected by a wire with a gold leaf electroscope. In front of the zinc plate and parallel with it stood a large screen of sheet zinc about 70 cm. broad and 60 cm. high. In the middle of this screen was a selenite window through which the rays of a Siemens arc lamp on the other side of it could fall on the plate. The system of plate and gold leaves was well insulated. In a day the divergence diminished by about $\frac{1}{4}$; while a research was in progress there was no noticeable change. Also if the arc lamp was in action while the window was covered with some suitable substance the insulation remained perfect.

If the plate and the electroscope, upon which latter the rays cannot fall, are negatively charged, then as soon as the light rays fall on the plate the gold leaves begin to come together rapidly; if they are positively charged the leaves at first sight show no motion at all and only after a long time will a more exact examination show a noticeable motion.

In order to obtain a quantitative measure of the rate at which the leaves come together, a scale was marked out on the glass plate of the Beetz electroscope and the displacements of the ends of the gold leaves were observed on this scale through a lens from a distant fixed point. In this way it was found, for example, that if the distance of the freshly cleaned zinc plate from the arc light was 70 cm. in the case of a negative charge, the divergence diminished in 5 seconds by 70 per cent, in 10 seconds the leaves had come together. When the charge was positive there was obtained a diminution of 10 per cent only in 60 seconds. The action can be observed up to very great distances, for example, 3 m. Its intensity depends very much on the size of the arc.

. . . .

From the experiments here described it appears that the phenomenon observed is particularly excited by the ultraviolet rays of light. Since magnesium also is effective the cause of the phenomenon cannot be electrostatic forces. An impairment of the insulation cannot explain the effect, because then both positively and negatively charged bodies would be affected in similar ways. The experiments on the absorption of the action by the interposition of bodies forbids us to look for the cause in material particles which are shot out from the arc; also these would be kept from the plate by the screen placed before it. Finally, that a rise of temperature caused by the rays cannot be used to explain the effect appears from this, that the red and infrared rays are inactive.

. . . .

PERRIN

Jean Perrin is professor of physical chemistry and director of the laboratory of physical chemistry at the University of Paris. He is distinguished for his study of Brownian movements, by which he showed that the law of equipartition of energy was followed by small particles suspended in liquids.

In the paper which follows, "Nouvelles propriétés des rayons cathodiques," taken from *Comptes Rendus*, Vol. 121, p. 1130, 1895, Perrin gives an account of the experiment by which he showed that the cathode discharge carried negative electric charges with it. A similar experiment was soon afterward carried out by Sir J. J. Thomson.

THE NEGATIVE CHARGES IN THE CATHODE DISCHARGE

I. Two hypotheses have been presented to explain the properties of the cathode rays.

Some, with Goldstein, Hertz, or Lenard, think that this phenomenon, like light, results from vibrations of the ether, or even that it is light of short wave length. We then easily see that these rays might have a straight trajectory, excite phosphorescence, and act upon photographic plates.

Others, with Crookes or J. J. Thomson, think that these rays are formed of matter negatively charged and moving with great velocity. We then can easily understand their mechanical properties and also the way in which they bend in a magnetic field.

This latter hypothesis suggested to me some experiments which I shall present without troubling myself for the moment to consider if the hypothesis accounts for all the facts which are at present known, and if it alone can account for them.

Its partisans assume that the cathode rays are negatively charged; so far as I know, no one has demonstrated this electrification; I have therefore tried to determine if it exists or not.

II. To do this I made use of the laws of electric induction, by which we can recognize that electric charges are introduced into the interior of a closed conducting envelope and can measure them. I therefore arranged that cathode rays should enter a Faraday cylinder.

To do this I used the vacuum tube represented in Fig. 105.

ABCD is a metallic cylinder completely closed except that there is a small opening α in the middle of the face *BC*.

This acts as the Faraday cylinder. A metallic wire sealed at *S* in the wall of the tube connects this cylinder with an electroscope.

EFGH is a second metallic cylinder, which is always connected with the ground and pierced with two little openings at β and γ. It protects the Faraday cylinder from external influences.

Finally, at about 0.10m in front of *FG*, is placed the electrode *N*.

The electrode *N* serves as cathode. The anode is the protecting cylinder *EFGH*; a pencil of cathode rays then penetrates into the Faraday cylinder. Invariably this cylinder becomes charged with negative electricity.

The vacuum tube can be placed between the poles of an electromagnet.

When this magnet is excited, the cathode rays are deflected and no longer enter the Faraday cylinder: then this cylinder does

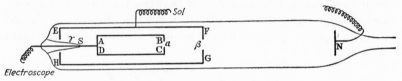

F_IG_. 105.

not become charged; it becomes charged as soon as the magnet is no longer excited.

In short, the Faraday cylinder is charged negatively when the cathode rays enter it and only when they enter it: the cathode rays are therefore charged with negative electricity.

We may measure the quantity of electricity which the rays carry. I have not finished this investigation, but I may give an idea of the order of magnitude of the charges obtained, by saying that for one of my tubes, at a pressure measured by 20 microns of mercury and for one interruption of the primary of the coil, the Faraday cylinder received enough electricity to charge a capacity of 600 C.G.S. units to 300 volts.

III. Since the cathode rays are charged negatively, the principle of the conservation of electricity leads us to look somewhere for the corresponding positive charges. I believe that I have found them in the very region where the cathode rays are formed, and have proved that they travel in the opposite sense by falling down on the cathode.

To verify this hypothesis it is sufficient to use a hollow cathode pierced with a small opening through which a part of the positive electricity attracted to it can enter. This electricity can then act on a Faraday cylinder inside the cathode.

The protecting cylinder *EFGH* with its opening β satisfies these conditions; I therefore used it this time as cathode, and the electrode *N* as anode.

The Faraday cylinder was then invariably charged with positive electricity.

The positive charges were of the order of magnitude of the negative charges formerly obtained.

Thus at the same time as negative electricity is radiated from the cathode, positive electricity proceeds toward the cathode. I have sought to find if this positive flux forms a second system of rays absolutely symmetrical with the first.

IV. For this purpose I constructed a tube somewhat like the former one, (Fig. 106) except that between the Faraday cylinder and the opening β there is placed a metallic diaphragm pierced with an opening β', so that the positive electricity which enters

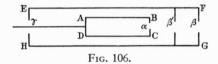

Fig. 106.

through β cannot act on the Faraday cylinder unless it also traverses the diaphragm β'. Then I repeated the former experiments.

When *N* was cathode the cathode rays emitted from it passed without difficulty through the two openings β and β', and made the gold leaves of the electroscope diverge strongly. But when the protecting cylinder was cathode, the positive flow, which, from the preceding experiment, enters at β, does not produce a separation of the gold leaves, except at very low pressures. By substituting an electrometer for the electroscope, it can be seen that the action of the positive flux is real but very feeble, and that it increases when the pressure decreases. In a series of experiments at a pressure of 20μ it brought a capacity of 2000 C.G.S. units to 10 volts; and at a pressure of 3μ in the same time, it brought it to 60 volts.

By the use of a magnet we could completely suppress this action.

V. This group of results seems not to be easily reconciled with the theory which supposes that the cathode rays are an ultra-violet light. On the contrary they agree very well with the theory which considers them a material radiation. This theory, it seems to me, we can state as follows:

In the neighborhood of the cathode the electric field is sufficiently intense to break into pieces, or *ions*, some of the molecules of the residual gas. The negative ions move toward the region where the potential increases, acquire a considerable velocity, and form the cathode rays; their electric charge and therefore their mass (in the ratio of a valence-gramme for 100,000 coulombs) is easily measurable. The positive ions move in the opposite sense; they form a diffuse cloud that is affected by a magnet and is not a radiation properly so called.

THOMSON

Sir Joseph John Thomson was born on December 18, 1856, near Manchester. He studied at Owens College, Manchester, and at Trinity College, Cambridge. In 1884 he was appointed Cavendish professor of experimental physics in the University of Cambridge. In 1919 he was appointed research professor and was succeeded in the Cavendish professorship by Lord Rutherford. He has been master of Trinity College since 1919.

In the paper here given on "Cathode Rays," from the *Philosophical Magazine*, Vol. 44, Series 5, p. 293, 1897, Thomson gives the first account of his discovery of the electron and its properties.

THE ELECTRON

The experiments discussed in this paper were undertaken in the hope of gaining some information as to the nature of the Cathode Rays. The most diverse opinions are held as to these rays; according to the almost unanimous opinion of German physicists they are due to some process in the aether to which— inasmuch as in a uniform magnetic field their course is circular and not rectilinear—no phenomenon hitherto observed is analogous: another view of these rays is that, so far from being wholly aetherial, they are in fact wholly material, and that they mark the paths of particles of matter charged with negative electricity. It would seem at first sight that it ought not to be difficult to discriminate between views so different, yet experience shows that this is not the case, as amongst the physicists who have most deeply studied the subject can be found supporters of either theory.

The electrified-particle theory has for purposes of research a great advantage over the aetherial theory, since it is definite and its consequences can be predicted; with the aetherial theory it is impossible to predict what will happen under any given circumstances, as on this theory we are dealing with hitherto unobserved phenomena in the aether, of whose laws we are ignorant.

The following experiments were made to test some of the consequences of the electrified-particle theory.

Charge Carried by the Cathode Rays.

If these rays are negatively electrified particles, then when they enter an enclosure they ought to carry into it a charge of negative electricity. This has been proved to be the case by Perrin, who

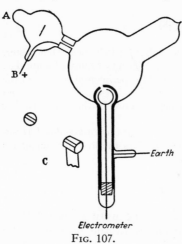

FIG. 107.

placed in front of a plane cathode two coaxial metallic cylinders which were insulated from each other: the outer of these cylinders was connected with the earth, the inner with a gold-leaf electroscope. These cylinders were closed except for two small holes, one in each cylinder, placed so that the cathode rays could pass through them into the inside of the inner cylinder. Perrin found that when the rays passed into the inner cylinder the electroscope received a charge of negative electricity, while no charge went to the electroscope when the rays were deflected by a magnet so as no longer to pass through the hole.

This experiment proves that something charged with negative electricity is shot off from the cathode, travelling at right angles to it, and that this something is deflected by a magnet; it is open, however, to the objection that it does not prove that the cause of the electrification in the electroscope has anything to do with the cathode rays. Now the supporters of the aetherial theory do not deny that electrified particles are shot off from the cathode; they deny, however, that these charged particles have any more to do with the cathode rays than a rifle-ball has with the flash when a rifle is fired. I have therefore repeated Perrin's experiment in a form which is not open to this objection. The arrangement used was as follows:—Two coaxial cylinders (Fig. 107) with slits in them are placed in a bulb connected with the discharge-tube; the cathode rays from the cathode *A* pass into the bulb through a slit in a metal plug fitted into the neck of the tube; this plug is connected with the anode and is put to earth. The cathode rays thus do not fall upon the cylinders unless they are deflected by a magnet. The outer cylinder is connected with the earth, the inner with the electrometer. When the cathode rays (whose path was traced by the phosphorescence on the glass) did not fall on the slit, the electrical charge sent to the electrometer when the induction-coil producing the rays was set in action was small and irregular; when, however, the rays were bent by a magnet so as to fall on the slit there was a large charge of negative electricity sent to the electrometer. I was surprised at the magnitude of the charge; on some occasions enough negative electricity went through the narrow slit into the inner cylinder in one second to alter the potential of a capacity of 1.5 microfarads by 20 volts. If the rays were so much bent by the magnet that they overshot the slits in the cylinder, the charge passing into the cylinder fell again to a very small fraction of its value when the aim was true. Thus this experiment shows that however we twist and deflect the cathode rays by magnetic forces, the negative electrification follows the same path as the rays, and that this negative electrification is indissolubly connected with the cathode rays.

When the rays are turned by the magnet so as to pass through the slit into the inner cylinder, the deflexion of the electrometer connected with this cylinder increases up to a certain value, and then remains stationary although the rays continue to pour into the cylinder. This is due to the fact that the gas in the bulb becomes a conductor of electricity when the cathode rays pass through it, and thus, though the inner cylinder is perfectly insulated when the rays are not passing, yet as soon as the rays pass through the bulb the air between the inner cylinder and the outer one becomes a conductor, and the electricity escapes from the inner cylinder to the earth. Thus the charge within the inner cylinder does not go on continually increasing; the cylinder settles down into a state of equilibrium in which the rate at which it gains negative electricity from the rays is equal to the rate at which it loses it by conduction through the air. If the inner cylinder has initially a positive charge it rapidly loses that charge and acquires a negative one; while if the initial charge is a negative one, the cylinder will leak if the initial negative potential is numerically greater than the equilibrium value.

Deflexion of the Cathode Rays by an Electrostatic Field.

An objection very generally urged against the view that the cathode rays are negatively electrified particles, is that hitherto no deflexion of the rays has been observed under a small electrostatic force, and though the rays are deflected when they pass near electrodes connected with sources of large differences of potential, such as induction-coils or electrical machines, the deflexion in this case is regarded by the supporters of the aetherial theory as due to the discharge passing between the electrodes, and not primarily to the electrostatic field. Hertz made the rays travel between two parallel plates of metal placed inside the discharge-tube, but found that they were not deflected when the plates were con-

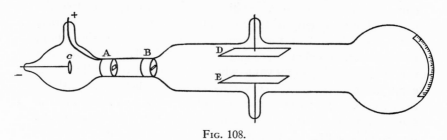

FIG. 108.

nected with a battery of storage-cells; on repeating this experiment I at first got the same result, but subsequent experiments showed that the absence of deflexion is due to the conductivity conferred on the rarefied gas by the cathode rays. On measuring this conductivity it was found that it diminished very rapidly as the exhaustion increased; it seemed then that on trying Hertz's experiment at very high exhaustions there might be a chance of detecting the deflexion of the cathode rays by an electrostatic force.

The apparatus used is represented in Fig. 108.

The rays from the cathode C pass through a slit in the anode A, which is a metal plug fitting tightly into the tube and connected with the earth; after passing through a second slit in another earth-connected metal plug B, they travel between two parallel aluminum plates about 5 cm. long and 2 broad and at a distance of 1.5 cm. apart; they then fall on the end of the tube and produce a narrow well-defined phosphorescent patch. A scale pasted on the outside of the tube serves to measure the deflexion of this patch. At high exhaustions the rays were deflected when the two aluminum plates were connected with the terminals of a battery of small storage-cells; the rays were depressed when the upper plate was connected with the negative pole of the battery, the lower with the positive, and raised when the upper plate was connected with the positive, the lower with the negative pole. The deflexion was proportional to the difference of potential between the plates, and I could detect the deflexion when the potential-difference was as small as two volts. It was only when the vacuum was a good one that the deflexion took place, but that the absence of deflexion is due to the conductivity of the medium is shown by what takes place when the vacuum has just arrived at the stage at which the deflexion begins. At this stage there is a deflexion of the rays when the plates are first connected with the terminals of the battery, but if this connexion is maintained the patch of phosphorescence gradually creeps back to its undeflected position. This is just what would happen if the space between the plates were a conductor, though a very bad one, for then the positive and negative ions between the plates would slowly diffuse, until the positive plate became coated with negative ions, the negative plate with positive ones; thus the electric intensity between the plates would vanish and the cathode rays be free from electrostatic force. Another illustration of this is afforded by what happens when the pressure is low enough to show the deflexion and a large difference of potential, say 200 volts, is established between the plates; under these circumstances there is a large deflexion of the cathode rays, but the medium under the large electromotive force breaks down every now and then and a bright discharge passes between the plates; when this occurs the phosphorescent patch produced by the cathode rays jumps back to its undeflected position. When the cathode rays are deflected by the electrostatic field, the phosphorescent band breaks up into several bright bands separated by comparatively dark spaces; the phenomena are exactly analogous to those observed by Birkeland when the cathode rays are deflected by a magnet, and called by him the magnetic spectrum.

A series of measurements of the deflexion of the rays by the electrostatic force under various circumstances will be found later on in the part of the paper which deals with the velocity of the rays and the ratio of the mass of the electrified particles to the charge carried by them. It may, however, be mentioned here that the deflexion gets smaller as the pressure diminishes, and when in consequence the potential-difference in the tube in the neighborhood of the cathode increases.

Magnetic Deflexion of the Cathode Rays in Different Gases.

The deflexion of the cathode rays by the magnetic field was studied with the aid of the apparatus shown in Fig. 109. The cathode was placed in a side-tube fastened on to a bell-jar; the opening between this tube and the bell-jar was closed by a metallic plug with a slit in it; this plug was connected with the earth, and was used as the anode. The cathode rays passed through the slit in this plug into the bell-jar, passing in front of a vertical plate of glass ruled into small squares. The bell-jar was placed between two large parallel coils arranged as a Helmholtz galvanometer. The course of the rays was determined by taking photographs of the bell-jar when the cathode rays were passing through it; the divisions on the plate enabled the path of the rays to be determined. Under the action of the magnetic field the narrow beam of cathode rays spreads out into a broad fan-shaped luminosity in the gas. The luminosity in this fan is not uniformly distributed, but is condensed along certain lines. The phosphorescence on the glass is also not uniformly distributed; it is much spread out, showing that the beam consists of rays which are not all deflected to the same extent by the magnet. The luminosity on the glass is crossed by bands along which the luminosity is very much greater than in the adjacent parts. These bright and dark bands are called by Birkeland, who first observed them, the magnetic spectrum. The brightest spots on the glass are by no means always the terminations of the brightest streaks of luminosity in the gas; in fact, in some cases a very bright spot on the glass is not connected with the cathode by any appreciable luminosity, though there may be plenty of luminosity in other parts of the gas. One very interesting point brought out by the photographs is that in a given magnetic field, and with a given mean potential-difference between the terminals, the path of the

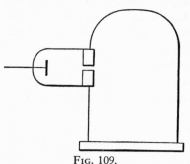

FIG. 109.

rays is independent of the nature of the gas. Photographs were taken of the discharge in hydrogen, air, carbonic acid, methyl iodide, i.e., in gases whose densities range from 1 to 70, and yet, not only were the paths of the most deflected rays the same in all cases, but even the details, such as the distribution of the bright and dark spaces, were the same; in fact, the photographs could hardly be distinguished from each other. It is to be noted that the pressures were not the same; the pressures in the different gases were adjusted so that the mean potential-differences between the cathode and the anode were the same in all the gases. When the pressure of a gas is lowered, the potential-difference between the terminals increases, and the deflexion of the rays produced by a magnet diminishes, or at any rate the deflexion of the rays when the phosphorescence is a maximum diminishes. If an air-break is inserted an effect of the same kind is produced.

In the experiments with different gases, the pressures were as high as was consistent with the appearance of the phosphorescence on the glass, so as to ensure having as much as possible of the gas under consideration in the tube.

As the cathode rays carry a charge of negative electricity, are deflected by an electrostatic force as if they were negatively electrified, and are acted on by a magnetic force in just the way in which this force would act on a negatively electrified body moving along the path of these rays, I can see no escape from the conclusion that they are charges of negative electricity carried by particles of matter. The question next arises, What are these particles? are they atoms, or molecules, or matter in a still finer state of subdivision? To throw some light on this point, I have made a series of measurements of the ratio of the mass of these particles to the charge carried by it. To determine this quantity, I have used two independent methods. The first of these is as follows:—Suppose we consider a bundle of homogeneous cathode rays. Let m be the mass of each of the particles, e the charge carried by it. Let N be the number of particles passing across any section of the beam in a given time; then Q the quantity of electricity carried by these particles is given by the equation

$$Ne = Q.$$

We can measure Q if we receive the cathode rays in the inside of a vessel connected with an electrometer. When these rays strike against a solid body, the temperature of the body is raised; the kinetic energy of the moving particles being converted into heat; if we suppose that all this energy is converted into heat, then if we measure the increase in the temperature of a body of known thermal capacity caused by the impact of these rays, we can determine W, the kinetic energy of the particles, and if v is the velocity of the particles,

$$\tfrac{1}{2}Nmv^2 = W.$$

If ρ is the radius of curvature of the path of these rays in a uniform magnetic field H, then

$$\frac{mv}{e} = H\rho = I,$$

where I is written for $H\rho$ for the sake of brevity. From these equations we get

$$\frac{1m}{2e}v^2 = \frac{W}{Q}$$

$$v = \frac{2W}{QI}$$

$$\frac{m}{e} = \frac{I^2Q}{2W}$$

Thus, if we know the values of Q, W, and I, we can deduce the values of v and m/e.

To measure these quantities, I have used tubes of three different types. The first I tried is like that represented in Fig. 108, except that the plates E and D are absent, and two coaxial cylinders are fastened to the end of the tube. The rays from the cathode C fall on the metal plug B, which is connected with the earth, and serves for the anode; a horizontal slit is cut in this plug. The cathode rays pass through this slit, and then strike against the two coaxial cylinders at the end of the tube; slits are cut in these cylinders, so that the cathode rays pass into the inside of the inner cylinder. The outer cylinder is connected with the earth, the inner cylinder, which is insulated from the outer one, is connected with an electrometer, the deflexion of which measures Q, the quantity of electricity brought into the inner cylinder by the rays. A thermo-electric couple is placed behind the slit in the inner cylinder; this couple is made of very thin strips of iron and copper fastened to very fine iron and copper wires. These wires passed through the cylinders, being insulated from them, and through the glass to the outside of the tube, where they were connected with a low-resistance galvanometer, the deflexion of which gave data for calculating the rise of temperature of the junction produced by the impact against it of the cathode rays. The strips of iron and copper were large enough to ensure that every cathode ray which entered the inner cylinder struck against the junction. In some of the tubes the strips of iron and copper were placed end to end, so that some of the rays struck against the iron, and others against the copper; in others, the strip of one metal was placed in front of the other; no difference, however, could be detected between the results got with these two arrangements. The strips of iron and copper were weighed, and the thermal capacity of the junction calculated. In one set of junctions this capacity was 5×10^{-3}, in another 3×10^{-3}. If we assume that the cathode rays which strike against

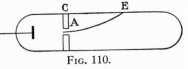

Fig. 110.

the junction give their energy up to it the deflexion of the galvanometer gives us W or $\frac{1}{2}Nmv^2$.

The value of I, i.e., $H\rho$, where ρ is the curvature of the path of the rays in a magnetic field of strength H was found as follows:— The tube was fixed between two large circular coils placed parallel to each other, and separated by a distance equal to the radius of either; these coils produce a uniform magnetic field, the strength of which is got by measuring with an ammeter the strength of the current passing through them. The cathode rays are thus in a uniform field, so that their path is circular. Suppose that the rays, when deflected by a magnet, strike against the glass of the tube at E (Fig. 110), then, if ρ is the radius of the circular path of the rays,

$$2\rho = \frac{CE^2}{AC} + AC;$$

thus, if we measure CE and AC we have the means of determining the radius of curvature of the path of the rays.

The determination of ρ is rendered to some extent uncertain, in consequence of the pencil of rays spreading out under the action of the magnetic field, so that the phosphorescent patch of E is several millimetres long; thus values of ρ differing appreciably from each other will be got by taking E at different points of this phosphorescent patch. Part of this patch was, however, generally considerably brighter than the rest; when this was the case, E was taken as the brightest point; when such a point of maximum brightness did not exist, the middle of the patch was taken for E. The uncertainty in the value of ρ thus introduced amounted sometimes to about 20 percent; by this I mean that if we took E first at one extremity of the patch and then at the other, we should get values of ρ differing by this amount.

The measurement of Q, the quantity of electricity which enters the inner cylinder, is complicated by the cathode rays making the gas through which they pass a conductor, so that though the insulation of the inner cylinder was perfect when the rays were off, it was not so when they were passing through the space between the cylinders; this caused some of the charge communicated to the inner cylinder to leak away so that the actual charge given to the cylinder by the cathode rays was larger than that indicated by the electrometer.

To make the error from this cause as small as possible, the inner cylinder was connected to the largest capacity available, 1.5 microfarad, and the rays were only kept on for a short time, about 1 or 2 seconds, so that the alteration in potential of the inner cylinder was not large, ranging in the various experiments from about .5 to 5 volts. Another reason why it is necessary to limit the duration of the rays to as short a time as possible, is to avoid the correction for the loss of heat from the thermo-electric junction by conduction along the wires; the rise in temperature of the junction was of the order 2°C.; a series of experiments showed that with the same tube and the same gaseous pressure Q and W were proportional to each other when the rays were not kept on too long.

Tubes of this kind gave satisfactory results, the chief drawback being that sometimes in consequence of the charging up of the glass of the tube, a secondary discharge started from the cylinder to the walls of the tube, and the cylinders were surrounded by glow; when this glow appeared, the readings were very irregular; the glow could, however, be got rid of by pumping and letting the tube rest for some time. The results got with this tube are given in the Table under the heading Tube 1.

The second type of tube was like that used for photographing the path of the rays (Fig. 109); double cylinders with a thermo-electric junction like those used in the previous tube were placed in the line of fire of the rays, the inside of the bell-jar was lined with copper gauze connected with the earth. This tube gave very satisfactory results; we were never troubled with any glow round the cylinders, and the readings were most concordant; the only drawback was that as some of the connexions had to be made with sealing wax, it was not possible to get the highest exhaustions with this tube, so that the range of pressure for this tube is less than that for Tube 1. The results got with this tube are given in the Table under the heading Tube 2.

The third type of tube was similar to the first, except that the openings in the two cylinders were made very much smaller; in this tube the slits in the cylinders were replaced by small holes, about 1.5 millim. in diameter. In consequence of the smallness of the openings, the magnitude of the effects was very much reduced; in order to get measurable results it was necessary to reduce the capacity of the condenser in connexion with the inner cylinder to .15 microfarad, and to make the galvanometer exceedingly sensitive, as the rise in temperature of the thermo-electric junction was in these experiments only about .5°C. on the average. The results obtained in this tube are given in the Table under the heading Tube 3.

The Tables in which the results of the several experiments are presented are omitted.

It will be noticed that the value of m/e is considerably greater for Tube 3, where the opening is a small hole, than for Tubes 1 and 2, where the opening is a slit of much greater area. I am of opinion that the values of m/e got from Tubes 1 and 2 are too small, in consequence of the leakage from the inner cylinder to the outer by the gas being rendered a conductor by the passage of the cathode rays.

It will be seen from these tables that the value of m/e is independent of the nature of the gas. Thus, for the first tube the mean for air is $.40 \times 10^{-7}$, for hydrogen $.42 \times 10^{-7}$, and for carbonic acid gas $.4 \times 10^{-7}$; for the second tube the mean for air is $.52 \times 10^{-7}$, for hydrogen $.50 \times 10^{-7}$, and for carbonic acid gas $.54 \times 10^{-7}$.

Experiments were tried with electrodes made of iron instead of aluminum; this altered the appearance of the discharge and the value of v at the same pressure, the values of m/e were, however, the same in the two tubes; the effect produced by different metals on the appearance of the discharge will be described later on.

In all the preceding experiments, the cathode rays were first deflected from the cylinder by a magnet, and it was then found that there was no deflexion either of the electrometer or the galvanometer, so that the deflexions observed were entirely due to the cathode rays; when the glow mentioned previously surrounded the cylinders there was a deflexion of the electrometer even when the cathode rays were deflected from the cylinder.

Before proceeding to discuss the results of these measurements I shall describe another method of measuring the quantities m/e and v of an entirely different kind from the preceding; this method is based upon the deflexion of the cathode rays in an electrostatic field. If we measure the deflexion experienced by the rays when traversing a given length under a uniform electric intensity, and the deflexion of the rays when they traverse a given distance under a uniform magnetic field, we can find the values of m/e and v in the following way:—

Let the space passed over by the rays under a uniform electric intensity F be l, the time taken for the rays to traverse this space is l/v, the velocity in the direction of F is therefore

$$\frac{Fe}{m} \frac{l}{v},$$

so that θ, the angle through which the rays are deflected when they leave the electric field and enter a region free from electric force, is given by the equation

$$\theta = \frac{Fe}{m} \frac{l}{v^2}$$

If, instead of the electric intensity, the rays are acted on by a magnetic force H at right angles to the rays, and extending across the distance l, the velocity at right angles to the original path of the rays is

$$\frac{Hev}{m} \frac{l}{v},$$

so that ϕ, the angle through which the rays are deflected when they leave the magnetic field, is given by the equation

$$\phi = \frac{He}{m} \frac{l}{v}.$$

From these equations we get

$$v = \frac{\phi}{\theta} \frac{F}{H}$$

and

$$\frac{m}{e} = \frac{H^2 \theta l}{F \phi^2}$$

In the actual experiments H was adjusted so that $\phi = \theta$, in this ease the equations become

$$v = \frac{F}{H},$$
$$\frac{m}{e} = \frac{H^2 l}{F \theta}$$

The apparatus used to measure v and m/e by this means is that represented in Fig. 108. The electric field was produced by connecting the two aluminum plates to the terminals of a battery of storage-cells. The phosphorescent patch at the end of the tube was deflected, and the deflexion measured by a scale pasted to the end of the tube. As it was necessary to darken the room to see the phosphorescent patch, a needle coated with luminous paint was placed so that by a screw it could be moved up and down the scale; this needle could be seen when the room was darkened, and it was moved until it coincided with the phosphorescent patch. Thus when light was admitted, the deflexion of the phosphorescent patch could be measured.

The magnetic field was produced by placing outside the tube two coils whose diameter was equal to the length of the plates; the coils were placed so that they covered the space occupied by the plates, the distance between the coils was equal to the radius of either. The mean value of the magnetic force over the length l was determined in the following way: a narrow coil C whose length was l, connected with a ballistic galvanometer, was placed between the coils; the plane of the windings of C was parallel to the planes of the coils; the cross section of the coil was a rectangle 5 cm. by 1 cm. A given current was sent through the outer coils and the kick a of the galvanometer observed when this current was reversed. The coil C was then placed at the centre of two very large coils, so as to be in a field of uniform magnetic force: the

596 A SOURCE BOOK IN PHYSICS

current through the large coils was reversed and the kick B of the galvanometer again observed; by comparing a and B we can get the mean value of the magnetic force over a length l; this was found to be

$$60 \times \iota$$

where ι is the current flowing through the coils.

A series of experiments was made to see if the electrostatic deflexion was proportional to the electric intensity between the plates; this was found to be the case. In the following experiments the current through the coils was adjusted so that the electrostatic deflexion was the same as the magnetic:

Gas	θ	H	F	l	m/e	v
Air..................	$\frac{8}{110}$	5.5	1.5×10^{10}	5	1.3×10^{-7}	2.8×10^9
Air..................	9.5/110	5.4	1.5×10^{10}	5	1.1×10^{-7}	2.8×10^9
Air..................	$\frac{13}{110}$	6.6	1.5×10^{10}	5	1.2×10^{-7}	2.3×10^9
Hydrogen.............	$\frac{9}{110}$	6.3	1.5×10^{10}	5	1.5×10^{-7}	2.5×10^9
Carbonic acid.........	$\frac{11}{110}$	6.9	1.5×10^{10}	5	1.5×10^{-7}	2.2×10^9
Air..................	$\frac{6}{110}$	5	1.8×10^{10}	5	1.3×10^{-7}	3.6×10^9
Air..................	$\frac{7}{110}$	3.6	1×10^{10}	5	1.1×10^{-7}	2.8×10^9

The cathode in the first five experiments was aluminum, in the last two experiments it was made of platinum; in the last experiment Sir William Crookes's method of getting rid of the mercury vapour by inserting tubes of pounded sulphur, sulphur iodide, and copper filings between the bulb and the pump was adopted. In the calculation of m/e and v no allowance has been made for the magnetic force due to the coil in the region outside the plates; in this region the magnetic force will be in the opposite direction to that between the plates, and will tend to bend the cathode rays in the opposite direction: thus the effective value of H will be smaller than the value used in the equations, so that the values of m/e are larger, and those of v less than they would be if this correction were applied. This method of determining the values of m/e and v is much less laborious and probably more accurate than the former method; it cannot, however, be used over so wide a range of pressures.

From these determinations we see that the value of m/e is independent of the nature of the gas, and that its value 10^{-7} is very small compared with the value 10^{-4}, which is the smallest value of this quantity previously known, and which is the value for the hydrogen ion in electrolysis.

Thus for the carriers of the electricity in the cathode rays m/e is very small compared with its value in electrolysis. The smallness of m/e may be due to the smallness of m or the largeness of e, or to a combination of these two. That the carriers of the charges in the cathode rays are small compared with ordinary molecules is shown, I think, by Lenard's results as to the rate at which the brightness of the phosphorescence produced by these rays diminishes with the length of path travelled by the ray. If we regard this phosphorescence as due to the impact of the charged particles, the distance through which the rays must travel before the phosphorescence fades to a given fraction (say $1/e$, where $e = 2.71$) of its original intensity, will be some moderate multiple of the mean free path. Now Lenard found that this distance depends solely upon the density of the medium, and not upon its chemical nature or physical state. In air at atmospheric pressure the distance was about half a centimetre, and this must be comparable with the mean free path of the carriers through air at atmospheric pressure. But the mean free path of the molecules of air is a quantity of quite a different order. The carrier, then, must be small compared with ordinary molecules.

The two fundamental points about these carriers seem to me to be (1) that these carriers are the same whatever the gas through which the discharge passes, (2) that the mean free paths depend upon nothing but the density of the medium traversed by these rays.

WIEN

Wilhelm (Willy) Wien was born on January 13, 1864 in Gaffken in East Prussia. He studied from 1882 to 1886 in Göttingen, Heidelberg, and Berlin. In 1890 he became assistant in the Physico-technical Establishment at Charlottenburg. After serving for a time in other universities as professor he became in 1900 professor of physics at Würzburg. In 1920 he became professor in Munich. He died on August 30, 1928. Wien rendered a great service to the development of the theory of radiation by his discovery of the displacement law and by his attempt to obtain a formula for the distribution of energy in the normal spectrum. For this work he was given a Nobel prize in 1911.

In the extract which follows from the paper "Untersuchungen über die elektrische Entladung in verdünnten Gasen," from the *Annalen der Physik und Chemie*, Vol. 65, p. 440, 1898, are described the experiments by which

Wien determined the properties of the canal rays. The first part of the paper contains a description of observations on the cathode rays, made with a Lenard tube, in which the cathode rays passed through an aluminium window into an extension of the tube where as high a rarefaction as possible was maintained. The magnetic and electrostatic deflections of a narrow cathode stream were observed and Wien concludes that it is proved by this investigation that the cathode rays which pass through the window carry with them strong negative charges. His measurements show that the particles of the stream have a velocity about equal to a third that of light, and that the ratio of the mass to the charge is 5×10^{-8}.

THE CANAL RAYS

. . . .

After the negative charge of the cathode rays was demonstrated, the thought came to me that the canal rays observed by Goldstein, which cannot be deflected appreciably by ordinary magnets, and which proceed backwards through a pierced cathode, might carry the positive charge. In this case it was not possible to shut out the field of observation completely from the discharge tube, because in spite of many attempts I could find no substance through which the canal rays will pass.

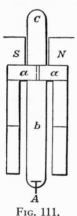

FIG. 111.

The description of the simple tube in which the experiments were tried and of some of the properties of the canal rays is omitted.

Hitherto it has not been known that the canal rays undergo magnetic or electrostatic deflection. After it has been proved that they carry positive charges, we may expect them to experience such deflections in the opposite direction to those of the cathode rays.

I was able to observe the electrostatic deflection in a simple way. A hole 2 mm. in diameter was bored in a metal plate (Fig. 111) and glass tubes were cemented to the plate on both sides. The anode was sealed into one of them, b, and in the other one, c, were placed two electrodes standing opposite each other, 5 cm. long and 0.5 cm. wide, at a distance apart of 1.7 cm. The tube was then placed in the zinc box and the metal plate was joined with the wall of the box and made the cathode. When the exhaustion was sufficient there came out from the hole a beam of canal rays which brought out a spot of fluorescence, of the well known yellow-green color, on the glass wall at a distance of 9 cm. from the plate.

This beam of rays passed between the two electrodes, and was deflected when the electrodes were brought to a difference of potential of 2000 volts by a high-tension accumulator; the deflection amounted to 6 mm. The stream was attracted by the negative electrode. When the exhaustion was great enough, it was possible, for a short time, to maintain the condition that the field between the electrodes remained constant. The narrow beam of canal rays increased the conductivity of the gas in the region so much that after a while the accumulator could discharge through it.

The proof of the magnetic deflection of the canal rays was less simple. The strong magnetic force of a Ruhmkorff electromagnet affected the discharge itself to such a degree that an accurate observation could not be made. And it was not possible to produce a long beam of canal rays such as would have permitted me to set up the discharge at some distance from the electromagnet. I did succeed in observing the deflection of the canal rays when I had a well defined beam by using a powerful ordinary horseshoe magnet. This deflection, however, was too small for measurement, and I therefore reached my object by using the following arrangement: The Ruhmkorff electromagnet carried pole pieces with faces 5 cm. long and 2 cm. wide. The tube c, 2.5 cm. wide, was placed between the poles, so that the lines of force of the magnetic field were perpendicular to the electric lines of force between the electrodes of the tube c. The glass tubes b, c, which have already been described, were cemented on an iron plate a, 11 cm. in diameter and 2.5 cm. thick. A hole 2 mm. in diameter was bored in the iron plate.

The iron plate was brought up to the pole pieces of the electromagnet. Around the tube b and in contact with the iron plate, there was placed an iron cylinder, whose walls were 2.5 cm. thick, surrounding the whole tube. In this way the magnetic forces which might act on b were very much weakened. Their strength could be determined from the deflection of the cathode rays if A was made the anode and a the cathode, while a was connected with the earth.

I first convinced myself that the canal rays which passed out toward c were not influenced if, when the iron cylinder was taken away, the cathode rays were deflected by an ordinary magnet as much as they were by the remaining magnetic force within the iron cylinder when the electromagnet was excited.

The iron cylinder was then put in place and the deflection of the canal rays observed when the electromagnet was excited. This deflection also amounted to 6 mm. and had the opposite direction to that of the cathode rays. The magnetic field was measured by the help of a Stenger bifilar galvanometer. It amounted to about 3250 C.G.S. In this case, as in the case of the electrostatic deflection, it appeared that the canal rays also consist of a mixture of rays which show different deflections.

From the numbers which have been given we obtain for the velocity of the canal rays 3.6×10^7 cm/sec., and for the ratio of the mass to the charge 3.2×10^{-3}.

ROENTGEN

Wilhelm Konrad Roentgen was born on March 27, 1845, in Lennep, Holland. He studied in Holland and at Zürich, where he took his doctor's degree in 1868. He was assistant to Kundt at Würzburg and Strassburg, professor of physics at Giessen, at Würzburg and from 1900 to 1920 at Munich. He died in Munich on February 10, 1923.

The paper which follows, "Ueber eine neue Art von Strahlen," is taken from the *Sitzungsberichte der Würzburger Physikalischen-Medicinischen Gesellschaft,* December, 1895. The translation by George F. Barker appeared in Harper's *Scientific Memoirs.* It contains the account of Roentgen's discovery of the rays which are known by his name, or more often now as X-rays.

THE ROENTGEN RAYS

First Communication

1. If the discharge of a fairly large induction-coil be made to pass through a Hittorf vacuum-tube, or through a Lenard tube, a Crookes tube, or other similar apparatus, which has been sufficiently exhausted, the tube being covered with thin, black card-board which fits it with tolerable closeness, and if the whole apparatus be placed in a completely darkened room, there is observed at each discharge a bright illumination of a paper screen covered with barium platino-cyanide, placed in the vicinity of the induction-coil, the fluorescence thus produced being entirely independent of the fact whether the coated or the plain surface is turned towards the discharge-tube. This fluorescence is visible even when the paper screen is at a distance of two metres from the apparatus.

It is easy to prove that the cause of the fluorescence proceeds from the discharge-apparatus, and not from any other point in the conducting circuit.

2. The most striking feature of this phenomenon is the fact that an active agent here passes through a black card-board envelope, which is opaque to the visible and the ultra-violet rays of the sun or of the electric arc; an agent, too, which has the power of producing active fluorescence. Hence we may first investigate the question whether other bodies also possess this property.

We soon discover that all bodies are transparent to this agent, though in very different degrees. I proceed to give a few examples: Paper is very transparent; behind a bound book of about one thousand pages I saw the fluorescent screen light up brightly, the printers' ink offering scarcely a noticeable hindrance. In the same way the fluorescence appeared behind a double pack of cards; a single card held between the apparatus and the screen being almost unnoticeable to the eye. A single sheet of tin-foil is also scarcely perceptible; it is only after several layers have been placed over one another that their shadow is distinctly seen on the screen. Thick blocks of wood are also transparent, pine boards two or three centimetres thick absorbing only slightly. A plate of aluminium about fifteen millimetres thick, though it enfeebled the action seriously, did not cause the fluorescence to disappear entirely. Sheets of hard rubber several centimetres thick still permit the rays to pass through them. Glass plates of equal thickness behave quite differently, according as they contain lead (flintglass) or not; the former are much less transparent than the latter. If the hand be held between the discharge-tube and the screen, the darker shadow of the bones is seen within the slightly dark shadow-image of the hand itself. Water, carbon disulphide, and various other liquids, when they are examined in mica vessels, seem also to be transparent. That hydrogen is to any considerable degree more transparent than air I have not been able to discover. Behind plates of copper, silver, lead, gold, and platinum the fluorescence may still be recognized, though only if the thickness of the plates is not too great. Platinum of a thickness of 0.2 millimetre is still transparent; the silver

and copper plates may even be thicker. Lead of a thickness of
1.5 millimetres is practically opaque; and on account of this
property this metal is frequently most useful. A rod of wood with
a square cross-section (20 × 20 millimetres), one of whose sides
is painted white with lead paint, behaves differently according as
to how it is held between the apparatus and the screen. It is
almost entirely without action when the X-rays pass through it
parallel to the painted side; whereas the stick throws a dark
shadow when the rays are made to traverse it perpendicular to the
painted side. In a series similar to that of the metals themselves
their salts can be arranged with reference to their transparency,
either in the solid form or in solution.

. . . .

6. The fluorescence of barium platino-cyanide is not the only
recognizable effect of the X-rays. It should be mentioned that
other bodies also fluoresce; such, for instance, as the phosphorescent calcium compounds, then uranium glass, ordinary glass,
calcite, rock-salt, and so on.

Of special significance in many respects is the fact that photographic dry plates are sensitive to the X-rays. We are, therefore,
in a condition to determine more definitely many phenomena,
and so the more easily to avoid deception; wherever it has been
possible, therefore, I have controlled, by means of photography,
every important observation which I have made with the eye by
means of the fluorescent screen.

In these experiments the property of the rays to pass almost
unhindered through thin sheets of wood, paper, and tin-foil is
most important. The photographic impressions can be obtained
in a non-darkened room with the photographic plates either in the
holders or wrapped up in paper. On the other hand, from this
property it results as a consequence that undeveloped plates
cannot be left for a long time in the neighborhood of the discharge-
tube, if they are protected merely by the usual covering of pasteboard and paper.

It appears questionable, however, whether the chemical action
on the silver salts of the photographic plates is directly caused by
the X-rays. It is possible that this action proceeds from the
fluorescent light which, as noted above, is produced in the glass
plate itself or perhaps in the layer of gelatin. "Films" can be
used just as well as glass plates.

I have not yet been able to prove experimentally that the X-rays are able also to produce a heating action; yet we may well assume that this effect is present, since the capability of the X-rays to be transformed is proved by means of the observed fluorescence phenomena. It is certain, therefore, that all the X-rays which fall upon a substance do not leave it again as such.

The retina of the eye is not sensitive to these rays. Even if the eye is brought close to the discharge-tube, it observes nothing, although, as experiment has proved, the media contained in the eye must be sufficiently transparent to transmit the rays.

7. After I had recognized the transparency of various substances of relatively considerable thickness, I hastened to see how the X-rays behaved on passing through a prism, and to find whether they were thereby deviated or not.

Experiments with water and with carbon disulphide enclosed in mica prisms of about 30° refracting angle showed no deviation, either with the fluorescent screen or on the photographic plate. For purposes of comparison the deviation of rays of ordinary light under the same conditions was observed; and it was noted that in this case the deviated images fell on the plate about 10 or 20 millimetres distant from the direct image. By means of prisms made of hard rubber and of aluminium, also of about 30° refracting angle, I have obtained images on the photographic plate in which some small deviation may perhaps be recognized. However, the fact is quite uncertain; the deviation, if it does exist, being so small that in any case the refractive index of the X-rays in the substances named cannot be more than 1.05 at the most. With a fluorescent screen I was also unable to observe any deviation.

Up to the present time experiments with prisms of denser metals have given no definite results, owing to their feeble transparency and the consequently diminished intensity of the transmitted rays.

With reference to the general conditions here involved on the one hand, and on the other to the importance of the question whether the X-rays can be refracted or not on passing from one medium into another, it is most fortunate that this subject may be investigated in still another way than with the aid of prisms. Finely divided bodies in sufficiently thick layers scatter the incident light and allow only a little of it to pass, owing to reflection and refraction; so that if powders are as transparent to X-rays as the same substances are in mass—equal amounts of material

being presupposed—it follows at once that neither refraction nor
regular reflection takes place to any sensible degree.   Experiments
were tried with finely powdered rock-salt, with fine electrolytic
silver-powder, and with zinc-dust, such as is used in chemical
investigations.   In all these cases no difference was detected
between the transparency of the powder and that of the substance
in mass, either by observation with the fluorescent screen or with
the photographic plate.

From what has now been said it is obvious that the X-rays
cannot be concentrated by lenses; neither a large lens of hard
rubber nor a glass lens having any influence upon them.   The
shadow-picture of a round rod is darker in the middle than at the
edge; while the image of a tube which is filled with a substance
more transparent than its own material is lighter at the middle
than at the edge.

. . . .

10. It is well known that Lenard came to the conclusion from
the results of his beautiful experiments on the transmission of the
cathode rays of Hittorf through a thin sheet of aluminum, that
these rays are phenomena of the ether, and that they diffuse
themselves through all bodies.   We can say the same of our rays.

In his most recent research, Lenard has determined the absorptive power of different substances for the cathode rays, and, among
others, has measured it for air from atmospheric pressure to
4.10, 3.40, 3.10, referred to 1 centimetre, according to the rarefaction of the gas contained in the discharge-apparatus.   Judging
from the discharge-pressure as estimated from the sparking
distance, I have had to do in my experiments for the most part
with rarefactions of the same order of magnitude, and only rarely
with less or greater ones.   I have succeeded in comparing by
means of the L. Weber photometer—I do not possess a better
one—the intensities, taken in atmospheric air, of the fluorescence
of my screen at two distances from the discharge-apparatus—
about 100 and 200 millimetres; and I have found from three
experiments, which agree very well with each other, that the
intensities vary inversely as the squares of the distances of the
screen from the discharge-apparatus.   Accordingly, air absorbs a
far smaller fraction of the X-rays than of the cathode rays.   This
result is in entire agreement with the observation mentioned above,

that it is still possible to detect the fluorescent light at a distance of 2 metres from the discharge-apparatus.

Other substances behave in general like air; they are more transparent to X-rays than to cathode rays.

11. A further difference, and a most important one, between the behavior of cathode rays and of X-rays lies in the fact that I have not succeeded, in spite of many attempts, in obtaining a deflection of the X-rays by a magnet, even in very intense fields.

The possibility of deflection by a magnet has, up to the present time, served as a characteristic property of the cathode rays; although it was observed by Hertz and Lenard that there are different sorts of cathode rays, "which are distinguished from each other by their production of phosphorescence, by the amount of their absorption, and by the extent of their deflection by a magnet." A considerable deflection, however, was noted in all of the cases investigated by them; so that I do not think that this characteristic will be given up except for stringent reasons.

12. According to experiments especially designed to test the question, it is certain that the spot on the wall of the discharge-tube which fluoresces the strongest is to be considered as the main centre from which the X-rays radiate in all directions. The X-rays proceed from that spot where, according to the data obtained by different investigators, the cathode rays strike the glass wall. If the cathode rays within the discharge-apparatus are deflected by means of a magnet, it is observed that the X-rays proceed from another spot—namely, from that which is the new terminus of the cathode rays.

For this reason, therefore, the X-rays, which it is impossible to deflect, cannot be cathode rays simply transmitted or reflected without change by the glass wall. The greater density of the gas outside of the discharge-tube certainly cannot account for the great difference in the deflection, according to Lenard.

I therefore reach the conclusion that the X-rays are not identical with the cathode rays, but that they are produced by the cathode rays at the glass wall of the discharge-apparatus.

13. This production does not take place in glass alone, but, as I have been able to observe in an apparatus closed by a plate of aluminum 2 millimetres thick, in this metal also. Other substances are to be examined later.

14. The justification for calling by the name "rays" the agent which proceeds from the wall of the discharge-apparatus I derive

in part from the entirely regular formation of shadows, which are seen when more or less transparent bodies are brought between the apparatus and the fluorescent screen (or the photographic plate).

I have observed, and in part photographed, many shadow-pictures of this kind, the production of which has a particular charm. I possess, for instance, photographs of the shadow of the profile of a door which separates the rooms in which, on one side, the discharge-apparatus was placed, on the other the photographic plate; the shadow of the bones of the hand; the shadow of a covered wire wrapped on a wooden spool; of a set of weights enclosed in a box; of a galvanometer in which the magnetic needle is entirely enclosed by metal; of a piece of metal whose lack of homogeneity becomes noticeable by means of the X-rays, etc.

Another conclusive proof of the rectilinear propagation of the X-rays is a pin-hole photograph which I was able to make of the discharge-apparatus while it was enveloped in black paper; the picture is weak but unmistakably correct.

15. I have tried in many ways to detect interference phenomena of the X-rays; but, unfortunately, without success, perhaps only because of their feeble intensity.

16. Experiments have been begun, but are not yet finished, to ascertain whether electrostatic forces affect the X-rays in any way.

17. In considering the question what are the X-rays—which, as we have seen, cannot be cathode rays—we may perhaps at first be led to think of them as ultra-violet light, owing to their active fluorescence and their chemical actions. But in so doing we find ourselves opposed by the most weighty considerations. If the X-rays are ultra-violet light, this light must have the following properties:

(a) On passing from air into water, carbon disulphide, aluminium, rock-salt, glass, zinc, etc., it suffers no noticeable refraction.

(b) By none of the bodies named can it be regularly reflected to any appreciable extent.

(c) It cannot be polarized by any of the ordinary methods.

(d) Its absorption is influenced by no other property of substances so much as by their density.

That is to say, we must assume that these ultra-violet rays behave entirely differently from the ultra-red, visible, and ultra-violet rays which have been known up to this time.

I have been unable to come to this conclusion, and so have sought for another explanation.

There seems to exist some kind of relationship between the new rays and light rays; at least this is indicated by the formation of shadows, the fluorescence and the chemical action produced by them both. Now, we have known for a long time that there can be in the ether longitudinal vibrations besides the transverse light-vibrations; and, according to the views of different physicists, these vibrations must exist. Their existence, it is true, has not been proved up to the present, and consequently their properties have not been investigated by experiment.

Ought not, therefore, the new rays to be ascribed to longitudinal vibrations in the ether?

I must confess that in the course of the investigation I have become more and more confident of the correctness of this idea, and so, therefore permit myself to announce this conjecture, although I am perfectly aware that the explanation given still needs further confirmation.

Second Communication, March 9, 1896.

Since my work must be interrupted for several weeks, I take the opportunity of presenting in the following paper some new phenomena which I have observed.

18. It was known to me at the time of my first publication that X-rays can discharge electrified bodies; and I conjecture that in Lenard's experiments it was the X-rays, and not the cathode rays, which had passed unchanged through the aluminium window of his apparatus, which produced the action described by him upon electrified bodies at a distance. I have, however, delayed the publication of my experiments until I could contribute results which are free from criticism.

These results can be obtained only when the observations are made in a space which is protected completely, not only from the electrostatic forces proceeding from the vacuum-tube, from the conducting wires, from the induction apparatus, etc., but is also closed against air which comes from the neighborhood of the discharge-apparatus.

To secure these conditions I had a chamber made of zinc plates soldered together, which was large enough to contain myself and the necessary apparatus, which could be closed air-tight, and which was provided with an opening which could be closed by a zinc door. The wall opposite the door was for the most part covered with lead. At a place near the discharge-apparatus, which was set up outside the case, the zinc wall, together with the lining of sheet-lead, was cut out for a width of 4 centimetres; and the opening was covered again air-tight with a thin sheet of aluminium. The X-rays penetrated through this window into the observation space.

I observed the following phenomena:

(a) Electrified bodies in air, charged either positively or negatively, are discharged if X-rays fall upon them; and this process goes on the more rapidly the more intense the rays are. The intensity of the rays was estimated by their action on a fluorescent screen or a photographic plate.

It is immaterial in general whether the electrified bodies are conductors or insulators. Up to the present I have not found any specific difference in the behavior of different bodies with reference to the rate of discharge; nor as to the behavior of positive and negative electricity. Yet it is not impossible that small differences may exist.

(b) If the electrified conductor be surrounded not by air but by a solid insulator, e.g. paraffin, the radiation has the same action as would result from exposure of the insulating envelope to a flame connected to the earth.

(c) If this insulating envelope be surrounded by a close-fitting conductor which is connected to the earth, and which, like the insulator, is transparent to X-rays, the radiation produces on the inner electrified conductor no action which can be detected by my apparatus.

(d) The observations noted under (a), (b), (c) indicate that air through which X-rays have passed possesses the power of discharging electrified bodies with which it comes in contact.

(e) If this is really the case, and if, further, the air retains this property for some time after it has been exposed to the X-rays, then it must be possible to discharge electrified bodies which have not been themselves exposed to the rays, by conducting to them air which has thus been exposed.

We may convince ourselves in various ways that this conclusion is correct. One method of experiment, although perhaps not the simplest, I shall describe.

I used a brass tube 3 centimetres wide and 45 centimetres long; at a distance of some centimetres from one end a part of the wall of the tube was cut away and replaced by a thin aluminium plate; at the other end, through an air-tight cap, a brass ball fastened to a metal rod was introduced into the tube in such a manner as to be insulated. Between the ball and the closed end of the tube there was soldered a side-tube which could be connected with an exhaust-apparatus; so that when this is in action the brass ball is subjected to a stream of air which on its way through the tube has passed by the aluminium window. The distance from the window to the ball was over 20 centimetres.

I arranged this tube inside the zinc chamber in such a position that the X-rays could enter through the aluminium window of the tube perpendicular to its axis. The insulated ball lay then in the shadow, out of the range of the action of these rays. The tube and the zinc case were connected by a conductor, the ball was joined to a Hankel electroscope.

It was now observed that a charge (either positive or negative) given to the ball was not influenced by the X-rays so long as the air remained at rest in the tube, but that the charge instantly decreased considerably if by exhaustion the air which had been subjected to the rays was drawn past the ball. If by means of storage cells the ball was maintained at a constant potential, and if the modified air was drawn continuously through the tube, an electric current arose just as if the ball were connected to the wall of the tube by a poor conductor.

(f) The question arises, How does the air lose the property which is given it by the X-rays? It is not yet settled whether it loses this property gradually of itself—i.e., without coming in contact with other bodies. On the other hand, it is certain that a brief contact with a body of large surface, which does not need to be electrified, can make the air inactive. For instance, if a thick enough stopper of wadding is pushed into the tube so far that the modified air must pass through it before it reaches the electrified ball, the charge on the ball remains unaffected even while the exhaustion is taking place.

If the wad is in front of the aluminium window, the result obtained is the same as it would be without the wad; a proof that it is not particles of dust which are the cause of the observed discharge.

Wire gratings act like wadding; but the gratings must be very fine, and many layers must be placed over each other if the modified air is to be inactive after it is drawn through them. If these gratings are not connected to the earth, as has been assumed, but are connected to a source of electricity at a constant potential, I have always observed exactly what I had expected; but these experiments are not yet completed.

.   .   .   .

BECQUEREL

Henri Becquerel was born on December 15, 1852, in Paris. He studied at the École Polytechnique, with which in 1875 he became connected as demonstrator. In 1895 he became professor in the same institution. He died on August 25, 1908, at Croisic in Brittany.

In the following short paper on "Sur les radiations émises par phosphorescence," from *Comptes Rendus*, Vol. 122, p. 420, 1896, is described the discovery of radioactivity from uranium.

The Radiation from Uranium

At a former meeting M. Ch. Henry announced that phosphorescent sulphide of zinc introduced in the path of the rays emanating from a Crookes tube increased the intensity of the radiations which passed through the aluminium.

Further M. Niewenglowski perceived that phosphorescent calcium sulphide emits radiations which pass through opaque bodies.

The same fact appears with several other phosphorescent bodies, and in particular with salts of uranium, of which the phosphorescence lasts only for a short time.

With the double sulphate of uranium and potassium, of which I possess crystals in the form of a thin transparent crust, I have made the following experiment:

I wrapped a Lumière photographic plate with bromized emulsion with two sheets of thick black paper, so thick that the plate did not become clouded by exposure to the sun for a whole day. I placed on the paper a plate of the phosphorescent substance, and exposed the whole thing to the sun for several hours. When I developed the photographic plate I saw the silhouette of the phosphorescent substance in black on the negative. If I placed between the phosphorescent substance and the paper a coin or a metallic screen pierced with an open-work design, the image of these objects appeared on the negative.

The same experiments can be tried with a thin sheet of glass placed between the phosphorescent substance and the paper, which excludes the possibility of a chemical action resulting from vapors which might emanate from the substance when heated by the sun's rays.

We may therefore conclude from these experiments that the phosphorescent substance in question emits radiations which penetrate paper that is opaque to light, and reduce silver salts.

. . . .

I particularly insist on the following fact, which appears to me exceedingly important and not in accord with the phenomena which one might expect to observe: the same encrusted crystals placed with respect to the photographic plates in the same conditions and acting through the same screens, but protected from the excitation of incident rays and kept in the dark, still produce the same photographic effects. I may relate how I was led to make this observation: among the preceding experiments some had been made ready on Wednesday the 26th and Thursday the 27th of February and as on those days the sun only showed itself intermittently I kept my arrangements all prepared and put back the holders in the dark in the drawer of the case, and left in place the crusts of uranium salt. Since the sun did not show itself again for several days I developed the photographic plates on the 1st. of March, expecting to find the images very feeble. The silhouettes appeared on the contrary with great intensity. I at once thought that the action might be able to go on in the dark, and I arranged the following experiment.

At the bottom of a box made of opaque cardboard, I placed a photographic plate, and then on the sensitive face I laid a crust of uranium salt which was convex, so that it only touched the emulsion at a few points; then alongside of it I placed on the same plate another crust of the same salt, separated from the emulsion by a thin plate of glass; this operation was carried out in the dark room, the box was shut, was then enclosed in another cardboard box, and put away in a drawer.

I did the same thing with a holder closed by an aluminium plate, in which I put a photographic plate and then laid on it a crust of uranium salt. The whole was enclosed in an opaque box and put in a drawer. After five hours I developed the plates, and the silhouettes of the encrusted crystals showed black, as in the former experiment, and as if they had been rendered phosphorescent by light.   In the case of the crust which was placed directly on the emulsion, there was a slightly different action at the points of contact from that under the parts of the crust which were about a millimeter away from the emulsion; the difference may be attributed to the different distances of the sources of the active radiation. The action of the crust placed on the glass plate was very slightly enfeebled, but the form of the crust was very well reproduced. Finally, in passing through the plate of aluminium, the action was considerably enfeebled but nevertheless was very clear.

It is important to notice that this phenomenon seems not to be attributable to luminous radiation emitted by phosphorescence, since at the end of one hundredth of a second these radiations become so feeble that they are scarcely perceptible.

. . . .

Emission of New Radiations by Metallic Uranium

Some months ago I showed that uranium salts emit radiations whose existence has not hitherto been recognized, and that these radiations possess remarkable properties, some of which are similar to the properties studied by M. Röntgen.   The radiations of uranium salts are emitted not only when the substances are exposed to light but when they are kept in the dark, and for more than two months the same pieces of different salts, kept protected from all known exciting radiations, continued to emit, almost without perceptible enfeeblement, the new radiations.   From the 3d. of March to the 3d. of May these substances were enclosed in a box of opaque cardboard.   Since the 3d. of May they have been in a double box of lead, which has never left the dark room.   A very simple arrangement makes it possible to slip a photographic plate under a black paper stretched parallel to the bottom of the box, on which rests the substances which are being tested, without exposing them to any radiation which does not pass through the lead.

In these conditions the substances studied continued to emit active radiation.

. . . .

All the salts of uranium that I have studied, whether they become phosphorescent or not in the light, whether crystallized, cast or in solution, have given me similar results.   I have thus been led to think that the effect is a consequence of the presence of the element uranium in these salts, and that the metal would give more intense effects than its compounds. An experiment made several weeks ago with the powdered uranium of commerce, which has been for a long time in my laboratory, confirmed this expectation; the photographic effect is notably greater than the impression produced by one of the uranium salts, and in particular by the sulphate of uranium and potassium.

Before publishing this result, I waited until our fellow member, M. Moissan, whose beautiful investigations on uranium have just been published, could put at my disposal some of the products which he had prepared. The results were still sharper and the impressions obtained on the photographic plate through the black paper were much more intense with crystallized uranium, with a casting of uranium, and with uranium carbide than with the double sulphate used as a check on the same plate.

The same difference appears again in the phenomenon of the discharge of electrified bodies. The metallic uranium provokes the loss of charge at a greater rate than its salts do.

. . . .

P. AND M. S. CURIE

Pierre Curie was born on May 15, 1859, in Paris. He was professor of physics at the Sorbonne. He died as the result of a street accident on April 19, 1906. His wife, Marie Sklodowska Curie, was born on November 7, 1867, in Warsaw. She worked with her husband in the laboratory of the Sorbonne and succeeded him after his death in 1906. She died on July 4, 1934.

The papers which follow are taken from *Comptes Rendus*, Vol. 127, pp. 175 and 1215, 1898. Their titles are practically the same, "Sur une substance nouvelle radio-active, contenue dans la pechblende" and "Sur une nouvelle substance fortement radio-active, contenue dans la pechblende." They contain the accounts of the discovery of polonium and of radium respectively. The second paper was prepared with the help of M. G. Bémont.

POLONIUM

Certain minerals containing uranium and thorium (pitch blende, chalcolite, uranite) are very active in emitting Becquerel rays. One of us has already shown that their activity is greater than that of uranium and thorium, and has expressed the opinion that this effect arises from some other very active substance contained in these minerals in small quantity.

The study of the compounds of uranium and of thorium has shown, in fact, that the property of emitting rays which make air conducting and which act on photographic plates, is a specific property of uranium and of thorium, which appears in all the compounds of these metals, being so much the more feeble as the proportion of the active metal in the compound is itself less. The physical state of the substances seems to be of altogether secondary importance. Various experiments have shown that if the substances are mixed with others their condition seems to have no effect except as it varies the proportion of the active body and the absorption produced by the inert substance. Certain causes, such as the presence of impurities, which have so great an effect on phosphorescence or fluorescence, are here altogether without effect. It therefore becomes very probable that if certain minerals are more active than uranium and thorium, it is because they contain a substance more active than these metals.

We have attempted to isolate this substance in pitch blende, and the experiment has confirmed our expectations.

Our chemical investigations have been guided by the tests made of the radiating activity of the products which were separated in each operation. Each product was placed on the plates of a condenser, and the conductibility acquired by the air was measured by an electrometer and a piezoelectric quartz, as in the work already referred to. We thus have not only an indication but a number which gives some measure of the richness of the product in the active substance.

The pitch blende that we analyzed was about two and a half times more active than uranium in our apparatus.

The details of the chemical methods employed are omitted.

By carrying on these different operations we obtained products which were more and more active. Finally, we obtained a substance whose activity is about 400 times greater than that of uranium.

We have attempted to discover among bodies which are already known if there are any which are radioactive. We have examined compounds of almost all the simple bodies; thanks to the kindness of several chemists we have received specimens of very rare substances. Uranium and thorium were the only ones which were evidently active, tantalum perhaps is very feebly so.

We believe, therefore, that the substance which we removed from pitch blende contains a metal which has not yet been known, similar to bismuth in its chemical properties.   If the existence of this new metal is confirmed, we propose to call it *polonium*, after the name of the native country of one of us.

.   .   .   .

Radium

Two of us have shown that by purely chemical processes we may extract from pitch blende a strongly radioactive substance.   This substance stands near bismuth in its chemical properties.   We have expressed the opinion that pitch blende perhaps contains a new element, for which we proposed the name polonium.

The researches which we have since carried on are in agreement with the first results obtained; but in the course of these researches we encountered a second substance also strongly radioactive and entirely different from the first in its chemical properties.   In fact, polonium is precipitated in acid solution by sulphuretted hydrogen; its salts are soluble in acids, and water precipitates them from these solutions; polonium is completely precipitated by ammonia.

The new radioactive substance that we have found presents the chemical aspects of almost pure barium; it is not precipitated either by sulphuretted hydrogen or by ammonium sulphide, or by ammonia; the sulphate is insoluble in water and in acids; the carbonate is insoluble in water; the chloride, very soluble in water, is insoluble in concentrated hydrochloric acid and in alcohol. Finally, it gives the spectrum of barium, which is easy to recognize.

We believe, nevertheless, that this substance, although for the most part consisting of barium, contains in addition a new element which gives it its radioactivity and which furthermore is very near barium in its chemical properties.   These are the reasons which speak in favor of this view.

(1) Barium and its compounds are not ordinarily radioactive; now, one of us has shown that radioactivity seems to be an atomic property, persisting in all the chemical and physical states of matter.   If we look at the thing this way, the radioactivity of our substance, which does not arise from barium, ought to be attributed to another element.

(2) The first substances which we obtained, in the state of hydrated chlorides, had a radioactivity 60 times greater than

that of metallic uranium.   (The radioactivity intensity is evaluated by the conductibility of the air in our apparatus.)   By dissolving these chlorides in water and precipitating a part of them by alcohol, the precipitated part is much more active than the part which remains dissolved.   By starting with this fact we may carry out a series of fractionations, from which we may obtain more and more active chlorides.   We have thus obtained chlorides which have an activity 900 times greater than that of uranium. We have been stopped by the lack of material, but from the progress of the operations we may assume that the activity would have been much more increased if we had been able to continue. These facts can be explained by the presence of a radioactive element of which the chloride is less soluble in alcoholic solution than is barium chloride.

(3) M. Demarçay has examined the spectrum of our substance, with a kindness for which we do not know how to thank him enough.   The results of his observations are presented in a special note which follows ours.   M. Demarçay has found in the spectrum a ray which seems not to belong to any known element.   This ray, which is scarcely visible in the chloride that is 60 times more active than uranium, becomes strongly marked in the chloride that was enriched by fractionation until its activity was 900 times that of uranium.   The intensity of this ray increases at the same time as the radioactivity, and this, we think, is a strong reason for attributing it to the radioactive part of our substance.

The various reasons which we have presented lead us to believe that the new radioactive substance contains a new element, to which we propose to give the name *radium*.

Index

A

Absolute measure of magnetic force, 519
Absolute time, space, and motion, 33
Acceleration, 2
Aepinus, 406
Air-pump, 80
Alembert, D', 55
Amontons, 128
Ampère, 114, 446
Andrews, 187
Anomalous dispersion, 381, 382
Arago, 324, 443
Atomic heat, 179

B

Balmer, 360
Barometer, 70, 73
Bartholinus, 280
Becquerel, H, 610
Bernoulli, Daniel, 247
Bernoulli, Jean, 48
Biot, 441
Biot and Savart's law, 441
Black, 134
Boltzmann, 262
Boyle, 84
Bradley, 337
Brown, 251
Brownian movements, 251

C

Cagniard de la Tour, 181
Cailletet, 192
Canal rays, 576, 598
Carnot, 220
Cathode discharge, 562, 564, 580
Cavendish, 105

Christiansen, 381
Clausius, 228
Collisions, laws of, 40
Coulomb, 97, 408
Critical temperature, 181, 187
Crookes, 563
Curie, M. S., 613
Curie, P., 547, 613
Currents, action of, on magnets, 437
actions between, 447
induced, 473

D

Davy, 161
Descartes, 50, 265
Diamagnetism, 503
Dimensions, theory of, 175
Double refraction, 280, 289
Du Fay, 398
Dulong, 178

E

Earth, density of, 105
Elasticity, law of, 93
modulus of, 95
of torsion, 98
Electric attractions and repulsions, 393
Electric charge on surface, 499
Electric conductors and nonconductors, 394
Electric convection, magnetic effect of, 538
Electric current, 421
Electric decomposition of water, 432
Electric discharge by light, 578
Electric force, law of, 408
Electric kite, 402
Electric radiation, 550

SE